CW00547316

# Biodiversity Economics

*Edited by*

Andreas Kontoleon, Unai Pascual
and Timothy Swanson

CAMBRIDGE UNIVERSITY PRESS
Cambridge, New York, Melbourne, Madrid, Cape Town, Singapore,
São Paulo, Delhi, Dubai, Tokyo, Mexico City

Cambridge University Press
The Edinburgh Building, Cambridge CB2 8RU, UK

Published in the United States of America by Cambridge University Press, New York

www.cambridge.org
Information on this title: www.cambridge.org/9780521866835



First published 2007

*A catalogue record for this publication is available from the British Library*

ISBN 978-0-521-86683-5 Hardback

# Contents

## Part IV. Managing agro-biodiversity: causes, values and policies

# Figures

# Tables

# Contributors

WIKTOR ADAMOWICZ, Professor & Canada Research Chair, Department of Rural Economy, University of Alberta, Canada.

RAFAT ALAM, PhD candidate, Department of Economics, University of Ottawa, Canada.

OFER BAHAT, Department of Environmental Science and Chemistry, University of Indianapolis, Ibillin, Israel.

EDWARD BARBIER, John S. Bugas Professor of Economics, Department of Economics and Finance, University of Wyoming, USA.

STEFAN BAUMGÄRTNER, Assistant Professor of Ecological Economics, Alfred Weber Institute of Economics, University of Heidelberg, Germany.

NIR BECKER, Professor in Economics, Department of Economics and Management Tel-Hai College, NRERC, Haifa University, Israel.

DORIS BEHRENS, Professor in Environmental Economics, Department of Economics, University of Klagenfurt, Austria.

EKIN BIROL, Research Fellow, Homerton College and Department of Land Economy, University of Cambridge, UK.

PETER BOXALL, Professor in Economics, Department of Rural Economy, University of Alberta, Canada.

ERWIN BULTE, Professor in Environmental Economics, Department of Economics, Tilburg University, The Netherlands.

VIVEK CHAUDHRI, Associate Professor, Department of Management, Monash University, Australia.

JEAN-PAUL CHAVAS, Professor of Agricultural Economics, Department of Agricultural & Applied Economics, University of Wisconsin-Madison, USA.

YAEL CHORESH, Researcher, Department of Natural Resources and Environmental Management, University of Haifa, Israel.

MICHAEL CHRISTIE, Assistant Professor, Institute of Rural Studies, University of Wales Aberystwyth, UK.

TOM DEDEURWAERDERE, Assistant Professor, National Foundation for Scientific Research, Belgium and Université catholique de Louvain, Belgium.

SALVATORE DI FALCO, Senior Research Fellow, Department of Agricultural and Resource Economics, University of Maryland, USA.

ADAM G. DRUCKER, Environmental Economist, Economics of Animal Genetic Resources Conservation Programme, International Livestock Research Institute (ILRI), Addis Ababa, Ethiopia and School for Environmental Research, Charles Darwin University, Australia.

PAUL FACKLER, Associate Professor, Department of Agricultural and Resource Economics, North Carolina State University.

DAVID FINNOFF, Assistant Professor, Department of Economics and Finance, University of Wyoming, USA.

BIRGIT FRIEDL, Assistant Professor in Economics, Department of Economics, University of Graz, Austria.

TIMO GOESCHL, Professor in Environmental Economics, Alfred Weber-Institute of Economics, University of Heidelberg, Germany.

ARTHUR HA, Senior Economist, Economics & Policy Research Branch, Department of Primary Industries, Victoria, Australia.

NICK HANLEY, Professor in Environmental Economics, Department of Economics, University of Stirling, UK.

MARC HELMER, Department of Agricultural Policy and Market Research, Justus Liebig University, Giessen, Germany.

PAULA HORNE, Senior Research Fellow, Finnish Forest Research Institute, Helsinki, Finland.

TONY HYDE, Institute of Rural Studies, University of Wales Aberystwyth, UK.

MOSHE INBAR, Associate Professor, Department of Biology, University of Haifa, Israel.

ANDREAS KONTOLEON, Assistant Professor in Environmental Economics, Department of Land Economy, University of Cambridge, UK.

VIJESH KRISHNA, PhD Candidate, Faculty of Agricultural Sciences, University of Hohenheim, Germany.

BRIAN LEUNG, Department of Biology & School of Environment, McGill University, Montreal, Canada.

DAVID LODGE, Professor in Conservation Biology, Department of Biological Sciences, University of Notre Dame, Notre Dame, USA.

ERIC LONSDORF, Research Associate, Lincoln Park Zoo, Alexander Center for Applied Population Biology, Chicago, USA.

KEVIN MURPHY, Institute of Rural Studies, University of Wales Aberystwyth, UK.

ERIK NELSON, PhD Candidate, Department of Applied Economics, University of Minnesota, USA.

ERNST-AUGUST NUPPENAU, Professor of Agricultural Economics, Department of Agricultural Policy and Market Research, Justus Liebig University Giessen, Germany.

UNAI PASCUAL, Assistant Professor in Environmental Economics, Department of Land Economy, University of Cambridge, UK.

DAVID W. PEARCE, Professor in Environmental Economics, University College London, London, UK.

CHARLES PERRINGS, Professor in Environmental Economics, Environment Department, University of York, UK.

ALEXANDER PFAFF, Associate Professor in Economics & International Affairs, School of International and Public Affairs and Department of Economics, Columbia University, USA.

STEVE POLASKY, Fesler-Lampert Professor of Ecological & Environmental Economics, Department of Applied Economics, University of Minnesota, USA.

NGUYEN V. QUYEN, Associate Professor, Department of Economics. University of Ottawa, Canada.

JUAN ROBALINO, PhD Candidate, Department of Economics, Columbia University, USA.

ARTURO SANCHEZ-AZOFEIFA, Associate Professor, Earth and Atmospheric Sciences Department, University of Alberta, Edmonton, Alberta, Canada.

JASON SHOGREN, Stroock Distinguished Professor of Natural Resource Conservation & Management, Department of Economics and Finance, University of Wyoming, USA.

ANDERS SKONHOFT, Professor in Environmental Economics, Department of Economics, NTNU, Trondheim, Norway.

MELINDA SMALE, Senior Economist, International Plant Genetic Resources Institute and International Food Policy Research Institute, Washington, DC, USA.

ANTHONY STARFIELD, Professor, Department of Ecology, Evolution and Behavior, University of Minnesota, USA.

GARY STONEHAM, Chief Economist, Economics and Policy Research Branch, Department of Primary Industries, Victoria, Australia.

LORIS STRAPPAZZON, Principal Economist, Economics Branch, Division of Agriculture, Department of Natural Resources and Environment, Victoria, Australia.

TIMOTHY SWANSON, Professor in Law and Economics, Department of Economics and School of Laws, University College London, UK.

YACOV TSUR, Professor of Agricultural Economics, Department of Agricultural Economics and Management, The Hebrew University of Jerusalem, Israel.

DAAN VAN SOEST, Associate Professor, Department of Economics and CentER, University of Tilburg, The Netherlands.

JANA VYRASTEKOVA, Associate Professor, Department of Economics and CentER, University of Tilburg, The Netherlands.

ROBERT WRIGHT, Professor in Economics, Department of Economics and Vice-Dean, Faculty of Management, University of Stirling, UK.

JOHN WARREN, Institute of Rural Studies, University of Wales Aberystwyth, UK.

AMOS ZEMEL, Professor in Economics, Department of Energy and Environmental Physics, The Jacob Blaustein Institute for Desert Research, Ben Gurion University of the Negev, Israel.

# Preface

The field of biodiversity economics, i.e. the analysis of the problems at the interface between the disciplines of economics and biology, probably has its origins primarily in the work of Colin Clark. Much of this early work looked at the exploitation of fisheries in the context of various institutional assumptions: open access, social planning, etc. Since these early efforts, the field of biodiversity economics has expanded in many different directions. It still concerns the analysis of the causes of resource overexploitation and decline, but also includes within its core the examination of the sorts of externalities involved (values) and the types of policies applied. In addition, and most crucially, the field now encompasses many resources other than simply marine resources: forests, wildlife, and even genetic resources (used in agriculture and pharmaceutical industries). The entire diversity of biological resources within the living world is now brought within the field of biodiversity economics.

All of these problems share a common aspect – the dynamic nature of biological resources. Biological resources are distinctive in that they live and grow and respond to other living things. This generates a common analysis across the entire discipline that focuses on how human societies interact with other living things and how management should take biological characteristics into consideration.

In this volume we provide a set of papers that demonstrates the application of this framework across the entire range of issues currently under consideration within this important field. We divide the volume into four sections, three representing the core areas of biodiversity economics and the last a demonstration of their application in a concrete context (agricultural biodiversity). In Part I, we commence with a set of eight papers comprising an examination of the causes of biodiversity loss. Then in Part II we turn to a section of five papers assessing the issues concerning the valuation of biodiversity. In Part III we examine the range of policies for biodiversity conservation. Finally, in Part IV, we include a case study on agricultural biodiversity: causes, values and policies. The volume as a whole serves as a demonstration of the means by which bio-economic analysis might be applied to the examination and evaluation of the problem of various forms of biodiversity losses.

The volume emanates from a collaborative effort undertaken by an interdisciplinary network of European scientists (known as BioEcon) working to advance economic theory and policy for biodiversity conservation. The BioEcon network has provided a platform for economists, lawyers and natural scientists from leading European academic and research institutions as well as members of prominent policy organisations to work together on advancing our understanding of the anthropogenic causes of biodiversity decline as well as on developing novel economic incentives for biodiversity conservation.[1] Over the past decade more institutions from all around the world have become involved in the network activities (such as its annual conference) while the network has provided the launching pad for many new researchers and research agendas in the field of biodiversity economics.[2] We hope that this volume will help to consolidate this relatively new field and continue to encourage new researchers and new research agendas in the area.

ANDREAS KONTOLEON, UNAI PASCUAL, TIMOTHY M. SWANSON

[1] The partners in BioEcon are: Alfred-Weber-Institute, University of Heidelberg, Germany; Center for Development Research, Department of Economics and Technological Change, University of Bonn, Germany; Centre for Economic Research, Tilburg University, Netherlands; Centre for Environment and Development Economics, Environment Department, University of York, UK; Centre for the Philosophy of Law, Université catholique de Louvain, Belgium; Department of Economics, Norwegian University of Science and Technology, Norway; Department of Economics, School of Oriental and African Studies, UK; Department of Economics, University College London, UK; Department of Land Economy, University of Cambridge, UK; Finnish Forest Research Institute, Vantaa Research Centre, Finland; Fondazione Eni Enrico Mattei, Italy; Laboratoire Montpellierain d'economie Theorique et Appliquee, Centre National de la Recherche Scientifique, Université Montpellier 1, France.

[2] Details of all network activities can be found at www.bioecon.ucl.ac.uk

# Acknowledgements

We are grateful to the European Commission for the initial funding of the BioEcon network under Framework V and for the guidance of Dr Martin Sharman in the development of the undertaking. We are also grateful to the European Commission, DEFRA-UK, and DIVERSITAS for their ongoing support and funding of the BIOECON annual conference. We are also grateful to several policy organisations that have been engaged in collaborative work with the BioEcon network and from which many of the chapters included in this volume have resulted. These include the IUCN-World Conservation Union, the International Food Policy Research Institute (IFPRI), the International Plant Genetic Resources Institute (IPGRI), the Organisation for Economic Co-operation and Development (OECD), the World Bank, Conservation International (CI), Resources for the Future (RFF), and the China Council for International Cooperation on Environment and Development (CCICED).

Lastly, we would like to dedicate the volume to the late Prof. David Pearce who has been an esteemed colleague, collaborator, teacher and friend to the contributors to this volume. Over the past thirty years Prof. Pearce has made several important conceptual and methodological contributions towards our understanding of the causes of biodiversity decline while he has been instrumental in popularising and establishing economic instruments for biodiversity conservation into major policy fora. The introductory chapter written specifically for this volume was sadly one of Prof. Pearce's last works. In this paper Prof. Pearce explores the strength and nature of societies' preferences for conserving biodiversity resources and finds that in many contexts actual conservation actions and budgetary outlays fall considerably short of the 'rhetoric' over how much we care about biodiversity. His insightful piece concludes by highlighting the importance of accurately valuing and accounting for biodiversity resources and services in public decision making, which constituted a recurrent and far-reaching policy message from his important body of work.

Every attempt has been made to secure permission to reproduce copyright material in this title and grateful acknowledgement is made to the authors and publishers of all reproduced material. In particular, we would like to acknowledge the following for granting permission to reproduce material from the sources set out below:

Chapter 7 originally published as 'Trade and Renewable Resources in a Second Best World: An Overview', Bulte, E. H. and Barbier, E. B., *Environmental and Resource Economics*, Vol. 30, No. 4, pp. 423–463, 2005. Reproduced with kind permission from Springer Science and Business Media.

Chapter 13 originally published as 'Multiple-use management of forest recreation sites: a spatially explicit choice experiment', Horne, P., Boxall, P. C. and Adamowicz, W. L., *Forest Ecology and Management*, Vol. 207, No. 1–2, pp. 189–199, 2005. Reproduced by permission of Elsevier.

Chapter 18 originally published as 'Conserving species in a working landscape: land use with biological and economic objectives', Polasky, S., Nelson, E., Lonsdorf, E., Fackler, P. and Starfield, A., *Ecological Applications*, Vol. 15, No. 4, pp. 1387–1401, 2005. Reproduced by permission of The Ecological Society of America.

# Foreword

I am delighted to see that biodiversity economics has become a discipline in its own right. Those of us who have been addressing the multiple dimensions of biodiversity have long sought better ways of incorporating economic thinking into our various challenges. Biodiversity loss is a serious preoccupation for the entire science of conservation biology, which has its own journal and scientific society, but it remains weak in delivering appropriate policy advice, largely because it is not able to demonstrate the economic implications of policy alternatives.

Other parts of the biodiversity community deal with what ultimately is an economic relationship, namely sustainable use. While the concept certainly has significant ethical dimensions, it more fundamentally deals with the costs and benefits of alternative management strategies, and these often will be based on economic principles. Is it more cost-effective to have safari hunting of rhinoceros, or photo safaris? How can economic calculations of sustainable off-take incorporate stochastic events, such as annual changes in rainfall (and thus productivity of vegetation)?

Others working on biodiversity focus on very specific issues, such as the impact of invasive alien species on natural ecosystems and human economies. Quantification of the negative impacts of these invasive alien species can help to convince policy-makers to design, implement and support appropriate measures to prevent such species from becoming established or to manage them efficiently once they have become part of an ecosystem. Biodiversity economics has much to contribute to the problem of invasive aliens, clearly demonstrating the suitability of alternative approaches to the problem.

I was also pleased to see the attention being given to the non-wild parts of biodiversity, here called 'agro-biodiversity'. The relationship between domesticated landscapes and the surrounding matrix has significant economic dimensions, as these non-domestic landscapes provide important ecosystem services to the agricultural lands. These include providing clean water, supporting pollinators, maintaining habitats for wild relatives of domesticated plants and animals (thereby providing genetic materials for the future), forming soils and ameliorating climate extremes. All of these have economic dimensions, and biodiversity economics has a key role to play in helping to develop appropriate incentive measures, such as systems of payment for ecosystem services, that are efficient and equitable as well as environmentally effective.

These are just a few of the topics where biodiversity economics is making important contributions. It is especially pleasing to see the breadth of institutions involved in BioEcon, demonstrating that biodiversity economics is built on a solid consensus of scholarly research.

I would like to close by paying homage to David Pearce, whose many contributions to biodiversity economics over the past few decades have been the foundation upon which so many other contributions have been built. His economics-based perspectives have helped to legitimise the arguments conservationists have been making for many decades, while also usefully challenging some of our cherished assumptions. His opening chapter well summarises many of the ideas that made his contributions so powerful to policy-makers and scientists alike. This is a worthy monument to his numerous contributions.

While biodiversity economics addresses issues such as valuation, incentives and tradeoffs, it is also apparent from this volume that much work remains to be done. This book is the best available account of the current state of the art in this important discipline. I have no doubt that the coming years will lead to even more dramatic progress in biodiversity economics. The future diversity of life on our planet depends on such progress.

JEFFREY A. MCNEELY
*Chief Scientist*
*IUCN-The World Conservation Union*
*Gland, Switzerland*

# Introduction

*Andreas Kontoleon, Unai Pascual, Timothy Swanson*

## 1 An introduction to biodiversity economics

Biodiversity economics examines the causes, values and policies associated with the problem of biodiversity decline. It usually involves a dynamic analysis of the living resources and hence it must often incorporate the facets of growth and responsiveness that are characteristics of such living things. In addition, it often focuses on the more esoteric forms of non-use values, such as real options and existence values, and the techniques available for quantifying them. Finally, it must also consider the manner in which decisions are taken in the context of such complicated dynamics and values. In sum, the field of biodiversity economics considers some of the interesting and complex dynamics within and between the social and natural worlds.

In this volume we attempt to categorise the various parts of this field under three headings: causes, values and policies. We then use a set of papers to demonstrate the meanings of these categories in this field and the development and extension of these concepts in this context. The intention is to demonstrate both the entire set of issues encompassed by the field of biodiversity economics and the manner in which frontier-level analysis and research is being undertaken within this realm.

Biodiversity economics is exciting and important work, as demonstrated by the various chapters within this volume. It encompasses both interesting topics and contexts (such as wildlife, forests and genetic resources) and important and complex problems (such as biological resistance, invasion and valuation). We set forth here a summary of the area, in the hopes of encouraging readers to pursue the chapters and the field in even greater detail.

## 2 The causes of biodiversity loss

Biodiversity loss is occasioned by many different factors, but three primary categories come to mind. First, there is the problem related to the continuing changing of land use across the globe and how this impacts upon the structure of ecosystems and their resident species diversity. This is a social problem to the extent that land use conversion is undertaken either without consideration of the value of biodiversity being lost in the process, or without consideration of the potential costliness of the continued conversion process in terms of undermining system resilience and stability. Second, there is the problem related to the ongoing expansion, emergence and integration of markets and states. In particular, existing trends of globalisation alongside the deepening of trade liberalisation have important but still insufficiently understood impacts on the stocks of living resources and the services that they provide to society. This is a social problem to the extent that it is difficult to invest in the maintenance of stocks when increased flows result from market integration; that is, it is often a problem of inadequate institutions and incentives. Third, there is the problem of the movement of some species (by societies) into the areas inhabited by others and the unintended or unforeseen impacts of these exotics. This is a problem determined by the biological character of some species, which are unable to inhabit areas in proximity to others, and the difficulty of internalising these biological characteristics into human decision making. Let us consider each of these distinct problems in turn.

### *Habitat conversion*

First, the problem of land use conversion concerns the difficulty of incorporating the values or potential costs of habitat conversions into social decision making. In this volume we examine three aspects of this problem, associated with institutions, externalities and potential hazards.

In Chapter 2, the question under consideration concerns the potential impact of institutions in contributing to the problem of habitat conversion. Edward Barbier considers the problem as one of potential for relatively uncontrolled or unmanaged resources to result in resource degradation and then tacit conversion. An example he cites is the mangrove fisheries of Thailand, which were managed under open access institutions, whereas the shrimp farms which replaced them were managed under private property rights. He argues that the form of institution applied to the resource determines the capacity of the resource to withstand exploitation, and those that continue are those which have more formal property rights applied.

He demonstrates the argument by an empirical analysis that examines the impact of a more formal institution as a source of 'friction' between the resource concerned and the market forces acting upon it. An open access resource suffers from exploitation immediately responsive to market pressures through its major signals (prices, wages), while a managed resource is conceived as one that responds (if at all) only with a lag. He finds that in the case of the Mexican *ejido* common landholding system, resource exploitation responded to market pressures with a lag while the Thai mangroves, mentioned above, responded with immediate effect. In short, Barbier is demonstrating that status quo open access regimes constitute unofficial policy for encouraging conversion of natural resources.

Another factor important in habitat conversion is the presence of externalities between various users. One user may perceive another as a potential competitor for its resources, in either the harvest or the marketing of the resource. In either case the incentive is for the resource user to respond to the existence of potential threats from other users with strategic overexploitation. This is simply the incentive to hoard the resource, or the disincentive to invest, by reason of potential competition. There are many reasons that such competition for resources and resource markets might exist, but the primary impact is the same as in the absence of adequate institutions: there is an incentive to convert the habitat to a use that might more easily be controlled. This might be the reason that many forests are converted to pasturelands, with the implied loss of biodiversity, simply because the residents of the forests are contestable whereas the converted cattle are not.

This hypothesis is examined in this volume in an empirical analysis by Robalino, Pfaff and Sanchez-Azofeifa (Chapter 3). They posit the case in which neighbouring users have impacts upon one another by reason of the relative balance of converted and non-converted goods and services provided within the market. They examine the nature of the interaction within the context of a spatially detailed dataset concerning two regions in Costa Rica. In their analysis they demonstrate the existence of such positive interaction effects in one region and its absence in the other. They argue that this difference could derive from something as formal as institutional differentiation, but also from something as informal as differing expectations within differing communities. So, their analysis emphasises the role of informal as well as formal institutions in the determination of the ultimate effect of externalities in resource conversion.

The final paper considering the problem of habitat conversion takes a very different approach. Chapter 4, by Tsur and Zemel, argues that the problem derives not from user-based externalities and the imperfection of local institutions but instead from the difficulty in internalising global-level externalities. They posit the problem as one in which the continuing conversion of habitat generates an increasing hazard on a global scale, one which must be endogenised by a global planner. This would be the case for example if the continuing conversion of arable lands to a few cultivated crops resulted in a production system that was susceptible to collapse. Then each individual conversion would enhance the prospect of collapse for the entire system which, if it happened, would be experienced as a collapse in welfare at the global level.

Tsur and Zemel examine the manner in which a social planner should act to manage for such a potential hazard from habitat conversion, and find three distinct cases. First, if there is a known threshold for the potential of collapse, then the social planner should simply bound conversion at that point and guide global conversion asymptotically towards that threshold. Second, if the threshold exists but its level is uncertain, then the social planner will provide a safety cushion against collapse in its conversion pathway. In short, with manageable uncertainty, there must be a precautionary cushion provided against the potential for collapse. Finally, if the threshold exists, is uncertain and exogenous (i.e. the conditions under which collapse will occur are not entirely within the control of the social planner), then under certain conditions it might be optimal to hasten the process (to ensure completion prior to exogenous collapse).

This analysis is an important example of the distinctive character of living resources. Some resources remain far more static and manageable (for example, exhaustible resources such as minerals) and so the process of control is much more manageable. Living resources provide elements of exogeneity and unpredictability to the problem that create complex approaches to decision taking.

### *Trade*

Trade also places pressure on biological resources. Many advocates argue that it is free trade that degrades biological diversity and that trade in diverse resources should be banned to halt its decline. Under very simple assumptions, this is true. It is straightforward to show that a resource-abundant economy with unmanaged resources will be subject to overexploitation (and resource decline) if opened up to trade with others. To the extent that biodiversity resources exist primarily within developing countries, with attendant resource abundance and institutional deficits, increased international trade must result in the decline of biodiversity resources.

The survey paper by Bulte and Barbier (Chapter 7) demonstrates the caveats to this simple observation. First, if we move away from this most simplistic caricature of North and South and towards a more realistic depiction in which both regions experience some institutional imperfections and some diverse resource endowments, then the impacts of trade are more ambiguous. It is possible both to achieve the economists' 'dream scenario' – in which the institutions and production patterns converge on the less resource-intensive and higher-quality institutions – and it is also possible to achieve the nightmare of the simplistic case set forth before. In general the outcomes are more mixed and less straightforward than the advocate's argument.

For Bulte and Barbier this points to the importance of trade in endogenising institutional development, so that those parts of the world with imperfect institutions are able to improve through interaction and exploitation. The incentives to engage in monitoring and enforcement, as well as any other institutional development, hinge upon the perceived investment-worthiness of the resources concerned. Bulte and Barbier argue that trade impacts crucially depend upon the institutions through which they are channelled.

In Chapter 8 Alam and Van Quyen provide a neat depiction of the manner in which trade interacts with other fundamental causes of biodiversity loss, namely population growth and habitat conversion. They posit the case in which biodiversity resources are under an exogenous threat by reason of the continuing expansion of human populations, and markets for biodiversity are thus crucial to the provision of incentives to avoid habitat conversion pressures. In this case, the absence of trade in biodiversity will necessarily lead to the decline of the resource, on account of demographic pressures, and hence a ban on trade is not a viable instrument in and of itself. The issues then come down to the nature of the trade instrument capable of channelling value and creating incentives to invest in biodiversity resources, and the capacity of these incentives to counterbalance the forces deriving from population pressures.

In sum, the chapters on trade survey a wide range of issues concerning habitat loss, resource exploitation and population pressure. The analysis by Bulte and Barbier emphasises the importance of institutions in channelling the pressures from trade into constructive purposes. The analysis by Alam and Van Quyen emphasises the necessity of providing trade as a counterbalance to the pressures from population. The papers together demonstrate the crucial manner in which the various drivers of biodiversity decline interact and the way in which management must consider them together.

### *Invasives*

A third important cause of biodiversity's decline is the introduction of unusual species within natural environments. This may happen purposefully (via the conversion of land uses described previously) or it may happen incidentally (via the spread of species attendant upon casual exchange and trade). The latter problem is the one considered under this heading. As trade patterns extend across the globe and the volume of trade increases, the spread of species beyond their natural domains continues to threaten many of the naturally resident species.

The problem of invasives is primarily a problem of monitoring and control. It concerns the need to expend resources to manage an externality attendant to otherwise casual activities, when the externality is both i) incidental to an otherwise unrelated activity (such as the carriage of zebra mussels in sea vessels' ballast water), and ii) primarily for the benefit of states other than the one that must undertake the control expenditures. Invasive species problems occur on many scales, but those that occur in the context of trade usually involve several, if not many, different states. Thus there is a significant public good and free rider facet to the invasive species problem.

In Chapter 5 Charles Perrings considers how such a problem is related to poverty in a survey paper that touches on many of the key issues concerning invasive species. Perrings finds that a crucial part of the invasive species' problem concerns the 'weakest link' nature of the public good, i.e. the state undertaking the least amount of management determines the quality and extent of the externality. This of course relates the invasives problem to the environmental Kuznets curve literature, examining the links between income and environmental management. Do poor countries necessarily undertake less environmental management and hence determine the quality and extent of the invasives problem? Perrings finds that the literature does not support the finding that poor countries must necessarily provide poor environmental management, but does promote the finding that the support of monitoring and enforcement in poor countries is important to the resolution of the problem. In general he expects that the invasives problem will become more prevalent as international trade increases and income expands, and that international cooperation will be important to its resolution.

A second paper on invasives by Finnoff, Shogren, Leung and Lodge (Chapter 6) concerns the choice of the appropriate instrument for the management of the problem: prevention or control. Control concerns the use of current information and current resources to minimise the negative impacts of invasives. Prevention involves investments in those assets most appropriate to minimising the problem over the future. The authors find that the choice between these two approaches depends on more than just their relative cost-effectiveness. It also depends on, first, the prevalent social discount rate and, second, the risk aversion of the decision maker. Basically, the two factors cut in opposite directions. The lower the social discount rate, the more that investment/prevention is preferred as the costs of discounting declines. In this case, the lower discount rate makes the case for prevention over control. However, in regard to risk aversion, the opposite conclusion results. Greater risk aversion in fact mitigates in favour of awaiting greater information prior to selecting action and so favours control over prevention. The paper makes clear that the problem of invasives concerns the control of a dynamic problem, in which the flow of information and the rate of return are critical elements of the solution.

Invasive species are clearly the most biological cause of the decline in biological diversity. It is the analysis and evaluation of the biological dynamics of interacting species that is at the core of the biology of this problem. The social dimensions are more concerned with the problem of enforcement/control and public good provision, and the combination of social and biological facets makes this a fascinating area for future work.

*Summary: causes of biodiversity loss*

The eight chapters in this part of the volume provide an interesting survey of the range of causes considered within the literature on biodiversity economics: habitat losses, trade and overexploitation, invasive species. These forces together account for much of the loss of biodiversity in the world today, and the authors here demonstrate the forms of analysis being brought to bear in the examination of this facet of the problem. They demonstrate that there are many economic facets to the problem of biodiversity loss, including i) the provision of public goods and development of institutions, ii) the valuation of externalities and iii) the choice and implementation of the appropriate instruments and controls. In the next two parts of this volume, we turn to the papers focusing on the latter two aspects of the biodiversity problem.

## 3 The value of biodiversity

In the introductory chapter to this volume, Prof. Pearce highlighted the importance of the monetary valuation of biodiversity for designing incentives that can induce optimal conservation efforts. In fact, valuation is important for all facets of public decision making that impact upon biodiversity resources. In particular, valuation plays an indispensable role in project appraisal and regulatory review (i.e. cost-benefit analysis), in the setting of environmental regulations (e.g. Pigouvian taxes), in the assessment of damages in liability cases (e.g. in oil spill cases), as a precursor for designing markets for biodiversity conservation (e.g. ecotourism, green forest products, etc.) and for green national accounting (e.g. accounting for the depreciation of national capital). Yet the accurate valuation of biodiversity resources and their services is hindered by their strong non-market and public good characteristics. This fundamental problem has spawned a vast literature examining both conceptual and technical issues on the valuation of biological resources. This research has provided insights of a wider scientific interest reaching beyond the field of biodiversity economics such as contributions to our understanding of individual preferences over public goods, the nature of altruism and quasi-option informational values, the problem of discounting as well as numerous important technical and econometric contributions.

In Part III of this volume we present two sets of papers with advances in this literature on the value of biodiversity resources and services. The first set includes two chapters addressing conceptual valuation issues, while the latter includes three chapters presenting developments in techniques of non-market valuation.

### *Concepts*

The first conceptual chapter by Goeschl and Swanson (Chapter 9) assesses the approaches for evaluating the informational value of genetic resources. The authors conceptualise the informational value of biological diversity analogous to that of a library containing all published written works. As resources are scarce, cost-benefit type decisions would need to be made over the optimal portfolio of books that should be preserved. The authors extend this logic to the management of the informational value of biodiversity and analyse three different approaches over how to manage this 'legacy library'. The first concerns how to construct the library richest in information given the opportunity costs. Here, the issue is one of taking a given budget and using it most effectively to maximise genetic variability, but without reference to any use or usefulness to humans. While the approach can be extended to encompass value dimensions other than distinctiveness, there are two cogent criticisms of this approach. The first is that the supply-side orientation overlooks that additional search investments have to be carried out to utilise the resources inherent in the legacy library. The other is that it is essentially static and thus places little emphasis on the potential values from biodiversity conservation as a means of solving problems important to human societies.

The second approach for maximising the informational value of biodiversity focuses on how to design the library to optimise the search for a given piece of information. Under such a search framework the problem is one of identifying the useful information as quickly as possible – the emphasis is upon the time and resources expended in the process of search for a solution to a specific problem. In this context, the existence of the biodiversity resource generates costs and benefits in equal measure, by being both information and obstacle. The most cogent criticism of this approach is that, although rooted in the idea of diversity as useful information, it also casts diversity as the major obstacle to its own usefulness. In a world in which the storehouse of genetic information is becoming ever smaller, it would seem that this is an interesting but potentially inadequate manner in which to cast the problem of biodiversity management.

The third approach for maximising the informational value of biodiversity centres on the problem of designing the library so as to provide the optimal stock of information to meet the demands arising from an endless stream of future unpredictable problems. In this context the authors explore the conditions under which today's decision should use current information, but provide an additional hedge against future uncertainties. This could imply the retention of larger libraries with a greater diversity of information. The most cogent criticism of this literature is that it implicitly assumes stagnant (or even regressive) technological progress and a lack of substitutability for biodiversity-sourced information.

These three very different approaches lead to very different answers regarding the design of the legacy library; however, taken together they provide insights into the value of the amount and quality of information contained within biological systems. The authors conclude that in the presence of irreversibilities, the current generation has the responsibility to make this decision over maximising the informational value of biodiversity by reference to the longer-term welfare of future generations, and that it may be dangerous to assume that technological change alone will be able to solve this problem.

The next conceptual paper, by Stefan Baumgärtner (Chapter 10), is concerned with the measurement of biodiversity as a precursor to its valuation. The author argues how measurement of biodiversity is subject to value judgements. He explores and compares the value judgements underpinning both the economic and ecological measurement of biological diversity. It is critically argued that there are systematic differences between these two approaches of biodiversity measurement. In doing so Baumgärtner makes two important advances to our understanding of measurement of biodiversity. First, he displays how the two types of biodiversity measures – the ecological and the economic – aim to characterise two very different aspects of biological systems. While the ecological measures describe the actual, and potentially unevenly distributed, allocation of species, the economic measures characterise the abstract list of species existent in the system. Second, he discusses how ecological and economic measures of biodiversity differ on account of fundamental differences in their philosophical perspective on biodiversity. The former follows a more 'conservative' view that can be traced to ideas by Leibniz and Kant, while the latter can be traced to a more 'liberal' perspective associated with the ideas of Descartes, Locke and Hume. These different value judgements lead to different valuation measures for biodiversity. Understanding these value judgements goes a considerable way to explaining commonly observed disagreements over the relative importance of biodiversity as well as over alternative conservation paths.

### *Techniques*

The section then turns to present frontier technical research on methods of non-market valuation. The papers selected for inclusion in this section of the volume make important technical and methodological advancements that contribute to fields such as experimental economics and applied micro-econometrics. Yet their main motivation (which is shared across all these papers) is to provide a direct contribution towards solving specific biodiversity conservation problems. That is, this set of papers provides an illustration of how current technical research in valuation can be purposefully pursued so as to enhance our understanding of how to design effective and efficient biodiversity conservation policies.

The first of these chapters, by Becker, Choresh, Inbar and Bahat (Chapter 11) develops an approach that combines the travel cost (TC) and contingent valuation (CV) methods for assessing the marginal value of an endangered species so as to assess the efficiency and cost-effectiveness of alternative wildlife conservation strategies. The authors apply this approach to assess the marginal values associated with the protection of the Griffon vulture in Israel via the development of alternative methods. In particular they utilise the result from the TC and CV to carry out cost-benefit analysis, both at the regional level with respect to assessing the welfare implications of one particular conservation activity (namely feeding stations) and at the national level with respect to assessing broader vulture conservation policy options. The former of these analyses was undertaken by comparing the costs of feeding stations to the estimated value of the marginal vulture derived from the two valuation methods. It was found that protecting vultures passes a national cost-benefit test and that feeding stations are economically viable in generating on average 0.23–2.12 vultures annually. In the latter analysis two additional policy issues were analysed: entrance fee policy and effort allocation. With respect to pricing policy it was shown that by charging the revenue-maximising entry fee level, policy-makers can generate a large increase in revenues compared with the current situation, but at the same time this will bring about a substantial welfare reduction. The region where the tradeoff between the revenues and welfare is relevant to the policy-makers was empirically identified. Further, it was shown how combining CVM and TCM results can provide insights as to the overall optimal allocation of vultures and visitors between 'competing' sites or nature reserves.

The wider policy contributions of this piece of research are the following. First, there are solid economic arguments to invest in 'charismatic' wildlife species, even if their population size is above the critical survival threshold level. Second, assuming we have good ecological appreciation of the cost-effectiveness of feeding stations, there is a welfare-enhancing rationale for differentiating efforts among different ecotourism sites. Finally, this research highlights the importance of creating a comprehensive database of critical survival threshold levels of different species, which would allow wildlife policy decisions to be made at the margin, which in turn would lead to a more efficient allocation of conservation funds and efforts.

The next chapter, by Christie, Hanley, Warren and Murphy (Chapter 12) uses the choice experiment (CE) method in an attempt to assess the value of different aspects of biodiversity *per se* instead of specific biodiversity *resources*. In doing so, the authors develop a novel methodological approach to the valuation of biodiversity. In particular, they draw heavily on ecological literature relating to the definition and measurement of biodiversity. This literature is then used to feed into the design of a CE which examines public values of various attributes of biodiversity in the UK. The difficulties involved with presenting complex and often new information in valuation studies had to be addressed. The authors make use of a novel way of conveying such information to respondents, in a manner which is consistent with ecological understandings of biodiversity. The attributes chosen for inclusion in the CE were familiarity of species, rarity of species, habitat quality (e.g. habitat restoration vs. new habitat creation) and state of ecosystem processes (e.g. ecosystem services, such as flood defence, have a direct impact on humans). The results suggest that respondents exhibited a high degree of understanding of the concepts used in the design of the CE study. Further, the authors find that the UK public exhibits strong preferences for the preservation of its biodiversity and that the nature of this value can be mostly classified in the passive non-use value category. Finally, the authors examine the relative importance placed on biodiversity attributes which provide insights as to what type of biodiversity conservation policies would be welfare enhancing for the UK public. For example, they find that the UK public would support policies that target rare and familiar species of wildlife while they would be unwilling to support policies that simply delay the time it takes for such species to become extinct. Further, there was evidence that the public would support policies that aimed both to protect and increase habitat, although this support was found to be weaker compared with that for the conservation of rare species. Finally, the choice experiment suggested that the public placed a higher priority on conservation efforts targeting ecosystem functions that *directly* affect humans and were less interested in other types of more indirect ecosystem services.

In the last chapter in this section (Chapter 13), Horne, Boxall and Adamowicz turn to the assessment of forest recreational and passive use values by employing a CE method. The novelty of their work lies in explicitly accounting for the impact of spatial dimensions in the supply of biodiversity goods and services. In an application of the CE method in Finland, the authors examine the values associated with specific municipal forest recreational locations and identify spatial preferences for biodiversity conservation.

The study shows how preferences for forest management at one site may be somewhat different than preferences for forest management over the set of recreation sites in a particular study area. This implies that forest management strategies should be best viewed over the entire system of spatial units, where the manager faces an option of varying levels of management intensity among the sites. Within this system, the manager could assign different management goals for each site, or integrate all management activities into a management system applied at all sites. Therefore, one important management attribute that should be considered is *variability* or *flexibility* in the management regime over the system of recreation sites. The study identifies the welfare impacts of altering the management regime. Further, by accounting for preference heterogeneity, their analysis allows for the examination of the characteristics of the 'winners' and 'losers' under alternative management regimes. The authors compare the results from a recreational site-selection model that included varying biodiversity levels across different sites with results from a model that used the average measures of biodiversity (such as species richness) across the entire system of sites. The comparative analysis clearly showed that the two models provided different policy conclusions, while the site-specific model was found to be statistically more efficient and to provide more information on the preferences for forest management. Comparison of the two models illustrates the benefit of including spatial information as a variable in understanding preferences for forest management.

The contributions made by these technical papers are indicative of some general patterns or trends in the literature on non-market valuation of biodiversity. In particular, there is considerable research in developing methods for combining different sources of data on preferences from different non-market valuation studies and/or methods. The paper by Becker *et al.* provides an illustration of this body of work by combining the TC with the CV method. Further, current research has focused on the valuation of biodiversity characteristics and services rather than biodiversity resources. The chapter by Christie *et al.* is one of the first attempts to quantify the value of ecologically coherent biodiversity measures using the CE method. One aspect of this research that has received considerable attention has to do with deciding on the optimal amount and quality of information that the researcher must convey to respondents partaking in stated preference studies that involve complex and largely unfamiliar biodiversity concepts. The contribution by Christie *et al.* provides an example of how such information can be harnessed from experts and then processed and conveyed to survey respondents.

In addition, researchers working on the valuation of biodiversity have acknowledged both the theoretical consistency and the policy importance of assessing marginal (as opposed to total) welfare impacts from changes in biodiversity levels. The work by Becker *et al.* exemplifies the importance of such marginal analysis for the design of optimal wildlife management strategies for the conservation of the Griffon vulture in Israel. Finally, a significant part of the biodiversity valuation 'scientific research programme' has been concerned with addressing and incorporating different levels and types of heterogeneity into the analysis. As shown in the paper by Horne *et al.* two such aspects that have received considerable attention are preference and spatial heterogeneity. With respect to the former, considerable econometric advances have been made in incorporating individual demographic and psychometric variables into random utility models. Regarding the latter, research has focused on incorporating location-specific characteristics (using spatial econometrics) into the assessment of both use and non-use values.

## 4 Policies for biodiversity conservation

Parts I and II of this volume help us understand better the fundamental institutional and economic factors determining the excessive depreciation rate of biodiversity and its associated social cost. However, this understanding would not be of much use if it is not used in the realm of policy-making. It is stressed throughout the volume that informal as well as formal institutions determine the fate of biodiversity conservation through changes in land use. This implies that investments in context-specific institutional assets (social capital) become important to adequately filter down the policies for the conservation of biodiversity resources. The level of precision and cost-effectiveness of conservation efforts is consequently greatly determined by the meso-economic environment.

Part III of this volume is concerned with policy issues and is divided in two complementary sets of papers. The first set focuses upon the design of innovative biodiversity contract mechanisms. The main characteristic of such contracts for the conservation of biological resources is their voluntary nature. That is, these three papers share a common concern which can be summarised by this question: how can voluntary agreements between providers of biological resources through conservation and beneficiaries of their valuable services be designed to yield cost-efficient conservation targets? Providers include land users with well-defined property rights to tracks of land in which conservation can be undertaken, and traditional local communities that manage land and its constituent biological assets as common pool resources. In addition, beneficiaries can either overlap with the community of resource managers, or belong to parts of society represented by government at different administrative scales. The answer to the above question logically depends both on the nature of the resources to be conserved, the institutional backdrop, and the nature of both providers and beneficiaries with their in-built strategies, such as free-riding. The next set of papers in this part attempts to move a step forward from the issue of policy design to the implementation realm. Therefore, the papers allow us to move closer in helping to find answers to another critical question: how can regulation be effectively implemented, in terms of precision (getting as close as possible to the target) and cost-effectiveness?

### *The role of voluntary agreements and contracts*

Departing from traditional market-based incentives, such as Pigouvian taxes and subsidies, a potentially fruitful way for conservation of biodiversity resources in private lands is by the direct creation of markets. In Chapter 14 Stoneham, Chaudhri, Strappazzon and Ha argue that this is by no means an easy task and that its design is challenging mainly because of problems associated with asymmetric information between the relevant parties. The causes of such asymmetries are obvious. Land owners (such as farmers) have a better appreciation than conservation programme administrators of the effects of undertaking conservation actions on their private economic decisions (e.g. on crop production).

Similarly, governments may hold information on the significance and scarcity of the biological resources found in private lands and this information may not be readily available or accessible to the individual landowners themselves. Against this backdrop, it is difficult to create markets for nature conservation. What is needed is the design of policy mechanisms that can reveal hidden information needed to develop markets or contracts between government and private landholders as primary stewards of biological resources and providers of environmental services through their conservation.

Gary Stoneham *et al.* demonstrate that it is feasible to create the supply side of a market for the conservation of biological resources by designing auction conservation contracts. With a defined budget, such auctions provide signals of the value of conservation translated into market prices which allow the allocation of financial resources to biodiversity conservation. The chapter also stresses the point that with a limited budget to recreate such markets for conservation by private land owners, flexible price auctions are more advantageous than fixed-price auctions as the former offer large cost savings to governments. Further, the chapter's case study, drawn from Victoria, Australia, is also helpful in highlighting several important auction and contract design complications that are likely to occur, including dealing with cases of multiple environmental outcomes, handling unforeseen site synergies and overcoming problems with revealing reserve prices.

Other forms of market creation have been heralded as potential means of biodiversity conservation, especially in situations in which land users manage land under diffuse and collective property right regimes, such as in most of the tropical regions of developing economies, and particularly in so-called biodiversity 'hot spots'. Retrospectively it can be said that one avenue that has possibly excessively been heralded as a panacea for conservation and development is that of bioprospection under the regulatory umbrella of the United Nations Convention of Biological Diversity (UNCBD). Dedeurwaerdere, Krishna and Pascual posit in Chapter 15 that under such a framework, contractual arrangements are the most usual institutional avenue to coordinate the different actors involved in bioprospection. Their key message is that there are important pitfalls of the actual biodiversity governance system arising from a too narrow and static notion of efficiency in the economics of regulation. The authors apply an evolutionary institutional perspective to a unique and widely acclaimed 'access and benefit sharing' (ABS) scheme from the Western Ghats, India.

The chapter first questions the validity of a narrow benefit-cost approach to approximate the social welfare loss from depreciating the genetic base of bioresources and the traditional knowledge which co-evolves with it. This criticism is then illustrated by focusing on the current ways that North–South bioprospection contracts and ABS agreements are designed. One important point that distils from the paper is that the value creation process of biological resources is a diffuse one occurring in adaptive and complex socio-ecological systems. By contrast, bioprospection contracts are almost entirely based on the assumption that the added value of bioresources arises just at the final stage of the full innovation chain. However, the chain also includes other nodes, such as the ecosystem itself recreating biological diversity and the contributions of the traditional knowledge of local communities. The paper thus calls for an institutional analysis of the full chain of the innovation process to assess the full potential benefits from bioprospection contracts.

The last of the chapters focusing on contract design is by Nuppenau and Helmer (Chapter 16). Their focus is on the design of compensation payment systems for waivers on ecologically unfavourable land use practices. The chapter presents a novel approach to spatial ecological-economic modelling, based on a principal (i.e. government) – agent (i.e. farmers) approach. This approach helps identify compensation payments schemes that are cost-effective, and in addition the analysis allows us to take into account the impacts of price policies on landscape structure and ecology following the impacts of such policies on farming intensity. The main question addressed is thus how to make compensation payments for conservation both more cost-efficient and targeted. In trying to answer such a question, the authors address a poignant issue, often downplayed in this literature: there is seldom a clear set of criteria to define what services landholders (e.g. farmers) should be compensated for, how payments ought to be organised and, as far as the outcome is concerned, how the issue of multiple land users can be addressed.

### *Conservation policy: implementation*

Part III of this volume also sets out the challenges of policy implementation for biodiversity conservation. The previous set of papers largely focuses on voluntary contract mechanisms for biodiversity conservation. But it can be argued that centralised intervention has been and is expected to remain an obvious element in the regulation of biodiversity. It is thus worth considering which strategies governments should follow to achieve optimal conservation levels, especially in the face of their administrative powers to create new norms and their disposable budgets on which the extent of monitoring activities largely depends.

The last four chapters in Part III portray the challenges facing government regulators. Chapter 17 by Vyrastekova and van Soest puts forward an interesting and very relevant question in the face of existing advantages and disadvantages of both decentralised (voluntary or informal) and centralised (coercive or formal) approaches to regulate the use of biological resources under common property. They contend that a mix of centralised and decentralised enforcement mechanisms may render outcomes which are superior to their individual counterparts. The idea is an appealing one from the point of view of conservation policy implementation. The government's powers can set the framework rules to exploit the self-regulatory mechanisms that can endogenously arise and be maintained by the community of resource use members itself. In their review chapter, the authors focus on past economic experiments to demonstrate, systematise and analyse the relative weaknesses and strengths of both formal and informal self-regulatory approaches. First, they contend that when regulations are likely to be poorly enforced by governments, implying a low probability of convicting someone breaking the norm, governments may often do better by not trying to impose any enforcement at all. This is a warning call about the potential hidden costs due to ill-implementation of regulatory strategies. For instance, one may think of the countless examples of increased strategic overexploitation of biological resources in newly created protected natural areas, largely due to an inadequate endowment of institutional assets and financial resources from local governments to enforce and effectively implement such protection. Second, self-regulation by natural resource users under common property may not be a panacea even when they have a stake in conserving the resources above what would be privately optimal. An often necessary, though not sufficient, condition for success is for such users to realise the welfare-enhancing effects of cooperation with respect to resource exploitation while not putting aside cooperation with respect to their other economic activities as well.

In Chapter 18 Polasky, Nelson, Lonsdorf, Fackler and Starfield acknowledge the likely counterproductive effects that may arise due to implementation problems of formally established natural protected areas to conserve habitats. The authors argue that given such institutional impediments, it is important to address conservation issues on lands outside such formally protected zones. Using a spatially explicit model they analyse the effects of alternative land-use patterns on species conservation and the economic returns of such a strategy. The paper is a clear example of the potential to integrate biological and economic models to search for efficient land-use patterns. Using a case study from Oregon (USA) with three typical land uses (managed forestry, agriculture and protected areas) they contend that thoughtful land-use planning that includes the possibility for recreating a simultaneous mosaic of different land uses can achieve very similar conservation objectives but at a lower cost compared with investments in exclusion zones (i.e. protected areas) with economic activity occurring only outside such reserves.

An interesting complement to the results by Polasky *et al.* is provided by Behrens and Friedl (Chapter 19). Their paper focuses on the implementation of another widely used policy approach to conservation, the so-called 'flagship approach', which involves efforts to conserve charismatic species such as the giant panda, the monk seal or the golden eagle. Here, the policy question revolves around the difficult choice between a 'flagship approach' and an 'ecosystem approach' to conservation of biological resources, the latter involving a more comprehensive approach targeted at protecting entire ecosystems. The choice becomes even more poignant when there are tradeoffs (or conflicts) between wildlife conservation and nature-based tourism.

The authors demonstrate that if policy implementation is thought to maximise intertemporal social welfare from both recreation and conservation subject to the natural links between species and their habitat, then in principle an optimal dynamic visitor control strategy can be found. Such a result is illustrated for the case of the golden eagle in the Eastern Alps. Moreover, the authors point out that the flagship approach to conservation may still be a good strategy for biodiversity conservation, thus underpinning efforts from large nature conservancy non-government organisations (NGOs) such as WWF, the global conservation organisation, which is increasingly targeting biodiversity conservation actions by using campaigns aimed at protecting a handful of charismatic species.

The closing paper of Part III by Anders Skonhoft (Chapter 20) sheds some light on optimal conservation implementation strategies focusing on another order of complexity of natural systems. This is the one associated with the species dynamics that occur when policy-makers choose to reintroduce species into specific habitats. Moreover, apart from emergent ecological dynamics, existing economic activities can also be affected in various ways. Skonhoft uses the example of the creation of severe conflicts when recolonised species are large carnivores like wolves and grizzlies that threaten livestock together with prey species which may have important consumptive values such as for food or hunting. Yet often, reintroduced or recolonised species can have similar consumptive values (e.g. hunting) in addition to *in situ* recreational or existence values. The main point made is that implementing species conservation strategies through recolonisation creates new complexities, both at the ecological level and in terms of generating new conflicting values that policy-makers need to take into account. Skonhoft analyses such complexities with an in-depth study of the recent recolonisation episode of the grey wolf in Scandinavia.

## 5 Managing agricultural biodiversity: causes, values and policies

The final part of the volume constitutes an illustration of frontier research work in the field of biodiversity economics as applied to *agricultural* biodiversity. The three papers chosen bring together many of the issues raised in the three previous parts of the volume over the *causes*, *values* and *policies* associated with managing dynamic biological resources.

The first of these papers by Di Falco and Chavas (Chapter 21) provides a demonstration of biodiversity economics through its analysis of the causes of *in situ* agro-biodiversity decline when farmers are faced with different types of environmental risk. In particular, the authors develop a theoretical and empirical framework for assessing the role of *in situ* crop biodiversity in productivity and environmental risk management. They start their analysis from the observation that much of the literature to date on the role of diversity in productivity and risk uses the stochastic production function suggested by Just and Pope. The adoption of this framework implies that risk effects are captured by the variance of yields. Di Falco and Chavas demonstrate how such a framework fails to capture the full extent of risk exposure, namely exposure to unfavourable downside risk (e.g. severe drought leading to crop failure). In order to illustrate empirically these issues, the authors present a case study that uses data from durum wheat farms from rainfed agriculture in drought-prone areas of Sicily, Italy. Their empirical analysis indicates that crop diversity has a potential beneficial role in supporting farm productivity and in managing environmental risk. The analysis also suggests that such diversity can reduce the variability of yields. However, the authors find that the effect of diversity on yield variance appears to vary with pesticide use. While both diversity and pesticides have the potential to reduce variance, they behave as substitutes in their risk-reducing effects. This finding suggests the presence of strong interaction effects between pest management, ecological management and risk management. Lastly, crop biodiversity was found to be positively correlated with the skewness of the distribution of crop yields. This indicates that diversity can help reduce downside risk exposure (such as the probability of crop failure). The analysis, therefore, concludes that when unfavourable climatic and agro-ecological conditions expose farmers to particular environmental risks, crop diversity may become an important asset for risk management.

The second paper of the concluding section by Birol, Kontoleon, and Smale (Chapter 22) provides an illustration of non-market *valuation* research as applied to agro-biodiversity. In particular the authors undertake one of the first attempts to assess the private use values associated with *in situ* agro-biodiversity conservation using a novel approach that combines stated and revealed choice data. Their study focuses on agro-biodiversity preserved in small-scale farms in Hungary. Such small plots, also known as 'home gardens', are frequently found in developing and transition economies where they are managed by farm households using traditional practices and family labour. Though these home gardens are believed to generate significant private benefits for farmers (e.g. enhanced diet quality, steady food supply) and public benefits for society at large (e.g. supporting long-term productivity advances in agriculture), the exact magnitude and nature of such values have not been adequately assessed. The study by Birol, Kontoleon and Smale contributes to this underdeveloped literature by focusing on estimating the private value to Hungarian farmers of agro-biodiversity conserved in home gardens. The authors use a 'data enrichment approach' that combines or fuses data from a CE model (a stated preference data) and a discrete-choice farm household model (revealed preference data). Their analysis suggests that data enrichment leads to a more accurate and robust estimation of the private value of agro-biodiversity in home gardens. The chapter concludes by discussing how the findings from this study can be used to identify those farming communities which would benefit most from agri-environmental schemes that support agro-biodiversity maintenance, at least public cost.

In the final paper of this volume, Smale and Drucker (Chapter 23) provide a systematic overview of the economic underpinnings of the policies for managing agro-biodiversity. Their analysis focuses on the economics of managing both crop and livestock genetic diversity. Their review discusses current research on the marginal value of such resources, their effects on productivity, vulnerability and efficiency in agriculture as well as economic factors that determine both the levels and targeting of *in situ* seed and animal breed conservation. Further, the authors systematise the state-of-the-art literature on the costs and benefits of *ex situ* plant and livestock diversity conservation. Lastly, the authors discuss the ways specific policies influence genetic resource conservation and sustainable use, as well as means for assessing conservation priorities.

Collectively the papers selected for inclusion in this final section demonstrate recent advancements and directions taken in the emerging field of biodiversity economics as applied to agro-biodiversity. The papers cut across the issues raised in the first three parts of the volume on the causes of agro-biodiversity decline, the values associated with plant and livestock diversity as well as the design and implementation of policies for their sustainable utilisation.

## 6 Conclusion to the volume

Biodiversity economics is a rapidly emerging field evolving at the interface of economics and ecology. It is a field that aims to explore the underlying anthropogenic causes of biodiversity decline, the possible incentive-based policies for addressing these causes, as well as the challenges for designing and implementing these policies. The field has expanded to address these issues at all levels of biodiversity from the genetic to the species to the ecosystem level. Beyond a fast-growing academic interest (such as that displayed by the development of the BioEcon research network), biodiversity economics has also acquired a central position in the work undertaken by major international environmental policy organisations such as the United Nations Environment Programme (UNEP) and the IUCN (the World Conservation Union). With the rapidly increasing anthropogenic pressures on biodiversity, as highlighted by the recent Millennium Ecosystem Assessment, this interest in the economics of biodiversity conservation is bound to intensify.

This volume provides an exposition of research at the frontier of this evolving field. The selected papers are mostly derived from the collaborative work undertaken within the BioEcon research network and focus on analysing the *causes* of biodiversity decline, the *values* associated with biodiversity, and the design and implementation of *policies* for the sustainable utilisation of biodiversity resources and services. The volume is intended to consolidate the field of biodiversity economics by offering an overview of the *current* advances in this area but also to offer an indication of its *future* intellectual trends and challenges. We hope that the volume can be used by researchers, graduate students and policy-makers as a springboard for the development of new research agendas in the area.

# 1 Do we really care about biodiversity?[1]

*David W. Pearce*

## 1 Introduction: the issue

The world community is allegedly very concerned about the fate of the world's biological diversity. Evidence for this concern arises from the ratification of various international treaties on biodiversity conservation. Among these are the truly global treaties: the International Convention for the Regulation of Whaling 1946; the International Convention for the Protection of Birds 1950; the Convention on Wetlands of International Importance (the 'Ramsar Convention') 1971; the Convention on International Trade in Endangered Species (CITES) 1973; the Convention on the Conservation of Migratory Species of Wild Animals 1979; and the Convention on Biological Diversity (CBD) 1992. Details of all these conventions and the various regional treaties can be found in Sands *et al.* (1994). Equally relevant treaties affecting biodiversity less directly are the Framework Convention on Climate Change (FCCC) 1992 and its first Protocol, the Kyoto Protocol of 1997; and the Convention to Combat Desertification 1994. Barrett (2003) lists over 300 international conventions relating to the environment in one form or another. Barrett states that only one treaty 'offers a comprehensive approach to biodiversity conservation. This is the Biodiversity Convention' (Barrett 2003, p. 350).

But just how serious is the world in respect of biodiversity conservation? We argue that the only true indicators of concern must relate to action taken. Rhetoric about the fate of the world's environments is politically cheap unless the electorate calls the politician to account. Action tends

[1] I am indebted to Cameron Hepburn of St Hugh's College, Oxford University, Stefano Pagiola of the World Bank, and Paul Jefferiss of the Royal Society for the Protection of Birds (RSPB), Sandy, UK, for comments and suggestions on earlier versions of this chapter. Paul Jefferiss of RSPB, Andrew Balmford of Cambridge University and Kirk Hamilton of the World Bank kindly supplied some of the references used in the paper. Finally, I am indebted to the audiences at several seminars at Cambridge University, Oxford University, University of Gothenburg, Sweden, and University College London for comments.

to be costly, despite the claims of some that so much of what can be done is 'win win', i.e. will pay for itself. Moreover, we need to measure action carefully. It might seem that negotiating a treaty is a sign of firm action. But many commentators now doubt that the international efforts on conservation are effective or, at least, are as effective as is claimed. In some cases, the design of treaties has been criticised as addressing the wrong problem or a problem of lesser importance – see, for example, Hutton and Dickson (2000) on CITES. The more crucial issue is the counterfactual, i.e. what would have happened if the treaties did not exist? While no-one can be sure in every case, the available evidence is consistent with the suggestion that most treaties achieve little more than the counterfactual. Thus, despite the wealth of treaty-making and national laws embodying the treaty intentions, Barrett declares that 'most treaties fail to alter state behaviour appreciably' (Barrett 2003, p. xi). Similarly, Sandler argues that 'many international treaties concerning the environment have merely codified actions that the ratifiers had already accomplished or were soon to achieve' (Sandler 1997, p. 213). Elsewhere, we have suggested that there is a global 'deficit of care' to resolve global warming problems (Pearce 2003).

In this chapter we try to measure the degree of care by measuring action taken, using two economic indicators: actual expenditures and stated, or implied, willingness to pay for biodiversity conservation. In so doing, we also try to resolve an apparent conundrum. A recent and widely discussed literature has suggested that the world's willingness to pay for ecosystem conservation generally runs into many trillions of dollars, suggesting that the world does recognise the importance of ecosystem services and is willing to pay for them. But when we look at the actual expenditures on ecosystem conservation, they appear to be measured in, at best, a few billions of dollars annually. How can willingness to pay and actual payments differ by several orders of magnitude?

A prior question is why measuring the degree of care matters. One answer is that problems will not be solved unless we are aware of their true extent. If the world spends too little solving the biodiversity problem, but believes it is spending enough, then the problem will not be resolved and there will be no incentive to spend the right amount until, perhaps, it is too late. False beliefs about the adequacy of the global effort to save biodiversity are simply encouraged by political rhetoric. In turn, politicians have an incentive to say they are doing a lot, while doing little. The rhetoric may get them re-elected. Spending the 'right' amount of taxpayers' money, however, may get them deselected because the implied tax burdens might not be acceptable. In short, the political system has in-built incentives for the truth not to be told. Only by seeking some measure of 'conservation effort' can we call the politicians to account.

In what follows we make some simplifying assumptions. First, we focus only on international conservation efforts. Partly this is because gathering the relevant data for biodiversity conservation efforts at the national level is too time-consuming. More importantly, the international focus is justified by the fact that a large part of the world's biological diversity resides in tropical countries where there are the least indigenous resources to conserve them. Hence, international flows of finance are one of the prime means of securing that conservation.

Second, we acknowledge that cash flows cannot be the whole story in terms of measuring the degree of care. Policy measures, for which associated cash flows may be difficult to identify and measure, will also affect conservation. But it might equally be argued that many policy measures actively encourage the destruction of biodiversity, e.g. the agricultural and industrial subsidy regimes employed primarily by rich countries. In other words, there may be as many 'bad' policies as there are 'good' ones from the standpoint of conservation.

Third, we will speak throughout of 'biodiversity' without dwelling too much on its definition. In its widest sense it refers to biological resources, while its proper sense would be confined to a measure of the diversity of those resources. Maximising the stock of biological resources is not the same as maximising diversity.

Fourth, we adopt the view that the 'right' amount of conservation effort is one where the marginal economic benefits from conservation just equal the marginal costs of conservation, i.e. the point where the net benefits of conservation are maximised. This is not a criterion of optimality that will appeal to many who do not like the economist's approach to these issues. But we argue this is an appropriate benchmark when trying to measure the degree of care since the economist's notions of costs and benefits relate directly to human preferences.

## 2 A diagrammatic construct

We begin with a diagram that tries to encapsulate the various flows of costs and benefits from biodiversity conservation. We will use the terms 'biodiversity conservation' and 'ecosystem conservation' interchangeably. We take an ecosystem to be broadly defined as 'a biotic community and its abiotic environment' (Krebs 1994, p. 12). All ecosystems generate flows of services to humankind and hence all ecosystem services have an economic value. The issue is, just how large is this value?

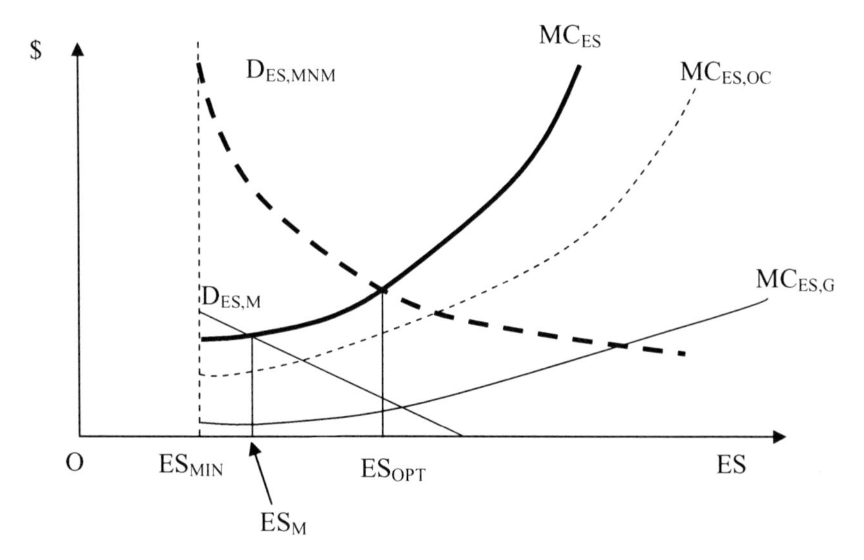


Figure 1.1. Stylised costs and benefits of ecosystem service provision

Figure 1.1 shows the relevant constructs. On the vertical axis we measure economic value in dollars. On the horizontal axis we measure the flow of ecosystem services (ES) which we assume can be conflated into a single measure for purposes of exposition.

The first construct is a demand curve for ecosystem services $D_{ES,M}$. This is a demand curve for the *commercial*, or *marketed*, services of ecosystems, i.e. those services that have associated with already established markets in which formal exchange takes place using the medium of money. Thus, if we have an ecosystem producing timber or fuelwood or wildmeat and, say, tourism, and if these products have markets, then the demand for these products would be shown by $D_{ES,M}$. Another name for a demand curve is a 'marginal willingness to pay' curve because the curve shows how much individuals are willing to pay for incremental amounts of the good in question, ES.

The second construct is another demand curve but this time for all ecosystem services, regardless of whether they currently have markets or not. This is $D_{ES,MNM}$ which is the demand curve for marketed (M) and non-marketed (NM) ecosystem services. There are various non-market services such as watershed protection, carbon sequestration and storage, scientific knowledge, the aesthetics of natural ecosystems, and so on. We know that $D_{ES,MNM}$ lies everywhere above $D_{ES,M}$. This is because, historically, ES have been abundant and hence there has been only a limited incentive for humans to establish property rights over them. As humans systematically expand their 'appropriation' of ecosystems, however, there is an incentive to establish property rights because ES become scarce relative to human demands on them. Humans have intervened in virtually all terrestrial ecosystems, especially as global population expands, appropriating around a third to a half of the net primary product (NPP) (Vitousek *et al.* 1986; Vitousek *et al.* 1997).[2] Nonetheless, a vast array of ES is not marketed, so there is a gap between $D_{ES,MNM}$ and$D_{ES,M}$.

We need to consider the shape of the two demand curves shown in Figure 1.1. Both are downward sloping, as we would expect. The more ES there are, the less humans are likely to value an *additional unit* of ES. We have no reason to suppose that ES are any different in this respect from other goods and services: they should obey the 'law of demand'. But notice what happens if we have a very low level of ES. Imagine a world with very few forests, very few unpolluted oceans, a much reduced stock of coral reefs, an atmosphere with a very much higher concentration of carbon dioxide and other greenhouse gases. In the limit, if there were no unpolluted oceans, no forests, extremely high concentrations of greenhouse gases, then the willingness to pay for one more unit of ES would be extremely high. Simply put, while a few may survive in some kind of artificial Earth bubble, humans would, by and large, disappear. For this reason, $D_{ES,MNM}$ bends sharply upwards as we go to points closer to the origin on the horizontal axis. Essentially, $D_{ES,MNM}$ is unbounded. There is some irreducible minimum ES below which marginal WTP would rise dramatically. This irreducible minimum corresponds to the kinds of limits that ecologists and others have tried to define for, say, climate change. For example, O'Neill and Oppenheimer (2002) set this limit at $+1^{\circ}$ C for long-run warming, well below the 'business as usual' level of warming.

Some suggest that at $ES_{MIN}$ the demand curve would become infinitely elastic (for example, see Turner *et al.* 2003). But as long as it is a (marginal) willingness to pay curve, this cannot strictly be correct since incomes and wealth would still be bounded. It is technically more correct to say that there is no meaning to the notion of economic value in the unbounded area of Figure 1.1.

As we shall see, this undefined region turns out to be rather important when we come to investigate the claims that the economic values attached to ES are extremely high.

In order to maintain ES of value to humans we know that certain costs are incurred. Figure 1.1 shows the first category of these costs as

[2] Net primary production is the energy or carbon fixed in photosynthesis less the energy (or carbon) used up by plants in respiration. NPP is like a surplus or a net investment after depreciation (what is required for maintenance of function).

$MC_{ES,G}$ – the marginal costs of managing ES. In the absence of any very strong evidence about the shape of $MC_{ES,G}$, we show it as a gently rising line. The second category of costs is of considerable importance and comprises the opportunity costs of providing ES. The assumption is that ES are best secured by conserving the ecosystems that generate them. This is not consistent with using the ecosystem for some other purpose, e.g. agriculture. Hence, a potentially significant cost of having ES is the forgone profits (more technically, the forgone social value) of the alternative use of the ecosystem. We refer to this as $MC_{ES,OC}$ – i.e. the marginal opportunity cost of ecosystem conservation. It is formally equivalent to the forgone net benefits of ecosystem conversion, i.e. 'development' as we tend to call it. The sum of $MC_{ES,G}$ and $MC_{ES,OC} = MC_{ES}$ gives us the overall marginal cost of conservation.

Figure 1.1 is obviously simplistic. For example, it ignores the possibility that ES might be largely maintained while serving some development function. Agro-forestry might be one example of this 'symbiotic' development. But in general, we know that there is a long-run trend towards ecosystem conversion, with the nature of the conversion meaning that many ES are lost. The diagram also ignores the possibility, realistic in practice, that the conversion process may be very inefficient. Ecosystems may be converted only for the development option not to be realised because of mismanagement of the conversion process or of the subsequent development. In what follows we ignore these qualifications in order to focus on the basic messages from the analysis.

Finally, Figure 1.1 shows us various points of interest. First, since the true *aggregate* costs of maintaining a given level of ES are given by the area under the overall $MC_{ES}$ curve,[3] and since the true global benefits of ES provision are given by the area under the $D_{ES,MNM}$ curve (assuming the demand curve shows 'true' willingness to pay – see later), the point $ES_{OPT}$ shows the economically optimal level of ES provision.

Second, any point to the left of $ES_{OPT}$ has benefits of ES (area under $D_{ES,MNM}$) greater than the overall costs of their supply. But all such points also have an interesting feature. Unless we arbitrarily confine attention to points between $ES_{MIN}$ and $ES_{OPT}$, all points to the left of $ES_{OPT}$ have either *infinite* total benefits or *undefined* total benefits, depending on how one wants to interpret the unbounded region.

Third, while $D_{ES,MNM}$ reflects the true global benefits of ES provision, it is not an 'operational' demand curve. This means that unless the WTP is captured by some form of market, or unless the evidence on WTP is

[3] Total cost is the integral of the marginal cost curve. Total benefits are given by the integral of the MWTP curve, i.e. the demand curve.

used to formulate some quantitative restrictions on ecosystem conversion (bans, restrictions on type of conversion, etc.), the demand curve that matters is curve $D_{ES,M}$. Figure 1.1 shows the real possibility that failure to reflect true WTP in actual markets results in a serious under-provision of ES. Here again we see the importance of the dual process of economic valuation (determining the location of $D_{ES,MNM}$) and capturing those values through forms of market creation.

## 3 Locating the current trend in ES provision

The next task is to gain some idea of where we might be in terms of Figure 1.1. Clearly, without detailed knowledge of the cost and benefit functions, we cannot be sure. But some of the evidence suggests strongly that things are getting worse, not better.

### *3.1 Historical land conversion*

First, we can be reasonably sure that we are moving leftwards in terms of the horizontal axis, provision of ES. The reason for this is that natural ecosystems have been converted to agriculture on a fairly systematic basis over very long periods of time. Indeed, the whole history of humankind is a history of land conversion. Richards (1990) estimates that between 1700 and 1980, the area of world forests and woodlands declined by 1.16 $10^9$ hectares, the area allocated to grassland and pasture stayed fairly constant, but that allocated to crops rose by 1.24 $10^9$ hectares. In short, cropland grew at the direct expense of forests and woodlands. One way of thinking about this process in terms of Figure 1.1 is to regard the $MC_{ES,OC}$ curve as being shifted upwards over time as population expands and the demand for food increases. This process is consistent with the 'human appropriation' estimates of Vitousek *et al.* discussed earlier.

### *3.2 The extinction record*

Second, while the evidence is more difficult to evaluate, many ecologists feel that species extinction rates are increasing. Ecologists usually determine whether extinction rates are high or not by comparing them with (a) past trends rates of extinctions over long periods of time, and (b) past episodes of mass extinctions. Thus, Pimm *et al.* (1995) argue that the background trend rate of 'natural' extinctions, based on the geological record, is 0.1–1.0 extinctions per million species years.[4] They suggest

[4] Thus, if on average species survive for 1 million years, natural extinction rates would be 1.0 extinctions per million species years (E/MSY). If average 'life' is 10 million years, the natural extinction rate would be 0.1 E/MSY.

that current extinction rates vary from 20 to 200 extinctions per million species years, orders of magnitude higher than past extinctions. Dirzo and Raven (2003) argue that recent recorded extinctions seriously understate actual extinctions, largely because of sampling errors, and concur with the longer-run estimates of Pimm *et al.* (1995). These estimates of extinctions are very uncertain, not least because there appears to be no consensus on just how many species there are in the first place, which is unsurprising when one considers the scale of the task that would be required to count them or infer their existence. Pimm *et al.* (1995) suggest a range of 10–100 million species. Stork (1999) suggests a more precise 'working' figure of 13 million species. Dirzo and Raven (2003) opt for approximately 7 million.

However, whereas the very long-run estimates derive from geological records, many of the more dramatic predictions are derived from species-area relationships, themselves controversial and generally thought to exaggerate loss rates (Stork 1999).[5] Analysis of actual recorded extinctions suggests that around 0.24 per cent of species have gone extinct since 1600 (Stork 1999). Such rates appear very low when compared with the more dramatic guesstimates using the other methods and especially the species-area relationship. Other evidence is consistent with the lower extinction rates (see, for example, van Kooten 1998).

The extinction record is, thus, still debated, but all that matters for current purposes is that a significant number of expert commentators believe that extinctions are increasing.

### *3.3 'Underfunding'*

If we know that the maintenance of existing *managed* natural and semi-natural ecosystems is under-funded, then we might also conclude that there is a leftwards move along the ES axis in Figure 1.1, at least as far as those ecosystems are concerned.[6] On the face of it, under-funding must mean that some ecosystems are not being maintained and hence must be being degraded. While there were extensive discussions about 'funding needs' at the Rio Earth Summit in 1992, the first comprehensive efforts to secure some insight into this issue are the important papers by James *et al.* (1999) and Balmford *et al.* (2003). James *et al.* (1999) estimate that there is serious under-funding of existing protected areas (PAs), primarily, but not exclusively, in developing countries. Expressing their estimated shortfall in expenditure as a fraction of the actual expenditure

[5] The species-area relationship takes the form $S = cA^z$ where S is species, A is area, and c and z are parameters. The value of z is usually taken to be around 0.3.

[6] Only a fraction of the world's natural and semi-natural ecosystems is managed.

(management costs only) shows shortfalls of 10 per cent in North America, developed Asia and Australasia. But shortfalls in Europe (somewhat surprisingly) are put at 50 per cent, Sub-Saharan Africa and Russia/East Europe at 100 per cent, 140 per cent in Latin America, 450 per cent in North Africa and the Middle East, and over 500 per cent in Asia. For the world as a whole, they put under-funding at around 40 per cent. Overall, then, the picture suggests under-funding on a major scale in developing countries, supporting the notion that existing ecosystems – as indicated by protected areas – are declining if not in area then in quality. If the world's managed areas are declining in terms of funding requirements, it seems reasonable to suppose that most of the rest – where, it will be recalled, human intervention is still dominant – will be in a worse state.

## 4 Reasons to be cheerful

Now consider some reasons why we might either be moving to the right in Figure 1.1 or, if we are losing ES, why it may not matter and could even be net beneficial.

### *4.1 Optimal ecosystem loss*

On the previous arguments, the record of:

- historical land conversion
- extinctions
- under-funding of protected areas

suggests that we are moving leftwards along the horizontal axis of Figure 1.1. However, such a finding, if correct, does not tell us whether we are to the right of $ES_{OPT}$ but moving left, or to the left of $ES_{OPT}$ and moving left. It may be that what we are failing to conserve is what we should not be conserving anyway. Potentially, we have some reason to be optimistic. The parallel argument in terms of global warming has been quite widely advanced. There is indeed, say some of the commentators, increased global warming, but it does not matter very much – e.g. see Lindzen (1994) – and the rate of return to alternative uses of the finance needed to combat global warming is higher – e.g. Lomborg (2004).

### *4.2 The property rights argument*

One argument advanced by 'free market' thinkers is that, as the appropriation of net primary product expands, so property rights to scarce ES will be established, markets will emerge and conversion will take place only if benefits exceed costs. In terms of Figure 1.1, $D_{ES,MNM}$ and $D_{ES,M}$

will gradually converge and the optimal amount of ES will come about. The non-market services will gradually be 'captured' by market creation. There is a range of views within this argument, from extreme free market positions in which the state has no role and biodiversity should be 'privatised' through to those who acknowledge the potentially large role that the private sector already plays in ecosystem management – see Anderson and Hill (1995) and Drake (1995).

Evidence in favour of such a process lies in the same international treaties outlined earlier. The FCCC and the Kyoto Protocol, for example, are ways of converting essentially open access rights to the global atmosphere and its ecological services into common property rights, with access being partially limited by agreements, at least for developed economies. Policy instruments such as carbon and energy taxes, and tradable carbon quotas, are the means by which some of this market creation works.

One problem with this view is that we have no guarantee this process will move fast enough to prevent serious loss of ES. One reason for doubt lies in the observation made above: it is not clear that such treaties do much to conserve ES (Pearce 2003; Barrett 2003, 2004). In other words, the common property provisions do not differ significantly in their effects compared with open access. Yet evidence suggests that those communities that escape resource degradation and overcome open access poverty are those that win the 'race' between environmental degradation and institutional adaptation to resource scarcity (Lopez 1998).

### 4.3 *Cornucopians*

Every now and again, some commentators suggest that things are certainly getting no worse and may be getting better. In the past, Simon (1981, 1986, 1995) and Simon and Kahn (1984) have suggested this, and in recent years Lomborg (2001) has echoed this view. Interestingly, Lomborg (2001) has attracted far more controversy than the earlier publications of Simon and Kahn, despite the fact that the messages, and the approach to the evidence, are very similar. Simon (1995), which is essentially an update of Simon and Kahn (1984), pays only limited attention to ecosystem protection and species loss, but does draw attention to the contrast between estimated rates of extinction of species and the recent historical record, which was noted earlier. Lomborg (2004) questions both the more alarming estimates of forest loss, including tropical forest loss, and the species extinction estimates. Part of the problem with all these contributions is that they tend to take the most pessimistic interpretations of ecosystem change and criticise them, a kind of 'straw man' approach.

Nonetheless, despite the controversy, any analysis that addresses the data as best it can needs to be taken seriously.

### 4.4 *The growth of protected areas*

Perhaps a more substantive piece of 'good news' is the expansion in the world's protected area system. Yellowstone National Park in the USA was the first PA, established in 1872. Globally, protected areas did not reach 1 million km$^2$ until just after the Second World War. Since then the growth has been fairly dramatic until, today, they cover some 18.8 million km$^2$ (Chape *et al.* 2003). Between 1962 and 2003 the area grew from 2.4 million km$^2$ to 18.8 million km$^2$, or around 0.4 million km$^2$ per annum.[7] Of course, this is not 'new' land but existing land with its use at least nominally proscribed to prevent its conversion to some other use, hopefully with biodiversity preserved if not encouraged.

A comparison can be made with converted land, although data for land conversion remain the subject of controversy. Most converted land is forest. The Food and Agriculture Organisation (FAO) (2001) estimates global net rates of deforestation of around 9 million hectares in the 1990s, or 0.23 per cent of total forest area. The World Resources Institute (Matthews 2001) disputes the figure, noting that FAO data include biodiversity-poor plantations as afforestation, offsetting natural forest loss. Net of plantation growth, annual losses are closer to 16 million hectares per annum, or 0.4 per cent per annum of forest cover, nearly double the FAO figure. A direct comparison therefore suggests that the world is designating protected areas at a rate of 40 million hectares per annum and deforesting land at perhaps 16 million hectares per annum, a net gain of 24 million hectares per annum.[8] If the world is protecting areas at a rate three to four times greater than rates of deforestation, this must be good news for biodiversity conservation, and *prima facie* evidence that we care. But there are some caveats.

First, PAs under IUCN Management Categories I–VI account for only 10 per cent of land in developing countries and 12 per cent in developed countries (World Resources Institute 2003).

Second, well over one half of all protected areas occur in nations where governance is weak (World Resources Institute 2003). Weak governance shows up as poor management and neglect and, in many cases,

[7] Note that the Greenland National Park, established in the 1970s, covers 97 million ha and the Great Barrier Reef Marine Park, established in the 1980s, covers 34 million ha.

[8] Actually, since the forest-loss figures are specific to the 1990s, the comparison should strictly be between the rate of PA formation and deforestation in that decade. The relevant figures would then be +59 m.ha − 16 m.ha = +43 m.ha.

corruption. Protected areas may therefore be 'paper parks', protected in name but not in reality (Whelan 1991). Van Schaik *et al.* (1997) document the poor state of many tropical rain forest reserves, showing how a combination of lack of resources, lack of commitment, lack of knowledge and what they call 'resource theft' places many of them in peril. In many cases this is unsurprising. Government involvement in protected areas arises precisely because market forces do not dictate that protection is the most privately beneficial use of the land. But governments have no comparative advantage in managing land for biodiversity. Where there are conflicts between protection and the conversion uses of the land, therefore, government is likely to lose out, or to become involved in rent capture procedures that involve its surrender of conservation. Furthermore, the available data on expenditures on PAs suggests major under-funding, as we discuss shortly.

Third, there is a real possibility that the better protected areas are biodiverse not because of protection effort but because the alternative use value of the land is low. The argument here is akin to that used by game theorists to explain why nations sign up to international agreements. Signing up is most likely when the nation in question has little or nothing to lose (Barrett 2003). Similarly, protection is more likely when the opportunity costs of protection are low. If we surmise that existing PAs will have low opportunity cost relative to any new ones, and the evidence for this is considered shortly, then it may well be that existing PAs would not have been damaged in any event.

What this suggests is that publicly owned and managed protected areas will be at risk wherever there is a high private economic value to the alternative use of the land, e.g. for agriculture or forestry. Governments are then either not able to resist encroachment and conversion, because they lack resources, or will actively connive in the conversion if there are rents to be gained for select groups and individuals. Where there is a low opportunity cost to conservation, the land will appear 'protected' when in reality the gazetting of the area makes little difference to its biodiversity status. Conservation would have occurred anyway. However, private ownership may succeed where government ownership fails (e.g. Langholz *et al.* 2000; Langholz and Lassoie 2001).

Finally, the protected areas movement is not one that, so far, has been well informed by an explicit balancing of costs and benefits to the nation in question. But as the demand for alternative uses of the land grows, especially for agriculture and human settlement, so a questioning of the national worth of protection will occur. Already, some of the results of this reappraisal suggest that nations may be better off sacrificing their protected areas – see, for example, Norton-Griffiths and Southey (1995).

Table 1.1 *Estimates of protected area costs (after James* et al. *1999)*

| Costs $ per ha. p.a. | Existing PAs | Extra PAs |
|---|---|---|
| Global costs | | |
| Management | 6.3 | 4.5 |
| Opportunity costs | n.a | 14.6 |
| *Total* | n.a | *19.1* |
| LDC costs | | |
| Management | 2.8 | 2.8 |
| Opportunity costs | 3.8 | 8.8 |
| *Total* | *6.6* | *11.6* |

## 5 Costs and benefits

Historical land conversion rates, the extinction record, evidence of 'under-funding' tells us that we are moving leftwards along the horizontal axis of Figure 1.1. The argument that things are at least getting no worse comes from the cornucopians, those who believe that property rights regimes react rapidly to changing scarcity, and the expansion of PAs. But none of the arguments tells us whether we are at or to one side of the economic optimum level. That is, we do not know whether the rates of loss are optimal or not. For that we need evidence on costs and benefits 'at the margin'.

### *5.1 The cost-benefit evidence: protected areas*

If we had some idea of the likely costs and benefits of expanded protected ecosystems, we would have some evidence to locate us to the right or left of the point $ES_{OPT}$ in Figure 1.1. Essentially, if the costs of 'new' protection exceed the benefits then we are to the right of $ES_{OPT}$ and if we have benefits in excess of costs then we are to the left of $ES_{OPT}$. The first piece of information needed concerns the costs.

James *et al.* (1999) looks at the world's protected areas and consider expenditures in current PAs and on a hypothetical expansion from the 13.2 million km$^2$ in 1999 to 20.6 million km$^2$. Converting their estimates to annual per hectare costs the picture appears as in Table 1.1.

Apart from the apparent reduction in global management costs as the PA area is expanded – they go down instead of up as might be expected – the picture is in keeping with Figure 1.1. One would expect opportunity costs to rise significantly as more ecosystems are conserved. The reason for this is that the 'low-cost' areas will tend to be protected first and, as the system is expanded, areas that have higher development potential will be converted. This prediction is also borne out by the later paper by Balmford *et al.* (2003) which shows that protection costs rise with an index of 'development'.

The World Bank (2002) has also costed the setting aside of just over 200 million hectares of *new* land in developing countries as protected forest areas. Its combined management plus opportunity cost estimates are very much higher than those in James *et al.* (1999) and amount to $93 per ha p.a, some eight times the James *et al.* estimates. Even allowing for the inclusion of some high-cost land acquisitions, 70 per cent of the land hypothetically considered in the Bank calculations is acquired at an opportunity cost of less than $50 per hectare. Whereas James *et al.* have opportunity costs of just under $9 per ha, the World Bank has opportunity costs of $83 and management costs of $10 per ha. It is not clear why the estimate should diverge so much. However, we make use of both sets of estimates shortly.

The extensive literature on environmental economics as it relates to ecosystem protection tends to focus on individual case studies. Moreover, while we have many studies of willingness to pay to conserve individual species and some habitats, it is hard to come by estimates of willingness to pay to conserve *diversity* as such. Finally, even if the focus of studies is on species and habitats, we have few *meta-studies* on which to base any consensus judgement. There are several 'global' assessments of the value of biodiversity, most of which, unfortunately, rest on serious errors of analysis. Accordingly, we reserve a special section for these studies shortly. While a full review of the evidence on willingness to pay has to take its turn for a later date, we briefly review some of the wider surveys of ecosystem values here, and also look at some estimates of global willingness to pay. Efforts to argue that the willingness to pay for ecosystem conservation outweigh the overall costs of conservation, based on individual case studies, can be found in Turner *et al.* (2003) and Balmford *et al.* (2002).

However, considerable caution is required in interpreting these reviews. First, published studies are more likely to report cases where benefits exceed costs, rather than vice versa: a 'censoring' effect is likely to be present. Second, some of the case studies utilise data from the illicit literature on ecosystem valuation such as that of Costanza *et al.* (1997). Another literature reports specific examples where markets have been created in an ecosystem's services, sometimes with the conclusion, stated or implied, that there are higher net benefits to ecosystem conservation than to the alternative 'development' option. This may be true, for example, of Costa Rica's Forest Law (Chomitz *et al.* 1998), although some authors express doubts about whether the Law passes a cost-benefit test.

Pearce (2003) and Pearce and Pearce (2001) survey the value of forest ecosystems. They conclude that the dominant economic value of forests lies in carbon storage and sequestration. Present values of $360 to 2,200 per hectare would more than compensate for many, though not all, conversion values for tropical forests. Genetic information for pharmaceuticals and agriculture probably has low per hectare value, perhaps a few dollars per hectare, although the debate on the appropriate procedures for valuing this information continues in the literature (for a review see Pearce *et al.* 2005, Chapter 12). Watershed protection, at $15 to 850 per hectare, and recreational values from near zero to $1,000 per hectare for unique forest areas, can be significant, but critically dependent on location. Non-timber forest products tend to be modest in terms of economic value relative to conversion values, but can be high relative to local community incomes. Since it is the former that tend to dictate conversion decisions, non-timber product values are unlikely to protect most forest areas, contrary to some of the early euphoria attached to these benefits. The somewhat gloomy finding is that, unless carbon is 'monetised', the economic values of tropical forests do not, at the moment, compete with alternative uses of the land in many cases. Put another way, implicit willingness to pay is not revealing high levels of care. Mobilising carbon values could change this for many areas of forest provided international markets are allowed to develop fully in carbon storage and sequestration. So far, efforts to do this have been very modest, with international negotiators on climate agreements finding too many reasons why stored carbon should not be the subject of the various flexibility mechanisms.

In addition to the use and indirect use values of forests, several authors have attempted to use stated preference techniques to secure some kind of 'global non-use value' for tropical forests. Kramer and Mercer (1997) use contingent valuation surveys to elicit US citizens' willingness to pay to conserve an extra 5 per cent of the world's tropical forests (taking 5 per cent as being already conserved in one form or the other). Their results suggest an annual per hectare valuation of about $4. Extended hypothetically to households in high-income countries, the value would rise to around $25 ha p.a.[9] The Kramer and Mercer study uses a 'one-off' payment. Horton *et al.* (2003) use a parallel approach to UK and Italian willingness to pay to protect the existing 5 per cent of (under-) protected areas in Amazonia. But their results contrast starkly with those

[9] The US has 91 million households. World high-income countries have 580 million households. The UK has just under 20 million households.

of Kramer and Mercer. For 'UK only' willingness to pay, the implied annual per hectare value is $48, producing a fund of $912 million *per annum*. If the UK result is extended to all industrialised countries, the implied 'fund' amounts to $26.8 billion and the per hectare value rises to a staggering $1,400 per annum, well in excess of the capital value, let alone the rental value, of most Amazonian land. Horton *et al.* (2003) argue that willingness to pay exceeds the costs of protection (calculated here as about $1 billion p.a. from the data in Horton *et al.*) 'by an enormous amount'. This would be true if the willingness to pay figures were credible. The authors themselves express doubts over their reliability. As we see shortly, it is also very hard to square these willingness to pay estimates with actual flows of funds for ecosystem protection.

Efforts at some kind of meta-analysis have been made with respect to wetlands. Woodward and Wui (2001) review thirty-nine studies and find values of $2 to $20,000 per hectare in 1990 prices, or say $3 to $30,000 in current prices. The average is around $3,000 per hectare. Brouwer *et al.* (1999) analyse contingent valuation studies only and present results in willingness to pay per household per year, making the study non-comparable and less useful than the Woodward and Wui study. The average willingness to pay in the Brouwer *et al.* study is around $40 per household per year in 1990 prices or, say, $60 per household per year in current prices. In turn, this might suggest $20 per person per year. If we imagine this sum was typical of all people over the age of fifteen and confine attention to the high-income countries of the world only, it would translate to about $14 billion per annum. However, as is well known, many economic valuation studies suffer from problems of aggregation across all goods. We cannot suppose that people would be willing to pay sums of this kind for wetlands conservation, plus another sum for tropical forests, and so on.

Overall, there are unquestionably contexts in which the inferred or stated willingness to pay for ecosystem conservation exceed the combined management costs of conservation plus the opportunity costs of conservation. But how far this is a general truth is open to serious question. Studies finding benefit-cost ratios greater than unity, and sometimes substantially greater than unity, may reflect 'censoring', the process whereby 'good news' is published and 'bad news' is not. Moreover, studies that do cost-benefit evaluations are often directed at ecosystems with fairly unique attributes. It is unwise to extrapolate from those studies to the far greater stock of ecosystems – this is perhaps one of the lessons of the literature on the value of genetic information in tropical forests. Finally, finding a willingness to pay is not the same as finding a value that can be captured and turned into cash flows. As is well known, only a fraction of overall willingness to pay, even when correctly estimated, can be converted to cash flows. Where this coefficient of capture is low, reliance has to be placed on decision makers measuring and understanding non-market values and using the information to establish regulatory frameworks that prevent ecosystem conversion.

What do we know about the relevant magnitudes of the costs and benefits of PAs? The answers for tropical forests appear to be that global WTP is anything from $25 ha p.a. (Kramer and Mercer) to $1,400 ha p.a. (Horton *et al.* 2003). Protection costs range from $7–12 ha p.a. (see Table 1.1) to $93 ha p.a. (World Bank). Given the doubts surrounding the extremely high value obtained by Horton *et al.* in geographically extending their WTP figures, the suspicion has to be that benefits do not automatically exceed costs, contrary to the optimistic findings in contributions such as Turner *et al.* (2003) and Balmford *et al.* (2002). This does not mean that those publications are reporting false case study conclusions, but simply that their findings may be far from general. But this comparison of costs and benefits is obviously fraught with difficulties. Perhaps the best that can be said is that they neither support nor disprove the notion that more global conservation passes a cost-benefit test.

### 5.2 *What do we actually spend on ecosystem conservation?*

Willingness to pay studies are the only way in which we can secure some idea of the economic 'worth' of marginal or discrete amounts of ecosystem services. Hence studies that seek this magnitude are wholly legitimate. But willingness to pay and actual payments are not the same thing. Whilst not denying the value of estimating areas under discrete ranges of $D_{ES}$ in Figure 1.1, finding out what we actually spend on ecosystem conservation should give us a 'reality check'. What follows is necessarily incomplete and constitutes a first attempt to estimate actual international flows of funds for ecosystem conservation.

#### 5.2.1 *Protected area costs*

First, we recall the estimates of James *et al.* (1999) that *actual* protected area expenditure is some $6 billion p.a. James *et al.* note that this figure excludes compensation that many feel should have been paid to those who have been displaced or deterred from converting or using the PA land. They estimate that at a further $5 billion. However, while there is clearly a strong moral (and economic) case for making such compensation payments, they are not in fact made. Hence the annual sum of relevance is $6 billion.

*5.2.2 Debt-for-nature swaps*

Debt-for-nature swaps (DfNSs) are one form of debt-for-development swaps and involve the purchase, usually by an international conservation organisation, but also by governments and even individuals, of developing countries' or transition countries' debt in the secondary debt market. Such debt is often quite heavily discounted, i.e. the redemption price is well below the face value, due to the market's realistic assessment of the prospects of repayment. In a DfNS, the purchaser of the secondary debt offers to give up the debt holding – usually by converting foreign exchange debt to domestic currency debt – in exchange for an undertaking by the debtor country government, usually through a local conservation NGO, to protect an environmentally important area, train conservationists, reduce pollution threats, etc. Some of the most celebrated debt swaps involving governments and NGOs are those under the Enterprise for the Americas Initiative (EfAI), established in 1990. Another significant government player in DfNSs is Switzerland, which set up a Swiss Debt Reduction Facility in 1991. DfNSs are clearly Coaseian bargains in which the indebted country has the property rights to a natural resource and accepts some attenuation of that right in exchange for payments by the beneficiaries of the resulting conservation. The involvement of at least the host government is necessary because rights are being attenuated and because issues of national sovereignty arise. But government involvement also helps reduce transactions costs. The involvement of lender governments is also clearly necessary where the debt is official debt.

A DfNS is an example of a Coaseian bargain (Coase 1960). Since the property rights to the environmental asset rest with the indebted country, and since the beneficiaries are environmentalists or the world as a whole, the beneficiaries are paying the host country not to convert the land in question or not to let it degrade. The essence of a Coaseian bargain is that the benefits derived from the payments made by the beneficiary must exceed the costs to the host country. It is the very fact that a bargain takes place at all that determines that a cost-benefit test is passed. If this is right, then the continuing existence of DfNSs confirms that, for some ecosystems at least, the benefits of conservation exceed the costs of conservation. Figure 1.2 shows this. In this case $MB_{D,H}$ shows the marginal benefits of land conversion, i.e. 'development', for the host country. $MEC_H$ shows the global externality imposed by this conversion of the rest of the world, W, due to the loss of biodiversity. This can also be interpreted as the marginal benefit to the world of *not* converting the land to development. Exercising its sovereign property rights, H will go to $-ES_H$ where its profits from converting the land are maximised. But the optimum is at $-ES_S$ (note that the horizontal axis measures the loss in ES, hence –ES). It pays

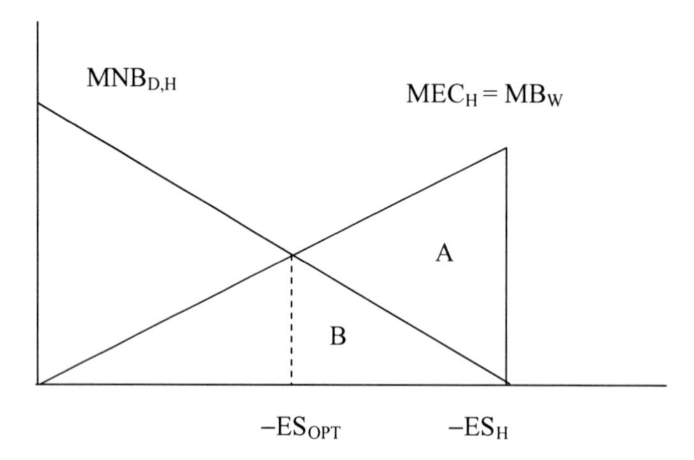


Figure 1.2. Beneficiary pays

the world to offer H any sum less than $MEC_H$ to prevent ecosystem loss, and it pays the host nation to accept any sum greater than $MNB_{D,H}$. Clearly there are gains from trade in moving from $-ES_H$ to $-ES_S$, area $(A + B)$ areas (B). The move passes a cost-benefit test.

Adapting data in Sudo (2003), Pearce computes the total flows of funds 1987–2003 under DfNSs. These are summarised in Table 1.2. Sudo (2003) estimates the funds leveraged by DfNSs, i.e. additional sums that 'piggy back' on the actual swap. The leveraging ratio is 1.9 for the overall portfolio of funds. The figures in Table 1.2 correspond to an annual flow of some $140 million. In order to find the 'implied price' of a hectare of conservation, Pearce and Moran (1994), following an earlier work by Ruitenbeek (1992), analysed some of the early DfNSs where information is available on payments and land area. They suggest that an implicit price of, at most, $5 ha is being paid for the 'average' swap. One can therefore argue that DfNSs have a conservation cost of up to $5 ha p.a., which is in keeping with the figures in Table 1.1. (Note that these are management costs – DfNSs appear typically to exclude land purchase.)

### *5.2.3 The GEF*

The Global Environment Facility (GEF) is a United Nations agency charged with meeting the 'incremental cost' of developing countries' provision of global environmental goods. The definition of incremental cost is treated rather broadly but is intended to reflect the additional costs a developing country would face if it switched from an activity that is justified in domestic terms only to one that has both a domestic and global

Table 1.2 *Debt-for-nature swaps – flow of funds 1987–2003*

| | Total face value of debt ($ million (rounded) | Total discounted value of debt ($ million, rounded) |
|---|---|---|
| Total excluding Poland (100 projects) | 1,943 | 582 |
| Total including Poland (102 projects) | 4,840 | 1,153 |
| Total including leverage | – | 2,190 |

Source: Pearce (2004), Sudo (2003)

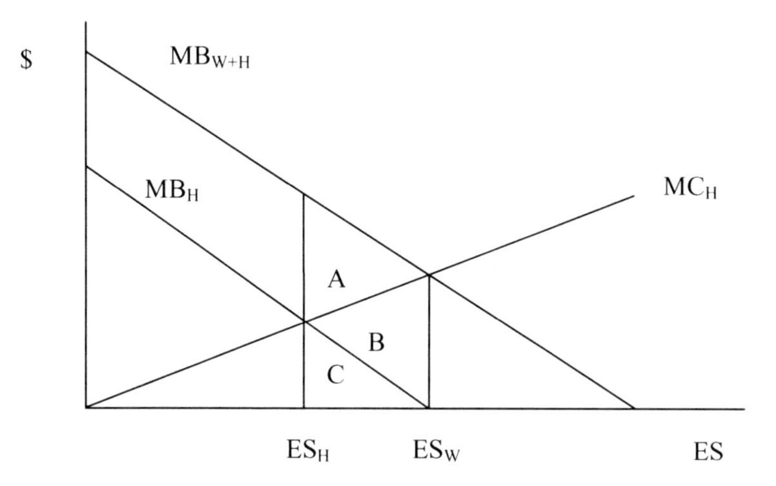


Figure 1.3. Incremental costs and the GEF

justification. As such, the GEF also fits the Coaseian model – Figure 1.3. $MB_{W+H}$ is the marginal benefit to the world plus the marginal benefit to the host nation of conservation. This time more conservation shows up as a move to the right along the horizontal axis. The marginal cost of conservation, $MC_H$, reflects the management costs of conservation plus the opportunity costs, i.e. the forgone development benefits. If we first ignore the additional global benefits of conservation in the host country, then the host nation will, if it optimises, go to $ES_H$. But the global optimum is at $ES_{H+W}$. The host nation has no incentive to go to the global optimum but will do so if it is compensated for the lost development benefits = area B + C. However, by going to the global optimum the host country secures some incremental benefit = area C. Hence there are two notions of incremental cost: gross incremental cost = area B + C and net

Table 1.3 *GEF-allocated funds and co-financing 1991–2002 ($ million)*

| | Climate change | Biodiversity | International waters | Ozone depletion | POPs | MFAs | Total |
|---|---|---|---|---|---|---|---|
| GEF | 1409 | 1486 | 551 | 170 | 21 | 210 | 3847 |
| Co-financing | 5000 | 2000 | n.a. | 67 | n.a | n.a | 7067 |
| Total | 6409 | 3486 | 551 | 237 | 21 | 210 | 10914 |

Source: GEF allocations from GEF *Annual Reports*. Co-financing estimates from GEF (2002)

Notes: MFAs = multi-focal areas such as land degradation. In 2002 land degradation was recognised as a separate focal area. POPs = persistent organic pollutants, approved as a focal area in 2001 and linked to the Stockholm Convention. Co-financing estimates for biodiversity and climate change are approximate and include expected sums. n.a. = not available but assumed to be zero or close to zero

incremental cost = area B. Whoever pays the residual element, area C, it is clear that gross benefit = area A + B + C exceeds incremental cost. A cost-benefit test is met.

The implementing agencies of the GEF were initially the World Bank, United Nations Environment Programme and the United Nations Development Programme, with various other agencies being given similar powers later on. The GEF was established in 1990 in a 'pilot phase', or GEF I, which lasted from 1991 to 1994, and its initial activities were unrelated to any international environmental conventions other than the Montreal Protocol on ozone layer depletion. Its coverage was biodiversity, climate change, ozone layer depletion and, curiously, 'international waters' – seas and lakes shared by two or more nations. The GEF soon took on the role of being the financing mechanism for the Framework Convention on Climate Change (1992), the Convention on Biological Diversity (1992) and the Stockholm Treaty on Persistent Organic Pollutants and the Convention to Combat Desertification.

Table 1.3 shows how much money the GEF allocated to its various 'focal areas' between 1991 and 2002. The crucial role of co-financing is revealed. Co-financing refers to the leverage that the GEF has on other funds outside the official Trust Fund.

Table 1.3 suggests that GEF funding has run at approximately $1 billion per annum. This certainly makes it the largest single source of market creation funding in the world. To facilitate comparison with other financial mechanisms, like has to be compared with like. Table 1.3 shows the comparison for biodiversity, although there are problems in separating out the biodiversity component in GEF expenditures because biodiversity is

Table 1.4 *Summary of flows of biodiversity conservation funds ($ million p.a.)*

| Debt-for-nature swaps | Protected areas | Costa Rica Forest Law[1] | Bioprospecting[1] | GEF biodiversity | GEF all areas | Bilateral aid |
|---|---|---|---|---|---|---|
| 140 | 6000 | 20 | Small | 315 | 1000 | 1000 |

Notes: 1 not discussed in this paper – see Pearce (2004)

often the beneficiary of non-biodiversity focal areas such as international waters. Table 1.4 shows some estimates for two other market-creation activities – payments to forest landowners by the Costa Rican government to encourage conservation, and bioprospecting, the payment by drug companies and others for genetic prospecting rights. Overall, GEF expenditures on biodiversity conservation appear to run at about $315 million p.a. but since a large part of 'international waters' expenditure is also biodiversity oriented, this sum could be raised to $365 million p.a. Finally, global warming control can also be seen in terms of protection of ES, in which case total ecosystem conservation expenditure rises to $950 million p.a.

### *5.2.4 Bilateral assistance*

Finally, we look at the available data on overseas development aid targeted at biodiversity. Under the Organisation for Economic Cooperation and Development (OECD) Creditor Reporting System (CRS), individual nations are meant to record biodiversity-related aid expenditures. OECD also records 'Rio Marker' expenditures, i.e. aid directed at achieving the Rio Earth Summit (1992) goals. These categories of expenditure tend to range much more widely than ecosystem conservation, for example including expenditures on water supply and agriculture. It is known that both sets of data are problematic (Lapham and Livermore 2003). Taking expenditures by the six main donors – USA, UK, France, Germany, Japan and the Netherlands – annual expenditures 1998–2000 were $38 million on the narrower definition and $240 million on the broader definition. An arbitrary compromise figure of $100 million is adopted here. Note that whereas we have a prima facie case for the GEF and DfNS expenditures passing a cost-benefit test, we cannot make any such presumption for bilateral assistance unless we can assume that assistance simply is not given without a cost-benefit analysis being carried out. This seems an unlikely assumption.

The bilateral aid figures also cast some light on the estimated willingness to pay figures in Horton *et al.* (2003). There it was suggested that the willingness to pay of UK citizens for conserving the 5 per cent existing protected areas of Amazonia would generate a fund of over $900 million p.a. Yet the OECD data suggest UK actual expenditure on overseas biodiversity aid at one-thirtieth of this sum, just $30 million p.a.

### *5.3 Conclusions of actual flows of funds*

Table 1.4 summarises the previous discussion. It is important to note that the totals cannot be added. This is because bilateral, GEF and DfNS expenditures will overlap with PA expenditures – there will be some double counting.

The caveats to Table 1.4 are fairly obvious. First, the coverage is incomplete and excludes, for example, domestic expenditures on 'own' ecosystem conservation. This item will be significant for developed economies. Second, the figures exclude opportunity cost payments or estimates of opportunity costs, whether paid or not. The focus here is on actual flows of funds so the exclusion of opportunity costs is justified in one sense. On another view, it could be argued that the world has implicitly 'paid' the opportunity costs of conservation because it has sacrificed those costs by adopting ecosystem conservation – we saw that the World Bank estimated these at $20 billion for an *expansion* of the PA system.

Despite these caveats, the message that tends to emerge is twofold: (a) actual expenditures on international ecosystem conservation appear to be remarkably small, and (b) they bear no resemblance to the willingness to pay figures obtained in the various stated preference studies. At best, the world spends perhaps $10 billion annually on ecosystem conservation. As others have noted – e.g. James *et al.* (2001) – these sums are trivial in relation to what the world actually spends on subsidising economic activity, perhaps $1 trillion p.a., 1,000 times as much. Unfortunately, while such comparisons demonstrate the 'affordability' of a massively expanded ecosystem conservation programme, they also raise the complex question of why the world prefers to spend its money on subsidies that damage the environment rather than saving the environment in the first place. One might also contrast the $10 billion spent on conservation with the suggested 'required' budget for an effective system. James *et al.* (2001) put this budget at $300 billion p.a.: again, affordable but dramatically at odds with what we do.

## 6 A curious literature: the value of everything

The previous conclusion is a gloomy one. If there is a major problem of ecosystem services loss – and this seems likely – the world exhibits precious little intent to solve it. There is one other comparison that has been made in the literature and this relates what we spend to the *total value of all ecosystem services*. For example, James *et al.* (2002) compare their 'required funds' estimate to the value of ecosystem services as estimated by Pimentel *et al.* (1997) and Costanza *et al.* (1997), noting that the sums would be a trivial 0.1 to 1.0 per cent of these total values. The implication is that these latter estimates of total value have meaning. Unfortunately, they do not. Despite this, they are widely quoted and the exercise has been repeated in various forms (e.g. Alexander *et al.* 1998).

The importance of the attempt by Costanza *et al.* (1997) to secure the aggregate willingness to pay for *all* ecosystem services derives in part from its publication in a distinguished science journal, *Nature*. The answer given by Costanza *et al.* to the question of the value of the world's ecosystems was $33 trillion per annum, with a confidence interval of $16–$54 billion. Confusingly, Costanza *et al.* also described the $33 trillion as a 'minimum' estimate, raising a question about the meaning of the range quoted. However, this issue turns out to be trivial since Costanza *et al.*'s figures do not have any meaning at all.[10]

Table 1.5 summarises the economic values estimated in Costanza *et al.* (1997), taking the central values only and condensing the categories of biome and ecological service. The first part of the table shows that coastal ecosystems provide around one-third of the economic value, with oceans, forests and wetlands also of major significance. The second part of the table indicates that nutrient cycling accounts for over half the value, with cultural and waste treatment values also being significant. Cultural value here refers to non-commercial activity such as gaining aesthetic, spiritual and educational pleasure. Nutrient cycles refer to the ways in which ecosystems acquire, process, store and recycle nutrients, while waste treatment refers to the role played by ecosystems in recovering and modifying nutrients.

A brief example of open oceans illustrates the procedure. Its cultural value (not shown separately in Table 1.5) is $76 per hectare. Multiplied

[10] There is an interesting issue for anyone interested in the sociology of the media and scientific publication. Only one of the authors of Costanza *et al.* is an economist, yet the article is explicitly about economics. The article was published in a science journal rather than an economics journal, with the refereeing process being clouded in some mystery. Efforts by very distinguished economists to refute the article by sending a reply to *Nature* were rebuffed by the editor of *Nature* on the grounds that Costanza had refuted the criticisms in an unpublished communication with the editor. *Nature* failed to publish any criticism of the article beyond an equally strange one-page follow-up in a later issue of the journal (*Nature* 1998) which failed to explain why the economists in question had taken the view they did, simply quoting 'sound bites' and reporting Costanza and some co-authors as justifying the article because it had provoked controversy. This criterion for publishing a paper in *Nature* appears not to be applied to other papers in that journal.

Table 1.5 *The alleged 'global value' of the world's ecosystems*

**(a) by biome:**

| Biome | | Economic value in $10^{12}$ \$ (1994) | |
|---|---|---|---|
| Marine: | open ocean | 8.4 | |
| | coastal | 12.6 | |
| *Total marine* | | | 21.0 |
| Terrestrial: | forests | 4.7 | |
| | grass/rangeland | 0.9 | |
| | wetlands | 4.9 | |
| | lakes/rivers | 1.7 | |
| | cropland | 0.1 | |
| *Total terrestrial* | | | 12.3 |
| *Total* | | 33.3 | |

**(b) by service (\$ 1994, $10^{12}$)**

| | Gas regulation | Disturbance regulation | Water regulation | Water supply | Nutrient cycling |
|---|---|---|---|---|---|
| Value | 1.3 | 1.8 | 1.1 | 1.7 | 17.1 |
| | Waste treatment | Food production | Cultural | Other | *TOTAL* |
| Value | 2.3 | 1.4 | 3.0 | 3.6 | 33.3 |

by the hectarage of the open oceans this gives \$2.5 trillion. The \$76 is derived as follows: in California coastal real estate commands a 'premium' of \$10 million $ha^{-1}$ over and above the value of similar non-coastal land and this premium is assumed to reflect the cultural values of living near the ocean. In Alabama the same differential is only \$0.5 million. Taking this as representative of developed country values, and measuring the coastal area in developed countries as 9.7 million hectares, we get (approximately) \$5–\$105 trillion as the cultural value of ocean land. This is a capital value so it needs to be amortised over the lifetime of the land, taken to be twenty years. When this amortised value is divided by the total ocean area and combined with a guesstimate for differential land values in developing countries, the resulting value is \$76 per ha. The example is sufficient to show the considerable risks in such an exercise. A few studies have been extrapolated in a process that economists call 'benefits transfer' or 'value transfer'. It is true that many economists engage in this activity, but perhaps with a little more attention being paid to the validity of the exercise. It is known that value transfer is subject to potentially very large errors even when local studies are extrapolated to other local areas. Engaging in this activity at the global level is obviously very problematic.

The Costanza *et al.* estimates have been the subject of severe criticism centring on (a) the illicit procedure of extrapolating local values to global estimates, (b) the equally illicit procedure of using studies which report marginal willingness to pay as if they are some kind of average willingness to pay, and (c) the failure to note that aggregate willingness to pay cannot exceed aggregate global income – at the time $33 trillion exceeded the best estimate of global world income by $5 trillion. Even a moment's reflection would reveal that if the world's oceans disappeared there would be no-one left to express a willingness to pay for anything! Severe criticisms of this work can be found in Pearce (1998) and Bockstael *et al.* (2000). The central points of criticism are as follows:

- As Figure 1.1 shows, if any meaning at all can be assigned to measuring the area under $D_{ES}$ it would be infinity. Others prefer to argue that the notion of human-oriented value has no meaning in the unbounded zone of Figure 1.1. As Toman puts it, Costanza *et al.*'s estimates are a 'serious underestimate of infinity'. The error is to use *marginal* valuations to compute the total value of something that, if supplied in zero quantities, would have infinite marginal value. Costanza's defence against this criticism, that he was not doing anything different to valuing gross national Product (GNP), the value of the total flow of goods and services at market prices, misses the point. As Bockstael *et al.* (2000) point out, economic valuation is about tradeoffs, not absolute measures. The value of anything is measured in terms of what has to be sacrificed if the good or service is to be obtained. The value attached to specified ecosystem services is therefore a value derived by holding all other features of the economy and, for that matter, features of all other ecosystems constant. Hypothetically 'removing' an ecosystem service without asking what would happen to everything else and then adding up the apparent economic values is simply illicit: it has no economic meaning. Such aggregates totally ignore the non-independence of ecosystems – i.e. their role as complements and substitutes for each other. Yet what the Costanza *et al.* article does is to treat each and every ecosystem as if it were separable for the purposes of aggregation. Apart from being bad economics, this is also bad ecology. Ecology, like economics, teaches that everything is interconnected. As major changes in ecosystems occur, so all prices (and incomes) would change, making the original measurement units irrelevant. In short, economic valuation procedures designed to measure small changes in quantities of goods and services cannot be used to measure totals.

- The resulting total value, $33 trillion, exceeded, at the time, the value of the world's GNP. Yet the estimates are allegedly derived from willingness to pay studies. Willingness to pay cannot exceed the world's GNP. As Bockstael *et al.* (2000) put it: 'If interpreted literally, it suggests that a family earning $30,000 annually would pay $40,000 each year for ecosystem protection.' The figure, they say, is 'absurd'.
- If one switched the basis of valuation from willingness to pay to secure ecosystem services to one of willingness to accept compensation to go without them, the absurdity of the Costanza *et al.* exercise is further underlined. Without all ecosystems there is no alternative state that would define willingness to accept: no-one would exist to accept the compensation.
- *Even if* the Costanza *et al.* (1997) estimates could be applied to the totality of things, the procedures adopted are incorrect in many cases. While some estimates are based on WTP studies, quite a few rest on 'replacement cost estimates'. This means that the loss of something is valued at the cost of replacing it. This would be valid if and only if individuals were willing to pay that sum to replace the lost service. But if this was the case then one should use the WTP figure in the first place. Using the cost of replacing something to measure the WTP for it implies, logically, that the benefit-cost ratio of restoring any loss, wherever and whenever it occurs, is at least one. No losses would ever be justified. This obviously cannot be the case.
- In terms of Figures 1.1 and 1.2, what would the Costanza *et al.* estimates measure, assuming they adopted the correct methodology? It appears that what the authors *thought* they were measuring was an area such as OPQES', i.e. a minimum estimate of the total area under $D_{ES}$. But, even if $D_{ES}$ could be construed as the demand curve for *all* ES, we have seen that it is necessarily unbounded. The remarks above about the meaning of prices and values in economics now serve to underline the fact that $D_{ES}$ is not really defined across such major changes in the scale of ES as envisaged in the Costanza *et al.* paper.

It seems clear that the main 'driver' for the Costanza *et al.* paper, and a similar one by Pimentel *et al.* (1997), is the motivation to demonstrate to the world at large that the conservation of ecosystem services matters. That is a laudable goal, but it cannot excuse the publication of scientifically meaningless analysis. In the event, some of the authors have actually repeated the mistakes (e.g. Sutton and Costanza 2002), others have quoted them approvingly, and the relevant journals have not repented their decision to publish.

## 7 Conclusions: what have we learned?

The purpose of this chapter has been to explore the extent to which political rhetoric on the importance of biodiversity conservation is backed by action. The rationale for trying to probe this issue is the suspicion that we are doing far too little whereas the rhetoric is suggesting we are doing a lot. If we are being misled – deliberately or inadvertently, in one sense it does not much matter which – then we may have a serious problem of misallocation of resources. We cannot find out unless we track down the flows of funds and relate them to willingness to pay. In short, what are the costs and benefits of global conservation?

The indicators of action we have chosen have been fairly narrowly defined as actual expenditures or willingness to pay. The paper quite deliberately excludes other indicators of the degree of concern for biodiversity. Overall, the contention is that conservation policy is not backed by financial commitments. Game theory teaches that many policies, and especially international policies that appear to transcend national interests, may achieve little when compared with the counterfactual.

The next problem is to define the costs and benefits of ecosystem conservation and to identify, if possible, where we currently stand. The costs are clearly the costs of ecosystem management plus the forgone opportunity costs of the 'development' that might otherwise take place. A simplifying assumption, that all ecosystem conversion is successful in developmental terms, was made. In practice we know that a significant fraction of such conversions results in neither development nor conservation. The benefits are measured by the economic value of ecosystem services as measured by the world's willingness to pay for them. It was noted that this willingness to pay is not a constant, will vary with the quantity and quality of the ecosystem service in question, and cannot be defined when some subset of ecosystem services goes below some threshold. We noted that some ecosystem services are already subject to market forces, but most are not. While the process of conferring property rights on currently 'open access' resources is developing, we have no guarantee that it will move fast enough to prevent serious degradation of ecosystem services. Indeed, we have significant evidence that this process of institutional change is moving all too slowly.

Various indicators were discussed which demonstrate that the dynamic process is very probably one of ecosystem service loss, not gain. Others might suggest a more optimistic story and that, at least, things are not getting any worse. Neither the pessimists nor the optimists, however, can tell us whether we are at, above or below some economic optimum in terms of ecosystem services. Most ecologists are probably convinced that we are well below the optimum and this view is no doubt shared by quite a few economists. But demonstrating this is hard since data that compare costs and benefits for the same thing are difficult to come by. Micro cost-benefit studies may well suffer from censoring – the studies that show net benefits get reported, those that show net costs do not. Moreover, some of the claims about citizen willingness to pay for ecosystem conservation look decidedly dubious when compared with actual expenditures. Needless to say, willingness to pay estimates *should* exceed actual expenditures (by the amount of consumer surplus), but it is hard to believe some of the reported differences. A comparison of the costs of protected areas and their benefits also raises some doubts, not about specific studies but about the generality of the view that conservation benefits exceed conservation costs.

A partial picture of actual expenditures on conservation was produced by looking at protected area expenditures, bilateral biodiversity assistance, expenditures by the Global Environment Facility and debt-for-nature swaps. For the GEF and DfNSs we have an a priori reason to assume benefits exceed costs: both are examples of Coaseian bargains and such bargains should not take place unless a cost-benefit test is met. The overall impression is that ecosystem conservation expenditures are to be measured in a few billions of dollars, certainly not hundreds of billions. Several apparent conundrums ensue.

First, why are actual expenditures so far below apparent willingness to pay? This question is not explored in any detail here. It could be that willingness to pay studies exaggerate real intent. It could be that the element of 'surplus' of willingness to pay over actual expenditure is huge. It could be that we are under-recording actual expenditures by significant amounts – certainly we have not tried to measure domestic expenditures on 'own' conservation. Maybe there are other explanations too.

The next conundrum arises from a comparison of actual expenditures (and, for that matter, willingness to pay estimates) and some of the more celebrated estimates of the economic value of the world's ecosystems. The former seem to be measured in billions of dollars, the latter in trillions of dollars. But studies that attempt to measure the *total* value of all ecosystem services are more than flawed – they are arbitrary. What they do is to take valuation techniques designed to value small (marginal) or discrete changes in ecosystem services and fallaciously apply them to the totality of systems. It is perhaps significant that these estimates are produced (in the main) by non-economists writing in science journals.

There is no pleasure in reporting the *suspicion* that, despite all the rhetoric, the world does not care too much about biodiversity conservation. Maybe the efforts of economists and ecologists will force a change of policy in the future. But the proper place to begin is with an honest appraisal of just how little we do. Hopefully, others will show that we do more than suggested in this chapter, or that we can have an expectation that a lot more will be done in the future.

## REFERENCES

Alexander, A., List, J., Margolis, M. and d'Arge, R. 1998. A method for valuing global ecosystem services. *Ecological Economics*. **27**. 161–170.

Anderson, T. and Hill, P. (eds.). 1995. *Wildlife in the Marketplace*. Lanham, MD: Rowman and Littlefield.

Balmford, A., Bruner, A., Cooper, P., Costanza, R., Farber, S., Green, R., Jenkins, M., Jefferiss, P., Jessamy, V., Madden, J., Munro, K., Myers, N., Naeem, S., Paavaola, J., Rayment, M., Rosendo, S., Roughgarden, J., Trumper, K. and Turner, R. 2002. Economic reasons for conserving wild nature. *Science*. **297**. 950–953.

Balmford, A., Gaston, K., Blyth, S., James, A. and Kapos, V. 2003. Global variation in terrestrial conservation costs, conservation benefits, and unmet conservation needs. *Proceedings of the National Academy of Sciences*. **100** (3). 1046–1050.

Barrett, S. 2003. *Environment and Statecraft: the Strategy of Environmental Treaty Making*. Oxford: Oxford University Press.

Barrett, S. 2004. Kyoto plus. In Helm, D. (ed.). *Climate Change Policy*. Oxford: Oxford University Press.

Bockstael, N., Freeman, A. M., Kopp, R., Portney, P. and Smith, V. K. 2000. On valuing Nature. *Journal of Environmental Science and Technology*. **34** (8). 1384–1389.

Brouwer, R., Langford, I., Bateman, I. and Turner, R. K. 1999. A meta-analysis of wetland contingent valuation studies. *Regional Environmental Change*. **1** (1). 47–57.

Chape, S., Blyth, S., Fish, L., Fox, P. and Spalding, M. 2003. *2003 United Nations List of Protected Areas*. Gland, Switzerland: International Union for the Conservation of Nature, and Cambridge, UK: World Conservation and Monitoring Centre

Chomitz, K., Brenes, E. and Constantino, L. 1998. *Financing Environmental Services: The Costa Rican Experience and its Implications*. Washington DC: World Bank, mimeo. kchomitz@worldbank.org

Coase, R. 1960. The problem of social cost. *Journal of Law and Economics*. **3**. 1–44.

Costanza, R., d'Arge, R., de Groot, R., Farber, S., Grasso, M., Hannon, B., Limburg, K., Naeem, S., O'Neill, R., Paruelo, J., Raskin, R., Sutton, P. and van den Belt, M. 1997. The value of the world's ecosystem services and natural capital. *Nature*. **387**. 253–260.

Dirzo, R. and Raven, P. 2003. Global state of biodiversity and loss. *Annual Review of Environment and Resources*. **28**. 137–167.

Drake, V. 1995. *Dealing in Diversity: America's Market for Nature Conservation*. Cambridge: Cambridge University Press.

Food and Agriculture Organisation. 2001. *Forest Resources Assessment 2000.* Rome: FAO.

Horton, B., Colarullo, G., Bateman, I. and Peres, C. 2003. Evaluating non-user willingness to pay for large-scale conservation programs in Amazonia. A UK/Italian contingent valuation study. *Environmental Conservation.* **30** (2). 139–146.

Hutton, J. and Dickson, B. (eds.). 2000. *Endangered Species, Threatened Convention: the Past, Present and Future of CITES.* London: Earthscan.

James, A., Gaston, K. and Balmford, A. 1999. Balancing the Earth's accounts. *Nature.* **401**. 323–324.

James, A., Gaston, K. and Balmford, A. 2001. Can we afford to conserve biodiversity? *BioScience.* **51** (1). 43–52.

Kramer, R. and Mercer, E. 1997. Valuing a global environmental good: US residents' willingness to pay to protect tropical rain forests. *Land Economics.* **73**. 196–210.

Krebs, C. 1994. *Ecology. 4th Edition.* Menlo Park, CA: Addison-Wesley.

Langholz, J., Lassoie, J., Lee, D. and Chapman, D. 2000. Economic considerations of privately owned parks. *Ecological Economics.* **33**. 173–183.

Langholz, J. and Lassoie, J. 2001. Perils and promise of privately owned protected areas. *Bioscience.* **51**. 1079–1085.

Lapham, N. and Livermore, R. 2003. *Striking a Balance. Ensuring Conservation's Place on the International Biodiversity Assistance Agenda.* Washington DC: Conservation International.

Lindzen, R. 1994. On the scientific basis for global warming scenarios. *Environmental Pollution.* **83**. 125–134.

Lomborg, B. (ed.). 2004. *Global Crises, Global Solutions.* Cambridge: Cambridge University Press.

Lopez, R. 1998. Where development can or cannot go: the role of poverty-environment linkages. In B. Pleskovic and J. Stiglitz (eds.). *Annual World Bank Conference on Development Economics 1997.* Washington DC: The World Bank Press. 285–306.

Matthews, E. 2001. *Understanding the Forest Resources Assessment 2000.* Washington DC: World Resources Institute.

*Nature.* 1998. Audacious bid to value the planet whips up a storm. *Nature.* **395**. 430.

Norton-Griffiths, M. and Southey, C. 1995. The opportunity costs of biodiversity conservation in Kenya. *Ecological Economics.* **12**. 125–129.

O'Neill, B. and Oppenheimer, M. 2002. Dangerous climate impacts and the Kyoto Protocol. *Science.* **296**. 1971–1972.

Pearce, D. W. 1998. Auditing the Earth. *Environment.* **40** (2). March. 23–28.

Pearce, D. W. 2003. Will global warming be controlled? Reflections on the irresolution of humankind. In Pethig, R. and Rauscher, M. (eds.). *Challenges to the World Economy: Festschrift for Horst Siebert.* Berlin: Springer Verlag. 367–382.

Pearce, D. W. 2004. Environmental market creation: saviour or oversell? *Portuguese Economic Journal.* **3** (2). 115–144.

Pearce, D. W. and Moran, D. 1994. *The Economic Value of Biodiversity.* London: Earthscan.

Pearce, D. W. and Pearce, C. 2001. *The Value of Forest Ecosystems*. Montreal: Convention on Biological Diversity. www.biodiv.org/doc/publications/cbd-ts-04.pdf

Pearce, D. W., Atkinson, G. and Mourato, S. 2005. *Cost-Benefit Analysis and the Environment: Recent Developments*. Paris: Organisation for Economic Cooperation and Development.

Pimentel, D., Wilson, C., McCullum, C., Huang, R., Dwen, P., Flack, J., Tran, Q., Saltman, T. and Cliff, B. 1997. Economics and environmental benefits of biodiversity. *BioScience*. **47**. 747–757.

Pimm. S., Russell, G., Gittleman, J. and Brooks, T. 1995. The future of biodiversity. *Science*. **269**. 347–350.

Richards, J. 1990. Land transformation. In B. Turner, W. Clark, R. Kates, J. Richards, J. Mathews and W. Meyer. *The Earth as Transformed by Human Action*. Cambridge: Cambridge University Press. 163–178.

Ruitenbeek, J. 1992. The rainforest supply price: a tool for evaluating rainforest conservation expenditures. *Ecological Economics*. **6** (1). 57–78.

Sandler, T. 1997. *Global Challenges: an Approach to Environmental, Political and Economic Problems*. Cambridge: Cambridge University Press.

Sands, P., Tarasofsky, R. and Weiss, M. (eds.). 1994. *Documents in International Environmental Law. Volumes I and II*. Manchester: Manchester University Press.

Simon, J. 1981. *The Ultimate Resource*. Oxford: Blackwell.

Simon, J. 1986. *Theory of Population and Economic Growth*. Oxford: Blackwell.

Simon, J. 1995. *The State of Humanity*. Oxford: Blackwell.

Simon, J. and Kahn, H. 1984. *The Resourceful Earth*. Oxford: Blackwell.

Stork, N. 1999. The magnitude of global biodiversity and its decline. In J. Cracraft and F. Grifo (eds.). *The Living Planet in Crisis: Biodiversity, Science and Policy*. New York: Columbia University Press. 3–32.

Sudo, T. 2003. *A Study of International Financial Instruments for Global Conservation: Debt for Nature Swaps and the Clean Development Mechanism*. Master's Thesis in Environmental and Natural Resource Economics, University College London.

Sutton, P. and Costanza, R. 2002. Global estimates of market and non-market values derived from nighttime satellite imagery, land cover, and ecosystem service valuation. *Ecological Economics*. **41**. 509–527.

Turner, R. K., Pavavola, J., Cooper, P., Farber, S., Jessamy, V. and Georgiou, S. 2003. Valuing nature: lessons learned and future research directions. *Ecological Economics*. **46**. 493–510.

van Kooten, C. 1998. Economics of conservation biology: a critical review. *Environmental Science and Policy*. **1**. 13–25.

van Schaik, C., Terborgh, J. and Dugelby, B. 1997. The silent crisis: the state of rain forest nature preserves. In R. Kramer, C. van Schaik and J. Johnson (eds.). *Last Stand: Protected Areas and the Defense of Tropical Biodiversity*. Oxford: Oxford University Press. 64–89.

Vitousek, P., Ehrlich, P., Ehrlich, A. and Matson, P. 1986. Human appropriation of the products of photosynthesis. *BioScience*. **36**. 368–373.

Vitousek, P., Mooney, H., Lubchenco, J. and Melillo, J. 1997. Human domination of Earth's ecosystems. *Science*. **277**. 494–499.

Whelan, T. (ed.). 1991. *Nature Tourism: Managing for the Environment*. Covelo, CA: Island Press.

Woodward, R. and Wui, Y-S. 2001. The economic value of wetland services: a meta-analysis. *Ecological Economics*. **37**. 257–270.

World Bank. 2002. *Costing the 7th Millennium Development Goal: Ensure Environmental Sustainability*. Environment Department and Development Economics Research Group. Washington DC: World Bank. (Restricted.)

World Resources Institute. 2003. *World Resources 2002–2004. Decisions for the Earth – Balance, Voice and Power*. Oxford: Oxford University Press.

*Part I*

# Causes of biodiversity loss

*Section A*

# Land conversion

# 2 The economics of land conversion, open access and biodiversity loss

*Edward B. Barbier*

## 1 Introduction

In developing economies, especially those without oil and natural gas reserves, the most important source of natural wealth is agricultural land. In these economies, the agricultural land base is expanding rapidly through conversion of forests, wetlands and other natural habitat (Barbier 2005). During 1980–1990 over 15 million hectares of tropical forest were cleared annually and the rate of deforestation averaged 0.8 per cent per year (FAO 1993). Although over 1990–2000 global tropical deforestation slowed to less than 12 million ha per year, or an annual rate of 0.6 per cent, this trend reflects less deforestation mainly in Latin America and Asia. Forest clearing increased over 1990–2000 in Africa to over 4.8 million ha annually, or 0.8 per cent per year. Whereas deforestation has declined in Tropical South America, Central Africa and Southeast Asia, it has risen significantly in Tropical Southern, West and East Sahelian Africa (FAO 2001).

López (1998a, 1998b) identifies most of sub-Saharan Africa, parts of East and Southeast Asia and the tropical forests of South America as regions with 'abundant land' and open access resource conditions that are prone to agricultural expansion. This expansion is mainly due to the high degree of integration of rural areas with the national and international economy as well as population pressures. The poor intensification of agriculture in many tropical developing countries, where use of irrigation and fertiliser is low, is also an important factor (FAO 1997, 2003).

The trend of rapid agricultural land expansion in tropical regions is of major concern to the problem of biodiversity loss because of the implications for forest conversion. Tropical forest ecosystems are the most species-rich environments. Although they cover less than 10 per cent of the world's surface they contain 90 per cent of the world's species (UNEP 2002). In order to develop an indicator of trends in the stock of biodiversity, the United Nations Environment Program-World Conservation Monitoring Centre (UNEP-WCMC) in cooperation with the

World Wide Fund for Nature (WWF) created the Living Planet Index (UNEP-WCMC 2000). The index is derived from trends in the size of wild populations of species in three habitats – forest, freshwater and marine ecosystems. The prevailing trend of all three indices between 1970 and 1990 was downward and the forest index displayed a decline of approximately 12 per cent, predominately in tropical species, which is consistent with deforestation trends in tropical forest regions. Other predictions of global extinction rates are projected on the basis of estimates of species richness in tropical forests combined with actual and projected deforestation trends. These calculations suggest a rate of species loss of around 1–5 per cent per decade (WCMC 1992).

If conversion and alteration of tropical forest habitats and ecosystems are by far the most important factors underlying global biodiversity loss, then one must ask: what are the principal causes behind the expansion of agricultural land in developing countries? Many recent economic analyses of tropical deforestation and land conversion have emphasised the important role of institutional factors (Brown and Pearce 1994; Kaimowitz and Angelsen 1998; van Kooten *et al.* 1999; Barbier and Burgess 2001a; Barbier 2004). One key institutional factor, the prevalence of open access conditions and poorly defined property rights in land frontier regions, is routinely cited as a major contributing factor to excessive tropical agricultural land expansion. On the positive side, there is ample evidence in developing economies that more secure rights over natural resources, particularly land, will lead to incentives for increased investments in resource improvements and productivity (Feder and Onchan 1987; Feder and Feeny 1991; Besley 1995; Bohn and Deacon 2000). There is also counterevidence that tenure insecurity in tropical forest frontier regions will also create the incentives for agricultural land conversion (Barbier and Burgess 2001b). Finally, several studies emphasise the rent-dissipation effect of poorly defined property rights, including the breakdown of traditional common property rights regimes, in developing countries (Bromley 1989, 1991; Ostrom 1990; Baland and Plateau 1996; Alston *et al.* 1999; Deacon 1999).

As open access conditions and ill-defined property rights are thought to be important factors driving agricultural land expansion and forest conversion in developing countries, there needs to be developed an adequate economic model of forest land conversion under open access that can be empirically tested. The purpose of this chapter is to illustrate one such land conversion model at the country case study level, following the approach of Barbier (2002) and Barbier and Cox (2004). The model presented (Sections 3 and 4) is based on the behaviour of an economic actor which converts open access lands. Two versions of the model are developed to contrast the role of formal and informal institutions (e.g. legal ownership rules versus traditional common property rights regimes) as constraints on the land conversion decision. Following the perspective on institutions developed by North (1990), the model demonstrates that the equilibrium level of land cleared will differ under conditions of no institutional constraints – i.e. the *pure open access* situation – compared with conditions where effective institutions exist to control and thus raise the cost of land conversion. The model is then applied to two case studies. The first (Section 5) investigates expansion of agricultural planted area in Mexico at the state level and over the 1960–1985 period before implementation of the North American Free Trade Agreement (NAFTA) reforms. The second case study (Section 6) deals with mangrove conversion for shrimp farming in Thailand's coastal provinces between 1979 and 1996.

## 2 Institutional constraints and forest conversion

In many tropical regions a key factor influencing deforestation is the lack of effective property rights and other institutional structures that limit access and conversion of forest land. In the absence of formal ownership rules, traditional common property regimes in some forested regions have also proven to be effective in controlling the open access deforestation problem (Bromley 1989, 1991; Larson and Bromley 1990; Ostrom 1990; Baland and Platteau 1996; Richards 1997; Gibson 2001). In short, formal and informal institutions can influence the process of forest loss by imposing increased costs of conversion on farmers who clear forestland.

In this chapter we are concerned with analysing the role of formal and informal institutions as constraints on the conversion of forestland to agriculture in developing countries. The perspective on institutions adopted here follows the approach of North (1990), who defines institutions as 'humanly devised constraints that shape human interaction' and which 'affect the performance of the economy by their effect on the costs of exchange and production'. One can therefore formalise the relationship between institutional constraints and forestland conversion by smallholders using a simple agricultural household model. Because institutions raise the cost of land clearing, more land should be converted under pure open access.[1] In turn, the existence of institutional constraints prevents the adjustment of the stock of converted land to the long-run

[1] However, formal and informal institutions governing agricultural land ownership and expansion are not uniform across developing countries or even within the same country (Baland and Platteau 1996; Burger *et al.* 2001; López 1998b). As will become clear, the following analysis is capable of only assessing two equilibrium situations: one with the

equilibrium desired by agricultural households, which is the amount of land that could be cleared under open access.

The next two sections develop the two versions of the formal model of agricultural land conversion, under open access conditions and under institutional constraints governing conversion. We develop the model with the assumption that the economic agent undertaking land conversion is an agricultural household seeking to add to its existing cropland area at the expense of freely available forested land. This model is directly applicable to the case study of Mexico, where maize land expansion by peasant farmers was the main cause of forest loss in the pre-NAFTA era. However, as the Thailand case study illustrates, the same model can be applied to other processes of land conversion under open access situations, such as mangrove deforestation by commercial shrimp farms seeking to expand aquaculture areas.

## 3 A pure open access model of land conversion

The following model of land conversion is based on an approach similar to that of Panayotou and Sungsuwan (1994), López (1997, 1998a) and Cropper *et al.* (1999). The model and the two case studies of Mexico and Thailand appear in Barbier (2002) and Barbier and Cox (2004).

Assume that the economic behaviour of all $J$ rural smallholder households in the agricultural sector of a developing country can be summarised by the behaviour of a representative $j^{th}$ household. Although the representative household is utility-maximising, it is a price taker in both input and output markets. Farm and off-farm labour of the household are assumed to be perfect substitutes, such that the opportunity cost of the household's time (i.e. its wage rate) is determined exogenously. The household's behaviour can be modelled recursively, in the sense that the production decisions are made first and then the consumption decisions follow (Singh *et al.* 1986).

In any time period, $t$, let the profit function of the representative agricultural household's production decisions be defined as:

$$\max \pi(p, w, w_N) = \max_{N_j, x_j} pf(x_j, N_j) - wx_j - w_N N_j \tag{1}$$

where the variable inputs include cleared land by the $j^{th}$ household, $N_j$, and a vector, $x_j$, of other $i, \ldots, k$ inputs (e.g. labour, fertiliser, seeds)

presence of institutions that are effective in controlling or limiting forest conversion and one where open access prevails. Although it may be the case that a state with effective institutions may cause less deforestation than a state without, it is also possible that the transition path from a pure open access situation to a state in which effective institutions are established may result in increased deforestation during this transition period. For some possible examples, see López (1998b).

used in production of a single agricultural output. The corresponding vector of input prices is $w$, and $p$ is the price of the farm output. Finally, $w_N$ is the rental 'price' of land. If the household clears its own land from freely accessible forest, then this is an implicit price or opportunity cost (Panayotou and Sungsuwan 1994). However, if the household purchases or rents additional cleared land from a market, then $w_N$ would be the market rental price of land (Cropper *et al.* 1999).

Utilising Hotelling's lemma, the derived demand for cleared land by the $j^{th}$ household, $N_j$, is therefore:

$$N_j = N_j(p, w, w_N) = -\partial\pi/\partial w_N, \quad \partial N_j/\partial w_N < 0, \quad \partial N_j/\partial p > 0 \tag{2}$$

As (2) is homogeneous of degree zero, it can also be rewritten as a function of relative prices, using one of the input prices, $w_i$, as a numeraire:

$$N_j = N_j(p/w_i, w/w_i, w_N/w_i), \quad \partial N_j/\partial(w_N/w_i) < 0, \quad \partial N_j/\partial(p/w_i) > 0 \tag{3}$$

Equations (2) and (3) depict the derived demand for cleared land by the representative $j^{th}$ household. Assuming a common underlying technology for all rural households engaged in land clearing allows us to aggregate either relationship into the total demand for converted land by all $J$ households. To simplify the following analysis, we will work primarily with the derived demand relationship (2).

In aggregating the demand for cleared land across all $J$ agricultural households, it is important to consider other factors that may influence the aggregate level of conversion, such as income per capita, population and economy-wide policies and public investments.[2] Thus, allowing $Z$ to represent one or more of these exogenous factors and $N$ the aggregate demand for cleared land, the latter can be specified as:

$$N = N(p, w, w_N; Z) \tag{4}$$

As rural households generally provide their own supply of cleared land, $N$, one can view this type of supply as a kind of 'production' of cleared

[2] For reviews of relevant empirical studies, see Barbier (2004), Barbier and Burgess (2001a), Brown and Pearce (1994), Kaimowitz and Angelsen (1998), van Kooten *et al.* (1999). The assumptions as to how exogenous 'macroeconomic' factors influence the aggregate demand for cleared land vary across the different studies. For example, to derive the aggregate demand for cleared land, Cropper *et al.* (1999) multiply the household demand by the total number of agricultural households. The latter is assumed to be endogenously determined, with one of its explanatory variables being non-agricultural income. In contrast, Barbier and Burgess (1997) and Panayotou and Sungsuwan (1994) simply assume that the aggregate demand equation for cleared land includes both a population variable and income per capita as additional exogenous factors in the demand relationship.

land governed by the following conditions. The source of the cleared land (i.e. forested areas) is an open access resource, so that land is cleared up to the point where any producer surpluses (rents) generated by clearing additional land are zero.[3] The principal input into clearing land is labour, $L$, which is paid some exogenously determined wage rate, $w_L$, and the production function is assumed to be homogeneous. This production of cleared land may also be affected by a range of exogenous factors, $\alpha$, that may influence the accessibility of forestland available for conversion, including roads, infrastructure and closeness to major towns and cities.

Thus one can specify a cost function, based on the minimum cost for the rural household of producing a given level of cleared land, $N$, for some fixed levels of $w_L$ and $\alpha$, as:

$$C_j = C_j(w_L, N; \alpha) \tag{5}$$

Under open access conditions, each household will convert forest area up to the point where the total revenues gained from converting $N_j$ units of land, $w_N N_j$, equal the total costs represented by (5). If a farming household clears its own land, then $w_N$ is now the household's implicit 'rental' price, or opportunity cost, of utilising additional converted land. However, as the household is essentially supplying land to itself, then in equilibrium the implicit price ensures that the household's costs of supplying its own land will be equated with its derived demand for converted land. Then for the $j^{th}$ representative household the following cost conditions for supplying its own cleared land must hold:

$$w_N = c_j(w_L, N_j; \alpha), \quad \partial c_j/\partial w_L > 0, \quad \partial c_j/\partial N_j > 0,$$
$$\partial c_j/\partial \alpha < 0, \quad j = 1, \ldots, \mathcal{J}. \tag{6}$$

The right-hand side of (6) is the average cost curve for clearing land, which may be increasing with the amount of land cleared as, among other reasons, one must venture further into the forest to clear more land (Angelsen 1999). Equation (6) therefore represents the equilibrium 'own' supply condition for households exploiting a pure open access resource (Freeman 2003, Chapter 9). That is, in equilibrium, the household's implicit price for cleared land will be equated with its per unit costs of

[3] The assumption that open access conditions prevail in 'accessible' forest areas implies that, if there are any rents or producer surpluses generated from clearing land, then others attracted by these profits will enter the forest to clear land as well. In equilibrium, any rents will then be dissipated and thus each individual will clear land up to the point where total revenues equal total costs. This assumption is common in bioeconomic models of unregulated open access resources, in particular fisheries (see Freeman 2003, Chapter 9 and Heal 1982 for reviews).

forest conversion, thus ensuring that any rents from clearing are dissipated. Together with the household's derived demand for converted land (2), equation (6) determines the equilibrium level of land clearing by the household as well as its implicit price. Although the latter variable is not observed, it is possible to use (2) and (6) to solve for the reduced form equation for the equilibrium level of cleared land. Substituting (6) for $w_N$ in equation (2), and then rearranging to solve for $N_j$, yields:

$$\begin{aligned} N_j &= N_j(p, w, w_N(w_L, \alpha)), \\ dN_j/dw_L &= \partial N_j/\partial w_L + \partial N_j/\partial w_N \partial w_N/\partial w_L, \\ dN_j/d\alpha &= \partial N_j/\partial w_N \partial w_N/\partial \alpha > 0. \end{aligned} \tag{7}$$

Aggregating (7) across all $J$ households in a province or region that convert their own land, and including exogenous factors $Z$, leads to a reduced form relationship for the aggregate equilibrium level of cleared land:[4]

$$N^* = N(p, w_I, w_L; \alpha, Z), \quad dN/dp > 0, dN/d\alpha > 0, \tag{8}$$

where the wage rate, $w_L$, is now distinguished from the vector of prices for inputs other than labour, $w_I$. The amount of land converted should increase with the price of output and the accessibility of forest land. However, the impact of a change in the wage rate or other input prices is ambiguous.[5]

## 4 Institutional constraints and land conversion

In the case of deforestation, effective formal and informal institutions may limit the ability of smallholders and others to obtain and convert forestland, thus increasing the costs of clearing compared with pure open access conditions. For example, it is straightforward to demonstrate that,

[4] Note that if the household-derived demand relationship (3) was used with (6) to solve for the reduced form level of land conversion, $N_j^*$, then the aggregate land conversion relationship (8) would be specified in relative prices, i.e.

$$N^* = N(p/w_i, w/w_i, w_N/w_i; \alpha, Z)$$

[5] In the case of the impacts of a change in the wage rate on land clearing, the ambiguity of the impacts arises because of two possible counteracting effects. First, a higher wage rate should make it more costly for the household to convert more land area, thus reducing the equilibrium amount of land converted. However, labour is also used in agricultural production, and if land and labor are substitutes, then a higher wage rate may also increase the use of converted land in production. Whether the equilibrium level of cleared land will increase or decrease in response to a rise in the wage rate will depend on the relative magnitude of these two effects. See Barbier and Cox (2004) for further details.

if private or common property institutions enable individuals to optimally manage forest resources to supply converted land, then not only are producer surpluses being generated but also the costs of supplying converted land will always be higher than under conditions of open access supply.

Even though the conditions for establishing effective private or common property regimes to manage resources optimally in developing countries are stringent (Baland and Platteau 1996; Ostrom 2001), it is unlikely that these conditions are met in many remote, frontier forest areas prone to agricultural conversion (Barbier and Burgess 2001b). Nevertheless, in some regions and countries, the presence of formal and informal institutions may not have led to optimal management of the supply of converted land from the forests, but they may have controlled open access exploitation by restricting land clearing and increasing the costs of conversion. If institutional constraints on forest conversion in developing countries do operate in this way, then it is straightforward to extend the model of pure open access conversion to incorporate such impacts. This in turn can yield a 'testable' hypothesis of the effectiveness of institutional constraints on deforestation.

Let $\beta$ represent some impact of institutions on the costs of clearing land. If the presence of such effects increases the average costs of clearing, then it should follow that:

$$c_j(w_L, N_j; \alpha, \beta) > c_j(w_L, N_j; \alpha). \tag{9}$$

Due to the institutional constraints, $\beta$, the per unit costs of land clearing are now higher compared with pure open access conditions. Defining $N^I$ as the aggregate amount of land cleared under the presence of such constraints, then from (6)–(8):

$$N^* > N^I = N^I(p, w, w_L; \alpha, \beta, Z). \tag{10}$$

The equilibrium amount of cleared land will be lower when institutional constraints are present compared with the pure open access situation.[6]

The above relationships can be used to develop a simple empirical test of whether institutional constraints may be affecting the level of agricultural-related deforestation. If $D_t$ is the rate of deforestation caused by agricultural conversion over any time period $(t-1, t)$, then by definition $D_t = N_t - N_{t-1}$. That is, deforestation is equal to the change in the amount of agricultural land cleared and cultivated by farmers. However, equation (10) indicates that, if over the time period $(t-1, t)$ institutional constraints are present, then the rate of deforestation under

[6] The reduced-form level of land conversion when institutional constraints are present, $N^I$, can also be specified in terms of relative prices, i.e. $N^I = N^I(p/w_i, w/w_i, w_N/w_i; \alpha, \beta, Z)$.

these constraints will be less than under pure open access conditions, i.e. $D_t^I = N_t^I - N_{t-1} < D_t^* = N_t^* - N_{t-1}$. Adjustment in the level of agricultural conversion will be slower if institutional constraints raise the costs of clearing land. Assuming that the difference in the respective deforestation rates can be accounted for by some adjustment parameter, $\delta$, it therefore follows that:

$$D_t^I = N_t^I - N_{t-1} = \delta(N_t^* - N_{t-1}) = \delta D_t^* \qquad 0 \leq \delta \leq 1 \qquad (11)$$

Equation (11) is a basic partial adjustment model. It allows for a straightforward test of whether institutional impacts on the costs of land clearing, $\beta$, are restricting agricultural land expansion *without* having to specify the relationship between $\beta$ and the amount of land cleared, $N_t^I$. For example, if $\delta = 1$, this implies that institutional impacts, $\beta$, are having a negligible impact on land conversion, i.e. the actual level of land conversion is equivalent to the level under pure open access conditions, $D_t^*$. In contrast, $\delta = 0$ indicates that institutional constraints on land conversion are absolutely binding and land use change is not responding to any of the factors influencing the supply and demand for cleared land, i.e. $N_t^I = N_{t-1}$. Values of $\delta$ within these two extremes will indicate the degree to which institutional impacts, $\beta$, on the costs of land clearing are 'constraining' the rate of forest conversion.

Substituting equation (8) into (11) yields the partial adjustment model for cleared land. For purposes of estimation, a linear version of this model is assumed:

$$N_t^I = \delta[\gamma_0 + \gamma_1 p_t + \gamma_2 w_t + \gamma_3 w_{Lt} + \gamma_4 \alpha_t + \gamma_5 Z_t] + \lambda N_{t-1} + \delta\mu_t, \qquad (12)$$

where $\lambda = 1 - \delta$ and $\mu_t$ is the error term.

Alternatively, employing the relative price specification (3):

$$N_t^I = \delta\left[\gamma_0 + \gamma_1 \frac{p_t}{w_{it}} + \gamma_2 \frac{w_t}{w_{it}} + \gamma_3 \frac{w_{Lt}}{w_{it}} + \gamma_4 \alpha_t + \gamma_5 Z_t\right] + \lambda N_{t-1} + \delta\mu_t, \qquad (13)$$

A regression of either (12) or (13) will yield estimated coefficients $\delta_{\gamma k}$, which depict the adjusted impacts of the explanatory variables on land conversion under the presence of institutional constraints. The adjustment parameter $\delta$ can be calculated from the estimated value of $\lambda$. The latter value can in turn be used to derive the $\gamma_k$ coefficients, which indicate the impacts of the explanatory variables under open access conditions. The regression estimates will therefore yield a direct test of the hypothesis that the presence of formal or informal institutional controls on land clearing will restrict land expansion and thus the rate of deforestation. That is, if $\gamma = (1 - \delta) > 0$, then effective institutional constraints on land clearing will reduce the rate of deforestation due to agricultural land expansion.

The next section discusses the application of the above model to the case of agricultural land expansion in Mexico during the pre-NAFTA reform era, 1960–1985.

## 5 Agricultural land expansion in pre-NAFTA Mexico[7]

Until the early 1990s, one of the most enduring pieces of land tenure legislation in Mexico had been Article 27 of the 1917 Mexican Constitution (Brown 1997; Cornelius and Myhre 1998). Article 27 had established communal land ownership – the *ejido* – as the principal agrarian institution in rural Mexico. The *ejido* provided a framework for collectively managed, community-based land ownership. Although individual use rights of land could be assigned through a collective decision made by the community, the use rights could not be rented, sold or mortgaged. By 1991 there were 29,951 *ejidos* in Mexico accounting for 55 per cent of the land area (Jones and Ward 1998). Estimates suggest that as much as 70 per cent of the total forest area of 49.6 million hectares was owned by *ejidos*.

Over the period 1989–1994, Mexico implemented a series of major rural reforms aimed at transforming its agricultural sector to promote private investment and growth (Appendini 1998). The main impetus for such reforms was Mexico's participation in NAFTA, although the removal of agricultural subsidies started after the 1982 debt crisis. One of the most significant NAFTA reforms was the 1992 revisions to Mexico's land tenure legislation, as enshrined in Article 27 of the 1917 Mexican Constitution.

As the *ejido* system of land management is widely believed to have been a major factor in controlling deforestation in pre-NAFTA Mexico, there are major concerns that the removal of this system of control may spur greater deforestation (Richards 1997; Gibson 2001; Sarukhán and Larson 2001). Substantial forest conversion did occur over the pre-NAFTA era, ranging from 400,000 to 1.5 million ha per year, and mainly in tropical areas (World Bank 1994). A major cause of this deforestation was the expansion of agricultural and livestock production, largely by poor rural farmers seeking new land (Barbier and Burgess 1996; Deininger and Minten 1999). Road building and timber extraction

[7] The following case study is based on Barbier (2002).

may also have contributed through 'opening up' new areas of forest for encroachment by these activities.

A panel analysis conducted by Barbier and Burgess (1996) found that prior to the NAFTA reforms, the majority of agricultural production in Mexico was essentially low input and extensive in land use, which appears to characterise much of *ejido*-based smallholder cultivation across Mexico (World Bank 1989; Brown 1997; Cornelius and Myhre 1998). A more recent study of deforestation over the 1980–1990 period rejected the hypothesis that *ejido* ownership of agricultural land led to greater deforestation, leading the authors to conclude that there is little evidence that widespread communal land ownership over 1980–1990 promoted increased forest conversion (Deininger and Minten 1999). To the contrary, the authors suggest that *ejido*-based communities, 'valuing the safety net provided by such arrangements, have developed forms of organization capable of overcoming the "tragedy of the commons"' (Deininger and Minten 1999, p. 334).

In sum, although forest conversion to agriculture did occur during the pre-NAFTA reform era, the prevalence of *ejido* collective management of agricultural and forested lands may have deterred deforestation somewhat. During this period, such institutional constraints may have led to a lower rate of deforestation than if the remaining forested areas were under pure open access. Thus an analysis of the agricultural land expansion that occurred in Mexico during the pre-NAFTA reform period makes a relevant case study for examining the effectiveness of institutional constraints on deforestation. Such an analysis was implemented by Barbier (2002), with equation (13) chosen as the specification for the reduced-form land conversion relationship.

The partial adjustment relationship (13) was estimated through a dynamic panel analysis of longitudinal data for planted agricultural area. This was applied across the thirty-one states of Mexico, plus the Federal District, and included the 1960, 1970, 1980 and 1985 time periods.[8] The latter periods coincide with the era of pre-rural reforms in Mexico, when agricultural policies were fairly stable and *ejido* ownership of agricultural and forested lands was most prevalent (Brown 1997; Appendini 1998; Jones and Ward 1998).

In the dynamic panel analysis of (13), the dependent variable, $N^l$, was agricultural area planted, which was also lagged one time period to obtain $N_{t-1}$. The relative price variable, $p/w_i$, was represented by the ratio for guaranteed maize prices to fertiliser prices, and the relative wage variable, $w_L/w_i$, by the ratio of rural wage rates to fertiliser prices. Unfortunately,

[8] See Barbier (2002) for further details of the specific panel analysis approach.

Table 2.1 *Mexico – random effects estimation of agricultural land expansion, 1960–1985*

| | Dependent variable: agricultural area planted ('000 ha) | | | |
|---|---|---|---|---|
| Explanatory variables[a] | Adjusted parameter estimates $(\delta\gamma_k)^a$ | Adjusted elasticity estimates | Unadjusted parameter estimates $(\gamma_k)$ | Unadjusted elasticity estimates |
| Maize-fertiliser price ratio (Mexican pesos (MEX$)/metric ton) | 628.71 (2.995)** | 1.1477 | 4,765.12 | 8.6986 |
| Rural wage-fertiliser price ratio (MEX$/day per MEX$/kg) | −10,512 (−1.986) * | −0.4642 | −79,673 | −3.5184 |
| Population ('000 persons) | 0.025 (2.501)** | 0.0943 | 0.19 | 0.7146 |
| Income per capita (MEX$/population) | −0.0049 (−1.540) | −0.0938 | −0.04 | −0.7106 |
| Road density (Km/ha) | −23.724 (−1.593) | −0.0617 | −179.81 | −0.4676 |
| Lagged agricultural area planted (Lagged one period, initial period 1960) | 0.8681 (17.612)** | – | – | – |
| Constant | −210.56 (−2.012) * | – | – | – |

Estimated $\delta = 1 - \lambda = 0.1319$

Notes: [a]t-ratios are indicated in parentheses. **Significant at 1 per cent level. *Significant at 5 per cent level

Source: Barbier (2002)

lack of data on other input prices used in agricultural production precluded the inclusion of a variable for other relative input prices, $w/w_i$.[9] Exogenous economic and policy factors, *Z*, that might also affect land clearing included population and income per capita. Exogenous factors influencing the accessibility of forested lands, $\alpha$, were represented by road density.

Table 2.1 indicates the results for the random effects model, which was the preferred regression. The maize price-fertiliser ratio, population and lagged planted area are highly significant and lead to an increase in agricultural land area. The ratio of rural wage rates to fertiliser prices is also significant at the 5 per cent level. As expected, an increase in this ratio

[9] In fact, land, labor and fertilisers were the predominant inputs in smallholder, mainly land-extensive and rainfed agriculture across Mexico during the pre-NAFTA period (World Bank 1989).

leads to a fall in planted area. However, neither income per capita nor road density is a significant factor in explaining agricultural expansion. The negative sign of the latter variable suggests that it may be reflecting the rapid growth of urbanisation in many states rather than indicating greater accessibility to frontier forest areas.

As noted, the coefficient on lagged agricultural area, *MAAP*(-1), is both highly significant and relatively large (i.e. $\lambda = 0.868$). This implies that the null hypothesis that effective institutional constraints may have restricted the rate of land expansion cannot be rejected for the 1960–1985 period in Mexico. The presence of *ejido* communal ownership of agricultural and forest lands may have exerted some degree of control on land conversion in pre-NAFTA Mexico, thus slowing down the pace of deforestation compared with pure open access conditions.[10]

The possible impacts of this effect are indicated by a comparison of the 'adjusted' and 'unadjusted parameter' and elasticity estimates depicted in Table 2.1. As the table shows, the adjusted responses of planted area to the key explanatory factors are significantly lower than the unadjusted estimates. This is particularly striking for the three significant variables in the regression: the maize-fertiliser price ratio, the wage-fertiliser price ratio and population. For example, the maize-fertiliser price ratio clearly had the largest impact on agricultural land use in pre-NAFTA Mexico. However, whereas the adjusted elasticity indicates that a 10 per cent increase in the price ratio over this period caused an 11.5 per cent increase in agricultural area planted, the 'unadjusted' response would have been an 87 per cent increase in agricultural land use. Compared with pure open access conditions, *ejido* land management may therefore have mitigated considerably the incentives for farmers to convert forestland to agriculture in response to any increases in the maize-fertiliser price ratio during the pre-NAFTA period. Similar comparisons can be made for the influence of the wage-fertiliser price ratio and population on planted area. The adjusted response to a 10 per cent rise in the wage-fertiliser price ratio over 1960–1985 was a fall in agricultural area of 4.6 per cent. In contrast, the unadjusted response would have been a decrease in agricultural land use of 35.2 per cent. A 10 per cent increase in population leads to an adjusted 0.9 per cent rise in planted area, whereas the unadjusted increase is 7.1 per cent.

[10] There is limited anecdotal evidence that, in some areas, the *ejido* system may have controlled deforestation better than property-owning alternatives. In Chiapas, a controlled comparison of an *ejido* with a neighbouring community of property-owning individuals revealed that the former was characterised by fewer inequalities in wealth and land holdings, greater community solidarity and fewer social problems (Brown 1997). Since the 1950s, the *ejido* community had also experienced less land use change and expansion and more stable land ownership patterns.

These regression results are also consistent with the theoretical model of smallholder land conversion. By far the largest significant influence on agricultural land expansion in pre-NAFTA Mexico was the maize-fertiliser price ratio, followed by the wage-fertiliser price ratio and then population. As noted, smallholder farming in Mexico over 1960–1985 was characterised by low agricultural productivity, predominantly maize-based and dependent largely on unskilled farm labour and land as its main inputs (World Bank 1989; Brown 1997; Cornelius and Myhre 1998). Although subsidies helped to increase the use of fertilisers among farmers, these inputs tended to be under-utilised, especially by poorer smallholders. Thus a rising maize-fertiliser price ratio would effectively represent greater returns to smallholder production, leading to more land being converted and brought into cultivation. Equally, an increasing population would mean more farming households and labourers, causing a further increase in the demand for agricultural land. Finally, a rise in the price of labour relative to fertiliser reduces both cultivated area and land conversion, suggesting that land is being substituted for labour in cultivation.

Although changes in the maize-fertiliser price ratio, the wage-fertiliser price ratio and population are important factors influencing forest conversion in pre-NAFTA Mexico, Table 2.1 indicates that such impacts may have been mitigated by the effective controls on land use and ownership by *ejido* collective management compared with pure open access conditions. The key issue is, of course, whether or not the 1989–1994 rural reforms in Mexico – and principally the 1992 reforms of *ejido* land ownership – have affected any such institutional constraints on land conversion in the post-NAFTA era.

As summarised by Barbier (2002), there remains a degree of institutional control of forest conversion by smallholders in rural Mexico. Forested lands continue to be held and managed collectively by *ejidos* and there is very little evidence that the parcelling of communal agricultural land into individual plots has resulted so far in greater levels of forest conversion. Nevertheless, the widespread changes in institutional arrangements ushered in by the 1992 land tenure reforms are likely to have some influence on the rate of forest land conversion, although it may be some time yet before the effects on conversion can be detected. In addition, other NAFTA reforms and structural changes in the Mexican economy and agricultural sector have affected agricultural land conversion (Barbier and Burgess 1996). The latter incentive effects could be considerable and may make it difficult to determine the impacts of the recent institutional changes on land conversion. Possibly the greatest concern for the future is what might happen to forested lands if more and more *ejidos* are dissolved or become increasingly ineffective in managing land collectively. Although legally the forest land will revert to state ownership, public authorities may have a great deal of difficulty enforcing control of forest conversion. Throughout Latin America, the inability of central and regional governments to control illegal land clearing, squatting and land ownership disputes in remote frontier areas is not an encouraging precedent. If this occurs on a large enough scale, the open access model of land conversion by smallholders may become a more appropriate description of the process of deforestation in rural Mexico.

## 6 Shrimp farm expansion and mangrove loss in Thailand[11]

The issue of coastal land conversion for commercial shrimp farming is a highly debated and controversial topic in Thailand. Frozen shrimps are one of the country's major export products, earning more than $1.6 billion each year, and the government has encouraged these exports (Tokrisna 1998; Barbier and Sathirathai 2004). Yet expansion of shrimp exports has caused much devastation to Thailand's coastline and has impacted other valuable commercial sectors, such as fisheries.

Thailand's coastline is vast, stretching for 2,815 km, of which 1,878 km is on the Gulf of Thailand and 937 km on the Andaman Sea (Indian Ocean) (Kaosa-ard and Pednekar 1998). In recent decades, the expansion of intensive shrimp farming in the coastal areas of Southern Thailand has led to rapid conversion of mangroves (Barbier and Sathirathai 2004). In the mid-1990s the annual loss was estimated to be around 3,000 ha per year (Sathirathai 1998).

Although mangrove conversion for aquaculture began in Thailand as early as 1974, the boom in intensive shrimp farming through mangrove clearing took off in 1985 when the increasing demand for shrimps in Japan pushed up the border-equivalent price to $100 per kilogram (kg) (Barbier and Sathirathai 2004). For example, from 1981 to 1985 in Thailand, annual shrimp production through aquaculture was around 15 thousand metric tons (KMT), but by 1991 it had risen to over 162 KMT and by 1994 to over 264 KMT (Kaosa-ard and Pednekar 1998).

Shrimp farm area expanded from 31,906 ha to 66,027 ha between 1983 and 1996. A more startling figure is the increase in the number of farms during that period, from 3,779 to 21,917. In general, this reflects a rapid shift from more extensive to more small-scale, intensive and highly productive aquaculture systems of on average 2–3 ponds, with each pond comprising up to 1 ha in size (Kongkeo 1997; Tokrisna 1998; Goss *et al.*

[11] The following case study is based on Barbier and Cox (2004).

2001). However, much of the semi-intensive and intensive shrimp farming in Thailand is short term and 'unsustainable', i.e. water quality and disease problems mean that yields decline rapidly and farms are routinely abandoned after 5–6 years of production (Flaherty and Karnjanakesorn 1995; Dierberg and Kiattisimkul 1996; Thongrak *et al.* 1997; Tokrisna 1998; Vandergeest *et al.* 1999).

Although shrimp farm expansion has slowed in recent years, unsustainable production methods and lack of know-how have meant that more expansion still takes place every year simply to replace unproductive and abandoned farms. Estimates of the amount of mangrove conversion due to shrimp farming vary, but recent studies suggest that up to 50–65 per cent of Thailand's mangroves have been lost to shrimp farm conversion since 1975 (Dierberg and Kiattisimkul 1996; Tokrisna 1998). In provinces close to Bangkok, such as Chanthaburi, mangrove areas have been devastated by shrimp farm developments (Raine 1994). More recently, Thailand's shrimp output has been maintained by the expansion of shrimp farming activities to the far southern and eastern parts of the Gulf of Thailand and across to the Andaman Sea coast (Flaherty and Karnjanakesorn 1995; Sathirathai 1998; Vandergeest *et al.* 1999).

Moreover, conversion of mangroves to shrimp farms is irreversible. Without careful ecosystem restoration and manual replanting efforts, mangroves do not regenerate, even in abandoned shrimp farm areas. In Thailand, most of the estimated 11,000 or more ha of replanted areas over 1991–1995 have occurred on previously unvegetated tidal mudflats (Lewis *et al.* 2000). Such 'afforestation' efforts have been strongly criticised as being ecologically unsound (Stevenson *et al.* 1999; Erftemeijer and Lewis 2000). However, more recent efforts at mangrove replanting in Southern Thailand have focused on ecological restoration of mangrove areas destroyed by both legal and illegal shrimp ponds, although the total area restored is very small relative to the natural mangrove forest area that has been converted (Lewis *et al.* 2000). Currently in Thailand there is no legal requirement that shrimp farm owners invest in replanting and restoring mangroves once farming operations have ceased and the ponds are abandoned. Shrimp farming does not necessarily have to pose any environmental threat, provided that waste water from the farm has been treated before being released. In addition, it is possible to design shrimp aquaculture systems in coastal areas that do not involve removal of vegetation and areas naturally fed by tidal conditions. However, the establishment of these farm systems is too expensive for the type of small-scale pond operations found in much of Thailand, which are dependent on highly intensive and untreated systems through rapid conversion of mangrove and coastal resources (Thongrak *et al.* 1997; Tokrisna 1998).

Much of the financial investment in coastal shrimp farms is from wealthy individual investors and business enterprises from outside the local community (Flaherty and Karnjanakesorn 1995; Goss *et al.* 2000, 2001). Although some hiring of local labour occurs, in the past shrimp farm owners have tended to hire Burmese workers as their wage rates are much lower.

Ill-defined property rights have accelerated the rapid conversion of mangroves to shrimp farms in Thailand. Historically, this has been a common problem for all forested areas in the country (Feder *et al.* 1988; Feeny 1988, 2002; Thomson *et al.* 1992). Although the state, through the Royal Forestry Department, ostensibly owns and controls mangrove areas, in practice they are *de facto* open access areas onto which anyone can encroach. This has had three impacts on mangrove deforestation attributable to shrimp farms. First, the open access conditions have allowed illegal occupation of mangrove areas for establishing shrimp farms, in response to the rising prices and profits from shrimp aquaculture (Barbier and Sathirathai 2004).[12] Second, in Thailand insecure property rights in cleared forest areas have been associated with under-investment in land quality and farm productivity (Feder *et al.* 1987, 1988; Feeny 1988; Thomson *et al.* 1992). The lack of tenure security for shrimp farms in Southern Thailand appears also to be a major factor in the lack of investment in improving productivity and adopting better aquaculture methods, leading to more mangrove areas being cleared than necessary (Barbier and Sathirathai 2004). Third, several studies have pointed out how open access forest lands in Thailand are more vulnerable to rapid deforestation and conversion to agricultural and other commercial uses as the development of roads and the highway network makes these lands more 'accessible' (Cropper *et al.* 1999; Feeny 2002). Similar problems exist for the open access coastal mangrove areas in Southern Thailand. In particular, the geographical 'spread' of shrimp farm expansion and accompanying mangrove deforestation has also proceeded from the more to less accessible areas: initially in the coastal provinces near Bangkok, spreading down the southern Gulf of Thailand coast towards Malaysia, and more recently beginning on the Andaman Sea coast (Raine 1994; Flaherty and Karnjanakesorn 1995; Sathirathai 1998; Vandergeest *et al.* 1999).

[12] This process has been a frequent occurrence historically on all of Thailand's forest lands, as noted by Feeny (2002, p. 193): 'In contrast to the creation of private property rights in crop land, the commercialization of forestry was associated with the creation of state property rights in forest lands. *De jure* state property was often, however, *de facto* open access. Illegal logging and the expansion of the area under cultivation in response to market opportunities and population growth led to rapid deforestation.'

Despite the lack of secure property rights and the frequently illegal occupation of mangrove areas, owners have an incentive to register their shrimp farms and converted land with the Department of Fisheries. The farms then become eligible for the preferential subsidies for key production inputs, such as shrimp larvae, chemicals and machinery, and for preferential commercial loans for land clearing and pond establishment (Tokrisna 1998; Barbier and Sathirathai 2004). Such subsidies inflate artificially the commercial profitability of shrimp farming, thus leading to more mangrove conversion, even though estimates of the economic returns to shrimp aquaculture in Thailand suggest that such conversion is not always justified (Sathirathai and Barbier 2001). Combined with insecure property rights, the subsidies also put further emphasis on shrimp aquaculture as a commercial activity for short-term exploitative financial gains rather than a long-term sustainable activity.

If shrimp farm expansion is a major cause of mangrove deforestation, then the resulting mangrove loss in any period, $r_t$, is directly related to the amount of land area converted by shrimp farms, i.e.

$$M_{t-1} - M_t = r_t = N_t - N_{t-1} \tag{14}$$

where $M$ represents mangrove area and $N$ is the amount of land cleared and used for shrimp farming. Equation (14) states that the land available for shrimp farming in the current period, $N_t$, equals the amount of productive land left over from a previous period $N_{t-1}$, plus any newly cleared land, $r_t$. Equally, the decline in mangroves between the current and previous periods, $M_{t-1} - M_t$, equals the amount of land newly converted for shrimp farming, $r_t$.

Equation (14) implies a direct link between mangrove deforestation and land conversion for shrimp farm area expansion, with the latter activity determined by the commercial profitability of aquaculture operations. For a relatively long time period $[t, t-1]$, it is possible to establish this link formally.[13] In equation (14), let $M_{t-1}$ represent the amount of mangrove area available in a previous period before much shrimp farming has occurred. Thus compared with the current period, $t$, in the previous period, $t-1$, mangrove area will be relatively abundant and very little of it will have been cleared for shrimp farming, i.e. $\frac{N_{t-1}}{M_{t-1}} \approx 0$. Thus dividing

[13] As noted above, the conversion of mangrove area by shrimp farms has been largely irreversible in Thailand. That is, even if unproductive shrimp farms are abandoned, mangrove systems cannot regenerate naturally on this land. Moreover, to date, very little replanting of mangroves has occurred on abandoned shrimp farm land, nor are shrimp farm owners required legally to undertake such replanting (Erftemeijer and Lewis 2000; Lewis *et al.* 2000; Stevenson *et al.* 1999).

equation (14) by $M_{t-1}$ we obtain:

$$\frac{M_{t-1} - M_t}{M_{t-1}} = \frac{N_t - N_{t-1}}{M_{t-1}} = \frac{N_t}{M_{t-1}}. \tag{15}$$

The left-hand side of (15) is a measure of the long-run proportionate change in mangrove area. It therefore represents a long-run indicator of *mangrove loss*. The right-hand side of (15) is the ratio of current shrimp farm area to mangrove area in a previous base period. It therefore represents a long-run indicator of *relative shrimp farm area expansion*.

Returning to the pure open access model of land clearing, recall equation (8), which defines an equilibrium reduced-form relationship between current shrimp farm area, $N_t^*$, and the output and input prices for shrimp farming, the accessibility of mangrove areas, and other economic and demographic factors:

$$N^* = N(p, w_I, w_L; \alpha, Z), \quad dN/dp > 0, \quad dN/d\alpha > 0, \tag{8}$$

Thus it follows from condition (15) that our long-run indicator of relative shrimp farm area expansion, $N_t/M_{t-1}$, will also be determined by equation (8). As equation (15) suggests that our long-run indicators of mangrove loss and shrimp farm expansion are equivalent, then our measure of long-run mangrove loss, $M_{t-1} - M_t/M_{t-1}$, is also determined by (8). Thus both indicators of mangrove loss and shrimp farm expansion can be estimated, using appropriate data for the shrimp output price, $p$, the wage rate, $w_L$, other input prices, $w_I$, the 'accessibility' of mangrove areas, $\alpha$, and other economic and demographic factors that may affect the mangrove clearing decision, $Z$.

Alternatively, if the household-derived demand relationship (3) was used with (6) to solve for the reduced-form level of land conversion, $N_j^*$, then the aggregate land conversion relationship (8) would be specified in relative prices, i.e.

$$N^* = N(p/w_i, w/w_i, w_N/w_i; \alpha, Z). \tag{16}$$

The relative price relationship for land conversion (16) was estimated through dynamic panel analysis across twenty-one coastal provinces of Thailand from 1979 to 1996.[14] As is clear from (15), to use either our mangrove loss or shrimp farm expansion indicators as a dependent variable requires first choosing an appropriate base year for mangrove area,

[14] Although data were collected for all twenty-two coastal provinces of Thailand for this period, only twenty-one coastal provinces were used in the analysis. As no mangrove area data were recorded by the Royal Forestry Department for the coastal province of Narathiwat, we excluded this province from the analysis. See Barbier and Cox (2004) for details.

$M_{t-1}$. We chose 1979 as the base year for two reasons. First, neither economic nor mangrove data in Thailand prior to that date were complete for all coastal provinces, and second, even though shrimp farming began prior to 1979, the major period of shrimp farm establishment and expansion in Thailand occurred over 1979 to 1996. Thus, the dependent variable for mangrove loss $(M_{t-1} - M_t t/M_{t-1})$ is the proportion of mangrove area cleared relative to the 1979 area of mangroves $[(M_{1979}\text{-}M_t)/M_{1979}]$ and the dependent variable for relative shrimp farm expansion $(N_t/M_{t-1})$ is the proportion of shrimp farm area in the current year relative to 1979 mangrove area $[S_t/M_{1979}]$.

Output price, $p$, in equation (16) was represented by the provincial price of shrimp in Thai baht/ton.[15] The two input prices chosen for $w_L$ and $w_I$, respectively, were the minimum provincial wage and the price of ammonium phosphate. The latter is a proxy for the price of feed used in shrimp aquaculture, with which it is highly correlated (Thongrak *et al.* 1997). To estimate (16), these output and input prices were expressed in terms of relative prices with respect to the minimum wage. The distance of each province from Bangkok was included as the measure of the 'accessibility' of provincial mangrove resources, $\alpha$. Finally, several exogenous factors, $Z$, were chosen to represent both economic effects and demographic changes at the provincial level that might influence mangrove conversion: gross provincial product (GPP) per capita, population growth, and the number of shrimp farms per total provincial land area.[16] Table 2.2 shows the results of random effects estimations for the mangrove loss regression and for two versions of the shrimp farm expansion regressions, one with shrimp farm density and one without.[17]

[15] In the regressions, all price variables as well as gross provincial product per capita are expressed in local currency (Thai baht) and in real terms (1990 values), using the gross domestic product (GDP) deflator for Thailand.

[16] The exchange rate and real interest rate were also included as additional exogenous variables in the analysis. However, these variables are not represented at the provincial level. Neither variable was significant and their inclusion distorted the original regression results. Both variables were therefore dropped from the final regressions. Population growth was used instead of population density as the latter was highly correlated with GPP and shrimp farm density. See Barbier and Cox (2004) for further details.

[17] The general approach advocated for panel analysis was followed in estimating equation, and in all cases the one-way random effects models performed best. Log-log and semi-log forms of the regression were also tested but the linear form performed best. Inclusion of the variable for average distance of each province from Bangkok in the models meant that any fixed effects regression would be collinear. We tested the models with and without this variable and for the possible endogeneity of the shrimp farm density variable in the regressions. The null hypothesis that shrimp farm density is an exogenous variable could be rejected for the mangrove loss regression but not for the shrimp farm estimation. However, the standard instrumental variable (IV) technique could not be employed to correct for the endogeneity of shrimp farm density in the latter estimation. The

Table 2.2 *Thailand – random effects estimation of mangrove loss and shrimp farm area expansion, 1979–1996*

| | % mangrove area cleared relative to 1979 $(M_{1979} - M_t)/M_{1979}$ | % shrimp farm area relative to 1979 mangrove area $S_t/M_{1979}$ | % shrimp farm area relative to 1979 mangrove area $S_t/M_{1979}$ |
|---|---|---|---|
| Shrimp price-wage ratio (Thai baht (B)/kg per B/day) | $4.081 \times 10^{-2}$ (5.524)** | $1.795 \times 10^{-1}$ (4.941)** | $2.089 \times 10^{-1}$ (3.769)** |
| Fertiliser price-wage ratio (B/kg per B/day) | $-2.620 \times 10^{-3}$ (−6.982)** | $-8.102 \times 10^{-3}$ (−7.244)** | $-9.031 \times 10^{-3}$ (−5.313)** |
| Distance of province from Bangkok (km) | $-5.013 \times 10^{-4}$ (−3.314)** | $-1.331 \times 10^{-3}$ (−2.033)* | $-1.681 \times 10^{-3}$ (−2.316)* |
| Population growth ( %/year) | $5.808 \times 10^{-7}$ (1.769)† | $5.915 \times 10^{-6}$ (2.431)* | $6.548 \times 10^{-6}$ (1.741)† |
| Gross provincial product per capita (B/person) | $8.466 \times 10^{-7}$ (2.428)* | $-2.875 \times 10^{-6}$ (−2.587)** | $-1.546 \times 10^{-6}$ (−0.919) |
| Shrimp farm density (Farms/km$^2$) | $1.071 \times 10^{-3}$ (0.945) | $2.380 \times 10^{-2}$ (5.086)** | – |
| CONSTANT | 0.773 (7.882)** | 1.536 (3.715)** | 1.780 (3.733)** |

Notes: t-ratios are indicated in parentheses
** Significant at 1% level. *Significant at 5% level. †Significant at 10% level
Source: Barbier and Cox (2004)

The results reported for mangrove loss in Table 2.2 show that all variables have the predicted signs. In addition, the only explanatory variable that has no significant impact on long-run mangrove loss in Thailand is shrimp farm density. The relative price of shrimp has a significant and positive effect on mangrove deforestation across the coastal provinces of Thailand, whereas mangrove loss declines for those coastal provinces that are further from Bangkok. A rise in the relative feed price has a significant and negative impact on long-run mangrove loss. Provincial economic development (represented by GPP) causes mangrove deforestation, as does population growth, although the latter variable is significant only at the 10 per cent level.

The two regressions for relative shrimp farm area expansion in Thailand vary little with respect to the sign and significance of the

IV technique in panel analysis requires using a two-stage fixed effects procedure, but unfortunately a fixed effects regression is incompatible with our preferred regression that includes the 'distance' variable. As an alternative, we therefore report two versions of our panel analysis of shrimp farm expansion in Table 2.2, one with the shrimp farm density variable and one without. See Barbier and Cox (2004) for further details.

coefficients of three main variables: relative shrimp price, relative feed price and the accessibility of mangrove areas. All three variables are significant and have the predicted signs (see Table 2.2). Shrimp farm area expansion increases with the relative price of shrimp, but declines with the relative feed price and for those coastal provinces further from Bangkok. Population growth is significant in explaining relative shrimp farm expansion in both regressions, but only at the 10 per cent level in the estimation that excludes shrimp farm density. Provincial economic development has a significant and negative impact on shrimp farm expansion in the regression that includes shrimp farm density, but is insignificant in the estimation without it. Finally, shrimp farm density appears to be a significant factor in shrimp farm expansion, but this variable might be endogenous in the regression.[18]

The panel analysis regressions of mangrove loss and relative shrimp farm area expansion reported in Table 2.2 are therefore consistent with the theoretical model of 'open access' land conversion developed above. Further insights into the causes of mangrove loss and shrimp farm expansion can be gained from the estimated elasticities, which are indicated in Table 2.3.

The variables with the largest impacts on mangrove loss are distance from Bangkok and the price of ammonium phosphate, followed by the minimum wage, shrimp price, gross provincial product per capita and population growth. In both regressions of relative shrimp farm area expansion, the variables with the largest effects are again distance from Bangkok and ammonium phosphate price, followed by the minimum wage and shrimp price.[19] In the estimation that includes shrimp farm density, the remaining impacts are attributed to GPP, shrimp farm density and population growth. In the estimation that excludes shrimp farm density, only population growth has a modest, but barely significant, impact on shrimp farm expansion. As expected, the effects of changes in the explanatory variables on relative shrimp farm expansion are always greater than on mangrove loss.[20]

[18] See previous note and further discussion in Barbier and Cox (2004).

[19] As Table 2.3 indicates, in the regression without shrimp farm density, the impact of the shrimp price on relative shrimp farm area expansion is slightly larger than the impact of the minimum wage rate.

[20] As noted previously, mangrove deforestation in Thailand has also resulted from tourism, agricultural, industrial and urban developments in coastal areas, and thus is not completely explained by mangrove clearing for shrimp farming. If economic activities other than shrimp farming are responsible for mangrove loss in the coastal areas of Thailand, this might explain why in Table 2.3 the elasticities for the explanatory variables are larger for the two versions with shrimp farm expansion as the dependent variable rather than the version with mangrove loss as the dependent variable.

Table 2.3 *Thailand – estimated elasticities for mangrove loss and shrimp farm area expansion, 1979–1996*

| Explanatory variables | % mangrove area cleared relative to 1979 $(M_{1979}-M_t)/M_{1979}$ | % shrimp farm area relative to 1979 mangrove area $S_t/M_{1979}$ | % shrimp farm area relative to 1979 mangrove area $S_t/M_{1979}$ |
|---|---|---|---|
| Shrimp price-wage ratio (Thai baht (B)/kg per B/day) | 0.158** | 0.402** | 0.468** |
| Shrimp price (B/kg) | 0.156** | 0.397** | 0.462** |
| Wage rate (B/day) | 0.302** | 0.421** | 0.450** |
| Fertiliser price-wage ratio (B/kg per B/day) | −0.460** | −0.824** | −0.918** |
| Fertiliser price (B/kg) | −0.445** | −0.796** | −0.887** |
| Distance of province from Bangkok (km) | −0.626** | −0.963* | −1.216* |
| Population growth (%/year) | 0.014† | 0.080* | 0.089† |
| Gross provincial product per capita (B/person) | 0.097* | −0.190** | −0.103 |
| Shrimp farm density (Farms/km$^2$) | 0.014 | 0.185** | – |

Notes: ** Significant at 1 % level. *Significant at 5 % level †Significant at 10 per cent level
Source: Barbier and Cox (2004)

Overall, these results reaffirm the hypothesis that the profitability of shrimp farming, coupled with 'open access' land conversion decisions, is a very important underlying cause of mangrove deforestation in Thailand. Intensive shrimp farming utilises a considerable amount of feed, the costs of which represent anywhere from 30–60 per cent of the total costs of shrimp aquaculture in various systems across Thailand (Kongkeo 1997; Thongrak *et al.* 1997; Tokrisna 1998). Thus it is not surprising that a change in the price of ammonium phosphate – our proxy for feed price – causes a relatively large impact on shrimp farm expansion and mangrove clearing.

As indicated in Table 2.3, if ammonium phosphate and thus feed prices across Thailand were to rise by 10 per cent, the relative decline in shrimp farm area would be 8–9 per cent and mangrove clearing would decrease by around 4.5 per cent. Our results indicate that shrimp farm area expansion and mangrove loss are also responsive to changes in the price of shrimp. As discussed above, expansion of shrimp farming in Thailand has occurred rapidly since 1985, which was when a rapid rise in world demand and prices for shrimp occurred. The elasticity estimates suggest that if the price of shrimp were to rise by 10 per cent, relative shrimp farm area would increase by 4–5 per cent and mangrove deforestation would expand by 1.6 per cent.

The analysis also confirms that the 'accessibility' of mangrove areas is an important determinant of mangrove clearing for shrimp farming in Thailand. This is an expected result, given that Bangkok is the major domestic market as well as the key port and terminus for both Thailand's export market and many regional domestic markets. In addition, many investors in shrimp farming operations are from outside the coastal provinces and in particular from Bangkok. The elasticity estimates suggest that coastal areas that are 10 per cent further from Bangkok have 10–12 per cent less relative shrimp farm area and have 6.3 per cent lower mangrove clearing rates. Distance from Bangkok appears to be an important factor determining the accessibility of coastal resources, the profitability of shrimp farming and therefore mangrove conversion. The historical pattern of mangrove loss in Thailand is consistent with this result. Mangrove deforestation began initially in the coastal provinces near Bangkok, spread down the southern Gulf of Thailand coast towards Malaysia and is now beginning on the Andaman Sea coast (Raine 1994; Flaherty and Karnjanakesorn 1995; Sathirathai 1998).

Table 2.3 indicates that the provincial minimum wage variable has a positive elasticity in the panel regressions. A 10 per cent rise in the rural minimum wage causes relative shrimp farm area to increase by over 4 per cent and mangrove clearing by 3 per cent.[21] As discussed above, our theoretical model would suggest that the amount of mangrove land converted should decrease with the cost of labour, which is the principal input involved in clearing operations, but this effect may be counteracted by an opposite impact of a rise in the wage rate on mangrove conversion, if land and labour are substitutes in shrimp farming. Our elasticity results suggest that this latter substitution effect might be the stronger influence. As the costs of labour use in production rise, shrimp farmers may be induced to move from more intensive aquaculture operations that employ relatively more labour than land to more semi-intensive and extensive systems that require relatively more land. For example, in Thailand extensive shrimp farms (5–7 ha) have average labour costs of only $36.1/ha, semi-intensive farms (3–4 ha) have labour costs of $96.6/ha and intensive farms

[21] By employing relative prices in each regression and using minimum wage as the numeraire, the impact of a rise in the wage rate will depend on the relative impacts of the shrimp price-wage ratio versus the fertiliser price-wage variables on the dependent variable. In all regressions the negative impact of the latter variable has the greater absolute effect, which is the reason why the elasticity associated with the minimum wage is positive.

(2–3 ha) have labour costs of $377.5/ha (Tokrisna 1998). Thus, a rise in wages may lead some shrimp farmers to expand shrimp farm area and switch to less intensive operations in order to save on overall labour costs (Goss *et al.* 2001).[22]

Shrimp farm expansion and mangrove loss may also be influenced somewhat by demographic pressures, such as provincial population change, although the significance of this impact is weak in two of the three regressions (see Tables 2.2 and 2.3). A 10 per cent rise in population growth will cause shrimp farm area to expand by 0.8–0.9 per cent and mangrove clearing also increases by 0.1 per cent.

A 10 per cent rise in gross provincial product per capita increases mangrove loss by about 1 per cent but the impact of GPP on relative shrimp farm area is less clear, given the possible problem of the endogeneity of shrimp farm density in the regressions of shrimp farm expansion (see Tables 2.2 and 2.3). As noted above, mangrove loss is increasingly occurring in coastal areas due to provincial economic development activities other than shrimp farming, such as urbanisation, agriculture, tourism and industrialisation (Dierberg and Kiattisimkul 1996; Tokrisna 1998).[23] Such coastal economic developments are likely to lead to increases in gross provincial product per capita while at the same time putting greater pressure on remaining mangrove areas.

To summarise, this case study provides strong evidence that our open access land conversion model applies to shrimp farm expansion and

[22] Despite the anecdotal and empirical evidence that higher wages may induce some substitution of land for labor in shrimp farm operations, thus leading to an increase in overall mangrove clearing, this interpretation of our results must be treated with some caution. Because of the lack of disaggregated data on shrimp farm operations across all provinces in Southern Thailand over 1976–1990 by type of technology – extensive, semi-intensive and intensive – we are unable to separate out the effects of wages on mangrove clearing by each type of farm. By employing the aggregated shrimp farm data in our analysis, we are essentially treating all three technologies as a single technology, which could lead to a misleading prediction about the likely effects of a wage change on land use. We are grateful to an anonymous reviewer for pointing out this possibility to us.

[23] Tokrisna (1998) provides some evidence of these changing trends in the rate of mangrove utilisation by various coastal economic activities. Before 1980 an average of 7 per cent of all mangrove areas in Thailand were converted to shrimp ponds. In 1986, the rate of mangrove conversion to shrimp ponds was estimated to be 30 per cent, but had declined to 17 per cent by 1994. In contrast, the rate of mangrove conversion due to other coastal economic activities, such as urbanisation, agriculture, tourism and industry, has increased rapidly from 15 per cent before 1980, 17 per cent in 1986 and 36 per cent by 1994. In terms of cumulative impacts on mangrove loss, over the entire 1979–1996 period, shrimp farming is still thought to have had the greatest effect, even though the rate of mangrove conversion due to shrimp aquaculture has tended to vary over this period. As reported above, estimates suggest that up to 50–65 per cent of Thailand's mangroves have been lost to shrimp farm conversion since 1975 (Dierberg and Kiattisimkul 1996; Tokrisna 1998).

mangrove loss in Thailand over 1979–1996. However, several recent developments could greatly influence the future impacts of shrimp farming on mangrove conversion in Southern Thailand.

First, the availability of new mangrove areas suitable for conversion to shrimp farming is becoming increasingly scarce. Of the 62,800 ha of mangrove areas considered suitable for shrimp farms in 1977, between 38 per cent and 65 per cent were already converted by shrimp farms between 1975 and 1993 (Dierberg and Kiattisimkul 1996). Thus expansion of shrimp farms is increasingly occurring on coastal land formerly used for rubber and palm plantations and, until the recent ban, in rice paddy areas.

Second, it is still too early to gauge the effect of the ban on shrimp farming in the rice and fruit growing areas in the central region of Thailand. One result is likely to be greater conversion of remaining areas of coastal mangrove forests, especially the remaining pristine mangrove on the Andaman coast. To prevent this from happening, however, recent policy initiatives have been proposed to promote the conservation of mangroves and the participation of local communities in their management (Sathirathai 1998). For example, the Royal Forestry Department is considering banning mangrove forest concessions and regulating the use of mangrove areas, particularly those affected by shrimp farming. Furthermore, new legislation on community management of forests has been introduced, which offers the hope that the right of local communities to protect mangrove forests may receive legal recognition. The motivation for this potential change in policy arises from the recognition that the economic benefits of mangroves to local communities may be substantial and could possibly even outweigh the returns to intensive shrimp farming that lead to mangrove conversion.

However, if Thailand is to become a model for reconciling shrimp farm production with coastal mangrove management, this study points to two clear policy recommendations beyond what is currently being considered by the government. First, there is an urgent need to address the main *institutional failure* concerning management of mangrove resources. The present law and formal institutional structures of resource management in Thailand do not allow coastal communities to establish and enforce their local rules effectively. Nor do the current formal institutions and laws provide the incentives necessary for local and other resource user groups to resolve conflicts among themselves. The result is that any effort to resolve such conflicts incurs high risk and management costs, which in turn make it even harder for the successful establishment of collaborative resource management systems by local communities. There is also a need to address the main *policy failure* at the heart of the economic incentives for excessive conversion of mangrove areas to shrimp aquaculture.

As long as government policies continue to subsidise shrimp farm establishment and production, this activity will remain financially profitable to the commercial investor. The result is that the commercial pressure to convert mangroves and other coastal land to shrimp farming will remain, even though the actual economic returns to such investments may not always justify such conversion (Sathirathai and Barbier 2001).

For example, a new institutional framework for coastal mangrove management in Thailand might contain the following features (Barbier and Sathirathai 2004, Chapter 12). First, remaining mangrove areas should be designated as conservation (i.e. preservation) and economic zones. Shrimp farming and other extractive commercial uses (e.g. wood concessions) should be restricted to the economic zones only. However, local communities which depend on the collection of forest and fishery products from mangrove forests should be allowed access to both zones, as long as such harvesting activities are conducted on a sustainable basis. Second, the establishment of community mangrove forests should also occur in both the economic and conservation zones. But the decision to allow such local management efforts should be based on the capability of communities to effectively enforce their local rules and manage the forest sustainably. Moreover, such community rights should not involve full ownership of the forest but be in the form of user rights. Third, the community mangrove forests should be co-managed by the government and local communities. Such effective co-management will require the active participation of existing coastal community organisations and will allow the representatives of such organisations to have the right to express opinions and make decisions regarding the management plan and regulations related to the utilisation of mangrove resources. Finally, the government must provide technical, educational and financial support for the local community organisations participating in managing the mangrove forests. For example, if only user rights (but not full ownership rights) are granted to local communities, the latter's access to formal credit markets for initiatives such as investment in mangrove conservation and replanting may be restricted. The government may need to provide special lines of credit to support such community-based activities.[24]

If successful, such local management policies might act as effectively combined formal and informal 'institutional constraints' on mangrove loss due to shrimp farm expansion in Thailand. As the model of land

[24] Other complementary policies may also be necessary to reduce the environmental damages associated with shrimp farming and other mangrove-converting activities, such as establishing concession fees and auctions for these activities, reducing subsidies for shrimp farming, introducing incentives for mangrove replanting, water pollution charges, and even environmental assurance bonds for large-scale developments. For further discussion see Barbier and Cox (2004) and Barbier and Sathirathai (2004, Chapter 12).

conversion developed in this chapter suggests, the result should be to slow down the rate of conversion. It may also lead to more efficient land use, including selection of the most appropriate mangrove areas for conversion to shrimp farms.

## 7 Final remarks

This chapter was concerned with analysing the role of formal and informal institutions as constraints on the conversion of forestland to agriculture within a developing country. Given that open access conditions and ill-defined property rights are thought to be important factors driving agricultural land expansion and forest conversion in developing countries, we have developed an economic model of forestland conversion under open access that is empirically tested.

The model demonstrates formally that the equilibrium level of land cleared will differ under open access compared with when effective institutions exist to control land conversion. Because institutions raise the cost of land clearing, more land should be converted to pure open access. This allows derivation of a simple test for the 'constraining' effect of institutions on conversion through a partial adjustment mechanism for the equilibrium level of cleared land.

The first case study applied the model to the expansion of agricultural planted area in Mexico at state level and over the 1960–1985 period before the NAFTA reforms were implemented. The results confirm that the presence of *ejido* communal land management acted as an 'institutional constraint' on deforestation due to maize land expansion. There is concern that as the *ejido* land management system weakens during the post-NAFTA era, its ability to control deforestation by smallholders may also suffer.

The second case study applied the model to mangrove conversion for shrimp farming in Thailand's coastal provinces over 1979–1996. The results suggest that the profitability of shrimp farming coupled with open access availability of mangrove areas in accessible coastal areas were powerful factors driving mangrove deforestation in Southern Thailand. The study illustrates that Thailand needs 'institutional constraints' to slow down mangrove loss in coastal areas, through combining effective local community and government management of the resource.

### REFERENCES

Alston, L. J., Libecap, G. D. and Mueller, F. 1999. *Titles, Conflict, and Land Use: the Development of Property Rights and Land Reform on the Brazilian Amazon Frontier*. Ann Arbor: The University of Michigan Press.

Angelsen, A. 1999. Agricultural expansion and deforestation: modelling the impact of population, market forces and property rights. *Journal of Development Economics*. **58**. 185–218.

Appendini, K. 1998. Changing agrarian institutions – interpreting the contradictions. In W. A. Cornelius and D. Myhre (eds.). *The Transformation of Rural Mexico: Reforming the Ejido Sector*. San Diego, CA: Center for US-Mexican Studies, University of California. 25–38.

Baland, J-M. and Platteau, J-P. 1996. *Halting Degradation: Is there a Role for Rural Communities?* Oxford: Clarendon Press.

Barbier, E. B. 2002. Institutional constraints and deforestation: an application to Mexico. *Economic Inquiry*. **40** (3). 508–519.

Barbier, E. B. 2004. Explaining agricultural land expansion and deforestation in developing countries. *American Journal of Agricultural Economics*. **86** (5). 1347–1353.

Barbier, E. B. 2005. *Natural Resources and Economic Development*. Cambridge: Cambridge University Press.

Barbier, E. B. and Burgess, J. C. 1996. Economic analysis of deforestation in Mexico. *Environment and Development Economics*. **1** (2). 203–240.

Barbier, E. B. and Burgess, J. C. 1997. The economics of tropical forest land use options. *Land Economics*. **73** (2). 174–195.

Barbier, E. B. and Burgess, J. C. 2001a. The economics of tropical deforestation. *Journal of Economic Surveys*. **15** (3). 413–432.

Barbier, E. B. and Burgess, J. C. 2001b. Tropical deforestation, tenure insecurity and unsustainability. *Forest Science*. **47** (4). 497–509.

Barbier, E. B. and Cox, M. 2004. An economic analysis of shrimp farm expansion and mangrove conversion in Thailand. *Land Economics*. **80** (3). 389–407.

Barbier, E. B. and Sathirathai, S. (eds.). 2004. *Shrimp Farming and Mangrove Loss in Thailand*. London: Edward Elgar.

Besley, T. 1995. Property rights and investment incentives: theory and evidence from Ghana. *Journal of Political Economy*. **103** (5). 903–937.

Bohn, H. and Deacon, R. T. 2000. Ownership risk, investment, and the use of natural resources. *American Economic Review*. **90** (3). 526–549.

Bromley, D. W. 1989. Property relations and economic development: the other land reform. *World Development*. **17**. 867–877.

Bromley, D. W. 1991. *Environment and Economy: Property Rights and Public Policy*. Oxford: Basil Blackwell.

Brown, K. and Pearce, D. W. (eds.). 1994. *The Causes of Tropical Deforestation: The Economic and Statistical Analysis of Factors Giving Rise to the Loss of the Tropical Forests*. London: UCL Press.

Brown, P. 1997. Institutions, inequalities, and the impact of agrarian reform on rural Mexican communities. *Human Organization*. **56** (1). 102–110.

Burger, J., Ostrom, E., Norgaard, R. B., Policansky, D. and Goldstein, B. D. (eds.). 2001. *Protecting the Commons: a Framework for Resource Management in the Americas*. Washington, DC: Island Press.

Cornelius, W. A. and Myhre, D. (eds.). 1998. *The Transformation of Rural Mexico: Reforming the Ejido Sector*. San Diego, CA: Center for US-Mexican Studies. University of California.

Cropper, M., Griffiths, C. and Mani, M. 1999. Roads, population pressures, and deforestation in Thailand. 1976–1989. *Land Economics*. **75** (1). 58–73.

Deacon, R. T. 1999. Deforestation and ownership: evidence from historical accounts and contemporary data. *Land Economics*. **75** (3). 341–359.

Deininger, K. W. and Minten, B. 1999. Poverty, policies and deforestation: the case of Mexico. *Economic Development and Cultural Change*. **47** (2). 313–344.

Dierberg, F. E. and Kiattisimkul W. 1996. Issues, impacts and implications of shrimp aquaculture in Thailand. *Environmental Management*. **20** (5). 649–666.

Erftemeijer, P. L. A. and Lewis, R. R. III. 2000. Planting mangroves on intertidal mudflats: habitat restoration or habitat conversion? In V. Sumantakul, *et al.* (eds.). *Enhancing coastal ecosystem restoration for the 21st Century. Proceedings of the Regional Seminar for East and Southeast Asian Countries: ECOTONE VIII*, Ranong and Phuket, Thailand, 23–28 May 1999. Bangkok, Thailand, Royal Forestry Department. 156–165.

Feder, G. and Feeny, D. 1991. Land tenure and property rights: theory and implications for development policy. *World Bank Economic Review*. **5** (1). 135–153.

Feder, G. and Onchan, T. 1987. Land ownership security and farm investment in Thailand. *American Journal of Agricultural Economics*. **69**. 311–320.

Feder, G., Onchan, T., Chalamwong, Y. and Hongladarom, C. 1988. Land policies and farm performance in Thailand's forest reserve areas. *Economic Development and Cultural Change*. **36** (3). 483–501.

Feeny, D. 1988. Agricultural expansion and forest depletion in Thailand, 1900–1975. In J. F. Richards and R. Tucker (eds.). *World Deforestation in the Twentieth Century*. Durham, NC: Duke University Press. 112–143.

Feeny, D. 2002. The co-evolution of property rights regimes for man, land, and forests in Thailand, 1790–1990. In J. F. Richards (ed.). *Land Property and the Environment*. San Francisco, CA: Institute for Contemporary Studies Press. 179–221.

Flaherty, M. and Karnjanakesorn, C. 1995. Marine shrimp aquaculture and natural resource degradation in Thailand. *Environmental Management*. **19** (1). 27–37.

Food and Agricultural Organization (FAO). 1993. *Forest Resources Assessment 1990: Tropical Countries*. Rome: FAO Forestry Paper 112.

Food and Agricultural Organization of the United Nations. 1997. *State of the World's Forests 1997*. Rome: FAO.

Food and Agricultural Organization of the United Nations. 2001. *Forest Resources Assessment 2000: Main Report*. Rome: FAO Forestry Paper 140.

Food and Agricultural Organization of the United Nations. 2003. *State of the World's Forests 2003*. Rome: FAO.

Freeman, A. M. III. 2003. *The Measurement of Environmental and Resource Values: Theory and Methods, 2nd ed.* Washington, DC: Resources for the Future.

Gibson, C. C. 2001. Forest resources: institutions for local governance in Guatemala. In J. Burger, E. Ostrom, R. B. Norgaard, D. Policansky and B. D. Goldstein (eds.). *Protecting the Commons: A Framework for Resource Management in the Americas*. Washington, DC: Island Press. 71–90.

Goss, J., Burch, D. and Rickson, R. E. 2000. Agri-food restructuring and Third World transnationals: Thailand, the CP Group and the global shrimp industry. *World Development*. **28** (3). 513–530.

Goss, J., Skladany, M. and Middendorf, G. 2001. Dialogue: shrimp aquaculture in Thailand: a response to Vandergeest, Flaherty and Miller. *Rural Sociology*. **66** (3). 451–460.

Heal, G. 1982. The use of common property resources. In V. K. Smith and J. V. Krutilla (eds.). *Explorations in Natural Resource Economics*. Baltimore, MD: Johns Hopkins University Press. 72–106.

Jones, G. A. and Ward, P. M. 1998. Privatizing the commons: ejido and urban development in Mexico. *International Journal of Urban and Regional Research*. **22** (1). 76–95.

Kaimowitz, D. and Angelsen, A. 1998. *Economic Models of Tropical Deforestation: A Review*. Bogor, Indonesia: Center for International Forestry Research.

Kaosa-ard, M. and Pednekar, S. S. 1998. Background report for the Thai Marine Rehabilitation Plan 1997–2001. Report submitted to the Joint Research Centre of the Commission of the European Communities and the Department of Fisheries, Ministry of Agriculture and Cooperatives. Bangkok: Thailand Development Research Institute.

Kongkeo, H. 1997. Comparison of intensive shrimp farming systems in Indonesia, Philippines, Taiwan and Thailand. *Aquaculture Research*. **28**. 89–796.

Kooten, van G. C., Sedjo, R. A. and Bulte, E. H. 1999. Tropical deforestation: issues and policies. Chapter 5 in H. Folmer and T. Tietenberg (eds.). *The International Yearbook of Environmental and Resource Economics 1999/2000*. Cheltenham: Edward Elgar. 199–248.

Larson, B. A. and Bromley, D. W. 1990. Property rights, externalities, and resource degradation: locating the tragedy. *Journal of Development Economics*. **33**. 235–262.

Lewis, R. R. III., Erftemeijer, P. L. A., Sayaka, A. and Kethkaew, P. 2000. Mangrove rehabilitation after shrimp aquaculture: a case study in progress at the Don Sak National Forest Reserves, Surat Thani, Southern Thailand. Mimeo. Mangrove Forest Management Unit, Surat Thani Regional Forest Office, Surat Thani, Thailand: Royal Forest Department.

López, R. 1997. Environmental externalities in traditional agriculture and the impact of trade liberalization: the case of Ghana. *Journal of Development Economic*. **53**. 17–39.

López, R. 1998a. Agricultural intensification, common property resources and the farm-household. *Environmental and Resource Economics*. **11** (3–4). 443–458.

López, R. 1998b. Where development can or cannot go: the role of poverty-environment linkages. In B. Pleskovic and J. E. Stiglitz (eds.). *Annual Bank Conference on Development Economics 1997*. Washington, DC: The World Bank. 285–306.

North, D. C. 1990. *Institutions, Institutional Change and Economic Performance*. Cambridge: Cambridge University Press.

Ostrom, E. 1990. *Governing the Commons: The Evolution of Institutions for Collective Action*. Cambridge: Cambridge University Press.

Ostrom, E. 2001. Reformulating the commons. In J. Burger, E. Ostrom, R. B. Norgaard, D. Policansky and B. D. Goldstein (eds.). *Protecting the Commons: A framework for Resource Management in the Americas*. Washington, DC: Island Press. 17–44.

Panayotou, T. and Sungsuwan, S. 1994. An econometric analysis of the causes of tropical deforestation: the case of Northeast Thailand. In K. Brown and D. W., Pearce (eds.). *The Causes of Tropical Deforestation: The Economic and Statistical Analysis of Factors giving Rise to the Loss of the Tropical Forests.* London: University College London Press.

Raine, R. M. 1994. Current land use and changes in land use over time in the coastal zone of Chathaburi province, Thailand. *Biological Conservation.* **67**. 201–204.

Richards, M. 1997. Common property resource institutions and forest management in Latin America. *Development and Change.* **28**. 95–117.

Sarukhán, J. and Larson, J. 2001. When the commons become less tragic: land tenure, social organization, and fair trade in Mexico. In J. Burger, E. Ostrom, R. B. Norgaard, D. Policansky, and B. D. Goldstein (eds.). *Protecting the Commons: a Framework for Resource Management in the Americas.* Washington, DC: Island Press. 45–70.

Sathirathai, S. 1998. Economic valuation of mangroves and the roles of local communities in the Conservation of the resources: case study of Surat Thani, South of Thailand. Final report submitted to the Economy and Environment Program for Southest Asia (EEPSEA), Singapore.

Sathirathai, S. and Barbier, E. B. 2001. Valuing mangrove conservation in Southern Thailand. *Contemporary Economic Policy.* **19** (2). 109–122.

Singh, I., Squire, L. and Strauss, J. (eds.). 1986. *Agricultural Household Models: Extensions, Applications and Policy.* Baltimore: Johns Hopkins University Press.

Stevenson, N. J., Lewis, III R. R. and Burbridge, P. R. 1999. Disused shrimp ponds and mangrove rehabilitation. In W. Dordrecht Streever (ed.). *An International Perspective on Wetland Rehabilitation,* The Netherlands: Kluwer Academic Publishers. 277–297.

Thomson, J. T., Feeny, D.H and Oakerson, R. J. 1992. Institutional dynamics: the evolution and dissolution of common property resource management. In Daniel W. Bromley (ed.). *Making the Commons Work: Theory, Practice, and Policy.* San Francisico, CA: Institute for Contemporary Studies Press. 129–160.

Thongrak, S., Prato, T. Chiayvareesajja, S. and Kurtz, W. 1997. Economic and water quality evaluation of intensive shrimp production systems in Thailand. *Agricultural Systems.* **53**. 121–141.

Tokrisna, R. 1998. The use of economic analysis in support of development and investment decision in Thai aquaculture: with particular reference to marine shrimp culture. A paper submitted to the Food and Agriculture Organization of the United Nations.

United Nations Environment Programme (UNEP) 2002. *Global Environment Outlook 3.* London: Earthscan Publications.

United Nations Environment Programme-World Conservation Monitoring Centre (UNEP-WCMC). 2000. *Global Biodiversity: Earth's Living Resources in the 21st Century.* Cambridge: World Conservation Press.

Vandergeest, P., Flaherty, M. and Miller, P. 1999. A political ecology of shrimp aquaculture in Thailand. *Rural Sociology.* **64** (4). 573–596.

World Bank 1989. *Mexico – Agricultural Sector Report*. Washington, DC: The World Bank.
World Bank 1994. *Mexico Resource Conservation and Forest Sector Review*. Washington, DC: The World Bank.
World Conservation Monitoring Centre (WCMC) 1992. *Global Biodiversity: Status of the Earth's Living Resources*. London: Chapman and Hall

# 3 Estimating spatial interactions in deforestation decisions

*Juan A. Robalino, Alexander Pfaff, Arturo Sanchez-Azofeifa*

## 1 Introduction

Ongoing decreases in the stock of tropical forest have long been a major concern, due to their implications for biodiversity loss and provision of ecosystem services. Ecological research also provides evidence that even if the stock is held constant, the spatial pattern of forest affects the level of services generated (McCoy and Mushinsky 1994; Twedt and Loesch 1999; Diaz *et al.* 2000; Parkhurst *et al.* 2002; Coops *et al.* 2004; Scull and Harman 2004). A highly fragmented forest made up of small patches may not provide the minimum habitat size that some organisms require. Thus it may offer less protection for species than the same amount of unfragmented forest. It is then important to understand the effects of human activities that fragment standing forest and, as a result, alter the size, the shape, and also the spatial arrangement of habitat. These properties of habitat affect extinction rates of local populations.

Standard economic models of rural land use (e.g. agriculture/forest frontiers) will generate predictions of spatial pattern down to the level of detail that their data permit. However, a focus on spatial pattern highlights a question these models do not address: are there spatial dynamics *per se*? If we look behind observed spatial correlation, do one's land-use choices actually have any causal impacts upon those made by one's neighbours? This chapter presents a model of such spatial interactions and then discusses a method to empirically test for their presence using observed deforestation behaviour. Their existence has implications for the stock of forest, its pattern and the effect of policies on forests.

Acknowledgements: we would like to thank Geoffrey Heal, Malgosia Madajewicz, Ernst Nuppenau, Cristian Pop-Eleches, Rajiv Sethi, Arthur Small, Chris Timmins and Eric Verhoogen for their valuable comments. We thank for generous support the following institutions: The Social Science and Humanities Research Council of Canada; Center for Economics, Environment and Society at the Earth Institute and the Institute of Social and Economic Research and Policy at Columbia University; NASA; Tinker Foundation Inc.; the National Center for Ecological Analysis and Synthesis; and the National Science Foundation. All errors are our own.

Such research builds upon a number of existing literatures. Empirical economic analysis of deforestation, one piece of the much broader literature on the economics of land use, has provided evidence of significant effects on land-use and land-cover outcomes of biophysical and socio-economic factors that one might expect to affect the relative profitability of competing land-use types (Panoyotou and Sungsuwan 1989; Rudel 1989; Stavins and Jaffe 1990; Cropper and Griffiths 1994; Pfaff 1999). Recently, much of this work has employed spatially specific data, making use of geographic information systems (GIS). Predictions are then more spatially detailed (Chomitz and Gray 1996; Geoghegan *et al.* 2001; Kok and Veldkamp 2001; Serneels and Lambin 2001; Walsh *et al.* 2001; Irwin and Bockstael 2002). In this chapter we add 'neighbour effects', or spatial interactions, to this strand of the literature.

Neighbour effects form one part of a broad set of studies of 'social interactions' between agents that need not be spatial. Two spatial examples of such interactions that concern land-use and agriculture are: (i) externalities in US residential development, which have been analysed both theoretically (Turner 2005) and empirically (Irwin and Bockstael 2002), and (ii) adoption of agricultural technologies that affects neighbours' adoption decisions (Case 1992). This chapter brings methods from the social interactions literature to tropical deforestation. The results of this blend can be integrated with many existing discussions of land use and habitat conservation.

For example, rules for selecting habitat for reserves have often been suggested from purely ecological perspectives focused upon where species may exist (Tubbs and Blackwood 1971; Kirkpatrick 1983; Cocks and Baird 1989; Polasky *et al.* 2000), but they could reflect land use as well. This has happened in the consideration of land costs that vary across sites (Ando *et al.* 1998; Polasky *et al.* 2000) and of the threat of clearing that also varies (Pfaff and Sanchez-Azofeifa 2004; Costello and Polasky 2004). This chapter suggests another land-use consideration for where public actors should focus, the spatial spillovers to neighbouring land-use decisions from reserves. Results concerning spatial interactions may suggest varying the intensity of such conservation actions over space.

Here we develop a model of neighbours' interactions that builds upon work by Brock and Durlauf (2001a). The key to the empirical application of this model is the use of instrumental variables to identify the magnitude of the effect individuals in a neighbourhood have on each other's choices.[1]

[1] Neighbourhoods and neighbours are defined in the following section. However, a neighbourhood can be seen as an area of land. Two individuals are neighbours if the land they manage is located in the same neighbourhood. While this paper uses a specific definition of neighbourhood and therefore neighbours, it is up to the researchers to redefine the concept of neighbourhood according to their needs.

We rely heavily upon the GIS for many calculations of the neighbours' characteristics, including the biophysical characteristics of neighbouring plots that serve as instruments.

The use of an instrumental variable addresses simultaneity and the presence of spatially correlated unobservable effects. Simultaneity rises when the explanatory variable not only affects but also is affected by the dependent variable.[2] In this case, neighbours' deforestation decisions (the explanatory variable) affect the individual's deforestation decision (the dependent variable) and simultaneously the individual's deforestation decision affects neighbours' deforestation decisions. Since the individual's decisions do not, though, affect the biophysical characteristics of neighbours' plots, using those characteristics to instrument for neighbouring deforestation choices addresses the simultaneity problem. The same reasoning applies to spatial correlation between the unobservable factors that affect an individual's decision and the unobservable factors that affect their neighbours' decisions.[3] A correlation between the two sets of unobservable factors implies correlation between individual and neighbours' behaviours that does not indicate causality. The same correlation does not, though, imply a correlation between an individual's choice and the biophysical characteristics of neighbours' plots. If the instrument is correctly chosen, it addresses these two major issues for the estimation of such interactions (Moffitt 2001).

If positive interactions exist in deforestation, as suggested by the example here that makes use of Costa Rican data, three important consequences arise. Neighbours' decisions reinforcing each other will generate more homogenous forest outcomes within neighbourhoods, i.e. highly fragmented forest patterns are less likely. Also, changed incentives to deforest in one location (e.g. from land policies) spill over to affect areas nearby. Finally, interactions generate the possibility that a given region could end up with significantly different deforestation outcomes (multiple equilibria) due simply to changes in beliefs about what neighbours will do. This depends upon the magnitude of the interactions.[4] Thus, projecting the effects of policies based on extrapolations from past equilibria could miss the possibility that a policy could induce another equilibrium. An agency could implement a policy with the expectation of small increases in deforestation, based on low clearing rates in the past equilibrium, and

[2] See Greene (2003) and Maddala (1983) for more details and examples.

[3] An example of a spatially correlated unobservable effect is the effect on deforestation decisions caused by a soil characteristic that is similar among neighbours, observable by all individuals but unobservable to the researcher.

[4] See Cooper and John (1988) and Brock and Durlauf (2001a) for the role of the magnitude of the interactions in the existence of multiple equilibria.

be surprised by its impact when changed beliefs about neighbours' deforestation behaviour amplifies the policy's impact. Another implication is that it may be desirable to intervene in the interest of a new equilibrium. Suboptimal equilibria can be maintained if individuals have self-fulfilling expectations of suboptimal actions by their neighbours, such as all clearing even though all would be better off by conserving.

This chapter is structured as follows. Section 2 describes a simple model of interactions in the context of deforestation, based on an equilibrium in beliefs about the neighbours' actions. Section 3 discusses empirical issues in measurement of interactions and the benefits of using an instrumental variable approach. Data requirements for analysing neighbours' interactions in deforestation decisions are discussed in section 4. Finally, results for two regions within Costa Rica, as well as discussion of how to obtain the equilibria once the parameters of the model are estimated, are presented in section 5.

## 2 A model of interactions in deforestation

Social interactions exist when an individual's decision is affected by decisions of other individuals. Models with social interactions can be divided into global interaction models and local interaction models (Brock and Durlauf 2001a; Glaeser and Scheinkman 2001). Global interaction models are those in which individuals are affected by the decisions of the entire population (see Brock and Durlauf 2001a) and local interaction models are those in which individuals are affected only by the decisions of neighbouring individuals (Schelling 1971; Blume 1993; Ellison 1993). This chapter addresses the modelling of local interactions in the context of deforestation, applying the concepts discussed by Brock and Durlauf (2001a).

Empirical economic models of land use without interactions, applied to deforestation, study how the relative profitability of agricultural and forest land uses is determined by a set of exogenous factors. Some of these models use continuous dependent variables such as county-level deforestation (Stavins and Jaffe 1990, Cropper and Griffiths 1994; Pfaff 1999; Pfaff and Sanchez-Azofeifa 2004). Other models use discrete dependent variables, as implied by the observation of binary plot-level deforestation decisions (Chomitz and Gray 1996; Geoghegan *et al.* 2001).

Here we develop a discrete dependent model with interdependent individual deforestation decisions. The model assumes a forested area divided into $n$ plots. Each plot is managed by one individual and each individual manages only one plot. Each individual faces a decision between conserving forest in the plot, $f$, or clearing the plot to engage in an alternative land use, $c$. In addition, decisions are assumed to be based on the maximisation of profits. Therefore, individuals clear their forest if the profit of any of the non-forest land uses is larger than the profits to be gained by conserving the forest.

Profits are affected by three different factors: a vector of observable plot characteristics, neighbours' deforestation decisions and a random profit shock. As in other standard deforestation models, the effect of the vector of the individual $i$'s observable characteristics, $x_i$, on profits depends on the action taken by the individual. High levels of rainfall in a plot, for example, increase profits if the plot is deforested and used for agriculture. However, it may decrease profits if the individual decides to conserve the forest for tourism activities. Tourists prefer visiting sunny areas with low levels of rainfall. Therefore, individual $i$ obtains $x_i\beta_c$ when he/she clears but he/she gets $x_i\beta_f$ when he keeps forest, where $\beta_c$ and $\beta_f$ are two vectors of parameters that linearly map plots' characteristics into profits.

Standard econometric deforestation models also allow for the existence of unobservable elements that affect profits and thus deforestation decisions. The random profit shock represents the magnitude by which $i$'s profits are affected by these characteristics in ways observed only by that same actor $i$.

Privately observed characteristics' effects on profits also depend on the action taken by $i$. For instance, each individual possesses skills in working the land that are unknown to the rest of the agents. A highly skilled individual would obtain greater profits if he decides to engage in agriculture but no particular gain if the decision is to conserve the forest. Therefore, individual $i$ receives an additional $\varepsilon_i(c)$ if clearing occurs but an additional $\varepsilon_i(f)$ if the forest is conserved.

Finally, neighbours' decisions also affect individual profits, unlike in standard empirical models of deforestation by individuals. The individual $i$'s neighbourhood is defined as the area, outside $i$'s plot, covered by forest within a distance $r$ of any point inside $i$'s plot. The set of $i$'s neighbours, $N_i$, contains the individuals with plots that intersect the neighbourhood of $i$. It can be assumed that neighbours' decisions affect the profits of clearing based on the fraction of the neighbourhood being deforested, $m_i$.

Furthermore, it can be assumed that neighbours' effects on individual $i$'s profits also depend on the action taken by $i$. If a fraction $m_i$ of the neighbourhood is cleared and $i$ also decides to clear, he receives $\rho_{cc}m_i$ for mimicking neighbours' behaviour and $\rho_{cf}(1 - m_i)$ for deviating from the neighbours' behaviour. If he conserves his plot of forest he gets $\rho_{fc}m_i$ for deviating from his neighbours' behaviour and $\rho_{ff}(1 - m_i)$ for mimicking his neighbours' behaviour. The parameters $\rho_{cc}$, $\rho_{cf}$, $\rho_{fc}$ and $\rho_{ff}$ map neighbours' deforestation decisions into profits.

However, deforestation decisions are simultaneous. Hence, individuals form beliefs or expectations about the fraction of the neighbourhood that his neighbours would deforest, $m_i^e$. Therefore, $i$ clears if expected profits of clearing, $\pi_i(c, m_i^e)$, are larger than the expected profits of conserving his forest, $\pi_i(f, m_i^e)$. Formally, the individual clears if

$$x_i\beta_c + \rho_{cc}m_i^e + \rho_{cf}\left(1 - m_i^e\right) + \varepsilon_i(c) > x_i\beta_f + \rho_{fc}m_i^e + \rho_{ff}m_i^e + \varepsilon_i(f). \quad (1)$$

If the distribution of the difference of the shocks, $\varepsilon_i(c) - \varepsilon_i(f)$, is independent, identically and normally distributed across individuals, the probability, $p_i \in [0, 1]$, that agent $i$ clears is:

$$p_i = \Phi\left(x_i\beta + (\rho_c + \rho_f)m_i^e - \rho_f\right) \quad (2)$$

where $\Phi$ represents the standard normal distribution function, $\rho_c$ represents $\rho_{cc} - \rho_{fc}$, $\rho_f$ represents $\rho_{ff} - \rho_{cf}$, and $\beta$ represents $\beta_c - \beta_f$.

Under rational expectations, individuals compute the probability of their neighbours' clearing, based on which they form beliefs about the fraction of their neighbourhood that will be deforested. Putting this formally,

$$m_i^e = \sum_{j \neq i} w_{ij}\Phi\left(x_i\beta + (\rho_c + \rho_f)m_j^e - \rho_f\right) \quad (3)$$

where $w_{ij}$ is the fraction of land managed by agent $j$ in $i$'s neighbourhood. The equilibrium in expectation is the vector, $(p_1, p_2, \ldots, p_n) \in [0, 1]^n$, that solves the set of equations:

$$p_i = \Phi\left(x_i\beta + (\rho_c + \rho_f)\sum_{j \neq i}\left(w_{ij}p_j\right) - \rho_f\right) \quad \forall i. \quad (4)$$

This set of equations has at least one solution, $p^*$ (Brock and Durlauf 2001b). A solution is an equilibrium that generates self-consistent beliefs. In equilibrium, all individuals' beliefs about their neighbours' actions equals their neighbourhood expected deforestation. Empirically, this allows the neighbourhood's actual deforestation to be used to estimate the interaction coefficient $\rho$ defined as: $\rho_c + \rho_f$.

In fact, there could be more than one vector of probabilities of deforestation that satisfy the system of equations (4). The number of equilibria depends on the magnitude of the interaction coefficient, $\rho$, and on the plots' observable characteristics.[5] Once the parameters have been

[5] Brock and Durlauf (2001b) and Brock and Durlauf (2001a), for instance, show how the magnitude of the interaction coefficient affects the number of equilibria assuming specific observable characteristics of the agents under a different assumption about the distribution function of the shocks.

estimated, computational procedures can be used to search for the number of vectors (equilibria) that satisfy the equations of the spatial deforestation model.

The potential existence of multiple equilibria has important implications. Different deforestation outcomes could arise in the same region. Given the irreversibility of deforestation decisions, such effects can last over time. There could also be a Pareto dominant equilibrium outcome. In such a case, decentralised decisions do not assure the best outcome and government intervention is then justified.

## 3 Estimation strategy

The identification of interaction effects has been widely discussed in economics (e.g. Brock and Durlauf 2001a; Glaeser and Scheinkman 2001; Moffitt 2001; Conley and Topa 2002; Bayer and Timmins 2003) and in modelling land use in particular (Irwin and Bockstael 2002). A number of alternatives have been proposed, but consensus is that the best solution depends on the application (Glaeser and Scheinkman 2001; Moffitt 2001).[6] Simultaneity and the presence of spatially correlated unobservable variables are among the most important sources of bias that should be addressed.

Simultaneity is present in the estimation of interaction coefficients in any application. If individual $i$ is affected by individual $j$, then individual $i$ also affects $j$'s decision (if $j$ belongs to $i$'s neighbourhood, $i$ must belong to $j$'s neighbourhood). This two-directional process biases the estimation. Without this potential bias being addressed, the estimate of the interaction coefficient would reflect not only the effect of agent $j$'s action on $i$'s decision but also the effect of $i$'s decision on $j$'s action.

Another critical issue is that only limited information in terms of individual and plot characteristics can be observed. Many other driving factors of deforestation end up in the errors of the regression equation. This is especially important since some of those other factors are also spatially correlated. The estimation, then, of the interaction term $\rho$ by only using the neighbourhood deforestation rate, $m_i$, is inconsistent. What appear to be effects of neighbouring choices on individuals' choices could be the result of spatial correlation between unobserved deforestation drivers.

Some estimation techniques can address simultaneity, others can address spatially correlated errors and some can address both.

[6] In each application, one specific econometric problem might be more severe than another. Therefore, the best strategy of estimation will vary according to the application and data availability.

### 3.1 *Spatial econometric approach*

Anselin's (1988) Spatial Autoregressive (SAR) model has been used for the estimation of local interactions. The SAR model deals with simultaneity by solving the econometric equation for the dependent variable present in the right- and left-hand sides of the equation and then estimating the non-linear resulting functional form of the parameter via maximisation of a likelihood function.[7]

Anselin's Spatial Error model addresses the correlation of the errors. However, this approach strongly depends on the assumption of the spatial relation of the errors. If the spatial error structure is not correctly determined, some of the unobservable variables will still affect the estimates of the interaction coefficient. Knowing the correct spatial structure of the unobservable variables that affect deforestation is by definition impossible. This is also true for the Anselin's General Spatial model that considers spatial correlation both of the errors and, as in the SAR, of the dependent variable.

### 3.2 *Instrumental variable approach*

Simultaneity and the presence of spatially correlated unobservable factors can be addressed using instrumental variable techniques (Evans *et al.* 1992; Moffitt 2001). The ideal instrumental variables are exogenous neighbours' characteristics that explain neighbours' deforestation decisions and that are not correlated to the unobservable shocks that affect individuals' deforestation decisions. If these conditions hold, the variation in the instruments can be used to infer the effects of neighbours' deforestation in individuals' deforestation decisions.

Using neighbours' characteristics to infer the interaction effect avoids simultaneity. In this case, the individual decision does not affect the exogenous neighbours' characteristic. Therefore, the feedback effect of the individual deforestation decision on neighbours' decisions does not affect the estimation.

Additionally, the instrumental variable approach addresses the effects of spatially correlated unobservable factors in the estimation. These factors do in part drive the deforestation decisions of neighbours, but do not affect their exogenous characteristics. By using exogenous neighbours' characteristics, the correct estimate of the interaction parameter can still be accomplished. This is true as long as the exogenous neighbours' characteristics are uncorrelated with the individual's unobservable shocks.

[7] The application of these models causes significant computational demands, limiting the possibility of using large data sets.

The key condition of the IV strategy is, therefore, that the instruments not be correlated with the unobservable factors that drive individuals' decisions. For example, the average of the neighbours' minimum distance to a local road can be used as instrument. This variable affects neighbours' deforestation decisions but also could reflect unobservable abundance of roads in the area, something which while unobserved may also directly lead the individual to clear forestland. This choice of instrument would reflect the effects of interactions and the direct effects of the abundance of roads in the interaction coefficient jointly, which clearly would bias the interaction coefficient.

Neighbouring ecological characteristics and topography may be uncorrelated to the unobservable individuals' characteristics. It follows that deforestation decisions may be affected by the individuals' own ecological and topographic characteristics, but not by their neighbours' ecological and topographic characteristics. These are the instruments that we use in the model that we show in this chapter in the context of Costa Rica.

However, there could still be unobservable variables that affect individuals' deforestation and that are correlated with such exogenous neighbouring characteristics. One response to such potential issues is to absorb possible unobservable plots' characteristics that could be correlated with the instrument in the deforestation equation itself with control variables. For instance, controlling for the density of local roads in the neighbourhood would reduce the bias when using neighbours' minimum distance to the roads as the instrument. In general, controlling for spatially explicit variables can minimise any possible correlation between the instrument and unobservable plot characteristics that directly affect plot deforestation.

### *3.3 Discreteness and the Two Stage Probit Least Squares*

The Two Stage Probit Least Squares (2SPLS) method (Maddala 1983) is available in order to implement instrumental variable techniques in a discrete dependent variable approach. This method involves two steps. The first step consists of regressing neighbours' deforestation on the instruments and the rest of exogenous variables that explain the individuals' deforestation decision. The second stage consists of using the predicted values of the first stage regression to estimate the interaction effect in the individual's deforestation equation.

In the first stage the instruments and exogenous individuals' characteristics[8] are used to predict neighbourhood deforestation using a linear

[8] Adding exogenous individuals' characteristics improves efficiency in the estimation.

specification,

$$m_i = \Pi_1 \sum_{j \in N_i} (w_{ij} \bar{x}_j) + \Pi_2 x_i + \mu_i \tag{5}$$

where $\Pi_1$ and $\Pi_2$ are the reduced-form coefficients to be estimated, $\bar{x}_j$ are the exogenous characteristics that affect the decision of only the individual $j$, for all $j$ in $N_i$ and does not affect the decision of individual $i$. Therefore, the instrument is

$$\sum_{j \in N_i} (w_{ij} \bar{x}_j), \tag{6}$$

which represents the value of the average exogenous characteristics in the neighbourhood of $i$.[9] The estimated reduced-form coefficients, $\hat{\Pi}_1$ and $\hat{\Pi}_2$ in equation (5), are used to predict neighbourhood deforestation, $\hat{m}_i$.

In the second stage, neighbourhood deforestation is substituted for using its predicted values. Then, the interaction coefficient $\rho$ can be estimated from

$$\Pr(y_i = 1) = \Phi(x_i \beta + \rho \hat{m}_i - \rho_f) \tag{7}$$

by standard likelihoods methods,[10] where the dependent variable, $y_i$, is discrete and reflects the observation of whether plot $i$ has been deforested, 1, or not, 0, in a specific period.

## 4 Data requirements and GIS

The estimation of the parameters of the model requires information on the individual's deforestation decision, $y_i$, the individual's vector of observable characteristics, $x_i$, and the individual's neighbourhood deforestation, $m_i$. Additionally, some of the observable characteristics of the individual's neighbours should satisfy certain conditions, discussed below, to construct the instrument. If these conditions hold, the instrument can be used to estimate correctly the interaction parameter.

Geographic information systems can be used to process spatially specific data to produce the variables required for the analysis. Recently, GIS has been used to analyse deforestation (see Chomitz and Gray 1996; Pfaff 1999; Kok and Veldkamp 2001; Serneels and Lambin 2001; Walsh *et al.* 2001; Irwin and Bockstael 2002).

The number of observations available for analysis is extremely large when using spatially explicit or pixel-level forest information. If

[9] Note that $i \notin N_i$.

[10] We follow the standard normalisation assumption that the variance of the privately observed shocks, $\sigma$, is one as in Brock and Durlauf (2001).

computations for either variable creation or estimation become difficult when using all of the point observations, a valid alternative is drawing random samples of pixels or locations from the maps. This can simplify the calculations and speed up these processes. This section discusses how the required variables are obtained, suggests what variables can be used to build the instrument, and conveys the role that GIS can play in this approach to estimation.

### *4.1 Deforestation decisions*

The object of study is the analysis of deforestation decisions in privately owned land during a period of time. Therefore, the analysis should be focused on those plots that are covered by forest. Furthermore, we exclude land within national parks as these are owned by the government and decisions about the management of the land in these areas are not based on individuals' profit calculations as assumed in the model. If plot $i$ has been deforested by the end of the period, $y_i$ the dependent variable is assumed to have value 1 but if the plot is still covered by forest at the end of the period, $y_i$, it is given the value 0.

Forest satellite pictures can be used to obtain deforestation dynamics (e.g. Chomitz and Gray 1996; Pfaff 1999; Kok and Veldkamp 2001; Serneels and Lambin 2001; Walsh *et al.* 2001). Data from Costa Rica are used to illustrate the estimation procedure. Forest satellite pictures taken in 1986 and 1997 and developed by the Tropical Scientific Center in Costa Rica are used to describe the presence of forest in 30 $m^2$ pixels across Costa Rica. In this study, 10,000 randomly drawn pixels across Costa Rica serve as plot observations to analyse deforestation. From these pixels, only those that are in privately owned forest are considered for the analysis. The dependent variable, in this case, is obtained as follows. Pixels covered by forest in 1986 that are deforested by 1997 are associated with value 1 and pixels covered by forest in 1986 that are still covered by forest in 1997 receive the value 0.

### *4.2 Neighbourhood deforestation*

The hypothesis being tested is whether the fraction of the neighbourhood that is deforested, $m_i$, affects the individual's deforestation decision, $y_i$. Therefore, the information about $m_i$ is required. One of the advantages of using GIS is that it is possible to calculate the actual fraction of the neighbourhood that is deforested during the period of study. Another alternative is using the randomly drawn sample of pixels in the neighbourhood and calculating the fraction that is deforested during the period. Brock and Durlauf (2001b) discuss the use of the sample to infer the fraction of neighbours that take a specific decision.

In order to calculate deforestation within the neighbourhood, the concept of neighbourhood should be well defined. Definitions of neighbourhoods and neighbours in the literature are as numerous as the type of interactions that have been studied. It is common to define neighbourhoods using political divisions such as provinces, counties or districts.[11] However, neighbourhoods can also be defined by distances alone, regardless of political boundaries. Here we follow the second approach. That is, neighbourhoods are defined based on distances and such an approach is used to estimate interactions in our Costa Rica example. More specifically, the neighbourhoods are defined as the areas covered by forest within a 10 km radius. Any two plots separated by a distance smaller than 10 km, covered by forest, are considered to be neighbouring plots. Figure 3.1, for instance, shows the location of a plot represented by a star, the 10 km radius neighbourhood, represented by the large circle, sampled neighbouring plots, represented by triangles, and the rest of the sampled plots (observations), represented by dots. Forest satellite pictures are used to calculate the deforested fraction of these neighbourhoods between 1986 and 1997 in Costa Rica.

### 4.3 *Observable drivers of deforestation*

Observable characteristics that are commonly used in deforestation models are those that describe the socio-economic conditions in the plot, such as population (Cropper and Griffiths 1994; Anderson *et al.* 2002), local wages (Pfaff 1999; Anderson *et al.* 2002), distance to markets (Pfaff 1999; Geoghegan *et al.* 2001; Serneels and Lambin 2001; Anderson *et al.* 2002), distance to roads (Chomitz and Gray 1996; Pfaff 1999; Serneels and Lambin 2001; Anderson *et al.* 2002; Geoghegan *et al.* 2001), and those that describe the ecological conditions in the plot, including vegetation type (Serneels and Lambin 2001; Pfaff and Sanchez-Azofeifa 2004), the slope of the terrain (Chomitz and Gray 1996) and soil type (Cropper and Griffiths 1994; Chomitz and Gray 1996; Pfaff 1999; Geoghegan *et al.* 2001).

An example of a set of plot characteristics that would control effectively for factors that might be correlated to the instrument and affect individuals' deforestation in Costa Rica is presented in Table 3.1. These variables are also calculated using GIS.

[11] Others have also defined neighbourhoods based on social connections. The literature on 'networks' also defines neighbourhoods from different perspectives.

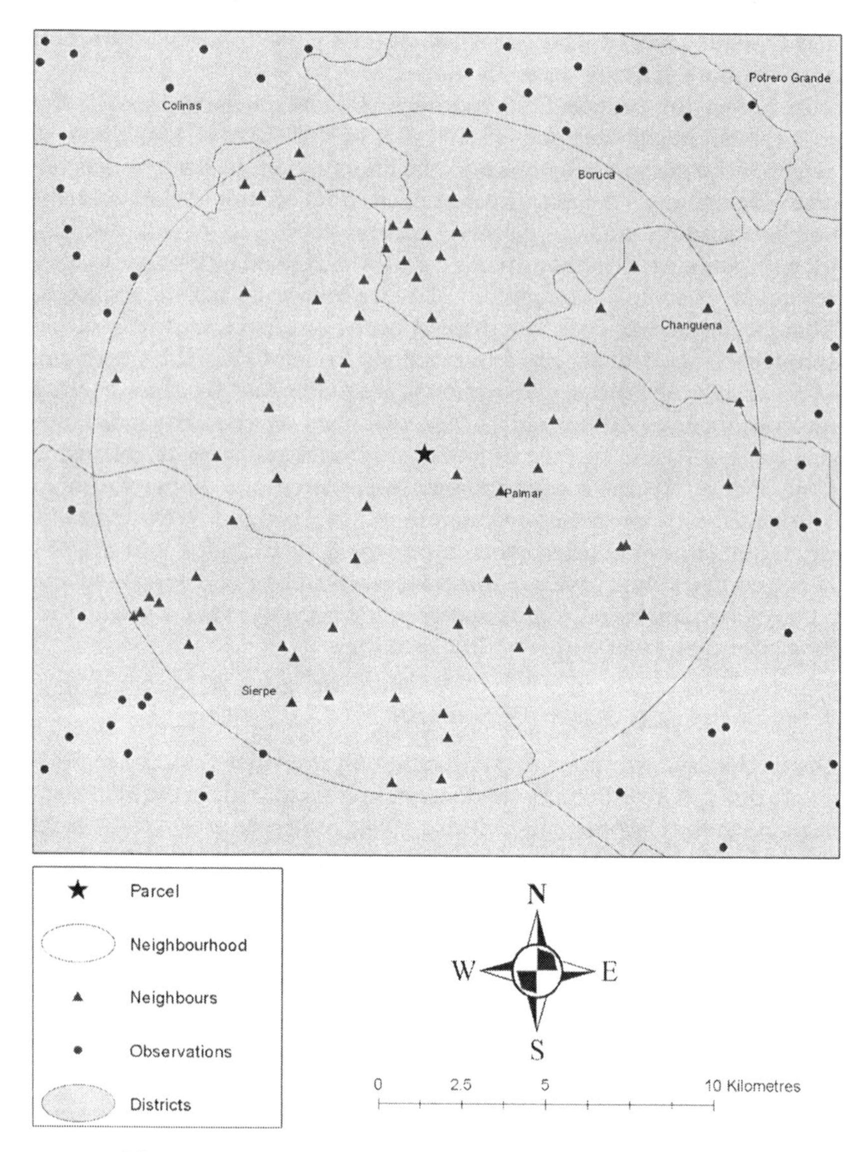


Figure 3.1. Illustration of the observations, neighbourhoods and neighbours

Table 3.1 *List of plot characteristics*

| Type | Characteristics | |
|---|---|---|
| Distances | Distance to the city (San José) | DSJ |
| | Distance to a port (Caldera) | DTC |
| | Distance to a port (Limon) | DTL |
| | Distance to national roads | DNR |
| | Distance to local roads | DLR |
| | Distance to sawmills | DTS |
| | Distance to schools | DSC |
| | Distance to cleared areas | DCA |
| | Distance to towns (county capital) | DMT |
| Natural characteristics | Slope of the terrain | SLO |
| | Life zones | LZ |
| Characteristics of areas around the plot | Length of national roads at 10 km radius | LNR |
| | Length of local roads at 10 km radius | LLR |
| | Number of sawmills at 10 km radius | NSM |
| | Number of towns at 10 km radius | NMT |
| | Number of schools at 10 km radius | NSC |
| | Percentage of cleared area at 10 km radius | CLP |

### 4.4 *Instrumenting neighbourhood deforestation*

As noted above, an instrument should satisfy two conditions. First, it should explain the neighbours' deforestation decisions. Second, it should not be correlated with unobservable characteristics that affect $i$'s decision. The first condition suggests that the characteristics from the vector $x_j$ that affect $j$'s decision, where $j$ represents those individuals in $i$'s neighbourhood, should be considered as instruments. However, not all of $j$'s observable characteristics can be used as instruments. Some of these characteristics, as discussed before, are correlated with unobservable characteristics that affect $i$'s deforestation decision, which violates the second condition.

Plots' characteristics determined by nature can satisfy these conditions. Natural characteristics reflect a source of exogenous variation that can be useful in identification of social interaction processes (e.g. Chaudhuri 1999; Munshi 2003).

Two proposed instruments are neighbours' slopes of the terrain and neighbours' ecological characteristics.[12] These are chosen as they do not affect the individuals' deforestation decisions directly. Moreover,

[12] The classification of the plots' ecological characteristics is based on Holdridge Life Zones.

individuals' deforestation decisions are affected by their own slopes and own ecological characteristics. Computing these instruments is a simple task. Each plot $i$ has a set of neighbouring plots. Each neighbouring plot has its characteristics, such as slope of the terrain or ecological characteristics. Therefore, the instruments can be easily computed by calculating the average of the neighbouring plot characteristics.

## 5 Results and equilibria

Two techniques are used to estimate the interaction parameter: standard probit and 2SPLS. The 2SPLS uses neighbours' slopes as the instrument for neighbours' deforestation. These techniques are applied to two different regions in Costa Rica shown in Figure 3.2. The regions were chosen based on their quantity of forest and their ecological importance. The area that was left out of the analysis does not have enough deforestation to perform the analysis.

Descriptive statistics of the characteristics of the plots in each region are shown in Table 3.2. The classification of the regions was accomplished by regrouping the government's planning sectors. Specifically, Region 1 contains Huetar Norte and Huetar Atlantica, Region 2 contains Brunca, the Central Area and the Central Pacific, and finally, the region left out is Chorotega. We divide Costa Rica into these two regions in order to test whether the level of interactions could vary across space. If so, that could generate different policy implications for the different areas. Regions are grouped according to their characteristics and location.

Estimates of the interactions parameter are presented in Table 3.3.[13] In Region 1, the standard probit estimate suggests that interactions are positive and significant. However, standard probit estimates are upward biased due to simultaneity and the presence of spatially correlated unobservable factors. Unbiased estimates can be found by using a 2SPLS technique. The 2SPLS estimates show insignificant neighbourhood effects. These two results show that if simultaneity and the presence of spatially correlated effects are not addressed in the empirical approach to measuring interactions, one might conclude wrongly that interactions in Region 1 exist when there is no evidence for that.

However, using 2SPLS can also lead the researcher to conclude that interactions exist. In Region 2, probit and 2SPLS estimates of the interaction parameter are positive and significant and their magnitude is similar.[14] Standard errors under the 2SPLS however are larger than the standard errors from the probit estimates. This difference arises as

[13] In the Appendix, complete regression results are presented.
[14] A statistical test cannot reject the null hypothesis that these estimates are equal.

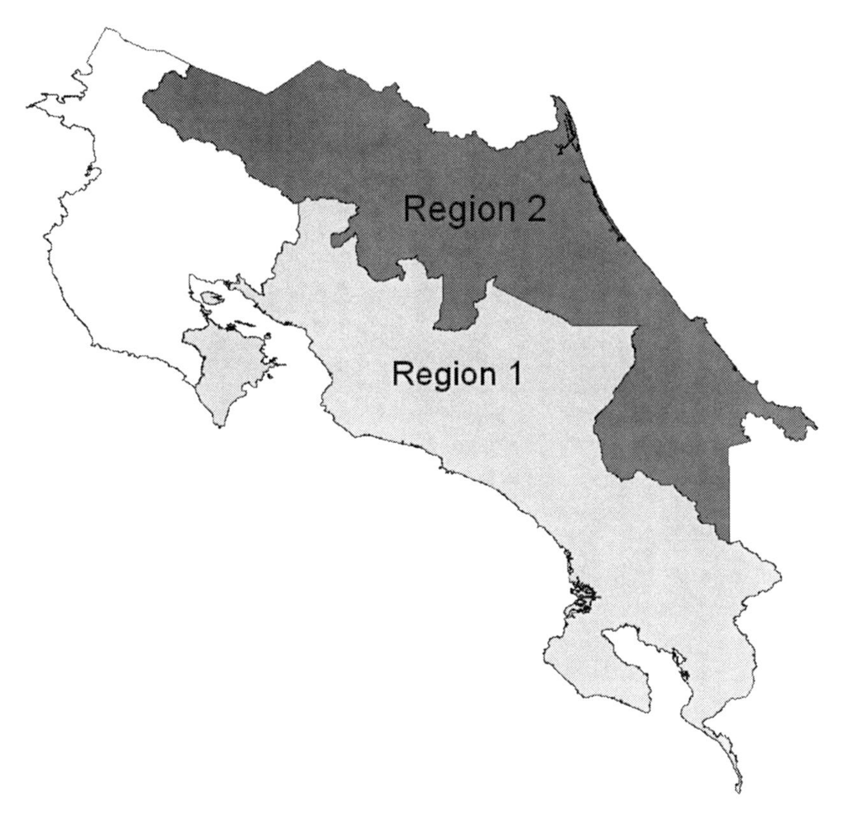


Figure 3.2. Region 1 and Region 2*

a consequence of the presence of simultaneity and spatially correlated unobservable effects. Moreover, the difference in the 2SPLS estimates between Region 1 and Region 2 shows that the presence of interactions might vary across regions.

Once the parameters of the model have been estimated, a numerical procedure can be used to search for the equilibrium outcome, $p^*$. The probabilities of deforestation in equilibrium can be computed by an iterative process. A set of initial beliefs, $p^{(1)}$, generates a second set of beliefs, $p^{(2)}$, using[15]

$$p_i^{(2)} = \Phi\left(x_i\hat{\beta} + \hat{\rho}\sum_{j\neq i}\left(w_{ij}\,p_i^{(1)}\right) - \hat{\rho}_f\right)\ \forall i \tag{8}$$

* Areas outside the regions do not have enough deforestaation in the sample

15 Note that $\rho_f$ cannot be identified from the model when among the set of individual's characteristics a constant term is present. The process can still go on since the estimated constant term would contain both effects.

Table 3.2 *Descriptive statistics for Region 1 and Region 2*

| | | Region 1 (obs. 637) | | | | Region 2 (obs. 810) | | | |
|---|---|---|---|---|---|---|---|---|---|
| Variables | | Mean | S. D. | Min | Max | Mean | S. D. | Min | Max |
| *Deforestation* | | | | | | | | | |
| Def. decision: $y$ | Dummy | 0.08 | 0.28 | 0 | 1 | 0.20 | 0.40 | 0 | 1 |
| Neighbours' def. | % | 0.07 | 0.07 | 0 | 0.57 | 0.20 | 0.13 | 0.01 | 0.70 |
| *Instrument* | | | | | | | | | |
| Neighbours' slopes | Degrees | 11.25 | 5.59 | 0 | 25.79 | 2.64 | 3.08 | 0 | 12.78 |
| *Controls* | | | | | | | | | |
| SLO | Degrees | 11.04 | 8.01 | 0 | 32.05 | 2.66 | 4.48 | 0 | 26.56 |
| Bad Life Zones | Dummy | 0.21 | 0.40 | 0 | 1 | 0.52 | 0.50 | 0 | 1 |
| Good Life Zones | Dummy | 0.17 | 0.37 | 0 | 1 | 0.01 | 0.11 | 0 | 1 |
| DSJ | Km | 89.06 | 53.38 | 0 | 223.04 | 77.57 | 34.83 | 1.48 | 174.48 |
| DTL | Km | 131.22 | 45.51 | 44.65 | 234.19 | 108.59 | 68.35 | 0.99 | 273.33 |
| DTC | Km | 126.12 | 74.35 | 9.75 | 278.94 | 119.25 | 39.94 | 35.91 | 239.69 |
| DLR | Km | 2.37 | 2.30 | 0.00 | 21.37 | 2.54 | 2.08 | 0.00 | 10.26 |
| DNR | Km | 3.87 | 3.77 | 0.01 | 31.09 | 5.17 | 4.33 | 0.01 | 22.08 |
| DTS | Km | 18.60 | 10.70 | 0.98 | 60.20 | 20.31 | 11.99 | 1.29 | 71.46 |
| DSC | Km | 12.29 | 7.42 | 0.32 | 53.65 | 19.39 | 11.24 | 0.99 | 54.79 |
| DCA | Km | 0.22 | 0.29 | 0.00 | 2.60 | 0.27 | 0.38 | 0.00 | 3.02 |
| DMT | Km | 15.50 | 7.99 | 0.76 | 54.82 | 27.44 | 15.13 | 1.89 | 71.46 |
| LNR | Km | 38.93 | 32.61 | 0 | 284.92 | 28.36 | 23.93 | 0 | 159.58 |
| LLR | Km | 52.88 | 41.12 | 0 | 248.91 | 50.00 | 34.47 | 0 | 234.05 |
| NSM | Units | 0.45 | 1.50 | 0 | 16 | 0.53 | 1.47 | 0 | 11 |
| NMT | Units | 0.36 | 0.81 | 0 | 9 | 0.19 | 0.61 | 0 | 7 |
| NSC | Units | 1.19 | 4.01 | 0 | 66 | 0.51 | 1.47 | 0 | 17 |
| CLP | % | 0.57 | 0.26 | 0 | 0.98 | 0.44 | 0.21 | 0 | 0.98 |

Table 3.3 *Estimates of the interactions parameter (ρ)*

| | Region 1 | | Region 2 | |
|---|---|---|---|---|
| | 2SPLS | Probit | 2SPLS | Probit |
| $\hat{\rho}$ | 0.40 | 6.49** | 3.31** | 3.06** |
| Standard errors | 12.28 | 1.37 | 1.31 | 0.50 |
| N | 637 | 637 | 810 | 810 |
| -Log likelihood | 147 | 147 | 353 | 353 |

** indicates significance at 99%

The iterative process consists in computing $(p^{(1)}, p^{(2)}, \ldots, p^{(k)})$, until $p^{(k)}$ equals $p^{(k+1)}$. The set of probabilities of deforestation, $p^{(k)}$, is an equilibrium because it satisfies the set of simultaneous equation (4). Formally,

$$p_i^{(k)} = \Phi\left(x_i\hat{\beta} + \hat{\rho}\sum_{j\neq i}\left(w_{ij}p_i^{(k)}\right) - \hat{\rho}_f\right) \quad \forall i. \tag{9}$$

This procedure finds only stable equilibria. It has been argued, though, that this type of equilibrium is more likely to be observed in the long run. This is because if the system is in an unstable equilibrium, then small changes in the beliefs of the agents can shift the system to a stable equilibrium. By increasing the number of initial conditions considered, the probability of finding all of the equilibria increases.

Using the set of equations (9), it can be seen that changes in the vector of individual characteristics $x_i$ affect the probabilities of deforestation of other individuals. This effect depends on the magnitude of the interaction, $\rho$. A change in characteristics of an individual $i$ affects the probability that $i$ clears, which in turn affects the probability that $j$ clears, given that $i$ is $j$'s neighbour. This second effect depends on the magnitude of the interaction coefficient, $\rho$. This example shows how policy interventions that affect only individual $i$ could end up affecting all of the individuals that have $i$ as a neighbour.

## 6 Conclusion

The dependency of the provision of ecosystem services on the stock and the spatial distribution of forest is leading researchers to focus on the spatial dynamics of forest. This chapter has discussed a method to empirically test one of the key factors that shape the stock and spatial pattern of forest: neighbours' interactions in deforestation decisions. The methodology applied, based upon the use of instrumental variables, could be used in different regions where different species reside but are threatened by deforestation and in other land-use contexts, such as settings where reforestation is occurring in cleared areas.

An illustration of the approach proposed here was presented for two regions in Costa Rica. In one of the regions, it is shown that there is no evidence for interactions using the instrumental variable approach, contradicting the result of a standard approach. In the other region, using instruments positive spatially reinforcing interactions are found.

Such interactions have important implications. Positive interactions reduce forest fragmentation within neighbourhoods and imply that policies which alter incentives to deforest in one location have spillover effects in neighbouring locations. Further, they create the possibility of multiple equilibria. The potential for multiple equilibria implies that projections of the effects of new policies which are based on extrapolations from past equilibria could be missing the possibility that a policy could induce another equilibrium.

Further research could focus on identifying impacts of spillover effects and multiple equilibria on the supply of environmental services. Ecological results or new research can link the quantity of forest and its spatial structure with the supply of environmental services. Analysis of overall impacts could be accomplished by generating simulations in an integrated model using interaction parameters of different magnitudes.

## REFERENCES

Anderson, L., Granger, C., Reis, E., Weinhold, D. and Wuder, S. 2002. *The Dynamics of Deforestation and Economic Growth in the Brazilian Amazon.* Cambridge: Cambridge University Press.

Ando, A., Camm, J., Polasky, S. and Solow, A. 1998. Species distributions, land values, and efficient conservation, *Science.* **279** (5359). 2126–2128.

Anselin, L. 1988. *Spatial Econometrics: Methods and Models.* Boston: Kluwer Academic Publishers.

Bayer, P. and Timmins, C. 2003. Estimating equilibrium models of sorting across locations. Yale University. Economic Growth Center. Discussion Paper No. 862. 1–31.

Blume, L. E. 1993. Statistical mechanics of strategic interaction, *Games and Economic Behaviour.* **5**. 387–424.

Brock, W. A. and Durlauf, S. N. 2001a. Discrete choice with social interactions. *Review of Economic Studies.* **68** (2). 235–260.

Brock, W. A. and Durlauf, S. N. 2001b. Interactions-based models. In J. J. Heckman. and E. Leamer (eds.). *Handbook of Econometrics.* Amsterdam: Elsevier. 3329–3371.

Case, A. 1992. Neighborhood influence and technological change. *Regional Science and Urban Economics.* **22** (3). 491–508.

Chaudhuri, S. 1999. Forward-looking behavior, precautionary savings, and borrowing constraints in a poor, agrarian economy: tests using rainfall data. *Working Paper 9899–10*. Columbia University.

Chomitz, K. M. and Gray, D. A. 1996. Roads, land use, and deforestation: a spatial model applied to Belize. *World Bank Economic Review*. **10**. 487–512.

Cocks, K. D. and Baird, I. A. 1989. Using mathematical-programming to address the multiple reserve selection problem – an example from the Eyre peninsula, South Australia. *Biological Conservation*. **49** (2). 113–130.

Conley, T. G. and Topa, G. 2002. Socio-economic distance and spatial patterns in unemployment. *Journal of Applied Econometrics*. **17** (4). 303–327.

Cooper, R. and John, A. 1988. Coordinating coordination failures in Keynesian models. *Quarterly Journal of Economics*. **103** (3). 441–463.

Coops, N. C., White, J. D. and Scott, N. A. 2004. Estimating fragmentation effects on simulated forest net primary productivity derived from satellite imagery. *International Journal of Remote Sensing*. **25** (4). 819–838.

Costello, C. and Polasky, S. 2004. Dynamic reserve site selection. *Resource and Energy Economics*. **26** (2). 157–174.

Cropper, M. and Griffiths, C. 1994. The interaction of population-growth and environmental-quality. *American Economic Review*. **84** (2). 250–254.

Diaz, J. A., Carbonell, R., Virgos, E., Santos, T. and Telleria, J. L. 2000. Effects of forest fragmentation on the distribution of the lizard psammodromus algirus. *Animal Conservation*. **3**. 235–240.

Ellison, G. 1993. Learning, local interaction, and coordination. *Econometrica*. **61** (5). 1047–1071.

Evans, W. N., Oates, W. E. and Schwab, R. M. 1992. Measuring peer group effects – a study of teenage behavior. *Journal of Political Economy*. **100** (5). 966–991.

Geoghegan, J., Villar, S. C., Klepeis, P., Mendoza, P. M., Ogneva-Himmelberger, Y., Chowdhury, R. R., Turner, B. L. and Vance, C. 2001. Modeling tropical deforestation in the southern Yucatan peninsular region: comparing survey and satellite data. *Agriculture Ecosystems and Environment*. **85**. 25–46.

Glaeser, E. and Scheinkman, J. 2001. Measuring social interactions. In S. N. Durlauf. and H. P. Young. (eds.). 83–132.

Greene, W. H., 2003. *Econometric Analysis*. Upper Saddle River, NJ: Prentice Hall.

Irwin, E. G. and Bockstael, N. E. 2002. Interacting agents, spatial externalities and the evolution of residential land use patterns. *Journal of Economic Geography*. **2** (1). 31–54.

Kirkpatrick, J. B. 1983. An iterative method for establishing priorities for the selection of nature reserves – an example from Tasmania. *Biological Conservation*. **25** (2). 127–134.

Kok, K. and Veldkamp, A. 2001. Evaluating impact of spatial scales on land use pattern analysis in Central America. *Agriculture Ecosystems and Environment*. **85**. 205–221.

Maddala, G. S. 1983. *Limited-dependent and Qualitative Variables in Econometrics*. Cambridge, New York: Cambridge University Press.

McCoy, E. D. and Mushinsky, H. R. 1994. Effects of fragmentation on the richness of vertebrates in the Florida scrub habitat. *Ecology*. **75** (2). 446–457.

Moffitt, R. 2001. Policy Interventions, low-level equilibria, social interactions. In S. N. Durlauf. and H. P. Young. (eds.). *Social Dynamics*. Cambridge, MA: The MIT Press. 45–82.

Munshi, K. 2003. Networks in the modern economy: Mexican migrants in the U.S. labour market. *Quarterly Journal of Economics*. **118** (2). 549–599.

Nelson, G. C. and Hellerstein, D. 1997. Do roads cause deforestation? Using satellite images in econometric analysis of land use. *American Journal of Agricultural Economics*. **79**. 80–88.

Panoyotou, T. and Sungsuwan, S. 1989. An econometric study of the causes of tropical deforestation: the case of northeast Thailand. *Development Paper 284*. Harvard Institute for International Development.

Parkhurst, G. M., Shogren, J. F., Bastian, C., Kivi, P., Donner, J. and Smith, R. B. W. 2002. Agglomeration bonus: an incentive mechanism to reunite fragmented habitat for biodiversity conservation. *Ecological Economics*. **41** (2). 305–328.

Pfaff, A. S. P. 1999. What drives deforestation in the Brazilian Amazon? Evidence from satellite and socioeconomic data. *Journal of Environmental Economics and Management*. **37** (1). 26–43.

Pfaff, A. S. P. and Sanchez-Azofeifa, G. A. 2004. Deforestation pressure and biological reserve planning: a conceptual approach and an illustrative application for Costa Rica. *Resource and Energy Economics*. **26** (2). 237–254.

Polasky, S., Camm, J. D., Solow, A. R., Csuti, B., White, D. and Ding, R. G. 2000. Choosing reserve networks with incomplete species information. *Biological Conservation*. **94** (1). 1–10.

Rudel, T. K. 1989. Population, development, and tropical deforestation – a cross-national-study. *Rural Sociology*. **54** (3). 327–338.

Schelling, T. C. 1971. Dynamic models of segregation. *Journal of Mathematical Sociology*. **1** (2). 143–186.

Scull, P. R. and Harman, J. R. 2004. Forest distribution and site quality in southern Lower Michigan, USA. *Journal of Biogeography*. **31** (9). 1503–1514.

Serneels, S. and Lambin, E. F. 2001. Proximate causes of land-use change in Narok District, Kenya: a spatial statistical model. *Agriculture Ecosystems and Environment*. **85**. 65–81.

Stavins, R. N. and Jaffe, A. B. 1990. Unintended impacts of public-investments on private decisions – the depletion of forested wetlands. *American Economic Review*. **80** (3). 337–352.

Tubbs, C. and Blackwood, J. 1971. Ecological evaluation of land for planning purposes. *Biological Conservation*. (3). 169–172.

Turner, M. A. 2005. Landscape preferences and patterns of residential development. *Journal of Urban Economics*. **57** (1). 19–54.

Twedt, D. J. and Loesch, C. R. 1999. Forest area and distribution in the Mississippi alluvial valley: implications for breeding bird conservation. *Journal of Biogeography*. **26** (6). 1215–1224.

Walsh, S. J., Crawford, T. W., Welsh, W. F. and Crews-Meyer, K. A. 2001. A multiscale analysis of LULC and NDVI variation in Nang Rong district, northeast Thailand. *Agriculture Ecosystems and Environment*. **85** (1–3). 47–64.

Table 3.A1 *Regression results: probit estimates and second-stage estimates from 2SPLS*

| | Dependent variable deforestation decisions 86–97 | | | |
|---|---|---|---|---|
| | Region 1 | | Region 2 | |
| | Probit | 2SPLS | Probit | 2SPLS |
| NDE ($\rho$) | 6.498 | 0.401 | 3.065 | 3.314 |
| | (1.378) | (12.28) | (0.508) | (1.312) |
| GLZ | −0.614 | −0.513 | 0.129 | 0.110 |
| | (0.285) | (0.280) | (0.178) | (0.199) |
| BLZ | 0.275 | 0.198 | −3.509 | −3.513 |
| | (0.270) | (0.260) | (38.05) | (62.50) |
| DSJ | 0.008 | 0.007 | 0.013 | 0.013 |
| | (0.009) | (0.012) | (0.008) | (0.008) |
| DLI | −0.016 | −0.016 | −0.012 | −0.010 |
| | (0.006) | (0.007) | (0.005) | (0.005) |
| DCA | −0.010 | −0.007 | −0.016 | −0.015 |
| | (0.006) | (0.012) | (0.009) | (0.009) |
| DLR | 0.016 | 0.036 | −0.019 | −0.007 |
| | (0.046) | (0.047) | (0.036) | (0.035) |
| DNR | 0.050 | 0.036 | −0.010 | −0.007 |
| | (0.043) | (0.042) | (0.021) | (0.021) |
| DTS | −0.016 | −0.011 | 0.008 | 0.007 |
| | (0.014) | (0.013) | (0.008) | (0.008) |
| DTH | −0.023 | −0.031 | −0.001 | −0.004 |
| | (0.024) | (0.023) | (0.007) | (0.007) |
| PTC | −2.806 | −2.582 | −1.613 | −1.592 |
| | (0.795) | (0.762) | (0.339) | (0.336) |
| DMT | 0.016 | 0.029 | 0.004 | 0.003 |
| | (0.021) | (0.034) | (0.006) | (0.006) |
| SDA | −0.053 | −0.053 | −0.006 | −0.005 |
| | (0.014) | (0.014) | (0.017) | (0.019) |
| LNR | 0.007 | 0.014 | −0.002 | −0.002 |
| | (0.006) | (0.012) | (0.005) | (0.004) |
| LLR | −0.003 | −0.002 | −0.001 | −0.001 |
| | (0.004) | (0.004) | (0.002) | (0.002) |
| NSM | −0.196 | −0.120 | −0.049 | −0.055 |
| | (0.119) | (0.125) | (0.056) | (0.056) |
| NMT | 0.193 | 0.028 | −0.007 | 0.011 |
| | (0.216) | (0.278) | (0.192) | (0.196) |
| NHS | −0.040 | −0.070 | 0.074 | 0.073 |
| | (0.055) | (0.076) | (0.065) | (0.063) |
| CLP | 1.314 | 1.050 | −0.160 | −0.416 |
| | (1.483) | (1.468) | (1.168) | (1.150) |
| CLP2 | −1.975 | −1.335 | 0.634 | 0.751 |
| | (1.384) | (1.472) | (1.127) | (1.092) |
| Constant | 1.738 | 1.386 | 0.892 | 0.673 |
| | (1.052) | (1.169) | (1.131) | (1.135) |

In parenthesis standard errors

Table 3.A2 *Regression results: first stage*

| Dependent variable is neighbours' deforestation 86–97 | | |
|---|---|---|
| | Region 1 | Region 2 |
| *Instrument* | | |
| Neighbours' slopes ($\Pi_1$) | −0.0019 | −0.0234 |
| | (0.0007) | (0.0017) |
| *Controls for efficiency* ($\Pi_2$) | | |
| GLZ | −0.0057 | −0.0764 |
| | (0.0074) | (0.0102) |
| BLZ | 0.0010 | 0.0151 |
| | (0.0078) | (0.0328) |
| DSJ | −0.0006 | −0.0013 |
| | (0.0002) | (0.0004) |
| DLI | 0.0000 | 0.0003 |
| | (0.0002) | (0.0003) |
| DCA | 0.0006 | 0.0010 |
| | (0.0001) | (0.0005) |
| DLR | 0.0011 | 0.0029 |
| | (0.0012) | (0.0021) |
| DNR | −0.0004 | 0.0036 |
| | (0.0010) | (0.0012) |
| DTS | −0.0002 | −0.0021 |
| | (0.0004) | (0.0004) |
| DTH | 0.0004 | −0.0009 |
| | (0.0005) | (0.0004) |
| PTC | −0.0139 | −0.0363 |
| | (0.0090) | (0.0104) |
| DMT | 0.0021 | −0.0019 |
| | (0.0005) | (0.0003) |
| SDA | 0.0000 | 0.0000 |
| | (0.0004) | (0.0010) |
| LNR | 0.0008 | −0.0009 |
| | (0.0001) | (0.0003) |
| LLR | 0.0000 | 0.0011 |
| | (0.0001) | (0.0001) |
| NSM | 0.0041 | −0.0011 |
| | (0.0027) | (0.0031) |
| NMT | −0.0150 | −0.0513 |
| | (0.0069) | (0.0112) |
| NHS | −0.0019 | 0.0006 |
| | (0.0012) | (0.0041) |
| CLP | −0.0011 | 0.0250 |
| | (0.0459) | (0.0645) |
| CLP2 | 0.0315 | 0.0500 |
| | (0.0430) | (0.0650) |
| Constant | −0.0066 | 0.3140 |
| | (0.0329) | (0.0636) |

In parenthesis standard errors

# 4 Resource exploitation, biodiversity loss and ecological events

*Yacov Tsur and Amos Zemel*

## 1 Introduction

We study the management of a natural resource that serves a dual purpose. First, it supplies inputs for human production activities and is therefore being exploited for beneficial use, however defined. Second, it supports the existence of other species. Large-scale exploitation competes with the needs of the wildlife populations and, unless controlled, can severely degrade the ecological conditions and lead to species extinction and biodiversity loss. Examples for such conflicts abound, including: (i) water diversions for irrigation, industrial or domestic use reduce in-stream flows that support the existence of various fish populations; (ii) reclamation of swamps and wetlands that serve as habitat for local plant, bird and animal populations and as a 'rest area' for migrating birds; (iii) deforestation reduces the living territory of a large number of species; (iv) intensive pest control may lead to the extinction of the pests' natural predators and eventually to the invasion of an immune pest species which is harder to control; (v) overgrazing reduces soil fertility and entails the destruction of natural vegetation over vast semi-arid areas in central Asia and sub-Saharan Africa, contributing to the process of desertification; and (vi) airborne industrial pollution falls as acid rain on lakes and rivers and interferes with freshwater ecosystems. In some of these examples the affected species may not contribute directly to human well-being, but their diminution or extinction entails a loss due to use and non-use values as well as the loss of option for future benefits such as the development of new medicines (Bird 1991; Littell 1992).

The global deforestation example illuminates the issue under consideration. Until recently, a rainforest area about the size of England was cleared each year (Hartwick 1992), leading to the extinction of numerous species (Colinvaux 1989). The biodiversity loss process often takes the form of a sudden collapse of the ecosystem, inflicting heavy damage and affecting the nature of future exploitation regimes. This is so because

Acknowledgements: financial support by the Paul Ivanier Center for Robotics Research and Production Management, Ben Gurion University of the Negev, is gratefully acknowledged.

ecosystems are inherently complex and their highly non-linear dynamics give rise to instabilities and sensitivity to various thresholds (see, for example, Mäler 2000; Arrow *et al.* 2003; Dasgupta and Mäler 2003). Moreover, ecosystems are often vulnerable to environmental events, such as forest fires, disease outbreaks, or invading populations, which are genuinely stochastic in nature. We refer to the occurrence of a sudden system collapse as an ecological event.

When the biodiversity loss process is gradual and can be monitored and controlled by adjusting exploitation rates, and/or when it involves a discrete ecological event whose occurrence conditions are *a priori* known, it is relatively simple to avoid the damage by ensuring that the event will never occur. Often, however, the conditions that trigger ecological events involve uncertainty and the corresponding management problems should be modelled as such. The present chapter characterises optimal resource exploitation policies under risk of occurrence of various types of events.

Impacts of event uncertainty on resource exploitation policies have been studied in a variety of situations, including emission-induced events (Cropper 1976; Clarke and Reed 1994; Tsur and Zemel 1996, 1998b; Aronsson *et al.* 1998; Fisher and Narain 2003), forest fires (Reed 1984; Yin and Newman 1996), species extinction (Reed 1989; Tsur and Zemel 1994), seawater intrusion into coastal aquifers (Tsur and Zemel 1995) and political crises (Long 1975; Tsur and Zemel 1998a). Occurrence risk typically leads to prudence and conservation, but may also invoke the opposite effect, encouraging aggressive exploitation in order to derive maximal benefit prior to occurrence (Clarke and Reed 1994).

Tsur and Zemel (1998b, 2004) trace these apparently conflicting results to different assumptions concerning the event occurrence conditions and the ensuing damage they inflict. An important distinction relates to the type of uncertainty. An event is called endogenous if its occurrence is determined solely by the resource exploitation policy, although the exact threshold level at which the event is triggered is not *a priori* known. This type of uncertainty is due to our partial ignorance of the occurrence conditions. It allows the avoidance of the occurrence risk by keeping the resource stock at or above its current state. Exogenous events, in contrast, are triggered by environmental circumstances that are genuinely stochastic and cannot be fully controlled by exploitation decisions. With this type of event, no exploitation policy is completely safe, although the managers can affect the occurrence hazard by adjusting the stock of the essential resource.

We show that the endogenous-exogenous distinction bears important implications for optimal exploitation policies and alters properties that are considered standard. For example, the optimal stock processes of renewable resources typically approach isolated equilibrium (steady) states. This feature, it turns out, no longer holds under endogenous event uncertainty: the equilibrium point expands into an equilibrium interval whose size depends on the expected event loss, and the eventual steady state is determined by the initial stock. In contrast, exogenous events maintain the structure of isolated equilibria and the effect of event uncertainty is manifest via the shift it induces on these equilibrium states.

In this chapter we avoid detailed exposition and mathematical derivations of optimal policies under uncertainty (these can be found in a number of cited papers, particularly Tsur and Zemel 2001, 2004). Our aim here is to explain the line of reasoning and present the main results characterising optimal exploitation policies under threats of ecological events.

## 2 Ecological setup

We consider the management of some environmental resource that is essential to the survival of an ecosystem (or of a key species thereof) and at the same time is exploited in various production processes. The stock $S$ of the resource can represent the area of uncultivated land of potential agricultural use, the water level at some lake or river or the level of cleanliness (measured, for example, by the pH level of a lake affected by acid rain or by industrial effluents). Without human interference, the stock dynamics is determined by the natural regeneration rate $G(S)$ (corresponding to groundwater recharge, to the decay rate of a pollution stock, or to the natural expansion rate of a forest area). The functional form of $G$ depends on the particular resource under consideration, but we assume the existence of some upper bound $\bar{S}$ for the stock, corresponding to the resource carrying capacity, such that $G(\bar{S}) = 0$ and $G'(\bar{S}) \leq 0$. With $x_t$ representing the rate of resource exploitation, the resource stock evolves with time according to

$$dS_t/dt \equiv \dot{S}_t = G(S_t) - x_t. \qquad (1)$$

Exploitation at a rate $x$ entails several consequences. First, it generates a benefit flow at the rate $Y(x)$ (from the use of land, water or timber or from the economic activities that involve the emission of pollutants), where $Y(x)$ is increasing and strictly concave with $Y(0) = 0$. Second, it bears the exploitation cost $C(S)x$, where the unit cost $C(S)$ is non-increasing and convex. Third, reducing the stock level (by setting $x > G(S)$) entails increasing the damage rate $D(S)$ inflicted upon the ecosystem that depends on the same resource for its livelihood. The damage function is assumed to decrease with $S$ and is normalised at $D(\bar{S}) = 0$. The net benefit flow is then given by $Y(x) - C(S)x - D(S)$.

Moreover, a decrease in the resource stock $S$ increases the probability of occurrence of an influential event of adverse consequences due to the abrupt collapse of the ecosystem it supports. In some cases the event is triggered when $S$ crosses an a priori unknown critical level, which is revealed only when occurrence actually takes place. Alternatively, the event may be triggered at any time by external effects (such as unfavourable weather conditions or the outburst of some disease). Since the resilience of the ecosystem depends on the current resource stock, the occurrence probability also depends on this state. We refer to the former type of uncertainty – that due to our ignorance regarding the conditions that trigger the event – as endogenous uncertainty (signifying that the event occurrence is solely due to the exploitation decisions) and to the latter as exogenous uncertainty. It turns out that the optimal policies are sensitive to the distinction between the two types of uncertainty.

Let $T$ denote the (random) event occurrence time, such that $[0, T]$ and $(T, \infty)$ are the *pre-event* and *post-event* periods, respectively. The benefit flow $Y(x) - C(S)x - D(S)$ defined above is the pre-event instantaneous net benefit. Let $\varphi(S_T)$ denote the post-event value at the occurrence time $T$, consisting of the value generated from the optimal post-event policy (discounted to time $T$) as well as of the immediate consequences of the event occurrence (see examples below).

An exploitation policy $\{x_t, t \geq 0\}$ gives rise to the resource process $\{S_t, t \geq 0\}$ via equation (1) and generates the expected present value

$$E_T\left\{\int_0^T [Y(x_t) - C(S_t)x_t - D(S_t)]e^{-rt}dt + e^{-rT}\varphi(S_T)\,|\,T > 0\right\} \tag{2}$$

where $E_T$ denotes expectation with respect to the distribution of $T$ and $r$ is the time rate of discount. The distribution of $T$ and the ensuing conditional expectation depend on the nature of the event and on the exploitation policy. Given the initial stock $S_0$, we seek the policy that maximises (2). We consider the reference case in which the event occurrence conditions are known with certainty and characterise the optimal policy. Uncertain endogenous and exogenous events are studied in sections 4 and 5, respectively.

## 3 Known events

Suppose that driving the stock to some *known* critical level $S_c$ triggers the collapse of the ecosystem and the loss of the species it supports, which entails a penalty $\psi > 0$ and prohibits any further decrease of the resource stock. The corresponding post-event value is $\varphi(S_c) = W(S_c) - \psi$, where

$$W(S) = [Y(G(S)) - C(S)G(S) - D(S)]/r \tag{3}$$

is the steady state value derived from keeping the extraction rate at the natural regeneration rate $G(S)$. The post-event value $\varphi$ thus accounts both for the fact that the stock cannot be further decreased (to avoid further damage) and for the penalty implied by the loss of biodiversity. Since the event occurs as soon as the stock reaches the critical level $S_c$, the event occurrence time $T$ is defined by the condition $S_T = S_c$ ($T = \infty$ if the stock is always kept above $S_c$).

Since $T$ is subject to choice, the conditional expectation in (2) can be ignored and the management problem becomes

$$V^c(S_0) = Max_{\{T,x_t\}} \left\{ \int_0^T [Y(x_t) - C(S_t)x_t - D(S_t)]e^{-rt}dt + e^{-rT}\varphi(S_T) \right\} \tag{4}$$

subject to (1), $x_t \geq 0$; $S_T = S_c$ and $S_0 > S_c$ given. Optimal processes associated with this 'certainty' problem are indicated with a '$c$' superscript. The event occurrence is evidently undesirable, since just above $S_c$ it is preferable to extract at the regeneration rate and enjoy the benefit flow $rW(S_c)$ associated with it rather than trigger the event and bear the penalty $\psi$. Thus, the event should be avoided, $S_t^c > S_c$ for all $t$ and $T = \infty$. The certainty problem, thus, can be reformulated as

$$V^c(S_0) = Max_{\{x_t\}} \int_0^\infty [Y(x_t) - C(S_t)x_t - D(x_t)]e^{-rt}dt \tag{5}$$

subject to (1), $x_t \geq 0$; $S_t > S_c$ and $S_0$ given. Thus, the effect of the certain event enters only via the lower bound on the stock level. This simple problem is akin to standard resource management problems and can be treated by a variety of optimisation methods (see, for example, Tsur and Graham-Tomasi 1991; Tsur and Zemel 1994, 1995, 2004). Here, we briefly review the main properties of the optimal plan.

We note first that because problem (5) is autonomous (time enters explicitly only through the discount factor), the optimal stock process $S_t^c$ evolves monotonically in time. The property is based on the observation that if the process reaches the same state at two distinct times, then the planner faces the same optimisation problem at both times. This rules out the possibility of a local maximum for the process, because the conflicting decisions to increase the stock (before the maximum) and decrease it (after the maximum) are taken at the same stock levels.

Similar considerations exclude a local minimum. Since $S_t^c$ is monotone and bounded in $[S_c, \bar{S}]$ it must approach a steady state in this interval. Using the variational method of Tsur and Zemel (2001), possible steady states are located by means of a simple function $L(S)$ of the state variable, denoted the evolution function, which measures the deviation of the objective of (5) from $W(S)$ due to small variations from the steady state policy $x = G(S)$ (see below). In particular, an internal state $S \in (S_c, \bar{S})$ can qualify as an optimal steady state only if it is a root of $L$, i.e. $L(S) = 0$, while the corners $S_c$ or $\bar{S}$ can be optimal steady states only if $L(S_c) \leq 0$ or $L(\bar{S}) \geq 0$, respectively.

For the case at hand, we find that the evolution function is given by

$$L(S) = (r - G'(S))\left\{\frac{-C'(S)G(S) - D'(S)}{r - G'(S)} - [Y'(G(S)) - C(S)]\right\} \tag{6}$$

When $Y'(0) < C(\bar{S})$, exploitation is never profitable. In this case $L(\bar{S}) > 0$ and the unexploited stock eventually settles at the carrying capacity level $\bar{S}$. The condition for the corner solution $L(S_c) < 0$ is obtained from (6) in a similar manner.

Suppose that $L(S)$ has a unique root $\hat{S}^c$ in $[S_c, \bar{S}]$ (multiple roots are discussed in Tsur and Zemel 2001). In this case, $\hat{S}^c$ is the unique steady state to which the optimal stock process $S_t^c$ converges monotonically from any initial state.

The vanishing of the evolution function at an internal steady state represents the tradeoffs associated with resource exploitation. Consider a variation on the steady state policy $x = G(\hat{S}^c)$ in which exploitation is increased during a short (infinitesimal) time period $dt$ by a small (infinitesimal) rate $dx$ above $G(\hat{S}^c)$ and retains the regeneration rate thereafter. This policy yields the additional benefit $(Y'(G(\hat{S}^c)) - C(\hat{S}^c))dxdt$, but decreases the stock by $dS = -dxdt$, which in turn increases the damage by $D'(\hat{S}^c)dS$, the unit extraction cost by $C'(\hat{S}^c)dS$ and the extraction cost by $G(\hat{S}^c)C'(\hat{S}^c)dS$. The present value of this permanent flow of added costs is given by $[D'(\hat{S}^c) + G(\hat{S}^c)C'(\hat{S}^c)]dS/(r - G'(\hat{S}^c))$. The effective discount rate equals the market rate $r$ minus the marginal regeneration rate $G'$ because reducing the stock by a marginal unit and investing the proceeds yields the market interest rate $r$ minus the loss in marginal regeneration $G'(S)$ (see, for example, Pindyck 1984). At the root of $L$ these marginal benefit and cost just balance, yielding an optimal equilibrium state.

While the discussion above implies that the stock process must approach $\hat{S}^c$, the time to enter the steady state is a choice variable. Using the conditions for an optimal entry time, one finds that the optimal extraction rate $x_t^c$ smoothly approaches the steady state regeneration rate $G(\hat{S}^c)$ and the approach of $S_t^c$ towards the steady state $\hat{S}^c$ is asymptotic, i.e. the optimal stock process will not reach the steady state at a finite time. These properties, as well as the procedure to obtain the full time trajectory of the optimal plan, are derived in Tsur and Zemel (2004).

When $L(S)$ obtains a root in $[S_c, \bar{S}]$, the constraint $S_t > S_c$ is never binding and the event has no effect on the optimal policy. However, with $S_c > \hat{S}^c$ the function $L(S)$ is negative in the feasible interval $[S_c, \bar{S}]$, hence no internal steady state can be optimal. The only remaining possibility is the critical level $S_c$, because the negative value of $L(S_c)$ does not exclude this corner state. The optimal stock process $S_t^c$, then, converges monotonically and asymptotically to a steady state at $S_c$. By keeping the process above the no-event optimal (i.e. the optimal policy without the constraint $S_T > S_c$), the event threat imposes prudence and a lower rate of extraction.

In this formulation the event is never triggered and the exact value of the penalty is irrelevant (so long as it is positive). This result is due to the requirement that the post-event stock is not allowed to decrease below the critical level. Indeed, this requirement can be relaxed whenever the penalty is sufficiently large to deter triggering the event in any case. The lack of sensitivity of the optimal policy to the details of the catastrophic event is evidently due to the ability to avoid the event occurrence altogether. This may not be feasible (or optimal) when the critical stock level is not *a priori* known. The optimal policy may, in this case, lead to unintentional occurrence, whose exact consequences must be accounted for in advance. In the following two sections we analyse the effect of uncertain catastrophic events on resource management policies.

## 4 Endogenous events

Here the critical level $S_c$ is imperfectly known and the uncertainty regarding the occurrence conditions is entirely due to our ignorance concerning the critical level rather than to the influence of exogenous environmental effects. The post-event value is specified, as above, $\varphi(S) = W(S) - \psi$.

Let $F(S) = Pr\{S_c \leq S\}$ and $f(S) = dF/dS$ denote the probability distribution and density functions of the critical level $S_c$ and denote by $q(S)$ the conditional density of occurrence due to a small stock decrease given that the event has not occurred by the time the state $S$ was reached:

$$q(S) = f(S)/F(S) \tag{7}$$

We assume that $q(S)$ does not vanish in the relevant range, hence no state below the initial stock can be considered *a priori* safe.

The distribution of $S_c$ induces a distribution on the event occurrence time $T$ in a non-trivial way, which depends on the exploitation policy. To see this notice that as the stock process evolves in time, the distributions of $S_c$ and $T$ are modified since at time $t$ it is known that $S_c$ must lie below $\tilde{S}_t = Min_{0 \leq \tau \leq t}\{S_\tau\}$ (otherwise the event would have occurred at some time prior to $t$). Thus, the distributions of $S_c$ and $T$ involve $\tilde{S}_t$, i.e. the entire history up to time $t$, which complicates the evaluation of the conditional expectation in (2). The situation is simplified when the stock process $S_t$ evolves monotonically in time, since then $\tilde{S}_t = S_0$ if the process is non-decreasing (and no information relevant to the distribution of $S_c$ is revealed), or $\tilde{S}_t = S_t$ if the process is non-increasing (and all the relevant information is given by the current stock $S_t$).

It turns out that the *optimal* stock process evolves monotonically in time. This property extends the reasoning of the certainty case above: if the process reaches the same state at two different times and no new information on the critical level has been revealed during that period, then the planner faces the same optimisation problem at both times. This rules out the possibility of a local maximum for the optimal state process, because $\tilde{S}_t$ remains constant around the maximum, yet the conflicting decisions to increase the stock (before the maximum) and decrease it (after the maximum) are taken at the same stock levels. A local minimum can also be ruled out even though the decreasing process modifies $\tilde{S}_t$ and adds information on $S_c$. However, it cannot be optimal to decrease the stock under occurrence risk (prior to reaching the minimum) and then increase it with no occurrence risk (after the minimum) from the same state (see Tsur and Zemel 1994 for a complete proof).

For a non-decreasing stock process it is known in advance that the event will never occur and the uncertainty problem reduces to the certainty problem (5). For non-increasing stock process the distribution of $T$ is obtained from the distribution of $S_c$ as follows:

$$\begin{aligned} 1 - F_T(t) &\equiv \Pr\{T > t | T > 0\} \\ &= \Pr\{S_c < S_t | S_c < S_0\} = F(S_t)/F(S_0). \end{aligned} \tag{8}$$

The corresponding density and hazard-rate functions are also expressed in terms of the distribution of the critical stock:

$$\begin{aligned} &\text{(a) } f_T(t) = dF_T(t)/dt = f(S_t)[x_t - G(S_t)]/F(S_0), \\ &\text{(b) } h(t) = \frac{f_T(t)}{1 - F_T(t)} = q(S_t)[x_t - G(S_t)]. \end{aligned} \tag{9}$$

Let $I(\cdot)$ denote the indicator function that obtains the value 1 when its argument is true and 0 otherwise. For non-increasing state process, the conditional expectation (2) can be expressed as

$$E_T\left\{\int_0^\infty [Y(x_t) - C(S_t)x_t - D(S_t)]I(T > t)e^{-rt}dt + e^{-rT}\varphi(S_T)|T > 0\right\}$$

Notice that $E_T\{I(T > t)|T > 0\} = 1 - F_T(t) = F(S_t)/F(S_0)$ and, using (9), the expectation of the second term gives $\int_0^\infty f_T(t)\varphi(S_t)e^{-rt}dt = \int_0^\infty f(S_t)[x_t - G(S_t)]\frac{\varphi(S_t)}{F(S_0)}e^{-rt}dt$. For non-increasing state processes the management problem becomes

$$V^{aux}(S_0) = \max_{\{x_t\}}\left\{\int_0^\infty \{Y(x_t) - C(S_t)x_t - D(S_t) + q(S_t)[x_t - G(S_t)]\varphi(S_t)\}\frac{F(S_t)}{F(S_0)}e^{-rt}dt\right\} \quad (10)$$

subject to (1), $x_t \geq 0$ and $S_0$ given. This problem is referred to as the *auxiliary* problem and the associated optimal processes are denoted by the superscript *aux*. Since we show below that the auxiliary problem is relevant for the formulation of the uncertain-endogenous-event problem only for stock levels above the root $\hat{S}^c$ of $L(S)$, we complement the constraints of (10) by the requirement $S_t^{aux} \geq \hat{S}^c$.

Formulated as an autonomous problem, the auxiliary problem also gives rise to an optimal stock process that evolves monotonically in time. Notice that at this stage it is not clear whether the uncertainty problem at hand reduces to the certainty problem or to the auxiliary problem, since it is not a priori known whether the optimal stock process decreases with time. We shall return to this question after the optimal auxiliary processes are characterised.

The evolution function corresponding to the auxiliary problem (10) is given by (Tsur and Zemel 2004)

$$L^{aux}(S) = [L(S) + q(S)r\psi]\,F(S)/F(S_0) \quad (11)$$

In (11), $L(S)$ is the evolution function for the certainty problem, defined in (6), and $q(S)$ is defined in (7). The event inflicts an instantaneous penalty $\psi$ (or equivalently, a permanent loss flow at the rate $r\psi$) that could have been avoided by the safe policy of keeping the stock at the level $S$. The second term in the square brackets of (11) gives the expected loss due to an infinitesimal decrease in stock. Moreover, $L^{aux}(\hat{S}^c) > 0$ at the lower bound $\hat{S}^c$ (since $L(\hat{S}^c) = 0$ and $q(\hat{S}^c)r\psi > 0$), implying that $\hat{S}^c$ cannot be an optimal equilibrium for the auxiliary problem.

The eventual steady state depends on the magnitude of the expected loss: for moderate losses, $L^{aux}$ vanishes at some stock level $\hat{S}^{aux}$ in the interval $(\hat{S}^c, \bar{S})$. We assume that the root $\hat{S}^{aux}$ is unique. Higher expected losses ensure that $L^{aux}(S) > 0$ for all $S \in (\hat{S}^c, \bar{S})$, leaving only the corner state $\hat{S}^{aux} = \bar{S}$ as a potential steady state. Thus, the optimal stock process $S_t^{aux}$ converges monotonically to $\hat{S}^{aux}$ from any initial state in $[\hat{S}^c, \bar{S}]$.

In order to characterise the optimal process $S_t^{en}$ under endogenous uncertain events, we compare the trajectories of the auxiliary problem with those obtained with the certainty problem corresponding to $S_c = 0$ (the latter can be referred to as the 'non-event' problem because the event cannot be triggered; see Tsur and Zemel 2004). The following characterisation holds:

(i) When $S_0 < \hat{S}^c$, the optimal certainty stock process $S_t^c$ increases in time. With event risk, it is possible to secure the certainty value by applying the certainty policy, since an endogenous event can occur only when the stock decreases. The introduction of occurrence risk cannot increase the value function, hence $S_t^{en}$ must increase. This implies that the uncertainty and certainty processes coincide ($S_t^{en} = S_t^c$ for all $t$) and increase monotonically towards the steady state $\hat{S}^c$.

(ii) When $S_0 > \hat{S}^{aux} > \hat{S}^c$, both $S_t^c$ and $S_t^{aux}$ decrease in time. If $S_t^{en}$ is increasing, it must coincide with the certainty process $S_t^c$, contradicting the decreasing trend of the latter. A similar argument rules out a steady state policy. Thus, $S_t^{en}$ must decrease, coinciding with the auxiliary process $S_t^{aux}$ and converging with it to the auxiliary steady state $\hat{S}^{aux}$.

(iii) When $\hat{S}^{aux} \geq S_0 \geq \hat{S}^c$, the certainty stock process $S_t^c$ decreases (or remains constant if $S_0 = \hat{S}^c$) and the auxiliary stock process $S_t^{aux}$ increases (or remains constant if $S_0 = \hat{S}^{aux}$). If $S_t^{en}$ increases, it must coincide with $S_t^c$ and if it decreases it must coincide with $S_t^{aux}$, leading to a contradiction in both cases. The only remaining possibility is to follow the steady state policy $S_t^{en} = S_0$ at all $t$.

To sum:

(a) $S_t^{en}$ increases at stock levels below $\hat{S}^c$.

(b) $S_t^{en}$ decreases at stock levels above $\hat{S}^{aux}$.

(c) All stock levels in $[\hat{S}^c, \hat{S}^{aux}]$ are equilibrium states of $S_t^{en}$.

The equilibrium interval is unique to optimal stock processes under uncertain endogenous events. Its boundary points attract any process initiated outside the interval while processes initiated within it must remain constant. This feature is evidently related to the splitting of the intertemporal exploitation problem into two distinct optimisation problems depending on the initial trend of the optimal stock process. At $\hat{S}^{aux}$, the expected loss due to occurrence is so large that entering the interval cannot be optimal even if under certainty extracting above the regeneration rate would yield a higher benefit. Within the equilibrium interval it is possible to eliminate the occurrence risk by not reducing the stock below its current level. As we shall see below, this possibility is not available for uncertain exogenous events and the corresponding management problem does not give rise to equilibrium intervals.

Endogenous uncertain events imply more conservative exploitation as compared with the certainty case. Observe that the steady state $\hat{S}^{aux}$ is a *planned* equilibrium level. In actual realisations, the process may be interrupted by the event at a higher stock level and the *actual* equilibrium level in such cases will be the realised occurrence state $S_c$.

A feature similar to both the certain event and the endogenous uncertain event cases is the smooth transition to the steady states. When the initial stock is outside the equilibrium interval, the condition for an optimal entry time to the steady state implies that extraction converges smoothly to the recharge rate and the planned steady state will not be entered at a finite time. It follows that when the critical level actually lies below $\hat{S}^{aux}$, uncertainty will never be resolved and the planner will never know that the adopted policy of approaching $\hat{S}^{aux}$ is indeed safe. Of course, in the less fortunate case in which the critical level lies above the steady state, the event will occur at finite time with the inflicted damage.

## 5 Exogenous events

Ecological events that are triggered by environmental conditions beyond the planners' control are termed 'exogenous'. Changing the resource stock level can modify the *hazard* of immediate occurrence through the effect of the stock on the resilience of the ecosystem, but the collapse event is triggered by stochastic changes in exogenous conditions. This type of event uncertainty has been applied for the modelling of a variety of resource-related situations, including nuclear waste control (Cropper 1976; Aronsson *et al.* 1998), environmental pollution (Clarke and Reed 1994; Tsur and Zemel 1998b) and groundwater resource management (Tsur and Zemel 2004). Here we consider the implications for biodiversity conservation. Under exogenous event uncertainty, the fact that a certain stock level has been reached in the past without triggering the event does not rule out occurrence at the same stock level some time in the future, as the exogenous conditions may turn out to be less favourable. Therefore, the mechanism that gives rise to the equilibrium interval under endogenous uncertainty does not work here.

As above, the post-event value is denoted by $\varphi(S)$ and the expected present value of an exploitation policy that can be interrupted by an event at time $T$ is given in (2). The probability distribution of

$T$, $F(t) = Pr\{T \le t\}$ is defined in terms of a stock-dependent hazard rate function $h(S)$ satisfying

$$h(S_t) = f(t)/[1 - F(t)] = -d\{\log[1 - F(t)]\}/dt, \tag{12}$$

such that

$$F(t) = 1 - \exp[-\Omega(t)] \quad \text{and} \quad f(t) = h(S_t)\exp[-\Omega(t)], \tag{13}$$

where

$$\Omega(t) = \int_0^t h(S_\tau)d\tau \tag{14}$$

With a state-dependent hazard rate, the quantity $h(S_t)dt$ measures the conditional probability that the event will occur during $(t, t + dt)$ given that it has not occurred by time $t$ when the stock level is $S_t$.

We assume that no stock level is completely safe, hence $h(S)$ does not vanish and $\Omega(t)$ diverges for any feasible stock process as $t \to \infty$. We further assume that $h(S)$ is decreasing, because a shrinking stock deteriorates the ecosystem conditions and increases the hazard for environmental collapse.

Given the distribution of $T$, (2) is evaluated by

$$\begin{aligned} E_T&\left\{\int_0^T [Y(x_t) - C(S_t)x_t - D(S_t)]e^{-rt}dt | T > 0\right\} \\ &= E_T\left\{\int_0^\infty [Y(x_t) - C(S_t)x_t - D(S_t)]e^{-rt}I(T > t)dt | T > 0\right\} \\ &= \int_0^\infty [Y(x_t) - C(S_t)x_t - D(S_t)]e^{-rt}(1 - F(t))dt \end{aligned}$$

and $E_T\{e^{-rT}\varphi(S_T)|T > 0\} = \int_0^\infty e^{-rt}\varphi(S_t)f(t)dt = \int_0^\infty e^{-rt}\varphi(S_t)h(S_t)(1 - F(t))dt$. Using (13), the biodiversity management problem is formulated as

$$\begin{aligned} V^{ex}(S_0) = \max_{\{x_t\}} \int_0^\infty &[Y(x_t) - C(S_t)x_t - D(S_t) \\ &+ h(S_t)\varphi(S_t)]e^{-rt-\Omega(t)}dt \end{aligned} \tag{15}$$

subject to (1), $x_t \ge 0$; $S_t \ge 0$ and $S_0$ given. Unlike the auxiliary problem (10) used above to characterise decreasing policies under endogenous events, problem (15) provides the correct formulation under exogenous events regardless of whether the stock process decreases or increases. We use the superscript '*ex*' to denote optimal variables associated with the exogenous uncertainty problem (15).

To characterise the steady state, we need to specify the value $W^{ex}(S)$ associated with the steady state policy $x^{ex} = G(S)$. Exogenous events may interrupt this policy, hence $W^{ex}(S)$ differs from value $W(S)$ defined in (3) to describe the value obtained from the steady state policy without occurrence risk. Under the steady state policy, (13) reduces to the exponential distribution $F(t) = 1 - \exp[-h(S)t]$, yielding the expected steady state value

$$W^{ex}(S) = W(S) - [W(S) - \varphi(S)]h(S)/[r + h(S)]] \qquad (16)$$

where the second term represents the expected loss over an infinite time horizon. The explicit time dependence of the distribution $F(t)$ of (13) renders formulation (15) of the optimisation problem non-autonomous. Nevertheless, the argument for the monotonic behaviour of the optimal stock process $S_t^{ex}$ holds, and the associated evolution function can be derived (see Tsur and Zemel 1998b), yielding

$$L^{ex}(S) = L(S) + d\{[\varphi(S) - W(S)]rh(S)/[r + h(S)]\}/dS. \qquad (17)$$

When the event corresponds to species extinction, it can occur only once since the loss is irreversible. If a further reduction in stock is forbidden, the post-event value is again specified as $\varphi(S) = W(S) - \psi$ and the second term of (17) simplifies to $-\psi h'(S)r^2/[r + h(S)]^2$. For decreasing hazard functions this term is positive and $L^{ex}(S) > L(S)$. Since $L(S)$ is positive below $\hat{S}^c$, so must $L^{ex}(S)$ be, precluding any steady state at or below $\hat{S}^c$. Thus, the root $\hat{S}^{ex}$ of $L^{ex}(S)$ must lie above the certainty equilibrium $\hat{S}^c$, implying more prudence and conservation compared with the policy free of uncertainty.

Biodiversity conservation considerations enter via the second term of (17) which measures the marginal expected loss due to a small decrease in the resource stock. The latter implies a higher occurrence risk, which in turn calls for a more prudent exploitation policy. Indeed, if the hazard is state-independent ($h'(S) = 0$), the second term of (17) vanishes, implying that the evolution functions associated with the problems with certain events and exogenous uncertain events are the same and the resulting steady states coincide. In this case, exploitation has no effect on the expected loss, hence the tradeoffs that determine the optimal equilibrium need not account for the biodiversity hazard, regardless of how severe it may be. For a decreasing hazard function, however, the degree of prudence (as measured by the difference $\hat{S}^{ex} - \hat{S}^c$) increases with the penalty $\psi$.

The requirement that the stock must not be further reduced following occurrence can be relaxed. For this situation, the post-event value is specified as $\varphi(S) = V^c(S) - \psi$, yielding a more complex expression for the evolution function, but the property $\hat{S}^{ex} > \hat{S}^c$ remains valid (Tsur and Zemel 1998b).

Another interesting situation involving exogenous events arises when the damaged ecology can be restored at the cost $\psi$. For example, the extinct population may not be endemic to the inflicted region and can be renewed by importing individuals from unaffected habitats. When restoration is possible, event occurrence inflicts the penalty but does not affect the hazard of future events. Under the steady state policy, then, one remains at the steady state also after occurrence and receives the post-event value $W^{ex}(S) - \psi$. With the fixed hazard rate $h(S)$, the exponential distribution for recurrent events yields the expected steady state value $W^{ex}(S) = W(S) - [W(S) - W^{ex}(S) + \psi]h(S)/[r + h(S)]$. Solving for $W^{ex}(S)$, we find that $W^{ex}(S) = W(S) - \psi h(S)/r$, reducing (17) to

$$L^{ex}(S) = L(S) - d[\psi h(S)]/dS. \quad (18)$$

When the event penalty $\psi$ depends on the stock, policy implications become more involved. Of particular interest is the case of *increasing* $\psi(S)$ and constant hazard, for which (18) implies more *vigorous* exploitation. An increasing penalty is typical for situations in which the damage is related to the uninterrupted value, which usually increases with the resource stock. This result is similar to the outcome of the 'irreversible' catastrophic events of Clarke and Reed (1994), which also give rise to exploitation policies that are less prudent than their certainty counterparts.

## 6 Concluding comments

Renewable resources are typically considered in the context of their potential contribution to human activities but they also support ecological needs that are often overlooked. This work examines implications of threats of ecological events for the management of renewable resources. The occurrence of an ecological event inflicts a penalty and changes the management regime. Unlike gradual sources of uncertainty (time-varying costs and demand, stochastic regeneration processes, etc.), which allow updating the exploitation policy in response to changing conditions, event uncertainty is resolved only upon occurrence, when policy changes are no longer useful. Thus, the expected loss must be fully accounted for prior to the event occurrence, with significant changes to the optimal exploitation rules.

In this chapter we have distinguished between two types of events that differ in the conditions that trigger their occurrence. An endogenous event occurs when the resource stock crosses an uncertain threshold level, while exogenous events are triggered by coincidental environmental conditions. We find that the optimal exploitation policies are sensitive to the type of the threatening events. Under endogenous uncertain events, the optimal stock process approaches the nearest edge of an equilibrium interval, or remains constant if the initial stock lies inside the equilibrium interval. The eventual equilibrium stock depends on the initial conditions. In contrast, the equilibrium states under exogenous uncertain events are singletons that attract the optimal processes from any initial stock. The shift of these equilibrium states relative to their certainty counterparts is due to the marginal expected loss associated with the events and serves as a measure of how much prudence it implies. In most cases, the presence of event threat encourages conservation, but the opposite behaviour can also be obtained.

A common feature to the types of events considered here is that information accumulated in the course of the process regarding occurrence conditions does not affect the original policy until the time of occurrence (see discussion of decreasing processes under endogenous events). In some situations, however, it is possible to learn during the process and continuously update estimates of the occurrence probability. This possibility introduces another consideration to the tradeoffs that determine optimal exploitation policies. In this case one has to account also for the information content regarding occurrence probability associated with each feasible policy. The investigation of these more complicated models is outside the scope of this chapter.

## REFERENCES


Aronsson, T., Backlund, K. and Löfgren. K. G. 1998. Nuclear power, externalities and non-standard Pigouvian taxes: a dynamical analysis under uncertainty. *Environmental and Resource Economics*. **11**. 177–195.

Arrow, K. J., Dasgupta, P. and Mäler, K.-G. 2003. Evaluating projects and assessing sustainable development in imperfect economies. *Environmental and Resource Economics*. **26**. 647–685.

Bird, C. 1991. Medicines from the forest. *New Scientist*. **17**. 34–39.

Clarke, H. R. and Reed, W. J. 1994. Consumption/pollution tradeoffs in an environment vulnerable to pollution-related catastrophic collapse. *Journal of Economic Dynamics and Control*. **18**. 991–1010.

Colinvaux, P. A. 1989. The past and future Amazon. *Scientific American*. **260**. 102–108.

Cropper, M. L. 1976. Regulating activities with catastrophic environmental effects. *Journal of Environmental Economics and Management*. **3**. 1–15.

Dasgupta, P. and Mäler, K.-G. 2003. The economics of non-convex ecosystems: introduction. *Environmental and Resource Economics*. **26**. 499–525.

Fisher, A. C. and Narain, U. 2003. Global warming, endogenous risk, and irreversibility. *Environmental and Resource Economics*. **25**. 395–416.

Hartwick, J. M. 1992. Deforestation and national accounting, *Environmental and Resource Economics*. **2**. 513–521.

Littell, R. 1992. *Endangered and other Protected Species: Federal Law and Regulations*. Washington, DC: The Bureau of National Affairs, Inc.

Long, N. V. 1975. Resource extraction under the uncertainty about possible nationalization. *Journal of Economic Theory*. **10**. 42–53.

Mäler, K.-G. 2000. Development, ecological resources and their management: A study of complex dynamic systems. *European Economic Review*. **44**. 645–665.

Pindyck, R. S. 1984. Uncertainty in the theory of renewable resource markets. *Review of Economic Studies*. **51**. 289–303.

Reed, W. J. 1984. The effect of the risk of fire on the optimal rotation of a forest. *Journal of Environmental Economics and Management*. **11**. 180–190.

Reed, W. J. 1989. Optimal investment in the protection of a vulnerable biological resource. *Natural Resource Modeling*. **3**. 463–480.

Tsur, Y. and Graham-Tomasi, T. 1991. The buffer value of groundwater with stochastic surface water supplies. *Journal of Environmental Economics and Management*. **21**. 201–224.

Tsur, Y. and Zemel, A. 1994. Endangered species and natural resource exploitation: extinction vs. coexistence. *Natural Resource Modeling*. **8**. 389–413.

Tsur, Y. and Zemel, A. 1995. Uncertainty and irreversibility in groundwater resource management. *Journal of Environmental Economics and Management*. **29**. 149–161.

Tsur, Y. and Zemel, A. 1996. Accounting for global warming risks: resource management under event uncertainty. *Journal of Economic Dynamics and Control*. **20**. 1289–1305.

Tsur, Y. and Zemel, A. 1998a. Trans-boundary water projects and political uncertainty. In R. Just and S. Netanyahu (eds.). *Conflict and Cooperation on Transboundary Water Resources*. Dordrecht, The Netherlands: Kluwer Academic Pub. 279–295.

Tsur, Y. and Zemel, A. 1998b. Pollution control in an uncertain environment. *Journal of Economic Dynamics and Control*. **22**. 967–975.

Tsur, Y. and Zemel A. 2001. The infinite horizon dynamic optimization problem revisited: A simple method to determine equilibrium states. *European Journal of Operational Research*. **131**. 482–490.

Tsur, Y. and Zemel, A. 2004. Endangered aquifers: groundwater management under threats of catastrophic events. *Water Resources Research*. **40**, W06S20 doi:10.1029/2003WR002168.

Yin, R. and Newman, D. H. 1996. The effect of catastrophic risk on forest investment decisions. *Journal of Environmental Economics and Management*. **31**. 186–197.

*Section B*

# Invasives

# 5 Pests, pathogens and poverty: biological invasions and agricultural dependence

*Charles Perrings*

## 1 Introduction

The problem addressed in this paper is the linkage between poverty and invasive alien species (IAS) – the introduction, establishment and spread of species outside of their original range. There are two main dimensions to the problem. One is the connection between poverty and the likelihood of the introduction, establishment or spread of invasive species. It includes the relation between poverty and strategies for the management of invasive species, investment in invasive species detection and control, and collaboration in international control measures. The second is the connection between poverty and the costs or benefits of invasions. This includes the links between invasive species, the structure of the economy, and poverty. It covers the relation between poverty and dependence on agriculture, wildlife utilisation, forestry and fisheries, and the importance of common property.

These two dimensions have been addressed in three generally distinct literatures. One is the literature on the costs of biological invasions. It is closely associated with the work of David Pimentel and colleagues, and comprises estimates of the more direct costs of invasive pests and pathogens in selected countries, including at least some developing countries (South Africa, India and Brazil). It also includes a longer-standing literature on the costs of various animal and plant pests and pathogens in agriculture, forestry and – to a lesser extent – fisheries. A second is on economics of invasive species. The research undertaken as part of Global Invasive Species Program (GISP) I was the first inquiry into this problem (Perrings *et al.* 2000). Since then a new literature on the economics of biological invasions has developed which looks at the efficient management of invasive species. As yet, this literature has not considered equity issues or the link between biological invasions and poverty, but it does address the factors that influence the probability of the introduction and spread of invasive species and the effectiveness of control. These can be related to poverty. A third literature considers the link between other kinds of environmental change and poverty, and includes both theoretical and empirical studies. It has tended to focus on particular areas of environmental change – especially pollution (air and water), habitat conversion (deforestation), water issues (water quality and water supply) and disease. However, this too can be used to say much about the link between the environmental changes associated with invasive species and poverty.

The chapter begins with the last of these – the general evidence for an empirical relation between poverty and environmental change, and between poverty and the primary source of IAS worldwide – the growth of trade, transport and travel. There are, by now, a number of surveys of the economics of biological invasions. Lovell and Stone (2005) have reviewed the literature on the economics of aquatic invasive species, while Evans (2003), Eisworth and Johnson (2002) have considered the literature on terrestrial systems – the latter in the context of a paper developing a general model for the management of invasive species. Stutzman *et al.* (2004) offer an annotated bibliography of economics of invasive plants.

There are three major points at issue in the economics of invasive species. The first is that the introduction, establishment and spread of potentially harmful alien species constitute an externality of international markets (international trade). In the absence of complete markets, the risk of biological invasions increases with the growth of trade. The second is that the control of invasive species is a public good at several different levels – national, regional and global. The provision of the public good requires the development of institutions that operate at the appropriate level and that can solve the free-rider problem at that level. This involves application of the subsidiarity principle to the development of governance mechanisms and international agreements. The third point at issue is the appropriate specification of the management problem and the evaluation of control options (where control subsumes interception, quarantine, eradication, containment and other management options).

The economics of the problem involves the identification of the source of the externality, estimation of its consequences for the welfare of people affected, and the development of mechanisms to ensure that resources committed to detection and control are commensurate. The methodological question is the following: given the set of prices, regulations, property rights and institutional conditions, how should the management problem be formulated and solved? It involves the identification and management of the risks and uncertainties associated with the introduction of novel species. It also involves the treatment of irreversible changes. When is it optimal to mitigate the risks of invasions (to take action that reduces the probability of invasions occurring) and when is it optimal to adapt (to take action that reduces the costs of invasions without affecting the probability that they will occur). Evans (2003) argues that economics has two major contributions to make to research on IAS. The first is to provide estimates of the impacts of invasions in order to improve both cost-effectiveness and efficiency of publicly funded IAS control programmes. The second is to develop economic sanitary and phytosanitary measures. This chapter addresses both questions.

The chapter is organised in six sections. Section 2 considers the general relationship between invasive species and poverty. Section 3 then evaluates the relation between trade and invasive species. That is followed by three sections on the evaluation of damage costs of IAS, the development of economic instruments to internalise IAS externalities, and expenditures on IAS control as a public good in poor economies and poor regions. A final section considers the decision tools available to inform mitigation and adaptation strategies, and relates these to the problem of uncertainty.

## 2 Poverty and environmental change

The linkages between poverty and environmental change have been widely studied, but it would be wrong to say that they are well understood. The Brundtland Report (WCED 1987) argued that there existed a causal connection between environmental change and poverty both within and between generations. A large literature has subsequently examined the empirical relation between per capita income (GDP or GNP) and environmental change. The Environmental Kuznets Curve literature stemmed from Grossman and Krueger's (1995) assessment of the environmental implications of Mexico's inclusion in the North American Free Trade Area, which showed that certain indicators of environmental quality first deteriorate and then improve as per capita incomes rise.

Within that literature the relation between per capita income and various other indicators of environmental change has subsequently been studied, using a range of databases and econometric approaches (see Stern 1998, 2004 for reviews of this literature). An inverted 'U'-shaped curve was found for the relation between per capita income and various atmospheric pollutants using both cross-sectional and panel data (Shafik 1994; Seldon and Song 1994; Cole *et al.* 1997; Stern and Common 2001), but the relation is by no means consistent. For some measures of environmental quality the relation with per capita income has been found to be monotonically increasing (e.g. carbon dioxide or municipal waste) or decreasing (e.g. faecal coliform in drinking water). For others it has been found to have more than one turning point. Moreover, even where the best fit is given by a quadratic function – the inverted 'U' – there are wide discrepancies in estimations of the turning point. This is the level of per capita income at which the particular measure of environmental quality starts to improve as per capita incomes rise. While some have chosen to interpret the Environmental Kuznets Curve as evidence that economic growth will, in some sense, take care of the environment, the consensus view is that there are no general rules to be drawn (Ekbom and Bojö 1999; Markandya 2000, 2001). The relation between changes in income and changes in the environment is complex, involving feedback effects in both directions.

Markandya's (2001) review of the literature on the relation between poverty, environmental change and sustainable development suggested that to the question 'does poverty damage the environment?', the answer was broadly 'no'. To the question 'does environmental degradation hurt the poor?', the answer was broadly 'yes'. Hence he concluded that while poverty alleviation would not necessarily enhance environmental quality and may in fact increase stress on the environment, environmental protection would generally benefit the poor. Of course, there are many caveats to this conclusion. Cutting the poor off from access to environmental resources by the establishment of protected areas without paying compensation is unlikely to improve their well-being.

The ambiguous nature of the statistical results on the linkages between poverty and the environment is reflected in the various case studies of environmental resource use in poor countries. For reasons that are well understood, the scarcity of commodities that satisfy basic needs such as water and fuelwood affects the poor more than the rich (Kumar and Hotchkiss 1988). So it is not at all surprising that environmental change which reduces the supply of basic goods held in common property should impact the poor. Where the case studies are less consistent is in the analysis of the relation between poverty, population growth, environmental change and institutions.

There are numerous studies of the effect of population growth – whether due to migration or fertility – on deforestation. Lopez and Scoseria (1996) found that in-migration to Belize from other Central American countries accounted for around a third of deforestation in that country. Population growth has similarly been implicated in environmental change in many other cases (De Janvry and Garcia 1988; Cleaver and Schreiber 1994; Lopez 1992; Lopez and Scoseria 1996). Sub-Saharan Africa has, however, provided some well-known counter-examples, where productivity increases that have accompanied population growth have more than compensated for any reduction in environmental resources (Pingali *et al.* 1987; Tiffen *et al.* 1994). Heath and Binswanger (1996), using the cases of Kenya and Ethiopia, argued that whether or not population growth had adverse effects on the environment depended on institutional conditions. There is also some evidence that the linkage between poverty, demography and environmental change is influenced by changes in household composition. Linde-Rahr's (2002) study of afforestation in Vietnam, for example, showed that in households with larger numbers of female members, tree planting was positively correlated with income, but that in households with larger numbers of male members the opposite was true.

What the differences between the various case studies have shown is that it is the determinants of household decisions on the use of environmental resources that matter. On the links between population growth and environmental change, Dasgupta's (1993, 2001) investigation of the connection between poverty, fertility decisions and environmental change concluded that both fertility decisions and the use made of environmental resources are strongly influenced by households' long-term security of income. Where poverty includes low expectations of secure future income, household responses include high fertility rates leading to increased pressure on the environment. This is especially marked where access to environmental resources is unregulated. Since this in turn increases uncertainty about future income, there is a positive feedback between poverty, fertility decisions and environmental degradation.

Another strand of the literature has addressed the link between poverty and the rate at which households discount the future (e.g. Perrings 1989; Chavas 2004). Building on the long-held observation that discount rates are not independent of income (Fisher 1930), these studies treat discount rates as endogenous. They find that if poverty causes people to ignore the longer-term consequences of their decisions, it also affects investment in conservation and environmental enhancement. Chavas's important (2004) paper shows that if the discount rate is endogenous and decreases in income, then in contrast to Markandya's view, poverty can contribute to environmental degradation. This is certainly consistent with at least some empirical findings on the topic (e.g. Pender 1996; Holden *et al.* 1998), although, as Markandya points out, the evidence remains mixed.

The linkage between poverty and growth has also been examined at a macro-economic level, where the evidence in the 1990s showed that declining public expenditures and a worsening distribution of income affected the ability of the poor to invest. In many cases, the rural poor were unable to respond to changing incentives while reductions in extension services and marketing support have further depressed rural incomes, particularly affecting rural women (Birdsall and Londoño 1997; Reed

Table 5.1 *Changes in inclusive wealth in China, India and sub-Saharan Africa, 1965–1996*

| | %ΔN | %ΔY/N | %ΔHDI | (dV/dt)/Y | %ΔV/N |
|---|---|---|---|---|---|
| | 1 | 2 | 3 | 4 | 5 |
| China | 1.7 | 6.7 | −0.2 | 0.100 | 0.8 |
| India | 2.1 | 2.3 | 2.2 | 0.08 | −0.1 |
| Sub-Saharan Africa | 1.7 | −0.2 | 0.9 | −0.028 | −3.4 |

Column 1: Average annual percentage rate of population growth, 1965–1996
Column 2: Average annual percentage rate change in per capita GNP, 1965–1996
Column 3: Average annual percentage rate of change in HDI, 1987–1997
Column 4: Genuine investment as a proportion of GDP, 1970–1993
Column 5: Average annual percentage rate of per capita wealth, 1970–1993
Source: Adapted from Dasgupta (2001)

1996). What made these findings disturbing was that many indicators of economic performance, including measures of trade growth, were moving in the opposite direction.

There has been substantial growth in capital flows and foreign direct investment (FDI) in the last decade in all income groups and all regions. However, both capital flows and FDI are much lower in areas where poverty is most persistent. This has implications for the resources committed to maintaining ecosystem services and the environmental assets from which such services are derived. The best current measure of this is the World Bank's adjusted net savings rate, which modifies conventional measures of net national savings by including changes not only in produced capital but also in human and natural or environmental capital. Adjusted net savings were originally defined as genuine savings (Hamilton and Clemens 1999; Hamilton 2000), or genuine investment (Dasgupta 2001; Arrow *et al.* 2003). It is a measure of the change in a country's wealth. Estimates of adjusted net savings are generally lower than other savings measures, reflecting the depreciation or degradation of environmental assets. Moreover, once population growth is taken into account, many regions of the world experienced negative changes in wealth per head during the last three decades of the twentieth century (Table 5.1).

Even regions that recorded strongly positive growth in conventional measures of economic performance, like India, recorded declining per capita inclusive wealth. In some regions the fall in the value of per capita

Table 5.2 *Depletion of natural capital, 2003*

| | Gross national savings % GNI | Net national savings % GNI | Adjusted net savings % GNI |
|---|---|---|---|
| World | 20.8 | 8.2 | 9.4 |
| Low income | 23.1 | 14.2 | 8.7 |
| Middle income | 27.9 | 17.8 | 10.1 |
| Low & middle income | 27.2 | 17.3 | 10 |
| East Asia & Pacific | 41.8 | 32.6 | 28.1 |
| Europe & Central Asia | 21.9 | 11.2 | 1.5 |
| Latin America & Caribbean | 19.5 | 9.2 | 5.3 |
| Middle East & N. Africa | 31.2 | 21.3 | −6.2 |
| South Asia | 24.9 | 15.9 | 13.8 |
| Sub-Saharan Africa | 16.9 | 6.3 | 0.7 |
| High income | 19.3 | 6.1 | 9.3 |
| Europe EMU | 21.3 | 7.5 | 11.6 |

Source: World Bank 2005. Global Economic Prospects, World Bank, Washington, DC

wealth was substantial. In sub-Saharan Africa, for example, annual per capita changes in wealth averaged −3.4 per cent between 1965 and 1996. In other words, Africans lost almost half of their wealth in that period (Dasgupta 2001).

Since the turn of the century the position has improved for many regions. In 2003 adjusted net savings were positive for all regions other than the Middle East (Table 5.2). However, they were close to zero in sub-Saharan Africa as a whole and were strongly negative in a number of countries.[1] In per capita terms, therefore, Africans were still getting poorer once changes in environmental stocks were taken into account.

Case studies of changes in inclusive wealth in particular countries have identified the policies and investment strategies that explain changes in national wealth. For example, Lange (2004) cites the contrasting cases of Botswana and Namibia. After independence in 1996 Botswana chose to reinvest the rents from the mining sector in building its capital stock. Namibia did not. The result is that whereas Botswana tripled per capita wealth in the last three decades of the twentieth century, Namibia's per capita wealth declined. In the 1980s Namibia's per capita wealth was 75 per cent greater than Botswana's; by the end of the 1990s it was only 33 per cent of Botswana's.

[1] In the Middle East adjusted net savings reflect the depletion of oil stocks. In sub-Saharan Africa, the worst performing countries are also oil-producing states (Nigeria −31.4; Angola −28.5; Congo −26.3) that are not reinvesting oil rents.

Changes in inclusive wealth are reflected in both recorded and projected poverty levels. Using the number of people living on less than either $1 or $2 per day as the criterion, the number of people in poverty increased in a number of developing regions in the last decade of the twentieth century, but current projections are that poverty will fall in all regions except sub-Saharan Africa in the next ten years (World Bank 2005).

The persistence of poverty in regions such as sub-Saharan Africa is also reflected in changes in the rural population. The linkage between poverty and rural activities has been well documented (Jazairy *et al.* 1992). Although the proportion of the population in rural areas has declined in every region due to the continuing movement of people from rural to urban areas, rural population growth remains positive in many low-income regions. It is highest in South Asia and sub-Saharan Africa, in both of which agricultural and forest-based employment account for a higher proportion of the labour force than elsewhere.

What does this mean for the linkages between poverty and biological invasions? There are three important points to make, each of which is considered later. First, if the resources committed to border inspection are positively correlated with GDP, the growth of trade increases the invasion risks of poor countries and poor regions disproportionately. That is, the risks of undetected species introductions will be higher. Second, if investment in the conservation of ecosystem services and the control of invasive species are also positively correlated with GDP, poor countries and poor regions may also be more invasible than rich countries and regions. That is, the risk that introduced species will be able to establish and spread will be higher in poor regions. Third, since invasive pests and pathogens primarily affect agriculture, forestry and fisheries, the greater dependence of poor producers on primary production makes them more vulnerable to the effects of biological invasions. In other words, the cost of invasive species will tend to impact more people in poor, resource-dependent economies than in rich economies, and will more directly affect their livelihoods.

## 3 Trade and invasive species

From an ecological perspective, any species introduced to an ecosystem beyond its 'home' range that establishes, naturalises and spreads is said to be invasive (Williamson 1996). From a policy perspective, however, the focus is generally on species whose home range lies beyond the national jurisdiction. In other words, the alien species that attract attention are those that are introduced as a consequence of international trade, transport or travel. This includes both species that are deliberately introduced as domesticated plants or animals and those that are introduced as an unintended by-product of the import of other goods and services – the so-called 'hitchhiker' species. The Office of Technology Assessment (OTA) (1993) estimated that four out of five invasive terrestrial weeds in the USA that had appeared during the twentieth century were introduced as by-products of the commodity trade. Although data are lacking on aquatic species, the proportion of invasive aquatic species that have been introduced by shipping is likely to be much higher. Many of the most famous examples of damaging species introductions, e.g. the zebra mussel (*Dreissena polymorpha*) and the Asian clam (*Corbicula fluminea*), are associated with ballast water exchange in ships. Their appearance is evidence of the failure of both international and domestic markets (Perrings *et al.* 2002; Margolis *et al.* 2005).

The precise relation between the growth in trade and invasive species is still unknown, reflecting the paucity of time series on species introductions to match the available time series on trade. Dalmazzone (2000) showed that economies that are more open tend to be more vulnerable to invasions. Small island states in particular are often geared to the production of primary products for export and are more dependent on imports than continental countries.[2] Since they are also ecologically more vulnerable to invasions than continental ecosystems, it follows that trade is not the only explanation for the success of introduced species in such economies. More recently, Levine and D'Antonio (2003) have considered merchandise trade as a predictor of invasions focusing on insects, plant pathogens and molluscs. They have also used the resulting model to predict increases in invasions over the next two decades. They conclude that trade-induced invasions will increase by between 3 and 61 per cent, depending on the model and the species.

What is beyond dispute is that species introductions increase with the volume of trade and that the frequency with which a species is introduced is positively correlated with the probability that it will establish (Enserink 1999). This means that the growth of trade, other things being equal, will increase the risk both of new introductions and of the establishment of introduced species (Lockwood *et al.* 2005).

Of course other things are not equal. What matters more to the ability to predict invasion risks than simply the volume of trade are the bioclimatic similarities between the ecosystems being connected, the nature of

[2] For example, the average percentage of merchandise imports as a share of the GDP, in the sample considered in Dalmazzone (2000), is 43 per cent for island countries, against an average 32 per cent for the whole sample and 26.8 per cent for continental countries.

the pathways (e.g. the time introduced species are in transit and their conditions during transit), the nature of the species themselves (e.g. traits, such as high plasticity, that make species invasive) and the invasibility of the ecosystems into which species are being introduced (e.g. the effects of fragmentation and biodiversity loss). However, given these conditions, an increase in propagule pressure due to an increase in the trade of goods will increase the risk of biological invasions. Furthermore, an increase in trade is expected to lead to habitat loss through conversion of land for agriculture, forestry and industry, with negative implications for biodiversity and the invisibility of ecosystems (Polasky *et al.* 2004). It follows that to understand the implications for biological invasions of changes in trade it is necessary to understand how the pattern of trade is changing as the volume of trade grows.

For the countries and regions where the world's poor live the trade that matters is trade in the products of agriculture, forestry and fisheries. Recent analyses of changes in the pattern of world commodity trade have pointed up a number of important features of agricultural trade. The first is that while the share of agriculture in global trade has been falling, it remains especially important to people in poverty precisely because poor people tend to live in rural areas and to derive their income from agriculture. The ratio of farm to non-farm income ranges from 40–80 per cent in low-income countries, but is around only 1 per cent in high-income countries. The fact that the decline in agricultural prices affects developing countries more than developed countries – world raw commodity prices declined by 6.6 per cent in the period 1990–2000, but developing country raw commodity prices declined by 15.2 per cent – also has implications for the rural poor (Aksoy 2005).

Aside from trade, transfers are also especially important in poorer regions. For example, grey leaf spot was first reported in South Africa in 1988. It has subsequently spread northwards into all the main maize-growing areas of Africa and its effect on yields has been such that it is now argued to pose a serious threat to food security (Rangi 2004). It was thought to have been introduced to the continent in US food aid shipments of maize during the drought years of the 1980s (Ward *et al.* 1999). Another example is parthenium weed from Mexico. This was first detected in Ethiopia in 1988 near food-aid distribution centres, implying that it had accompanied wheat grain distributed as food aid during the drought (GISP 2004). Since lower sanitary and phytosanitary standards apply to food aid, particularly emergency food aid, it may not be so surprising that the introduction and spread of potentially invasive species would follow the distribution of emergency relief.

A second important feature of agricultural trade in developing countries is that although it continued to grow at around 3.4 per cent in the 1990s, almost all the growth was accounted for by trade with other developing countries. More than 50 per cent of food imports in developing countries derives from other developing countries (Aksoy 2005). The World Bank reports that a major trend in the trading system involves the proliferation of bilateral and regional trade agreements (RTAs) and especially the proliferation of South–South RTAs (World Bank 2005). The number of RTAs has increased fourfold since 1990 and at the time of writing stands at over 230. Indeed, RTAs now account for nearly 40 per cent of world trade.

The development of RTAs is relevant to the problem of invasive species for three different reasons. The first is that many cover much broader issues than trade alone. An increasing number of RTAs address environmental issues. This is partly due to the limited scope for addressing environmental concerns in the General Agreement on Tariffs and Trade (GATT), but it is also in recognition of the fact that specific trade links involve specific environmental risks. The GATT does allow for actions in restraint of trade where human animal or plant life and health are threatened by trade. The Sanitary and Phytosanitary Agreement (SPA) provides the rules under which countries can do this, but allows individual countries some latitude. It encourages adoption of the standards set by the Codex Alimentarius Commission for food safety, the International Office of Epizootics for animal health, and the International Plant Protection Convention for plant health, but allows countries to choose their own level of protection (Jaffee and Henson 2005).

There is little doubt that this has been and continues to be used as a trade protection device. Large numbers of countries are ineligible to supply certain markets with a range of animal products and food crops because of restrictions based on threats to plant and animal health (Sumner 2003). A review of the complaints lodged by developing countries over the use of the SPA reveals a persistent set of concerns, including the overly restrictive and non-scientifically based measures by high-income countries for dealing with foot and mouth disease and bovine spongiform encephalopathy, and plant pests and pathogens, especially in the horticultural sector (Jaffee and Henson 2005). Nor is the SPA the only instrument used to restrict trade. Bacterial wilt (*Ralstonia solanacearum*) is listed in US law as a potential biological weapon. When it was found on a shipment of pelargonium cuttings, for example, it resulted in quarantine restrictions that have severely affected the horticultural trade in Kenya (Rangi, 2004).

A second reason why South–South RTAs are relevant to the problem of invasive species is precisely because they open up new trading opportunities between developing countries. What makes this interesting from the perspective of IAS is that the development of South–South trade brings about closer linkages between ecosystems in which bioclimatic conditions are broadly similar and therefore in which the risk that introduced species will establish, naturalise and spread is high. This aspect of the problem has not yet been investigated in the literature, but it is potentially an important risk factor.[3] There is evidence from NAFTA that the agreement has facilitated the spread of species within the NAFTA area that were introduced to a NAFTA country from some other country (Perrault *et al.* 2003). The promotion of agricultural trade between bioclimatically matching regions in which resources for the detection and control of potentially invasive species are weak must be a concern.

A third reason is that cooperation within RTAs may be an important part of the solution to biological invasion externalities and the free-rider problems attaching to the control of non-indigenous species. Schiff and Winters (2003) argue that if there are economies of scale or transboundary externalities, there is relatively little scope for market solutions to environmental problems and regional cooperation can provide the answer. A number of RTAs include environmental agreements. In many cases, these are designed to force compliance with environmental laws. So, for example, NAFTA has a Commission for Environmental Cooperation. Its role is to ensure that member states do not seek a trade benefit or attract inward investment by failing to comply with environmental laws. The US–Singapore Free Trade Agreement includes an environmental chapter requiring that both countries effectively enforce their environmental laws and including fines for non-compliance (World Bank 2005).

The same thing exists in developing country RTAs. The Southern Common Market (MERCOSUR), for example, includes an environmental working group charged with eliminating the use of environmental barriers to trade, promoting 'upward harmonisation' of environmental management systems and securing cooperation on shared ecosystems. Indeed, many of the main South–South RTAs – MERCOSUR, the Andean Pact, the Common Market for Eastern and Southern Africa

[3] The ratio between interception shares and import shares in any country gives a simple guide to the relative introduction risks attaching to different exporters. There are no data on this for developing countries, but a review of interception and trade data for the UK between 1996 and 2004 indicates the following ratios: Europe 0.85; Asia 4.25; Africa 0.91; North America 1.11; South America 1.29; Oceania 0.33. The riskiest source of imports was Asia, accounting for 17 per cent of all interceptions, but only 4 per cent of trade. The least risky was Oceania, accounting for 9 per cent of trade but only 3 per cent of interceptions.

(COMESA), the Southern African Development Community (SADC), the ASEAN Free Trade Area (AFTA) and the Caribbean Community (CARICOM) – include agreements on standards (World Bank 2005).

The regional scale is the appropriate level at which to manage environmental resources wherever the ecosystems affected are regional in extent. In marine systems, for example, the conservation of straddling or migratory stocks requires cooperation across the sea areas within which those stocks move. The conservation of such stocks is a regional public good and subsidiarity indicates that the right level of governance is the regional level. Similarly, the control of the introduction of potentially invasive species within a trading group should be regulated at the level of that group. Not only does this make it possible to ensure that the resources committed to control are commensurate with the collective benefits it offers, it also minimises transaction costs by reducing the number of participants to those with a real stake in the public good and builds trust by allowing repeated interaction between members over time (Sandler 2005).

## 4 Estimates of the damage costs of invasive species

The first estimate of the costs of invasive species by the Office of Technology Assessment of the US Congress (1993) considered the ecological and economic effects of harmful invasive species within the USA. It concluded that in the period from 1906, 59 per cent of all species introduced to the USA had caused economic or ecological damage and that the seventy-nine most harmful had caused damage of $97 billion over that period. Since then a number of papers by Pimentel and colleagues (Pimentel *et al.* 2000, 2001, 2005) have sought to update the OTA estimates and to extend them beyond the USA. The second of the Pimentel papers included estimates for three developed and three developing countries – South Africa, India and Brazil. To date this remains the most comprehensive summary of the control costs and lost output associated with invasive species in agriculture, forestry and fisheries in 'poor' countries.

The findings of Pimentel *et al.* (2001) are summarised in Tables 5.3 and 5.4. They represent a simple sum of various dollar estimates of annual damage costs in the countries concerned made over the preceding decade. Because of the way in which they were acquired, the numbers cannot be taken as a good approximation of net costs of species introductions in any of the countries concerned. There are no estimates of any benefits that may have accrued from the activities that led to the introduction of invasive species. The estimates of damage costs in the background literature are not made in any coherent way and are extremely patchy. The

Table 5.3 *Economic losses to introduced pests in crops, pastures and forests in the United States, United Kingdom, Australia, South Africa, India and Brazil (billion dollars per year)*

| Introduced pest | United States | United Kingdom | Australia | South Africa | India | Brazil | Total |
|---|---|---|---|---|---|---|---|
| Weeds | | | | | | | |
| Crops | 27.9 | 1.4 | 1.8 | 1.5 | 37.8 | 17.0[a] | 87.4 |
| Pastures | 6.0 | – | 0.6 | – | 0.92 | – | 7.52 |
| Vertebrates | | | | | | | |
| Crops | 1.0[b] | 1.2[c] | 0.2 | – | – | – | 2.4 |
| Arthropods | | | | | | | |
| Crops | 15.9 | 0.96 | 0.94 | 1.0 | 16.8 | 8.5 | 44.1 |
| Forests | 2.1 | – | – | – | – | – | 2.1 |
| Plant pathogens | | | | | | | |
| Crops | 23.5 | 2.0 | 2.7 | 1.8 | 35.5 | 17.1 | 82.6 |
| Forests | 2.1 | – | – | – | – | – | 2.1 |
| Total | 78.5 | 5.56 | 6.24 | 4.3 | 91.02 | 42.6 | 228.22 |

[a] Pasture losses included in crop losses
[b] Losses due to English starlings and English sparrows (Pimentel *et al.* 2000)
[c] Calculated damage losses from the European rabbit
Source: Pimentel *et al.* (2001)

Table 5.4 *Environmental losses to introduced pests in the United States, United Kingdom, Australia, South Africa, India and Brazil (billion dollars per year)*

| Introduced pest | United States | United Kingdom | Australia | South Africa | India | Brazil | Total |
|---|---|---|---|---|---|---|---|
| Plants | 0.148 | – | – | 0.095 | – | – | 0.243 |
| Mammals | | | | | | | |
| Rats | 19.000 | 4.100 | 1.200 | 2.700 | 25.000 | 4.400 | 56.400 |
| Other | 18.106 | 1.200 | 4.655 | – | – | – | 23.961 |
| Birds | 1.100 | 0.270 | – | – | – | – | 1.370 |
| Reptiles/Amph. | 0.006 | – | – | – | – | – | 0.006 |
| Fishes | 1.000 | – | – | – | – | – | 1.000 |
| Arthropods | 2.137 | – | 0.228 | – | – | – | 2.365 |
| Molluscs | 1.305 | – | – | – | – | – | 1.305 |
| Livestock diseases | 9.000 | – | 0.249 | 0.100 | – | – | 9.349 |
| Human diseases | 6.500 | 1.000 | 0.534 | 0.118 | – | 2.333 | 10.485 |
| Total | 58.302 | 6.570 | 6.866 | 3.013 | 25.000 | 6.733 | 106.484 |

Source: Pimentel *et al.* (2001)

findings are also inconsistent with the ecological literature in important respects – such as in the estimate of the proportion of introduced species that are 'harmful'. Nonetheless, it is interesting to consider the relative severity of the estimates in rich and poor countries.

Taking agricultural GDP in 1999 as the numeraire, the estimates reported in Table 5.3 indicate that invasive species caused damage costs equal to 53 per cent of agricultural GDP in the USA, 31 per cent in the UK and 48 per cent in Australia. By contrast, damage costs in South Africa, India and Brazil were, respectively, 96 per cent, 78 per cent and 112 per cent of agricultural GDP. Of course, there is considerable uncertainty about the Pimentel estimates given the ad hoc estimation methods. Public expenditure on invasive species control is not known in most countries, but where there are data it turns out to be very small relative to the Pimentel estimates. In the USA, for example, federal expenditure on invasive species in 1999 was less than $0.5 billion, i.e. 0.5 per cent of the estimated damage costs in agriculture. While there is almost certainly insufficient expenditure to counter the impact of invasive species, this also raises questions about the damage estimates themselves.

Nevertheless, if the relative values were of the right order of magnitude, the impact of invasive species on agriculture is significantly greater in developing than in developed countries. Furthermore, since agriculture accounts for a higher share of GDP in developing countries, the impact of invasive species on overall economic performance is proportionately even greater in developing countries. In India, for example, Pimentel's estimates imply that annual invasive species control and damage costs were 20 per cent of GDP in 1999, compared with less than 1 per cent in the USA. Pimentel *et al.*'s 2005 update of the US estimates added an estimate of the cost of weeds in lawns (without attempting to isolate non-indigenous from indigenous weeds, or to separate weed control and fertilisation), but otherwise reports similar figures (Pimentel *et al.* 2005).

There are a large number of case studies of the effects of particular invasive species, many of which focus on the USA (for a summary see Stutzman *et al.* 2004). Examples of invasive plants in the USA for which there exist cost estimates are leafy spurge (Bangsund *et al.* 1999), tansy ragwort (Coombs *et al.* 1996), yellow starthistle (Jetter *et al.* 2003) and tamarisk (Zavaleta 2000). A number of case studies of aquatic species have also been carried out, of which the impact of the zebra mussel, *Dreissena polymorpha*, on power stations is the best known (O'Neill 1997), but others include the effect of the green crab, *Carcinus maenas*, on the North Pacific Ocean fisheries (Cohen *et al.* 1995). Internationally, there have also been assessments of the role of the comb jelly, *Mnemiopsis leidii*, in changing the cost of fishing effort in the Black Sea (Knowler and Barbier 2000; Knowler 2005).

There are few case studies of individual invasive species in developing countries. Human diseases aside, invasive species that have the most direct effects on the livelihoods of the poor are those that impact agriculture, forestry and fisheries. The dominant crops grown in the poorer regions of the world are rice, maize, cassava, sorghum and millet. All are affected by invasive species – whether pests or pathogens. The range of effects includes the following:

- interference with crop growth through competition for light, water and nutrients
- allelopathy, or the production of toxins that inhibit the growth of other plants
- contamination of harvested crops
- provision of vectors for pests, pathogens, nemotodes and insects
- interference with harvesting
- requirement for additional cleaning and processing.

All of these have direct economic implications. Some increase the cost of production. Others reduce the value of harvested crops or result in their exclusion from international markets. The position is very similar with respect to animal pests and pathogens, foot and mouth disease being a good example.

Examples of pests and pathogens that have had particularly severe effects on crop yields in the world's poorest region, sub-Saharan Africa, include witchweed (*Striga hermonthica*), grey leaf spot (*Circosporda zeaemaydis*), the large grain borer (*Prostephanus truncatus*), cassava mealybug (*Phenacoccus manihoti*) and the cassava green mite (*Mononychellus tanajoa*) (Rangi 2004). Some of these species have been present for many decades, others are new arrivals. The larger grain borer was apparently introduced from south and central America during the 1970s. It was first detected in Tanzania in the late 1970s and is now established in east, central, south and west Africa. It primarily affects grain in storage, causing losses of up to 30 per cent within six months. Farrell and Schulten (2002) estimated that the income forgone as a result was in the order of $90 million for Tanzania alone.

The emergence of new agricultural pests has spurred the development of both new pesticides and alternative control measures, including biological control agents. For example, the cassava mealybug has been targeted by the parasitic wasp (*Epidinocarsis lopezi*), the cassava green mite by the mite (*Typhodromalus aripo*) and the large grain borer by the beetle (*Teretrisoma negrescens*) (Rangi 2004). Such biocontrol agents are themselves introduced species, with potential ecological consequences in addition to the control they exercise over the invasive pest.

There are far fewer studies of the impacts of invasive species on particular systems. Exceptions include the African Lakes and the South African fynbos. Kasulo (2000) analysed the ecological and socio-economic impact of invasive species in African lakes, focusing on introduced fish species and water weeds – the Nile perch (*Lates niloticus*), the Tanganyika sardine (*Limnothrissa miodon*) and water hyacinth (*Eichhornia crassipes*) into Lakes Victoria, Kyoga, Nabugabo, Kariba, Kivu, Itezhi-tezhi and Malawi. While the introduction of Nile perch had a major impact on the structure and profitability of fisheries, it is believed to have caused the extinction of numerous endemic species. The introduction of the Tanganyika sardine also benefited fisheries but had less dramatic impacts on the ecosystems of the lakes to which it was introduced. The water hyacinth, meanwhile, has proliferated in most African lakes. It has obstructed water passages and displaced native aquatic plants, fish and invertebrates by cutting out light and depleting dissolved oxygen. The weed is also believed to harbour disease-carrying organisms and has little potential for economic utilisation. Kasulo's (2000) estimate of the annual cost of the hyacinth in terms of its impact on fisheries in this group of lakes was $71.4.

The South African fynbos is affected by a number of invasive pinus, hakea and acacia species. By 2000 two-thirds of the fynbos area in the Western Cape had been significantly impacted. Damage costs include a reduction in biodiversity and, in particular, in species important for the international flower trade. But they also include a change in ecosystem functioning and hydrology. A number of studies have shown that fynbos mountain catchments are extremely valuable in terms of their water yield, and that the value of changes in water yields exceeded expected restoration costs (Higgins *et al.* 1997; van Wilgen *et al.* 1996; Turpie and Heydenrych, 2000; le Maitre *et al.* 2002). The result was a major control programme, the Working for Water Campaign, which had both restoration and poverty-alleviation goals. By 2004, the programme had cost in the region of $400 million and questions were being raised about its value relative to other development programmes. While the benefits of the programme in terms of employment and poverty alleviation are reasonably clear – the programme employed 24,000 people in 2000 – the environmental benefits aside from water flows are less easy to identify.

Turpie (2004) correctly points out that appropriate valuation of these benefits is needed to test the relative efficiency of conservation and development projects. Evaluation of the net benefits of resources committed to conservation and development projects includes a range of direct and indirect costs and benefits. It is easier to do in the case of control programmes for existing invasive species than it is for programmes designed to prevent the introduction of new, potentially invasive species. In fact, the cost-benefit ratio of successful control programmes for particularly harmful invaders can be surprisingly high (Hill and Greathead 2000), but this is like calculating the cost-benefit ratio of the purchase of a winning lottery ticket. The ex ante calculation involves uncertainty about the invasiveness of the species, the invasibility of the system being protected, the effectiveness of the control programme, and the responses of those whose life and livelihoods are affected by invasive species and their control.

To calculate the net benefit of restoration, control or eradication measures requires an evaluation not just of the damage or forgone output costs of invasive species and the cost of control but also of the benefits conferred by the invader or the activities that support the introduction or spread of the invader, and the distribution of those benefits. Most case studies of invasive species involve estimates of damage and control costs and do not deal with the benefits of the actions that lead to either the introduction, establishment or spread of invasive species. When those benefits are taken into account it is not always obvious that eradication or control is the optimal strategy. For example, siam weed (*Chromoleana odorata*) was introduced into Ghana in the 1960s and by the end of the century had spread to approximately 60 per cent of the land area. It has had major ecological effects. Nevertheless, a survey of users found that few would support its eradication since it confers significant benefits in terms of fuel, fibres, building materials and medicinal products (Rangi 2004). In semi-arid areas, mesquite (*Prosopis juliflora*) is a similar case. In South Africa, it has invaded the semi-arid Nama and succulent karoo biomes, and once again has had major ecological effects. In the more arid regions, however, it is highly valued for its capacity to provide a more reliable source of fuel and fibre than many native species in dry conditions (GISP 2004).

In many cases, control is exercised without explicit consideration of either damage costs or the benefits of the activities leading to the introduction, establishment or spread of invasive species. Certain pathogens, such as foot and mouth disease, are automatically eradicated whenever they appear without any cost-benefit calculation being made. In these circumstances it is still useful to consider the cost-effectiveness of control options. The literature on cost-effectiveness to date reflects a consensus that eradication is more cost-effective than control in most cases.

There is no consensus on the relative cost-effectiveness of eradication and prevention (through detection and interception in support of or red/black lists, or through quarantine).

The conclusion that many have drawn is that all conservation and development projects are location-specific, that the interactions between local people and ecological resources matter, and that it is important to understand the distribution of the costs and benefits of environmental change. If the people in locations where potentially invasive species appear are poor, their capacity to deal with the problem will be low. Borggaard *et al.* (2003), for example, note that cogon grass (*Imperata cylindrica*) has invaded shifting cultivation plots in many South and Southeast Asian countries. Since shifting cultivators are among the poorest members of those societies, the control measures needed to eradicate it (Johnson and Shilling 2003) are beyond their means. The problem is exaggerated by the migration or displacement of shifting cultivators since new farmers may be unaware of local conditions and so may not appreciate the extent of the problem (Adger *et al.* 2002). In these circumstances, it has been argued that the only effective strategies may be to manage for system resilience by adopting policies that enhance soil fertility, reduce clearing costs and increase the rate of forest recovery (Albers *et al.* 2006).

## 5 Invasion externalities: economic instruments of IAS control in poor countries

From an economic perspective, the problem of invasive species represents a classic market failure. Market prices of potentially invasive species do not reflect the costs they may impose on society, in part because many markets have been prevented from operating efficiently by agricultural policies and institutions. Agricultural tax/subsidy and price policies have increased the vulnerability of agro-ecosystems by reducing agro-biodiversity and by encouraging farm management regimes that leave agro-ecosystems open to invasion. Moreover, the lack of well-defined property rights in land and ecological services has discouraged people from taking action to control invading species. At the same time, the deregulation of both national and international markets has reduced both the surveillance of trade and the barriers to trade (Perrings *et al.* 2002).

A second point made by Perrings *et al.* (2002) is that there is a strong 'public good' element in the control of biological invasions. The benefits of quarantine, for example, are neither 'rival' nor 'exclusive'. If one extra person benefits from the protection offered by a quarantine policy it affects neither the cost of quarantine nor the benefits of quarantine to others. But because public goods are non-exclusive, any one person or any one country has a strong incentive to free-ride on the efforts of others. The implication of this is that if it were left to the market, there would be insufficient control of potentially invasive pests and pathogens. More importantly, the international control of many invasive species, such as infectious and communicable diseases, depends on the least effective provider – the weakest link in the chain (Sandler 1997). If control of an invasive species involves containment (or eradication) by all landowners, it will be only as good as the containment (or eradication) activities of the least effective landowner.

Since biodiversity conservation is at once a global, regional, national and local public good, it requires programmes of public investment that operate over a wide range of spatial and temporal scales (Perrings and Gadgil 2003). Moreover, a condition for an internationally efficient allocation of resources to conservation is that countries should be compensated for their contribution to the international public good. The incremental cost principle of the GEF implies just this, as does the CBD principle of the equitable sharing of the benefits of biodiversity conservation. Yet the structure of international markets and the rules governing international trade and investment mean that in practice those whose actions confer biodiversity benefits on others are seldom compensated. Equally, those whose actions impose biodiversity costs on others are seldom penalised.

It is not surprising therefore that the discussion has been dominated by instruments aimed at addressing the problems of externality and public goods. In the case of domestic markets, externalities can be addressed through a range of mechanisms extending from the assignment of property rights, through the use of market-based mechanisms such as taxes and subsidies, to simple regulatory measures supported by penalties for non-compliance. In international markets, where there is no sovereign authority, these options are not available. The choice of mechanisms open to any one country is limited by the bilateral and multilateral trade agreements to which it is party (Perrings *et al.* 2005). Much of the recent research on the economic problem of invasive species has accordingly focused on the options open to governments in these circumstances.

At a theoretical level, one of the main foci of this research has been the impact of tariffs, which concentrates on the interception of introduced species. Costello and McAusland (2003) consider the relationship between trade, tariffs and invasive agricultural pests. They show that the impact depends on the domestic agricultural price elasticity of imports. An increase in the tariffs always reduces the volume of trade and hence the rate of introductions, but the resulting stimulation of domestic agriculture increases the vulnerability of the sector to invasions.

Table 5.5 *Economic sanitary and phytosanitary instruments*

| National instruments | International instruments |
|---|---|
| Eradication/control charges | Risk-related tariffs |
| Monitoring charges | Inspection fees |
| Biodiversity maintenance fees | Fines/non-compliance penalties |
| Environmental bonds | Tradable risk permits |
| Risk-related land-use taxes | |
| Fines/non-compliance penalties | |
| 'Green box' agricultural support measures | |

McAusland and Costello (2004) then consider the degree to which non-tariff instruments, specifically inspections, may be used in combination with tariffs to achieve efficient control over alien species introductions. They show that the optimal mix of tariffs and inspections depends both on the rate to which imports are 'infected' by alien species and on the expected damage due to introductions that are not intercepted. Where the expected damage of unintercepted introductions is high, inspections dominate tariffs. But where the infection rate is high, tariffs dominate inspections.

More recently, Margolis *et al.* (2005) have applied the Grossman and Helpman (1994) model of tariff formation to the problem of international invasion externalities. They show that countries setting tariffs freely will indeed include expected damage cost of invasions into tariffs, but also that interest groups may set tariffs sub-optimally in order to introduce disguised protectionism.

Another novel instrument considered in the theoretical literature is tradable invasion risk permits. Horan and Lupi (2005) consider the introduction of aquatic invasive species in the Great Lakes, and propose the use of tradable invasion risk permits to allocate resources efficiently between risk-reducing options. Using the example of a small class of potentially invasive species from the Ponto-Caspian region, they show that the approach may offer efficiency gains over the more conventional regulations over ballast water exchange.

In practice, the instruments available to national governments are those admitted under the General Agreement on Tariffs and Trade, the Sanitary and Phytosanitary Agreement, the International Plant Protection Convention and related agreements (Shine *et al.* 2005) (see Table 5.5). The scope for using tariffs as a primary mechanism is strictly limited, and the only trade-related instruments involve defensive measures such as inspection and interception at ports of entry in support of black and white (or red, green and amber) lists, and combined with quarantine, confiscation and destruction.

There are few studies of the biodiversity impacts of economic measures permitted under current trade agreements. The Secretariat of the Convention of Biological Diversity has reviewed the effect of the Uruguay Round and its Agreement on Agriculture (URAA) on agro-biodiversity (SCBD 2005). It concludes that the reduction of 'amber box' (trade-distorting) measures under the URAA may have had positive effects on agro-biodiversity in countries where agriculture is highly intensive, but may have had negative effects in countries where agriculture is largely extensive and relies on traditional techniques. Moreover, this is particularly likely to have been the case in marginal lands of importance to biodiversity. The elimination of agricultural support schemes in these cases merely increases the rate at which soil nutrients are mined and reduces the resources committed to weed and pest control.

More interesting are the 'green box' (non-trade-distorting) measures designed to internalise externalities or to initiate payments for agricultural services. Many of these are redesigned agriculture support measures. They have largely been used by developed countries and in many cases appear to have been beneficial for agro-biodiversity. But the SCBD (2005) also notes that these have the potential to benefit biodiversity in developing countries where connected to wildlife or habitat conservation, or to the protection of traditional livestock strains and landraces. The report does not consider the problem of invasive species explicitly, but it would be consistent with this to suggest that green box agricultural support mechanisms that targeted invasion risks may be helpful in countries where agriculture is based on traditional landraces or livestock strains and on production methods that are vulnerable to the effects of invasive weeds, pests and pathogens.

The principle behind green box measures is that the (national) beneficiaries of environmental services provided by farmers should pay for these services. This implies either payments to farmers when their management practices confer benefits on society, or taxes when their management practices impose costs on society. In some cases this may imply the allocation of property rights. If invasive species increase the risks of fire, for example, the allocation of rights can create a market in fire risks. Where property rights are ill-defined, it may be easier to tax activities that lead to IAS risks. The problem then becomes one of determining the appropriate level of taxes. While the problem is straightforward in theory – the appropriate tax is equal to the marginal external damage cost of the activity – in practice this may be hard to calculate. Taxes are set at levels that lead to the desired behaviour. This means that they depend on the elasticity of producer responses. Since response elasticities are typically sensitive to income, poverty becomes an issue in designing economic sanitary and phytosanitary instruments. Short-run supply and demand elasticities can be extremely low among the poor (and even negative in the case of 'inferior' goods that provide essential life support). The relevant elasticities need to be understood and factored into the development of economic sanitary and phytosanitary instruments.

## 6 Public investment in invasive species control in low-income countries

The second element in the economic treatment of IAS is investment in IAS control (where this means detection, interception, eradication and control). International investment in the IAS problem is dominated by coordinated actions in response to particular threats such as SARS or Aids, or to bilateral or multilateral conservation and development projects that include an element of invasive species control. The best known of these is the South African Working for Water project, of which the control of IAS in the fynbos is a part. In general, however, lending for invasive species control is a very small share of World Bank lending for environmental and natural resource management (ENRM) projects. Overall ENRM lending has fallen substantially in the last decade, both in absolute terms and as a percentage of total lending. In 2002 it was less than a third of what it had been in 1994. Since then it has been improving, but is still only around 40 per cent of 1994 levels (Acharya *et al.* 2004).

Identifying the invasive species element in ENRM and linking this to the problem of poverty is not easy. The connection between poverty and public investments in invasive species control is most readily obtained by considering investment strategies in poor regions (e.g. sub-Saharan Africa). One problem is that invasive species are seldom explicitly identified. For example, a review of IDA, IBRD and GEF projects with a biodiversity element in the 1990s reported the Cape Peninsula biodiversity conservation project, but did not identify IAS as an element of that. Indeed, the only explicit reference to IAS was to a project for the eradication of IAS in Mauritius (MacKinnon *et al.* 2000). Invasive species are, however, a major component of the action plan of the environment initiative of the New Partnership for Africa's Development (NEPAD) (UNEP 2003).

The action plan notes that the impacts of invasive species are 'a major public policy' concern in many countries of Africa, affecting water supplies, fisheries, forestry, horticulture, trade and tourism. It also notes that they are a primary cause of biodiversity loss and ecosystem decline, that they exacerbate poverty and threaten the sustainability of development strategies (UNEP 2003).

The goal of the Programme Area on Prevention, Control and Management of Invasive Alien Species is stated to be 'to minimise the impact of IAS on the African continent's people, economies and ecological systems'. It proposes to use the same regional groupings referred to earlier – the East African Community, the Southern Africa Development Community and the Common Market for Eastern and Southern Africa – to regulate and control the introduction and spread of potentially invasive alien species, exploiting mechanisms identified at the 6th Conference of the Parties to the CBD. These include improvement in the capacity to undertake risk assessments, awareness raising and information provision, and development of the institutional capacity to manage IAS (UNEP 2003).

The foci for terrestrial systems under NEPAD are plant invaders in agriculture, forestry and rangelands in the Horn of Africa, sustainable management of invasive woody species in Southern Africa, and the control of the invasive Indian House Crow in Eastern Africa (UNEP 2003). While this maps reasonably well into IAS project funding from the IBRD, IDA and GEF, it meshes less well with the priorities identified in Rangi (2004), for whom the primary concern remains the effect of IAS on agriculture.

Part of the reason for this may be that while IAS are a major threat to food security, and while food security is the highest priority for many African governments, the linkage between them has not hitherto been made (Rangi 2004). Food security is certainly the first goal of poverty alleviation, which means that if IAS control is to be related to poverty alleviation, the natural foci are indeed IAS that affect the supply of food and water. Since many of the world's poor live in marginal, highly disturbed lands that are often the first to be colonised by invasive species, and since they do indeed exploit these species for food, fuel and fibre, it is not surprising that they sometimes have an ambivalent attitude to IAS control. One implication may be that invasive species should not be controlled. Another may be that IAS control should not be undertaken unless other measures have been put in place to compensate the poor for the loss of resources that results.

## 7 Factoring poverty into predictive modelling and management

Finally, an important feature of biological invasions is that they are, *ex ante*, highly uncertain. Williamson (1996) argues that this is because there are no general laws governing invasions. If so, it follows that it is extremely difficult to model the process and prediction of the population dynamics of a particular species in a particular habitat requires detailed study of that species in that habitat (Lawton 1999). Williamson (1996) claims that there are only two reasonably good predictors of the invasiveness of particular species: (a) a previous history of invasions by the same species, and (b) propagule pressure. Nevertheless, the assessment of the predictive capacity of models of the invasions process by the NAS Committee on the Scientific Basis for Predicting the Invasive Potential of Nonindigenous Plants and Plant Pests in the United States suggested that this may be too cautious and that there are other 'biological leads' that can be followed to improve predictability of invasiveness (NAS 2002). More importantly, these biological leads can be augmented by 'economic leads' – of which rural poverty is one.

Most recent work on the economics of IAS has involved the development of models of decision making under uncertainty, using a variant of either optimal control or stochastic dynamic programming in a bio-economic or ecological economic framework (Eisworth and Johnson 2002; Horan *et al.* 2002; Albers *et al.* 2006; Finnof *et al.* 2005; Knowler and Barbier 2005; Olson and Roy 2002; Perrings 2005). The quality of the data in these models is taken as given, although it is recognised that the risks confronting decision makers may not be independent of their actions.

The point has already been made that the nature and direction of pathways, the species that are likely to be introduced via those pathways and the frequency of introductions (propagule pressure) are all dependent on trade (and aid) flows. Hence changes in the structure, volume and value of trade will affect the probability that species from particular regions will be introduced into other regions. As the NAS (2002) puts it, China is likely to become a source of new invasive plants in the USA simply because of the growth in bilateral trade between them, the fact that they share similar physical and climatic conditions and have many related plant species. Nevertheless, the only recommendations they make on steps to take to improve the predictive capacity of models of biological invasions involve biotic and abiotic variables. Climate-matching models such as CLIMEX are recognised to be useful tools, and a range of research tasks is identified on, for example, host specificity among pathogens, the fate of biocontrol agents and the performance of US plants grown abroad. Nothing is said, however, about exploiting information on other factors that co-vary with biological invasions, such as trade or land use, or with the resources that are committed to detection and interception, eradication and control.

Perrings *et al.* (2002) make the point that the probability that a potentially invasive species is introduced, establishes and spreads depends on the strategy adopted to deal with invasive species. The main options are mitigation and adaptation. Mitigation is action to reduce the likelihood that a species will establish or spread. Adaptation is action that changes the cost of invasions, but does not affect the likelihood that they will occur. The choice between mitigation and adaptation strategies depends on their relative net benefits given the risk preferences of the decision maker and the (Shogren and Crocker 1999; Shogren 2000; Leung *et al.* 2002), but it also depends crucially on the capacity to predict the consequences of current actions. If it is not possible to affect the likelihood of invasions, then the only possible strategy is adaptation (Horan *et al.* 2002; Perrings 2005). The capacity to predict either invasiveness or invasibility allows decision makers to opt for a strategy of mitigation and this in turn changes the risk of invasions.

The likelihood of invasions depends on both the invasiveness of species and the invasibility of habitats. Both are influenced by socio-economic conditions. Invasiveness depends both on the properties of the organisms, resource flows (trade, transport and travel) and measures to detect and intercept introduced species. Invasibility depends on climatic and environmental conditions in the host system, but it also depends on the degree of habitat disturbance, fragmentation and simplification, on the openness of that system and on the effectiveness of control measures. Once again, these are influenced by socio-economic conditions. At the macro level, the openness of a country's economy, the composition of its trade flows, its regulatory regimes and the importance of agriculture, forestry or tourism all make it more or less vulnerable to invasions by alien species. So islands are susceptible to invasions partly because their native biodiversity is vulnerable, but also because they are typically very open. Dalmazzone (2000) observed that the average percentage of merchandise imports as a share of the GDP is 43 per cent for islands as against 27 per cent for continental countries.

At the micro level the invasibility of a habitat depends on land use and land management, including the management of alien species. In other words, the risks of biological invasions are endogenous (Shogren, 2000; Finnoff *et al.* 2005). So the habitat disturbance associated with the migration of shifting cultivators into new lands in Southeast Asia has been associated with the spread of cogon grass (Borggaard *et al.* 2003). Information of this kind may be used to improve the predictability of models. In the South African case, for example, models to predict the spread of IAS include at least some data on land use, but turn out to be quite sensitive to the modelling approach employed. Rouget *et al.* (2003) found that between 27 per cent and 32 per cent of land untransformed by agriculture in the fynbos and the renosterveld might be expected to be invaded, depending on whether rule-based or statistical modelling techniques are used. By adding models of the allocation of resources by resource users it should be possible to improve the capacity to predict changes in the invasibility of such habitats. Moreover, by adding models of trade and land use it should be possible to improve the capacity to predict both the introduction and spread of species. This, in turn, will make it possible to mitigate invasion risks. A strong positive correlation between trade volumes and the establishment of potentially invasive species has been shown for particular species of birds and fish, and there is some evidence of a correlation between the volume of all trade and general invasion risks (Dalmazzone 2000; Levine and D'Antonio 2003). Since poverty is positively correlated with many of the risk factors related both to the invasibility of ecosystems and to the weakness of detection and control measures, it should prove possible to factor it into predictive models of biological invasions.

## REFERENCES

Acharya, A., Aparatha-Hemantha, A. and Tsutsui, E. 2004. The environment and natural resources management portfolio. *The World Bank Group – Environment Matters* (June). 32–33.

Adger, W. N., Kelly, P. M., Winkels, A., Luong Quang Hai and C. Locke 2002. Migration, remittances, livelihood trajectories and social resilience. *Ambio* **31**. 358–366.

Aksoy, M. A. 2005. The evolution of agricultural trade flows. In M. A. Aksoy and J. C. Beghin (eds.). *Global Agricultural Trade and Developing Countries*. Washington, DC: World Bank. 17–36.

Albers, H., Goldbach, M. J. and Kaffine, D. 2006. Implications of agricultural policy for species invasion in shifting cultivation systems. *Environment and Development Economics*. **11**. 429–452.

Arrow, K. J., Dasgupta, P. and Mäler, K.-G. 2003. Evaluating projects and assessing sustainable development in imperfect economies. *Environmental and Resource Economics*. **26**. 647–685.

Bangsund, D. A., Leistritz, F. L., and Leitch, J. A. 1999. Assessing economic impacts of biological control of weeds: the case of leafy spurge in northern great plains of the United States. *Journal of Environmental Management*. **56**. 35–53.

Birdsall, N. and Londoño, J. L. 1997. Asset inequality matters: an assessment of the world bank's approach to poverty reduction. *American Economic Association Papers and Proceedings*. **87** (2). 32–37.

Borggaard, O. K., Gafur, A. and Peterson, L. 2003. Sustainability appraisal of shifting cultivation. *Ambio*. **32**. 118–123.

Chavas, J.-P. 2004. On impatience, economic growth and the environmental Kuznets curve: a dynamic analysis of resource management. *Environmental and Resource Economics*. **28**. 123–152.

Cleaver, K. M. and Schreiber, A. G. 1994. *Reversing the Spiral: The Population, Agriculture, and Environment Nexus in Sub-Saharan Africa*. World Bank, Washington DC.

Cohen, A. N., Carlton, J. T. and Fountain, M. C. 1995. Introduction, dispersal and potential impacts of the green crab *Carcinus maenas* in San Francisco Bay. California: *Marine Biology*. **122** (2). 225–237.

Cole, M. A., Rayner, A. J. and Bates, J. M. 1997. The environmental Kuznets curve: an empirical analysis. *Environment and Development Economics*. **2** (4).

Coombs, E. M., Radtke, H., Isaacson, D. L. and Snyder, S. 1996. Economic and Regional Benefits from Biological Control of Tansy Ragwort, *Senecio Jacobaea*, in Oregon. In V. C. Moran and J. H. Hoffmann (eds.). *International Symposium on Biological Control of Weeds*. University of Cape Town: Stellenbosch. 489–494.

Costello, C. and McAusland, C. 2003. Protectionism, trade and measures of damage from exotic species introduction. *American Journal of Agricultural Economics*. **85** (4). 964–975.

Dalmazzone, S. 2000. Economic Factors affecting vulnerability to biological invasions. In C. Perrings, M. Williamson and S. Dalmazzone (eds.). *The Economics of Biological Invasions*. Cheltenham: Edward Elgar. 17–30.

Dasgupta, P. 1993. *An Inquiry into Well-being and Destitution*. Oxford: Oxford University Press.

Dasgupta, P. 2001. *Human Well-Being and the Natural Environment*. Oxford: Oxford University Press.

De Janvry, A. and Garcia, R. 1988. *Rural Poverty and Environmental Degradation in Latin America: Causes, Effects and Alternative Solutions*. S 88/1/L.3/Rev.2. Rome: IFAD.

Eisworth, M. E. and Johnson, W. S. 2002. Managing nonindigenous invasive species: insights from dynamic analysis. *Environmental and Resource Economics*. **23**. 319–342.

Ekbom, A. and Bojö, J. 1999. *Poverty and Environment: Evidence of Links and Integration into the Country Assistance Strategy Process*. World Bank Discussion Paper No. 4, Environment Group, Africa Region, Washington, DC.

Enserink, M. 1999. Biological invaders sweep in. *Science*. **285**. 1834–1836.

Evans, E. A. 2003. Economic dimensions of invasive species. *Choices*, June.

Farrell, G. and Schulten, G. M. M. 2002. Large grain borer in Africa: a history of efforts to limit its impact. *Integrated Pest Management Review*. 7. 67–84.

Finnof, D., Shogren, J. F., Leung, B. and Lodge., D. 2005. The importance of bioeconomic feedback in invasive species management. *Ecological Economics*. **52** (3). 367–382.

Fisher, I. 1930. *The Theory of Interest*. New York: Macmillan.

Global Invasive Species Program (GISP). 2004. *Africa invaded: the growing danger of invasive alien species*. CapeTown: GISP.

Grossman, G. M. and Helpman, E. 1994. Protection for sale. *American Economic Review*. **89** (5). 1135–1155.

Grossman, G. M. and Krueger, A. B. 1995. Economic growth and the environment. *Quarterly Journal of Economics*. **110**. 353–377.

Hamilton, K. 2000. *Sustaining Economic Welfare: Estimating Changes in Per Capita Wealth*. Policy Research Working Paper 2498. Washington, DC: The World Bank.

Hamilton, K. and Clemens, M. 1999. Genuine savings rates in developing countries. *World Bank Economic Review*. **13**. 333–356.

Heath, J. and Binswanger, H. 1996. Natural resource degradation effects of poverty and population growth are largely policy induced: the case of Colombia. *Environment and Development Economics*. **1** (1). 65–84.

Higgins, S. I., Azorin, E. J., Cowling, R. M. and Morris, M. H. (1997). A dynamic ecological-economic model as a tool for conflict resolution in an invasive alien-plant, biological control and native-plant scenario. *Ecological Economics*. **22**. 141–154.

Hill, G. and Greathead, D. 2000. Economic evaluation in classical biological control. In C. Perrings, M. Williamson and S. Dalmazzone (eds.). *The Economics of Biological Invasions*. Cheltenham: Edward Elgar. 208–223.

Holden, S. T., Shiferaw, B. and Wik, M. 1998. Poverty, market imperfections and time preferences: Of relevance for environmental policy? *Environment and Development Economics*. **3** (1). 105–130.

Horan, R. D., Perrings, C., Lupi, F. and Bulte, E. 2002. Biological pollution prevention strategies under ignorance: the case of invasive species. *American Journal of Agricultural Economics*. **84** (5). 1303–1310.

Horan, R. D. and Lupi, F. 2005. Economic incentives for controlling trade-related biological invasions in the Great Lakes. *Ecological Economics*. **52** (3). 289–304.

Jaffee, S. M. and Henson, S. 2005. Agro–food exports from developing countries: the challenges posed by standards. In M. A. Aksoy and J. C. Beghin (eds.). *Global Agricultural Trade and Developing Countries*. Washington, DC: World Bank. 91–114.

Jazairy, I., Almagir, M. and Panuccio, T. 1992. *The State of World Rural Poverty*. London: IT Publications for IFAD.

Jetter, K. M., DiTomaso, J. M., Drake, D. J., Klonsky, K. M., Pitcairn, M. J. and Sumner, D. A. 2003. Biological control of yellow starthistle. In D. A. Sumner (ed.). *Exotic Pests and Diseases: Biology and Economics for Biosecurity*. Ames, Iowa: Iowa State University Press. 225–241.

Johnson, E. R. R. L. and Shilling, D. G. 2003. Cogon grass: *Imperata cylindrica (L.) Palisot*. Plant Conservation Alliance–Alien Plant Working Group. Washington, DC: National Park Service.

Kasulo, V. 2000. The impact of invasive species in African lakes. In C. Perrings, M. Williamson and S. Dalmazzone (eds.). *The Economics of Biological Invasions*. Cheltenham: Edward Elgar. 183–207.

Knowler, D. 2005. Reassessing the costs of biological invasion: *Mnemiopsis leidyi* in the Black Sea. *Ecological Economics*. **52** (2). 187–200.

Knowler, D. and Barbier, E. B. 2000. The economics of an invading species: a theoretical model and case study application. In C. Perrings, M. Williamson and S. Dalmazzone (eds.). *The Economics of Biological Invasions*. Cheltenham: Edward Elgar. 70–93.

Knowler, D. and Barbier, E. B. 2005. Importing exotic plants and the risk of invasion: are market based instruments adequate? *Ecological Economics*. **52** (3). 341–354.

Kumar, S. K. and Hotchkiss, D. 1988. *Consequences of Deforestation for Women's Time Allocation, Agricultural Production and Nutrition in the Hills of Nepal.* Research Report No. 69. Washington, DC: IFPRI.

Lange, G.-M. 2004. Wealth, natural capital, and sustainable development: contrasting examples from Botswana and Namibia. *Environmental and Resource Economics.* **29**. 257–283.

Lawton, J. 1999. Are there general laws in ecology? *Oikos.* **84**. 177–192.

Le Maitre, D. C., van Wilgen, B. W., Gelderblom, C. M., Bailey, C., Chapman, R. A. and Nel, J. A. 2002. Invasive alien trees and water resources in South Africa: case studies of the costs and benefits of management. *Forest Ecology and Management.* **160**. 143–159.

Leung, B., Lodge, D. M., D. Finnoff, Shogren, J. F., Lewis, M. A. and Lamberti, G. 2002. An ounce of prevention or a pound of cure: bioeconomic risk analysis of invasive species. *Proceedings of the Royal Society of London, Biological Sciences.* **269** (1508). 2407–2413.

Levine, J. and D'Antonio, C. 2003. Forecasting biological invasions with increasing international trade. *Conservation Biology.* **17** (1). 322–326.

Linde-Rahr, M. 2002. *Household economics of agriculture and forestry in rural Vietnam.* PhD Thesis, Department of Economics. University of Göteborg.

Lockwood, J. L., Cassey, P. and Blackburn, T. 2005. The role of propagule pressure in explaining species invasions. *Trends in Ecology and Evolution.* **20** (5). 223–228.

Lopez, R. 1992. Environmental degradation and economic openness in LDCs: The poverty linkage. *American Journal of Agricultural Economics.* **74**. 1138–1145.

Lopez, R. and Scoseria, C. 1996. Environmental sustainability and poverty in Belize: A policy paper. *Environment and Development Economics.* **1** (3). 289–308.

Lovell, S. J. and Stone, S. 2005. *The Economic Impacts of Aquatic Invasive Species: A Review of the Literature.* Working Paper No. 05–02. Washington, DC: U.S. Environmental Protection Agency National Center for Environmental Economics.

MacKinnon, K., Chow, N., Esikuri, E. and Platais, G. 2000. *Conserving and Managing Biodiversity in Dryland Ecosystems.* World Bank Environment Department Working Paper. Washington, DC: World Bank.

Margolis, M., Shogren, J. and Fischer, C. 2005. How trade politics affect invasive species control. *Ecological Economics.* **52** (3). 305–313.

Markandya, A. 2000. Poverty, environment and development. In A. Rose and L. Gabel (eds.). *Frontiers of Environmental Economics.* Cheltenham: Edward Elgar.

Markandya, A. 2001. *Poverty Alleviation and Sustainable Development Implications for the Management of Natural Capital.* Washington, DC: World Bank

McAusland, C. and Costello, C. 2004. Avoiding invasives: trade related policies for controlling unintentional exotic species introductions. *Journal of Environmental Economics and Management.* **48**. 954–977.

National Academy of Sciences (NAS). Committee on the Scientific Basis for Predicting the Invasive Potential of Nonindigenous Plants and Plant Pests in the United States. 2002. *Predicting Invasions of Non-Indigenous Plants and Plant Pests*. Washington, DC: National Academy Press.

Office of Technology Assessment (OTA) US Congress. 1993. *Harmful Non-Indigenous Species in the United States*. OTA Publication OTA-F-565. US Government Printing Office, Washington, DC.

Olson, L. J. and Roy, S. 2002. The economics of controlling a stochastic biological invasion. *American Journal of Agricultural Economics*. **84** (5). 1311–1316.

O'Neill, C. 1997. Economic impact of zebra mussels: Results of the 1995 zebra mussel information clearinghouse study. *Great Lakes Res. Review*. **3** (1). 35–42.

Pender, J. L. 1996. Discount rates and credit markets: theory and evidence from rural India. *Journal of Development Economics*. **50**. 257–296.

Perrault, A., Bennett, M., Burgiel, S., Delach, A. and Muffett, C. 2003. *Invasive Species, Agriculture and Trade: Case Studies form the NAFTA Context*. Montreal: North American Commission for Environmental Cooperation.

Perrings, C. 1989. An optimal path to extinction: poverty and resource degradation in the open agrarian economy. *Journal of Development Economics*. **30** (1). 1–24.

Perrings, C. 2005. Mitigation and adaptation strategies for the control of biological invasions. *Ecological Economics*. **52** (3). 315–325.

Perrings, C., Dehnen-Schmutz, K., Touza, J. and Williamson, M. 2005. How to manage biological invasions under globalization. *Trends in Ecology and Evolution*. **20** (5). 212–215.

Perrings, C., Williamson, M. and Dalmazzone, S. (eds.). 2000. *The Economics of Biological Invasions*. Cheltenham: Edward Elgar.

Perrings, C., Williamson, M., Barbier, E. B., Delfino, D., Dalmazzone, S., Shogren, J., Simmons, P. and Watkinson, A. 2002. Biological invasion risks and the public good: an economic perspective. *Conservation Ecology*. **6** (1). http://www.consecol.org/vol6/iss1/art1

Perrings, C. and Gadgil, M. 2003. Conserving biodiversity: reconciling local and global public benefits. In I. Kaul, P. Conceicao, K. le Goulven, and R. L. Mendoza (eds.). *Providing Global Public Goods: Managing Globalization*. Oxford: Oxford University Press. 532–555.

Pimentel, D., Lach, L., Zuniga, R. and Morrison, D. 2000. Environmental and economic costs of nonindigenous species in the United States. *Bioscience*. **50** (1). 53–56.

Pimentel, D., Zuniga, R. and Morrison, D. 2005. Update on the environmental and economic costs associated with alien-invasive species in the United States. *Ecological Economics*. **52**. 273–288.

Pimentel, D. David, McNair, S., Janecka, S., Wightman, J., Simmonds, C., O'Connell, C., Wong, E., Russel, L., Zern, J., Aquino, T. and Tsomondo, T. 2001. Economic and environmental threats of alien plant, animal and microbe invasions. *Agriculture, Ecosystems and Environment*. **84**. 1–20.

Pingali, P., Bigot, H. and Binswanger, P. 1987. *Agricultural mechanization and the evolution of farming systems in Sub-Saharan Africa*. Baltimore: Johns Hopkins.

Polasky, S., Costello, C. and McAusland, C. 2004. On trade, land-use and biodiversity. *Journal of Environmental Economics and Management*. **48**. 911–925.

Rangi, D. K. 2004. Invasive alien species: agriculture and development. Proceedings of a global synthesis workshop on biodiversity loss and species extinctions: managing risk in a changing world. IUCN. The World Conservation Union.

Reed, D. (ed). 1996. *Structural Adjustment, the Environment and Sustainable Development*. London: Earthscan.

Rouget, M., Richardson, D. M., Cowling, R. M., Lloyd, J. W. and Lombard, A. T. 2003. Current patterns of habitat transformation and future threats to biodiversity in terrestrial ecosystems of the Cape Floristic Region, South Africa. *Biological Conservation*. **112**. 63–85.

Sandler, T. 1997. *Global Challenges*. Cambridge: Cambridge University Press.

Sandler, T. 2005. Regional public goods and regional cooperation. Background Working Paper for the Task Force on Global Public Goods. Stockholm.

Schiff, M. and Winters, L. A. 2003. Regional Integration and Development, World Bank, Washington, D.C.

Secretariat of the Convention on Biological Diversity (SCBD). 2005. *The Impact of Trade Liberalisation on Agricultural Biological Diversity: Domestic Support Measures and their Effects on Agricultural Biological Diversity*. Montreal: SCBD.

Seldon, T. M. and Song, D. 1994. Environmental quality and development: is there a Kuznets curve for air pollution emissions? *Journal of Environmental Economics and Management*. **27**. 147–162.

Shafik, N. 1994. Economic development and environmental quality: an econometric analysis. *Oxford Economic Papers*. **46**. 757–773.

Shine, C., Williams, N. and Burhenne-Guilmin, F. 2005. Legal and institutional frameworks for invasive alien species. In H. A. J. Mooney, L. E. McNeely, P. J. Neville, J. K. Schei, and Waage, J. K. (eds.). *Invasive Alien Species: a New Sysnthesis*. Washington, DC: Island Press. 233–284.

Shogren, J. F. 2000. Risk reduction strategies against the ‘explosive invader.’ In C. Perrings, M. Williamson and S. Dalmazzone (eds.). *The Economics of Biological Invasions*. Cheltenham: Edward Elgar. 56–69.

Shogren, J. and Crocker, T. 1999. Risk and its consequences. *Journal of Environmental Economics and Management*. **37**. 44–51.

Stern, D. I. 1998. Progress on the environmental Kuznets curve. *Environment and Development Economics*. **3**. 381–394.

Stern, D. I. 2004. The rise and fall of the environmental Kuznets curve. *World Development*. **32** (8). 1419–1439.

Stern, D. I. and Common, M. S. 2001. Is there an environmental Kuznets curve for sulphur. *Journal of Environmental Economics and Management*. **41**. 162–178.

Stutzman, S., Jetter, K. M. and Klonsky, K. M. 2004. *An Annotated Bibliography on the Economics of Invasive Plants*. University of California, Davis, Agricultural Issues Center.

Sumner, D. A. (ed.). 2003. *Exotic Pests and Diseases: Biology and Economics for Biosecurity*. Ames, Iowa: Iowa State Press.

Tiffen, M., Mortimore, M. and Gichuki, F. 1994. *More People, Less Erosion, Environmental Recovery in Kenya*. New York: John Wiley.

Turpie, J. K. 2004. The role of resource economics in the control of invasive alien plants in South Africa. *South African Journal of Science*. **100** Jan/Feb. 87–93.
Turpie, J. K. and Heydenrych, B. J. 2000. Economic consequences of alien infestation of the cape floral kingdom's fynbos vegetation. In C. Perrings, M. Williamson and S. Dalmazzone (eds.). *The Economics of Biological Invasions*. Cheltenham: Edward Elgar. 152–182.
United Nations Environment Programme (UNEP). 2003. Action Plan of the Environment Initiative of the New Partnership for Africa's Development (NEPAD). Nairobi: UNEP.
van Wilgen, B. W., Cowling, R. and Mand Burgers, C. J. 1996. Valuation of ecosystem services: a case study from the fynbos, South Africa. *BioScience*. **46**. 184–189.
Ward, J. M. J., Stromberg, E. L., Nowell, D. C. and Nutter, F. W. 1999. Gray leaf spot: a disease of global importance in maize production. *Plant Disease* **83**. 884–895.
Williamson, M. 1996. *Biological Invasions*. London: Chapman and Hall.
World Bank, 2005. *Global Economic Prospects*. Washington, DC: World Bank.
World Commission on Environment and Development. 1987. *Our Common Future*. Oxford: Oxford University Press.
Zavaleta, E. 2000. The economic value of controlling an invasive shrub. *Ambio*. **29** (8). 462–467.

# 6 Prevention versus control in invasive species management

*David Finnoff, Jason F. Shogren, Brian Leung and David Lodge*

## 1 Introduction

As a leading cause of biodiversity loss and environmental damage, non-indigenous species can pose significant risks to society (see Mack *et al.* 2000, Lodge 2001). Managing these risks cost-effectively requires a consistent framework for bio-economic risk assessment. The economic theory of endogenous risk – merged with applied population ecology – provides such a framework (Shogren 2000; Leung *et al.* 2002). Endogenous risk captures the risk-benefit tradeoffs created by jointly determined ecosystem conditions, species characteristics and economic circumstances (Crocker and Tschirhart 1992; Settle *et al.* 2002). Endogenous risk theory stresses that management priorities depend crucially on both the *tastes* of the manager – his preferences over time and for risk bearing – and the *technology* of risk reduction – prevention, control and adaptation matter for optimal reduction strategies. Holding initial biological circumstances constant, managers with different preferences will likely make different choices on the mix of prevention and control. How different tastes affect technology choice, however, remains an open question in invasive species management.

This chapter investigates how manager types differentiated by preferences over time and over risk affect the optimal mix of prevention and control. The chapter advances our understanding on the behavioural underpinnings of risk-reduction strategies to control invasive species. Endogenous risk theory is a flexible tool that allows one to better understand the tradeoffs involved in changing the odds that good events are realised or in decreasing the severity of bad events if they are realised (Ehrlich and Becker 1972). The chapter also illustrates one approach to integrate economics and biology into a model to illustrate how humans affect nature and how nature affects humans. Capturing the dynamic feedback loops between the natural and social systems can be crucial for unbiased estimates of the key biological and economic parameters of interest (Finnoff *et al.* 2005).

Our analysis in this chapter proceeds in two steps. First, the problem is framed using a dynamic endogenous risk model that accounts for both biological and economic circumstances. In the framework the comparative statics on how changing tastes affect the technology mix are explored. Second, the model is implemented through an application of managing zebra mussels in a lake. Zebra mussels (*Dreissena polymorpha*) provide an illuminating example as their invasions currently cost US industries an estimated $100 million per year (Pimentel *et al.* 1999). Regional and federal governmental agencies and private producers faced with impacts (primarily power plants and water treatment facilities) continue to experiment with new control measures in an effort to maximise the benefits of zebra mussel control, and prevention of new infestations remains timely because zebra mussels are still expanding their range within North America (Bossenbroek *et al.* 2001). Zebra mussels have also been shown to cause substantial environmental impacts (Ricciardi *et al.* 1998; Lodge 2001). Using stochastic dynamic programming simulation, the impacts of preference changes on the mix of prevention and control, the probability of invasion and the overall welfare of the system are considered. Results suggest that an invasive species manager who is less risk averse and less myopic will likely invest more in prevention and less in control, which in turn requires less private adaptation by a firm, resulting in greater social welfare relative to a risk-averse, myopic manager.

## 2 Discrete dynamic endogenous risk framework

The classical models of choice under uncertainty underlie the theory of endogenous risk (von Neumann and Morgenstern 1944). Following Shogren (2000), consider a benevolent manager who allocates scarce resources to maximise expected social welfare subject to the risk of invasion. Herein, let the general circumstance of invasive species be seen as the management of an impure public 'bad'. Consider highly mobile invasive species with numerous transportation pathways, such that private citizens or firms cannot control the entry of the invasive into the overall system (e.g. zebra mussels entering into the great lakes in the ballast water of ships). Once established, the invader can cause adverse impacts. While private individuals or firms can only adapt to the invader, assume an overreaching governmental agency that acts as a benevolent manager. The manager can only partially control future invasions and growth of the invader through collective prevention and control strategies, given uncertainty in the 'kill function' (Feder 1979). These government actions provide a public good to private individuals who also respond to the invasion. Thus the framework casts the benevolent manager making optimal decisions given the risks of invasion and behaviour of private individuals, who also react to the consequences of invasion.

Private individuals (firms) are viewed as relatively myopic – they are relatively less farsighted than the benevolent manager. This restriction reflects the notion that firms make private decisions based on market discount rates, whereas the manager employs a rate based on social preferences. In general, the market discount rate is assumed to not exceed the social rate (e.g. Weitzman 1994). For tractability, assume that firms are completely myopic with a discount rate of zero. Lacking foresight, they take the state (as defined by the invasive species) as given and ignore any future repercussions of their behaviour. Consistent with myopic individuals, assume risk-neutral behaviour on the part of individuals to allow a focus on the effects of the manager's risk preferences.

In any period $t$, a representative individual (firm) maximises utility subject to their budget constraint taking the current state as given. Let states be defined by current period invader abundance $\theta_t$ (state variable). Invader abundances cause damages $D_t$, where monetised damages serve to diminish initial private wealth $M_t^p$. In response, individuals have costly strategies at their disposal and can adapt $Z_t^p$ to the invader. Adaptation (or self-insurance) accepts the direct damages and compensates in response to reduce the consequences of the damage.[1] This strategy refers to those options available to the individual that allow them to compensate for the realised damages. For example, if the individual in question is an individual power plant, zebra mussels clog coolant systems. The plant may be able to compensate/adapt to the damage inflicted by the mussels simply through employing factors of production and operate longer hours or burn more fuel than otherwise necessary.

Inserting the individual's budget constraint into their utility function yields the objective function,

$$\max_{Z_t^P} \quad U_t\left(M_t^P - D_t\left(Z_t^P;\theta^t\right) - C_t\left(Z_t^P\right)\right) \tag{1}$$

in which damages depend on given state $\theta_t$, where states are defined by current period invader abundance, which range from a minimum of zero before invasion or after complete eradication to any level of abundance within the system's carrying capacity. $C_t$ is the individual's cost function, assumed to be monotonically increasing in adaptation. The first-order condition for private adaptation is

$$-U'\left(M_t^P - D_t\left(Z_t^P;\theta^t\right) - C_t\left(Z_t^P\right)\right)\left[D_{Z_t^P} + C_{Z_t^P}\right] = 0 \tag{2}$$

[1] So that $D_Z < 0$ and if diminishing marginal effectiveness of $Z^P$ is assumed $D_{ZZ} > 0$ (Lichtenberg and Zilberman 1986).

Assuming an interior solution, time and state notation are suppressed and primes and subscripts indicate partial derivatives. As usual, condition (2) requires a balance of the marginal benefits of adaptation with its marginal costs such that the marginal damage reduction $D_{Z_t^P}$ is equal to the marginal cost of adaptation $C_{Z_t^P}$. Note that benefits arise from reduced consequences of damages given the adaptation response, and all benefits and costs (from the individual's viewpoint) accrue in the current period. Assuming the sufficiency conditions are maintained,[2] this condition determines the individual's optimal level of adaptation $\hat{Z}^P_{t,\theta_t}$ in any given period and state.

Given the individual's optimal choice, the benevolent manager maximises expected social welfare subject to the risk of invasion. Let social welfare be the discounted stream of intergenerational individual utility augmented by the costs of collective action. Unlike the firm, the manager considers the dynamics of the invasion process and can partially control entry and growth of the invader. The manager then directly influences the realised state $\theta_t$. The manager reduces the damages associated with invasion *in future periods* through either collective control or prevention. To result in damages, the invader must successfully traverse a number of interrelated processes: introduction, establishment and growth of the invader. Not all species that invade become established and not all established invaders cause damages (see Williamson 1996). Once a species establishes itself, let the system be considered invaded. After establishment, the invader can increase in abundance. It is the abundance that directly relates to damages. Unlike other forms of pollution, in which remedial efforts can have lasting effects, biological organisms can reproduce such that control efforts may be necessary in perpetuity.

To combat the risks of invasion and reduce the probability of damages, the resource manager can employ collective prevention $S$ to reduce the probability that invasion occurs at all. Once an invasion has occurred, they can collectively control $X$ to reduce the abundance and damages in the next period.

Let the risk of invasion be a multi-period compound lottery that reflects a separation in the probability of invasion in non-invaded states and transition probabilities in invaded states. Figure 6.1 presents a simplified view of a discrete invasion process for the first four periods of an invasion, $t$ through $(t+3)$.

[2] The second-order conditions require $U''(.)(D_{Z_t^P} + C_{Z_t^P})^2 - U'(.)(D_{Z_t^P Z_t^P} + C_{Z_t^P Z_t^P}) < 0$ which are satisfied by the assumptions of the model, namely that $U''(.) < 0$ and $D_{Z_t^P Z_t^P}, C_{Z_t^P Z_t^P} > 0$.

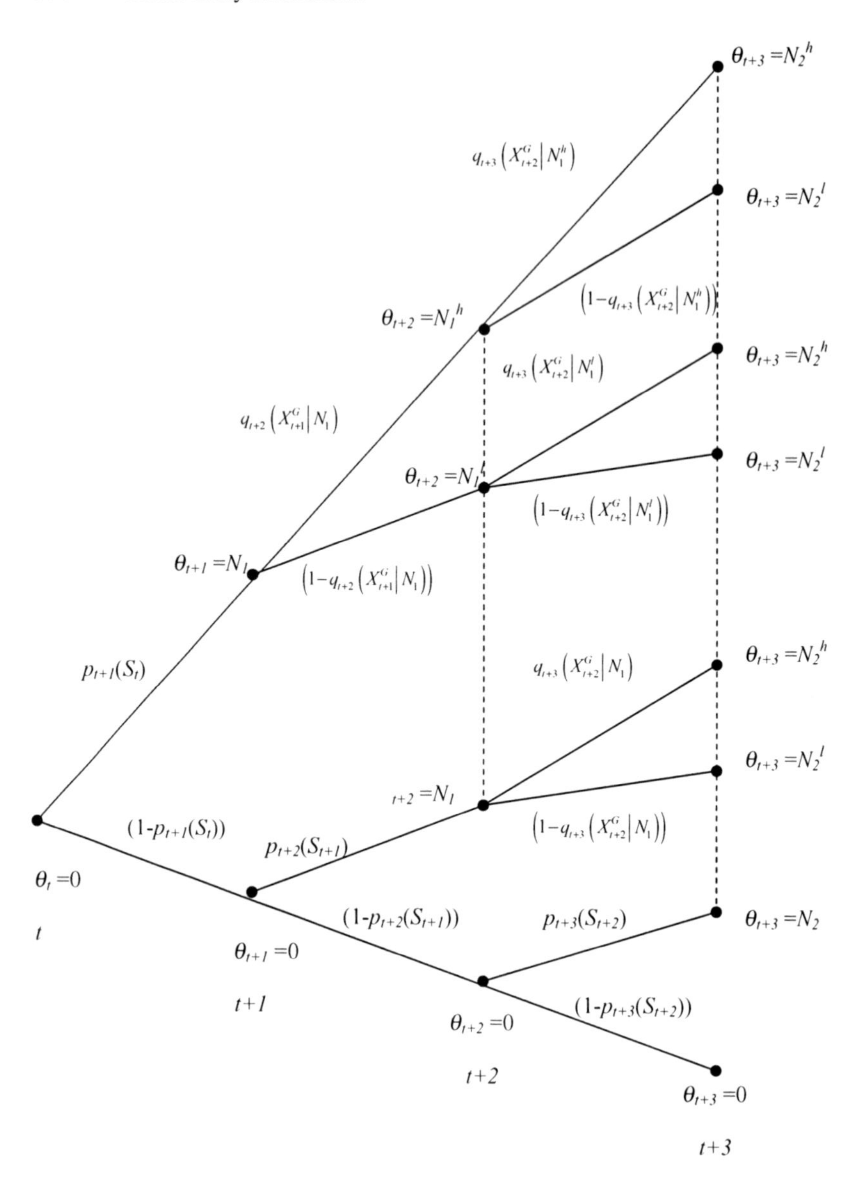


Figure 6.1. Schematic of the invasion process

In any time interval, there is only a single realised state. When forecasting the consequences of actions into the future, however, it is necessary to consider the probabilities of being in each possible state. For example, if the state of nature is uninvaded (current invader abundance $\theta_t = 0$) in the initial period, there is some probability of invasion $p_{t+1}(S_t)$, during the transition to $t + 1$. Let this be a diminishing function of collective prevention applied in $t$ such that $p_{t+1}(S_t)$ and $p_{t+1,S} < 0$, $p_{t+1,SS} > 0$, where the second set of subscripts indicate partial derivatives. If the invasion is successful the invaders become established ($\theta_{t+1} = N_1$) and cause damages in $(t + 1)$. If the invasion is not successful the invader does not become established ($\theta_{t+1} = 0$) and there is no damage.

In the transition to $(t + 2)$, in the non-invaded state the manager faces the threat of invasion (with probability $p_{t+2}$). But in the invaded state, they experience current period damages due to the abundance of the invader $N_I$ and face the threat of even larger damages in the subsequent period through growth of invaders (with probability $p_{t+2}$). Projected future actions include application of prevention $S_{t+1}$, and collective control measures $X_{t+1}$, as the realised state is not known with certainty. The probability of growth (transition probability) is conditioned on the abundance and would follow some population growth model. Collective control serves to reduce the reproducing invader population in $(t + 1)$ so that the magnitude of growth in the transition to $(t + 2)$ $q_{t+2}(X_{t+1}|N_1)$ depends on collective control and $q_{t+1,X} < 0$, $q_{t+1,XX} > 0$.[3]

If control measures are unsuccessful and the invader grows to a high level ($\theta_{t+2} = N_1^h$), there are damages, but if control is successful the invader's growth is halted and there are low (or zero) damages ($\theta_{t+2} = N_1^l$). But even if control is successful and there are low or no damages in $(t + 2)$, the biological population may grow and cause damages in future periods.

In our example, the manager takes current period damages as given and their employment of collective prevention and control is costly in the current period yet influences the invasion process in the subsequent period. The manager's strategies add to total costs, represented by augmenting the cost function to be $C_t(X_t, S_t, \hat{Z}_t^P)$, maintained as monotonically increasing in each argument.

The manager's objective is to maximise discounted social welfare over horizon $T$, where social welfare in $t$ is initial social wealth $M_t$ net of damages and the costs of invasion. In a discrete framework, write the stochastic dynamic programming equation (SDPE) as the summation of optimised

[3] In this format, both prevention and control are also referred to as *self-protection* or *mitigation* strategies (see Ehrlich and Becker 1972).

discounted welfare in year $t$ and all future years. Let $W$ be the maximum discounted expected social welfare from the perspective of initial period $t$, and $\alpha$ be a parameter reflecting the manager's absolute risk aversion. Periodic social welfare is $U_t$, an increasing ($U_t' > 0$) and strictly concave ($U_t'' < 0$) thrice-differentiable von Neumann-Morgenstern utility function. The SDPE is

$$W(\theta_t) = \max_{S_t, X_t} \quad U_t\left(M_t - D_t\left(\hat{Z}_t^P; N_t\right) - C_t\left(S_t, X_t, \hat{Z}_t^P\right); \alpha\right) + \rho E_t W(\theta_{t+1}; \alpha) \tag{3}$$

where current social welfare depends on damages due to current invader abundances, optimal private behaviour $\hat{Z}_t^P$, its cost, and the costs of collective action. Welfare in subsequent periods $t+1$ is discounted by factor $\rho^4$ and uncertain given random invasion, growth and damage. $E_t$ is the conditional expectation operator from the viewpoint of $t$. In the analytical model (not the simulation model) we consider a two-period model with a simplified case of three potential states in $(t+1)$: no invasion (and no damages if currently uninvaded), invaded and severe damages, and invaded and minimal damage. Expected welfare in $(t+1)$ is given by

$$E_t W(\theta_{t+1}) = p_{t+1}(S_t)\begin{bmatrix} q_{t+1}(X_t | N_t) U_{t+1}(G_{t+1}; \alpha) \\ +(1 - q_{t+1}(X_t | N_t)) U_{t+1}(H_{t+1}; \alpha) \end{bmatrix} + (1 - p_{t+1}(S_t)) U_{t+1}(H_{t+1}; \alpha) \tag{4}$$

Net incomes in $(t+1)$ are described by the following conventions:

$$G_{t+1} = M_{t+1} - D_{t+1}\left(\hat{Z}_{t+1}^P; N_t\right) - C_{t+1}\left(S_{t+1}, X_{t+1}, \hat{Z}_{t+1}^P\right)$$

$$H_{t+1} = M_{t+1} - C_{t+1}\left(S_{t+1}, X_{t+1}, \hat{Z}_{t+1}^P\right)$$

where $G_{t+1} < H_{t+1}$. As equation (4) demonstrates, odds exist $q_{t+1}$ that the invader grows rapidly in the transition to $(t+1)$ and causes damages only in the invaded state. If control measures are successful $(1 - q_{t+1})$, such that control is 100 per cent effective, the growth of the invader is halted with no damage. Note that the probability of growth and damage in the invaded state $q_{t+1}(X_t | N_t)$ is conditioned on the abundance in $t$, while damages in $(t+1)$ depend on the abundance of invader in $(t+1)$, $D_{t+1}(\hat{Z}_{t+1}^P; N_{t+1})$.

In $t$ the first-order condition for optimal collective prevention is given by

$$W_{S_t} = -U_t'(G_t; \alpha) C_{t,S_t} + \rho p_{t+1,S_t} q_{t+1}(U_{t+1}(G_{t+1}; \alpha) - U_{t+1}(H_{t+1}; \alpha)) = 0 \tag{5}$$

[4] The discount factor $\rho$ is related to the discount rate $r$ by $\rho = 1/(1+r)$.

where primes and subscripted variables indicate partial derivatives. Condition (5) requires the manager to employ prevention in $t$ up to the level in which the marginal costs of its current employment (first term) equals the discounted expected marginal benefits in the following period. The welfare gains are the result of a reduced probability of invasion and the increased chances of no damage in $(t + 1)$.

The first-order condition for collective control is in turn given by

$$\begin{aligned} W_{X_t} &= -U_t'(G_t;\alpha)C_{t,X_t} + \rho p_{t+1} q_{t+1,X_t}(U_{t+1}(G_{t+1};\alpha) \\ &\quad - U_{t+1}(H_{t+1};\alpha)) = 0 \end{aligned} \tag{6}$$

which requires collective control to be employed in period $t$ up to the level that equates the marginal cost of control in the current period to the discounted expected marginal benefits of control in the subsequent period. The marginal benefits result from a reduced chance of growth in the invaded state in $(t + 1)$.[5]

## 3 Comparative statics: discounting and risk aversion

It would seem apparent that a manager's choice of prevention and control depends in part on his preferences for time and for risk bearing. A manager with a high discount rate and high risk aversion is likely to make different risk-reduction decisions than if he/she was far-sighted and risk-neutral. Herein the influence of these postulates over a manager's prevention and control choices are explored.

*Discounting*

Let's first consider the comparative statics for time preferences. For computational simplicity, the discount factor is employed – the inverse of the discount rate. First consider prevention. Using the first-order conditions (4) and (5), the implicit function theorem, and assuming the Hessian matrix $H$ is negative definite, the comparative static for prevention yields

$$\frac{\partial S}{\partial \rho} = \frac{\overbrace{-\{EMBP_\rho\}W_{xx}}^{Direct\ \ Effect} + \overbrace{\{EMBC_\rho\}W_{sx}}^{Indirect\ \ Effect}}{|H|} \tag{7}$$

The first term in the numerator on the right-hand side is the direct effect of the discount factor on prevention; the second term is the indirect effect. The sign and magnitude of the direct effect depends on how a change in

[5] Note that throughout we assume the second-order sufficiency conditions are maintained for optimal collective decision making, with a negative definite Hessian matrix $H$ such that $W_{X_t X_t} < 0$, $W_{s_t s_t} < 0$ and $|H| = W_{X_t X_t} W_{s_t s_t} - (W_{X_t s_t})^2 \geq 0$.

the discount factor affects the expected marginal benefits of prevention (*EMBP*) or,

$$\frac{\partial EMBP}{\partial \rho} \equiv EMBP_{\rho} = p_{t+1,S_t} q_{t+1}(U_{t+1}(G_{t+1};\alpha) - U_{t+1}(H_{t+1};\alpha)) \tag{8}$$

*EMBP* are increasing in the discount factor (decreasing in the discount rate). Given $W_{xx} < 0$ by the second-order conditions, the direct effect of the discount factor on prevention is positive – a greater discount factor (i.e. a lower discount rate) implies more prevention.

The second term is the indirect effect. The effect is a function of a change in the discount factor on the expected marginal benefits of *control* (*EMBC*)

$$\frac{\partial [EMBC]}{\partial \rho} \equiv EMBC_{\rho} = p_{t+1} q_{t+1,X_t}(U_{t+1}(G_{t+1};\alpha) - U_{t+1}(H_{t+1};\alpha)) \tag{9}$$

*EMBC* are also increasing in the discount factor (decreasing in the discount rate). Given this, the indirect effect can either accentuate or attenuate the positive direct effect depending on the relationship $W_{sx}$. The term is given by

$$W_{SX} = U_t''(G_t;\alpha)C_{t,S_t}C_{t,X_t} - U_t'(G_t;\alpha)C_{t,S_tX_t} + \rho p_{t+1,S_t} q_{t+1,X_t} \times \begin{pmatrix} U_{t+1}(G_{t+1};\alpha) \\ -U_{t+1}(H_{t+1};\alpha) \end{pmatrix} \tag{10}$$

The first term on the right-hand side is negative given the assumptions over utility and cost functions. A reasonable assumption finds $C_{xs} = 0$ and the second term disappears. The third term can be shown to be negative as both $p_S$ and $q_X < 0$ and the term in parentheses being negative. $W_{sx}$ is therefore negative, such that the indirect effect attenuates the direct effect, perhaps to the point of reversing the sign – a higher discount rate increases prevention.[6]

How the discount factor affects control is summarised by

$$\frac{\partial X}{\partial \rho} = \frac{\overbrace{-\{EMBC_{\rho}\}W_{ss}}^{Direct\quad Effect} + \overbrace{\{EMBP_{\rho}\}W_{xs}}^{Indirect\quad Effect}}{|H|} \tag{11}$$

Again the first term is a direct effect of the discount factor on control, which is positive ($EMBC_{\rho} > 0$ and $W_{ss} < 0$). Again, a larger discount

[6] This reliance on relative magnitudes of direct and indirect effects reveals the need for more and better species-specific data on the underlying technology of invasive risk reduction.

factor (rate) implies more (less) control today. The second term is an indirect effect. Since $EMBC_\rho > 0$, the sign of the indirect effect again depends on $W_{xs}$, demonstrated to be negative above. The indirect effect of the discount factor (rate) is negative and the total effect depends on relative magnitudes.

Reiterating, a striking feature of the model is the equivalence between the direct effects of each strategy and the indirect effect of the other. But the relationship is not 1:1. This is because the direct effects are weighted by the other strategy's own effects on their employment ($W_{xx}$ and $W_{ss}$) and the indirect effects weighted by cross effects ($W_{xs}$ and $W_{sx}$). The key is to understand the relative magnitudes of the direct and indirect effects. If indirect effects are negligible as would be the case for small $W_{xs}$, and/or one of $EMBC_\rho$ or $EMBC_\rho$ being small in relation to the other, the direct effects of *both* strategies could dominate (and be negative in the discount rate). But if $W_{xs}$ is large and/or $EMBC_\rho$ and $EMBC_\rho$ are relatively similar in magnitude, the indirect effects become important. For example, if the direct effect on prevention is stronger than the direct effect on control, then an increase in the discount rate reduces prevention (through dominant own direct effect) and increases levels of control (through dominant indirect effect).

### *Risk aversion*

A similar procedure is followed to explore how differing levels of risk aversion affect prevention and control. In general, the direct and indirect effects of increased risk aversion yield indefinite comparative statics for two reasons. First, the direct impacts now depend on both the rate of change in the marginal utility of income and the rate of change in utility, which complicates the basic economic intuition that a more risk-averse manager should use more of the safer strategy, while shying away from the risky strategy (Briys and Schlesinger 1990). Second, the indirect effect again attenuates the direct effect, perhaps to the point of reversing the relationship. To see this, first consider the effect of changes in risk aversion on prevention:

$$\frac{\partial S}{\partial \alpha} = \frac{\overbrace{-\{EMBP_\alpha - MCP_\alpha\}W_{xx}}^{\textit{Direct Effect}} + \overbrace{\{EMBC_\alpha - MCC_\alpha\}W_{sx}}^{\textit{Indirect Effect}}}{|H|} \tag{12}$$

The first term in the numerator is the direct effect, which has three parts: change in the discounted expected marginal benefits of prevention (*EMBP*), $\frac{\partial[EMBP]}{\partial\alpha} \equiv EMBP_\alpha$; net of the change in current marginal opportunity costs of prevention ($MCP$), $\frac{\partial[MCP]}{\partial\alpha} \equiv MCP_\alpha$; and $W_{sx}$. Making generalisations now is problematic because $EMBC_\alpha$ depends on the rate of change in *utility* for a change in risk aversion, $U_\alpha(.) = \frac{\partial U(.)}{\partial\alpha}$ and because $MCP_\alpha$ depends (in part) on the rate of change in the *marginal utility of income* for a change in risk aversion, $U'_\alpha(.) = \frac{\partial U'(.)}{\partial\alpha} = \frac{\partial MU_I}{\partial\alpha}$. While the signs of both utility effects move in unison and could be either positive or negative (or zero, which implies no changes),[7] the most plausible case is positive when welfare is normalised across levels of risk aversion.[8]

For a positive utility effect, the sign of $EMBP_\alpha$ depends on the utility effect (positive in this case $U_\alpha(.) > 0$ ) such that

$$EMBP_\alpha = \rho p_{S_t} q_{t+1}(U_{t+1,\alpha}(G_{t+1};\alpha) - U_{t+1,\alpha}(H_{t+1};\alpha)) \tag{13}$$

and $EMBP_\alpha$ is negative following the maintained assumptions. Further, if $H$ is taken as the maximum wealth achieved (i.e. that with no adverse impacts of the invader) and is therefore equal to the normalisation level of wealth used,[9] then $U_{t+1,\alpha}(G_{t+1};\alpha) - U_{t+1,\alpha}(H_{t+1};\alpha) > 0$ delivering a negative $EMBP_\alpha$. Given $U_\alpha(.) > 0$ the $MCP$ is positive,

$$MCP_\alpha = U_{t,\alpha}{}'(G_t;\alpha)C_{t,S_t} \tag{14}$$

i.e. $MCP_\alpha > 0$. This implies a dollar saved today is more valuable to the manager. Therefore, since $EMBP_\alpha < 0$ and $MCP_\alpha > 0$ is subtracted, the direct effect is negative. This suggests that a more risk-averse manager actually has a direct incentive to reduce prevention. This result arises because the opportunity costs of prevention ($MCP_\alpha > 0$) have increased and because the expected future benefits have decreased ($EMBP_\alpha < 0$). A more risk-averse manager directly reacts unfavourably towards prevention because it is a riskier strategy relative to the control strategy – prevention acts to reduce the probability of invasion, while control acts to reduce the chance established invaders grow and cause damage. Risk-averse managers directly opt for less of the riskier strategy.[10]

[7] See the review in Eeckhoudt and Hammitt (2004).

[8] Normalising welfare for levels of risk aversion implies that $W(M) = M$, for all utility representations, where $M$ is the maximum wealth achieved in a system. Normalisation allows one to compare differing levels of risk aversion across a common metric.

[9] Such that as $H$ approaches the maximum achievable wealth $U_{t+1,\alpha}(H_{t+1};\alpha) \rightarrow 0$.

[10] In contrast, in the less likely case of negative utility effects, i.e. $U'_\alpha(.) < 0$ and $U_\alpha(.) < 0$, which implies $MCP_\alpha < 0$. The net in parentheses in (10) is negative and $EMBP_\alpha > 0$. Now the direct effect is positive, more risk aversion leads to more prevention. Richer managers are less impacted by an increase in risk aversion and they are still willing to use the riskier strategy.

Now consider the indirect effect, the second term in the numerator of (12). The indirect effect depends on the effect on discounted expected marginal benefits of *control* $EMBC\ \frac{\partial[EMBC]}{\partial\alpha} \equiv EMBC_\alpha$, the effect on current marginal opportunity costs of *control* $MCC\ \frac{\partial[MCC]}{\partial\alpha} \equiv MCC_\alpha$, and the cross effect $W_{xs}$ (demonstrated above to be negative). As in the direct effect, the $EMBC_\alpha$,

$$EMBC_\alpha = \rho p_{t+1} q_{t+1,X_t}(U_{t+1,\alpha}(G_{t+1};\alpha) - U_{t+1,\alpha}(H_{t+1};\alpha)) \quad (15)$$

depends on $U_\alpha(.)$ while the $MCC_\alpha$, i.e.

$$MCC_\alpha = U'_{t,\alpha}(G_t;\alpha)C_{t,X_t} \quad (16)$$

depends on $U'_\alpha(.)$. Again, for the most plausible case of a positive utility effect $MCC_\alpha > 0$. The $EMBC_\alpha$ depends (as with the direct effect) on the net $U_{t+1,\alpha}(G_{t+1};\alpha) - U_{t+1,\alpha}(H_{t+1};\alpha)$ which is positive if $H$ is taken as the maximum wealth, such that $EMBC\alpha < 0$ as $q_{t+1,X} < 0$ . The indirect effect is therefore positive and serves to attenuate the direct effect of risk aversion on prevention.[11]

Now consider how greater risk aversion affects control, which is summarised as

$$\frac{\partial X}{\partial \alpha} = \frac{\overbrace{-\{EMBC_\alpha - MCC_\alpha\}W_{ss}}^{\text{Direct Effect}} + \overbrace{\{EMBP_\alpha - MCP_\alpha\}W_{sx}}^{\text{Indirect Effec}}}{|H|} \quad (17)$$

Again the direct effect is the first term, the indirect the second. Note the correspondence between the direct (indirect) effect for prevention and the indirect (direct) effect for control.

Consider first the direct effect. For managers with $U'_\alpha(.) > 0$ and $U_\alpha(.) > 0$, it was shown above that $EMBC_\alpha < 0$ and $MCC_\alpha > 0$ so that the direct effect is negative. As a manager becomes more risk averse, he has a direct incentive to decrease control. More risk aversion increases his marginal gains of getting an extra dollar today (at an increasing rate) – this implies the opportunity cost of spending a dollar today has increased. That is, with increased risk aversion, a manager now puts more weight on both the extra dollar spent on control today and the extra dollar gained in the future. Money means more to him, today and into the future. He is tempted to save his dollar today – but if the marginal benefit gained in the future is big enough, he will increase control.[12]

[11] If $U_\alpha(.) < 0$, then $MCC_\alpha < 0$, $EMBC_\alpha > 0$ and the indirect effect is negative, which will attenuate the positive direct effect in this case.

[12] For managers with $U'_\alpha(.) < 0$ and $U_\alpha(.) < 0$, $MCC_\alpha < 0$ and $EMBC_\alpha > 0$, so that the direct effect is positive. A more risk-averse rich manager increases control.

The indirect effect of risk aversion on control follows the implications of the direct effect on prevention. For a positive utility effect $U'_\alpha(.) > 0$, $U_\alpha(.) > 0$ and the most plausible scenario finds $EMBP_\alpha < 0$, $MCP_\alpha > 0$, and coupled with $W_{sx} > 0$ the indirect effect is positive and attenuates the direct effect.[13]

In summary, increasing risk aversion alters both the marginal opportunity cost of spending money on prevention and control today and the expected marginal benefits in the future. If the change in both utility and the marginal utility of income for a change in risk aversion is positive, the direct effects on both strategies are negative, while the indirect effects are positive and attenuate the direct effects. The reverse holds in the case of a negative change in utility and the marginal utility of income for a change in risk aversion, and the indirect effects continue to attenuate the direct effects. The overall sign remains ambiguous and will be determined by the relative magnitude of the direct and indirect effects. These magnitudes in turn are influenced by the marginal opportunity cost of spending money on prevention and control today, the expected marginal benefits in the future, and the relative magnitudes of the own and cross effects $W_{ss}$, $W_{xx}$, and $W_{xs}$. If the cross effect is small, the indirect effects will be minor, perhaps allowing the direct effects of *both* strategies to dominate. But if there is a high degree of connection, the indirect effects could well matter. Making precise statements about sign and magnitude of the direct and indirect effects is likely to depend on specific applications for specific species.

## 4 Application: zebra mussel invasion

### *4.1 Empirical model*

To illustrate how tastes and technology interact in invasive species management a specific case is employed in numeric simulation. First, the underlying biology of the invasion process is specified. Second, human and biological components are integrated using a stochastic dynamic programming (SDP) model (Bellman 1961). The SDP framework allows the explicit incorporation of uncertainty into the decision process and provides the flexibility to incorporate jointly mediated biological and economic behaviour.

[13] Again the situation changes when $U'_\alpha(.) < 0$ and $U_\alpha(.) < 0$, $EMBP_\alpha > 0$ and $MCP_\alpha < 0$. The indirect effect of risk aversion on control flips to being negative and continues to attenuate the direct effect, positive in this case.

In the applied model there is a finite set of states $i\{i = 0, 1, \ldots\ldots n\}$ and time $t\{t = 0, 1, \ldots, T\}$. Let states be defined by discrete levels of population abundance $N_{it}$ for each period $t$. Let the manager maximise discounted social welfare over the time horizon $T$. From the perspective of any particular state and period, assume the state variable $N_t$ is known before the manager makes decisions over controls $S_t$, $X_t$ and private firms make decisions over adaptation $Z_t^P$ (state subscripts are suppressed). These choices define social welfare as $W(D(Z_t; N_t), C(S_t, X_t, Z_t))$ for that period and state, a function of the damages $D$ caused by the abundance of the invader, and the costs of $S_t$, $X_t$ and $Z_t$. Future social welfare is uncertain because of the underlying stochastic ecological process governing transitions between states.

Transitions between states over time through population growth are Markov and governed by $N_{t+1} = f(\varepsilon_t, N_t, X_t)$, where $\varepsilon_t$ represents stochastic population growth. Control effort in period $t$ serves to lower the reproducing population in $t$, whereas prevention effort in period $t$ reduces the probability of invasion in $(t + 1)$.

Following Leung *et al.* (2002), the ecology of the invasion process is captured as a multi-state compound lottery. A continuum of states $N_{it}$ is allowed between 0 (unsuccessful invasion) and the carrying capacity $K$. Assume a clear differentiation in the points of contact between prevention, control and adaptation and the ecological system as described above.

The probability of invasion is specified as

$$p_{i,t+1}^a = p^b e^{-\Lambda s_{i,t}} \tag{18}$$

where $p_{i,t+1}^a$ is the realised probability of invasion in the following period. $p_{i,t+1}^a$ depends on the baseline probability of invasion $p^b$ and the manager's prevention effort $S_{i,t}$ in the current period. Parameter $\Lambda$ reflects the efficacy of mitigation efforts and $e$ is the exponential function.

Given an invasion in $(t + 1)$, the probability of growth $q_{i,t+1}$ depends on initial population $N_{i,t+1}^b$, which in turn depends on collective control efforts in the preceding period $X_{i,t}$ and stochastic population growth (from random variable $\varepsilon_{i,t}$). The process proceeds in two stages. First, in *period t*, collective control reduces the abundance of reproducing invaders (i.e. the kill function) during the transition to $(t + 1)$, hence

$$N_{i,t}^a = N_{i,t}^b e^{-\upsilon X_{i,t}} \tag{19}$$

where $N_{i,t}^a$ are the residual of initial invaders $N_{i,t}^a$ that survive control measures and may growth, and $\upsilon$ is a parameter describing the effectiveness of adaptation. The accompanying stock growth uncertainty from random variable $\varepsilon_t$ occurs through the logistic expression

$$N^b_{i,t+1} = N^a_{i,t} + rN^a_{i,t}\left(1 - \frac{N^a_{i,t}}{K}\right) + \varepsilon_{i,t} \tag{20}$$

$K$ is the invader's carrying capacity and $r$ the invader's intrinsic growth rate. Together (19) and (20) dictate the transition process/growth probabilities $q_{i,t+1}$. Combining (18), (19) and (20) defines the transition process.

For any given state and period, assume social welfare is a function of social net wealth $SW$. $SW$ consists of the (private) net income of a representative producer adversely impacted by an invasion, inclusive of collective expenditures on prevention and control. The resource manager takes the producer's optimal choices as given in the determination of optimal collective prevention and control. The producer hires factors of production labour $L$ and capital $K$ in the production of output $Q$. It is through excessive employment of these factors (in comparison with no establishment) that firms are able to adapt to the consequences of an invasion (such that $Z(L, K)$). Suppressing state and period subscripts, social welfare is

$$SW = \left[P_Q\hat{Q}(N) - C_L\hat{L}(N) - C_K\hat{K}(N)\right] - C_S S - C_X X \tag{21}$$

where hats indicate variables endogenous to the firm. Invaders cause damages directly to the firm, reflected in these variables through the functional notation (note collective strategies are also ultimately a function of invader abundance). $P_Q$ is the (constant) price of $Q$, $C_L$ is the wage rate, $C_K$ the rental rate of capital, $C_S$ stands for the per unit cost of preventative measures and $C_X$ reflects the per unit adaptation costs. Following Lichtenberg and Zilberman (1986), the productivity of damage adaptation strategies is captured through a Cobb-Douglas production function:

$$Q = \alpha L^a K^{b1} D(N^b)^c \tag{22}$$

where $\alpha$, $a$, $b1$ and $c$ are parameters and $D(N^b)$ a damage function relating the impacts of the invader population to monetary damages. The exponential specification of $D$ is modified so that it depends on the initial invader abundance $N^b$ and parameter $\lambda$:

$$D(N) = 1 - e^{-\frac{\lambda}{N^b}} \tag{23}$$

Equation (17) says that greater abundances of invaders increase the damage they cause, deviating $D$ from its uninvaded magnitude of unity.

For the application the private firm is taken to be an electricity generator, who is assumed to be risk neutral (although risk aversion on the part of the resource manager is considered). Given the regulated environment of the electric power industry such that firms must satisfy all the demand they face at regulated rates (Christensen and Green 1976), output levels are taken by the firm as exogenous.[14] Also assume firms hire inputs of production in an optimal fashion from perfectly competitive input markets. Exogenous output and input prices and endogenous factor employment make the dual formulation appropriate in the determination of optimal factor employment.[15] Adaptation effort $Z_t$ depend on the damages caused by the invader, and represented by additional factors the firm hires to compensate for the damages of the invader (given an exogenous output level).

The SDP framework allows for the inclusion of a wide range of resource manager risk perceptions. Social welfare (from the viewpoint of the resource manager) is characterised through a von Neumann-Morgenstern utility function and employs a functional form which allows a range of risk preferences – risk neutrality to increasing relative risk and decreasing absolute risk aversion (Holt and Laury 2002). The form of the utility function is as follows:

$$U(SW) = \frac{1 - e^{-\alpha_{sw} SW^{(1-r_{sw})}}}{\alpha_{sw}} \tag{24}$$

which exhibits risk neutrality when parameter $\alpha_{sw}$ approaches zero. The function captures increasing relative risk aversion and decreasing absolute risk aversion when both parameters $\alpha_{sw}$ and $r_{sw}$ are positive.

[14] An accompanying characteristic of the industry is that it possesses several inputs that are less variable than others, or quasi-fixed inputs. Additions and removals of generation assets typically require long periods of time, while the amount of electricity generated can vary substantially within the short run. While power generators may be able to hire variable inputs optimally, they may be in a temporary disequilibrium with respect to quasi-fixed inputs. While it would be preferable to incorporate these inputs into the analysis as demonstrated in Brown and Christensen (1981), Caves *et al.* (1981), Berndt and Hesse (1986) and Sickles and Streitwieser (1998), given the additional complexity their inclusion would force and data unavailability we are forced to investigate only short-run production.

[15] In any given state, optimal choices are

$$\hat{L} = \alpha^{\frac{-1}{a+b1}} \left[\frac{aC_K}{b1C_L}\right]^{\frac{b1}{a+b1}} \left[\frac{Q}{D(N^b)^c}\right]^{\frac{1}{a+b1}} \quad \text{and} \quad \hat{K} = \alpha^{\frac{-1}{a+b1}} \left[\frac{b1C_L}{aC_K}\right]^{\frac{a}{a+b1}} \left[\frac{Q}{D(N^b)^c}\right]^{\frac{1}{a+b1}}$$

Table 6.1 *Firms in the sample*

| Firm | Firm |
|---|---|
| Central Illinois Public Service Co. | Indiana Michigan Power Co. |
| Commonwealth Edison | Indianapolis Power and Light Co. |
| Illinois Power Company | Union Electric Co. |

*Parameters*

Parameters employed in the simulations follow from the specification of equations (18)–(24). Ecological parameters were selected to represent a generic invasion process. Following the hypothetical example of Leung *et al.* (2002), consider a generic zebra mussel invasion of a lake and its impact on a representative electricity-generation facility. Given the focus in this work on the importance of risk and temporal attitudes of the decision maker's decision process, observed data are employed in the parameterisation of the economic components to make the magnitudes of change in the results somewhat reasonable. Data on a small set of large electric utilities in the great lakes region (see Table 6.1) were collected from generators' filings with the Federal Energy Regulatory Commission (FERC)[16] to construct Table 6.2.

Variables are firm and year specific and measured at the level of the plant.[17] All monetary variables are deflated (with base year 1982) using the Consumer Price Index.[18] Data exist on $Q$, $L$, total revenues, expenses and wages. Determining capital proved to be challenging so we use capital as a measure of all inputs not related to labour, and is a broad aggregate.[19] Prices (evaluated at the sample mean) were found by dividing the expense total by the corresponding real total (see Table 6.3).[20] In

[16] Form 1 filings by the generators, by firm and year (1994–2000) were accessed from the RIMS web site http://rimsweb2.ferc.fed.us/form1viewer/. These Form 1 data were augmented with additional data accessed from historical FERC Form 423 filings – http://www.ferc.fed.us/electric/f423/F423annual.htm

[17] Plant-level data are obtained from firm-level data by simply dividing firm data by number of plants. Unfortunately we have only a single observation of number of plants for each firm, but given the quasi-fixed nature of these assets this is a reasonable restriction.

[18] US Department of Labour, Bureau of Labour Statistics – ftp://ftp.bls.gov/pub/special.requests/cpi/cpiai.txt

[19] Further complicating the issue is that so many components of an electricity-generation plant are quasi fixed and lumpy in their investment. To avoid these complications, we focus solely on annual capital inputs. These are measured as an aggregation of utility fossil fuel inputs, the summation of total BTUs from coal (in thousand tons), oil (in thousand barrels) and natural gas (in thousand MMBtu) consumption to construct our measure of annual capital inputs.

[20] The cost of zebra mussel control was set at $1.6 million per control event (consistent with data from large power plants, Leung *et al.* 2002), which includes costs of molluscicide and reduced production during treatment.

Table 6.2 *Variables in the sample*

| Variable | Definition | Source | Mean |
|---|---|---|---|
| $Q_t$ | Total output (MWH[a]) | Sales to ultimate customers | 2,440,937 |
| $L_t$ | Labour inputs (number of employees) | Number of employees | 434 |
| $K_t$ | Capital inputs (BTUs) | Inferred as the summation of Utility fuel BTUs: calculated as the product of the quantity of fuel (oil (1,000 tons), oil (1,000 barrels) and natural gas (1,000 MMBtu)) and the fuel-specific BTU content for each firm | 13,496,686 |
| $TR_t$ | Total revenues ($) | Total sales of electricity | 185,261,805 |
| $TC_{L,t}$ | Total labour costs ($) | Total salaries and wages | 29,775,675 |
| $TC_{K,t}$ | Total capital costs ($) | Capital expenses[b] | 78,847,876 |

[a] Mega Watt Hours.
[b] Where capital expenses are the residual of total electric ops and main exps net of total salaries and wages.

Table 6.3 *Prices*

| Variable | Definition | Calculated value |
|---|---|---|
| $P_Q$ | Price of output | 47.48 |
| $C_L$ | Wage rate | 4.29 |
| $C_K$ | Rental rate of capital | 3.64 |
| $C_X$ | Per unit cost of adaptation effort | 1.6 |

the calculations and calibrations, all monetary variables were scaled by millions of inflation-adjusted dollars and real variables also scaled for computational simplicity.[21] All remaining economic variables and baseline parameters went through a calibration procedure (see Table 6.4).

For the manager's utility function, the baseline parameters reflect risk neutrality. Across cases of risk aversion, for the initial level of risk aversion a value for $r_{sw}$ as estimated by Holt and Laury (2002) was arbitrarily employed. Their value of $\alpha_{sw}$ was increased by a factor of ten due to differences in baseline wealth in this study, and welfare is normalised for all levels of risk aversion. Together, these parameters represent the case of increasing relative risk aversion and decreasing absolute risk aversion, consistent with observed data in Holt and Laury (2002).

[21] Q and K scaled in millions and L in hundreds.

Table 6.4 *Parameters*

| Parameter | Definition | Baseline value | Parameter | Definition | Baseline value |
|---|---|---|---|---|---|
| $\alpha_{sw}$ | Utility function parameter | 0.029* | $c$ | Production function parameter | 0.416 |
| $r_{sw}$ | Utility function parameter | 0.269* | $\Lambda$ | Efficiency of mitigation effort | 2.303 |
| $C_S$ | Per unit cost of mitigation effort | 0.1 | $v$ | Efficiency of adaptation effort | 2.303 |
| $\alpha$ | Production function parameter | 0.641 | $\lambda$ | Damage function parameter | 660 |
| $A$ | Production function parameter | 0.161 | $K$ | Invading species carrying capacity | 1000 |
| $b1$ | Production function parameter | 0.423 | $r$ | Invading species intrinsic growth rate | 1 |

* Value from Holt and Laury (2002), not used in the baseline

No direct data exist for per unit control costs. In the baseline simulation it is assumed that the social planner employs a mix of control and prevention efforts and a sensitivity test used to develop a reasonable value. All production function parameters for equation (22) were based on the assumption that all firms in the sample minimise costs subject to their specified production function. Employing the necessary conditions, the definition of the production function, imposing constant returns to scale on the production function and data from Table 6.2, $\alpha$, $a$, $b1$ and $c$ were determined.

For ecological parameters, the baseline probability of invasion $p^b$ extrapolates the monthly value used in Leung *et al.* (2002) into an annual value of 0.0828. The level of the efficacy of prevention efforts, $\Lambda$, was found from manipulation of equation (18) and the assumption that a unit of prevention reduces the probability of invasion by 90 per cent. An identical procedure was followed to find $v$. A reasonable value for $\lambda$ followed from equation (24), the (assumed representative, along with $r$) invader carry capacity $K$, and the assumption that if the invader population were to achieve its carrying capacity, economic production would be reduced to 50 per cent of its non-damaged levels with all other variables held constant.

### 4.2 *Results*

Four key results emerge from the numerical simulations that examine how alternative levels of risk aversion and discounting affect the optimal

Table 6.5 *Risk preference structures*

| | Parameters | | Arrow-Pratt coefficients | |
|---|---|---|---|---|
| Risk preferences | $\alpha_{sw}$ | $r_{sw}$ | Absolute risk aversion | Relative risk aversion |
| Risk neutrality (RN) | $0.26 \times 10^{-6}$ | $0.269 \times 10^{-6}$ | 0 | 0 |
| Risk aversion 1 (RA1) | 0.029 | 0.269 | 0.3594 | 1.7970 |
| Risk aversion 2 (RA2) | 0.29 | 0.269 | 0.9299 | 4.6497 |
| Risk aversion 3 (RA3) | 0.39 | 0.269 | 1.1485 | 5.7428 |

mix of prevention and control, the probability of invasion and the welfare of the system. Overall, seventeen simulations were run – two baselines and fifteen variations. The first baseline was a lower bound case of 'no prevention-control'; the second run replicated the Leung *et al.* (2002) baseline case of risk neutrality and no discounting.[22] The remaining fifteen runs were over three increasing levels of risk aversion ($\alpha_{sw}$) and three alternate discount rates ($\delta$).[23] Table 6.5 presents the set of risk-preference parameters, including the Arrow-Pratt measures of absolute and relative risk aversion; the alternative discount rates were 0 per cent, 3 per cent, 5 per cent and 15 per cent.

Our primary task is to investigate the comparative static effects of risk aversion and discounting on the optimal mix of prevention and control. Our first result is somewhat counterintuitive, yet consistent with the comparative statics.

**Result 1**. *While ambiguities exist, in general an increase in either the discount rate or degree of risk aversion (holding the other constant) causes prevention to fall and control to increase. Managers with greater preferences towards today decrease investment in prevention and increase investment in control of invasive species. Regarding the comparative statics, this result arises because the direct effect dominates the indirect effect for prevention and vice versa for control.*

*Support*: Figure 6.2 presents the mean annual levels of prevention for alternative levels of time preference (discount rate) and risk aversion.[24] Overall, Figure 6.2 suggests that prevention falls given an increase in either the rate of time preference or the degree of risk aversion. Panel (a) shows for each level of risk aversion, increasing the discount rate lowers

22 We add the feature of exogenous demand for the representative producers good.

23 From the baseline, we chose three increasing levels of risk aversion and two rates of time preference based on criteria requiring an observable change in behaviour for successive increases in risk aversion and discounting, all else equal.

24 The annual levels are the values in each state weighted by the probability of being in that state.

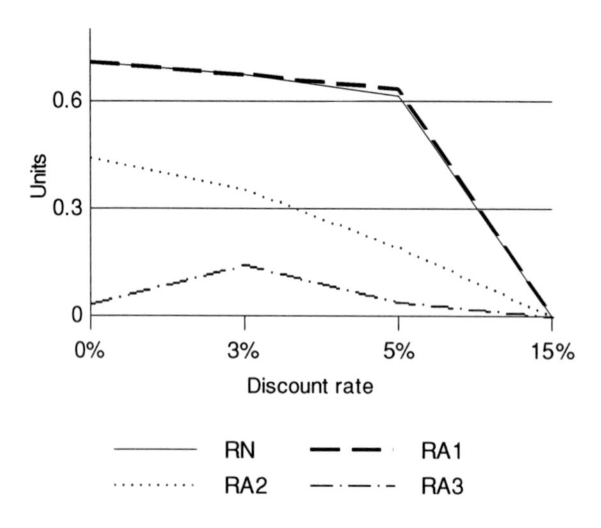


(a) Mean annual collective prevention

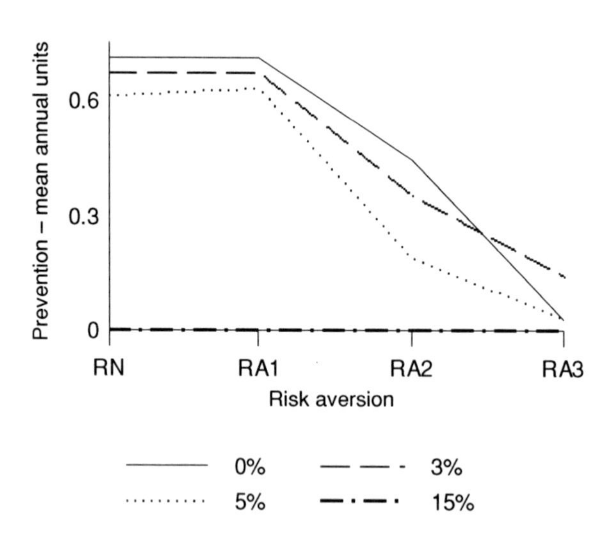


(b) Mean annual collective prevention

Figure 6.2. The influence of risk aversion and discounting on collective prevention

prevention (apart from the highest level). For the highest level of risk aversion, low discount rates lead initially to a slightly higher mean level of prevention, but as the discount rate is increased further, prevention falls. Panel (b) illustrates a similar pattern for mean prevention given an increase in the degree of risk aversion. In most cases, greater discounting leads to less prevention. The exception is the extreme discount rate.

Figure 6.3 shows the impacts on control fall in the opposite direction. Panel (a) shows that a higher discount rate leads to more control across degrees of risk aversion. The exception is the highest risk aversion, which decreases with the change from 0 per cent to 3 per cent and increases for each further consecutive change. In panel (b) for each discount rate, increasing risk aversion increases control.

In interpreting the results of Figures 6.2 and 6.3, note they are annual means. The impacts of changes in the rate of time preference and the degree of risk aversion serve to alter both the magnitudes of each strategy and their timing. Increases in the discount rate shift the period when prevention is abandoned towards the present (for the case of a risk-neutral decision maker with an increase in the discount rate from 0 per cent to 5 per cent see Figure 6.4, panel a which displays the time path of prevention and control). Lower levels of prevention in latter periods cause the probability of invasion to increase (panel a, Figure 6.5), which is followed by larger populations (panel b). The resulting damages require increased levels of adaptation (the results for capital employment are representative of those for labour and displayed in Figure 6.6, panels a and b) and prompt more control in following periods (Figure 6.4a).

In terms of the comparative statics, given our parameterisation the direct effects confirm the comparative static results and find a negative relationship between the discount rate and prevention and control. The indirect effects attenuate the direct effects, and the response to prevention dominates such that its indirect effect overwhelms the direct effect on control. The dominance of the response to prevention follows from prevention being a less effective strategy, only reducing the chance of invasions. Control reduces the chance of growth in all states. Prevention is the riskier strategy, making its direct effect more responsive to a change in the interest rate. With a dominant direct effect on prevention, its indirect influence over control overwhelms control's direct effect, so that control increases with an increase in the interest rate.

The consequences of an increase in the degree of risk aversion are similar, as shown for an increase from risk neutrality (RN) to the second level of risk aversion (RA2) in Figure 6.4b, for a zero discount rate. The increase in risk aversion serves both to terminate prevention in earlier periods and to delay the date of implementation. This serves to increase

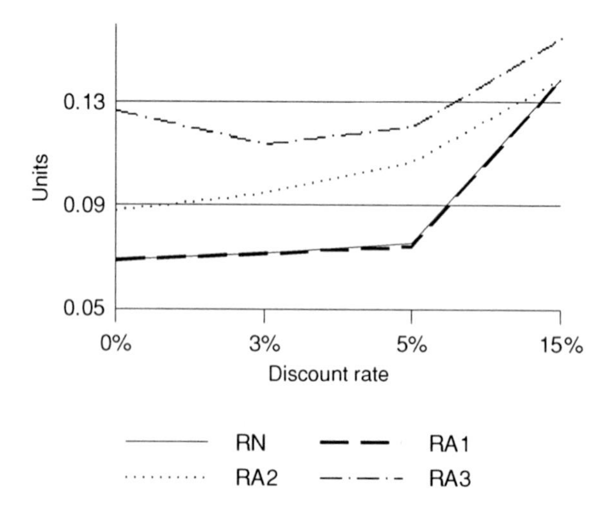


(a) Mean annual collective control

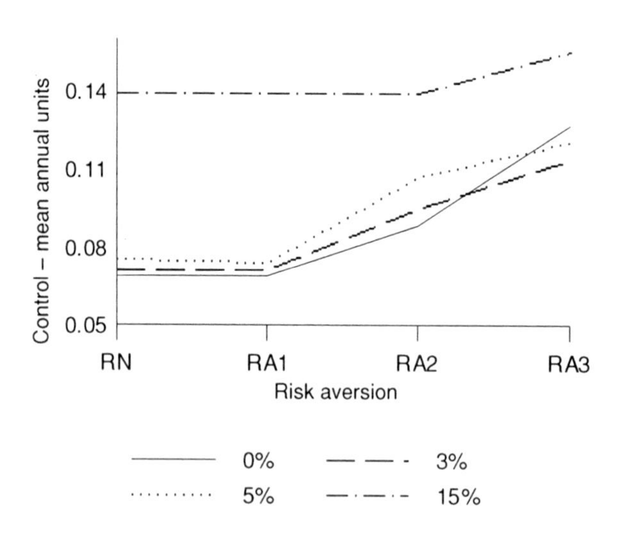


(b) Mean annual collective control

Figure 6.3. The influence of risk aversion and discounting on collective control

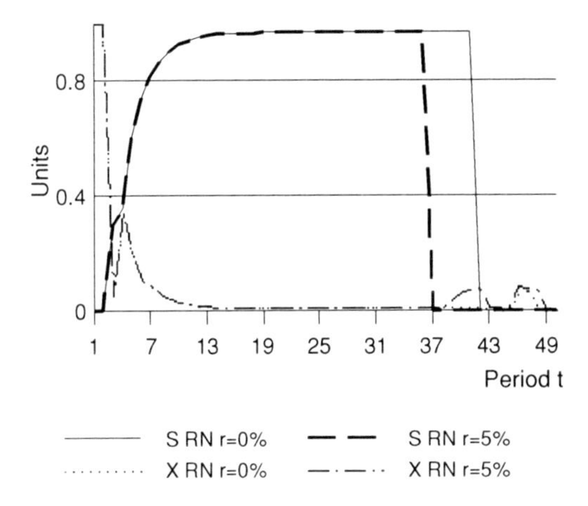


(a) Collective prevention (S) and control (X)

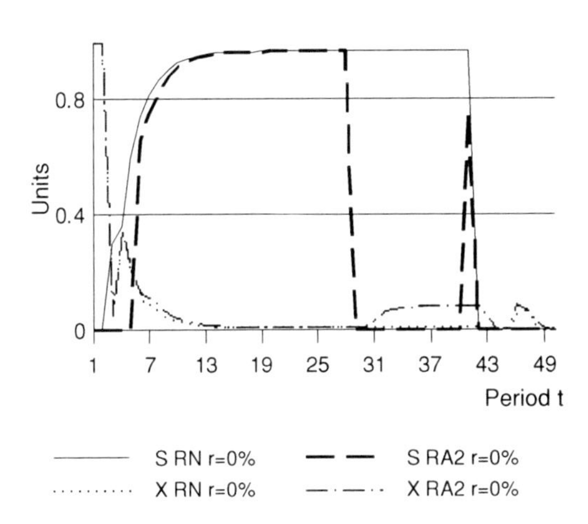


(b) Collective prevention (S) and control (X)

Figure 6.4. Dynamic effects of risk aversion and discounting on collective variables

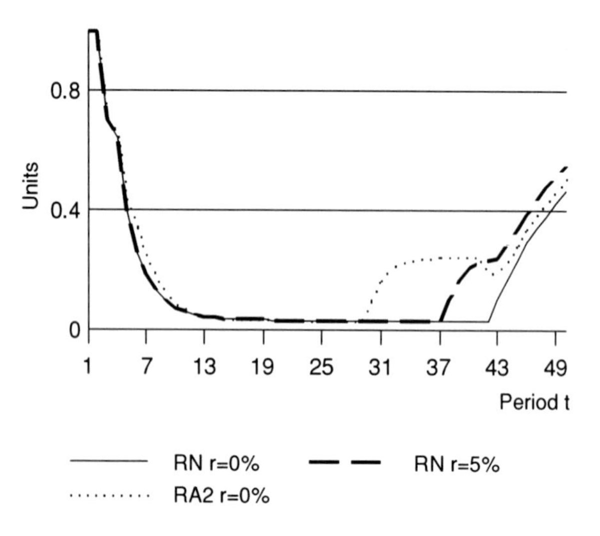


(a) Probability of invasion

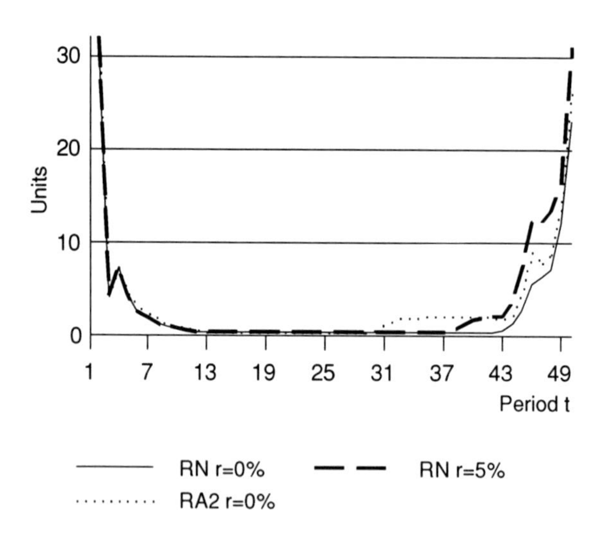


(b) Invader abundance

Figure 6.5. Dynamic effects of risk aversion and discounting on biological variables

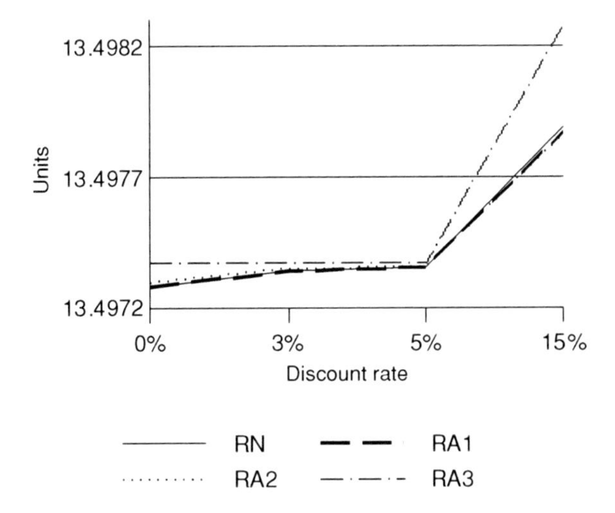


(a) Mean annual private adaptation

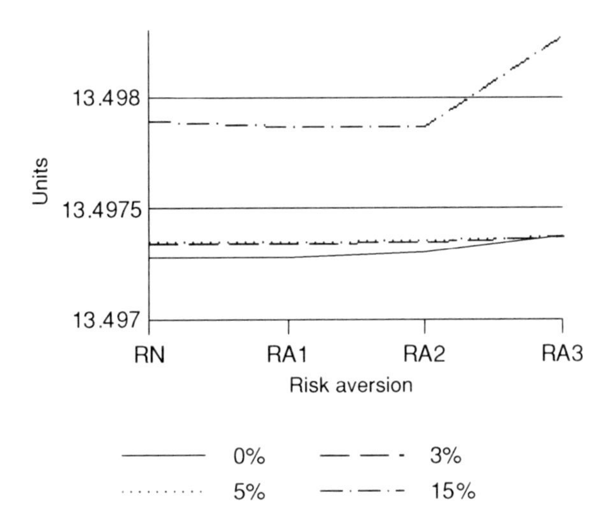


(b) Mean annual private adaptation

Figure 6.6. The influence of risk aversion and discounting on private adaptation

the probability of invasion at the end and beginning of the planning horizon, with resulting population, adaptation and lagged control increases (Figures 6.5a, 6.5b and 6.4b).

From the comparative statics, the direct effects of risk aversion on prevention and control are negative as the opportunity costs of current expenditures have increased and expected future benefits have decreased. Following similar reasoning as with discounting, indirect effects attenuate the direct effects. As prevention is a riskier strategy relative to the control, risk-averse managers opt for less of the riskier strategy and the direct effect on prevention dominates so that prevention is decreased and control is increased.

The ambiguities that emerge in our simulations do not contradict the findings of the literature that increased risk aversion does not necessarily lead to more mitigation (Dionne and Eeckhoudt 1985; Briys and Schlesinger 1990; Lee 1998). This occurs because mitigation affects probabilities not utility, which implies the convexity of the comparative statics is not guaranteed. In our multi-stage, compound lottery application, both prevention and control are forms of mitigation, although control is similar in many respects to self-insurance-cum-protection following Lee (1998), in which control serves to reduce the probability of establishment and therefore to reduce damages of established invaders.

**Result 2.** *In general, a manager with greater discount rates or greater risk aversion (holding the other constant) causes firms' adaptation to increase.*

*Support*: Changes in factor employment capture how discounting and risk aversion affect adaptation (Figure 6.6 presents the results for capital employment and are representative of those for labour). Panel a illustrates that for all degrees of risk aversion, increasing the discount rate increases adaptation. The impacts of risk aversion on adaptation are similar in b but not as responsive (for our arbitrary changes in risk aversion). Noting the complimentarity between adaptation and collective control (Figure 6.6 with Figure 6.3), these results are largely a function of the impact of prevention and control on the probability of invasion (Figure 6.7) and invader population (Figure 6.8).

**Result 3.** *A manager with greater discount rates or greater risk aversion increases the probability of invasion and invader populations.*

*Support*: The upper panels (a) demonstrate that the probability of invasion and invader population increases for all degrees of risk aversion as the discount rate is increased. The lower panel displays a similar effect for each level of the discount rate as the degree of risk aversion is increased. The difference between the two is that the mean population

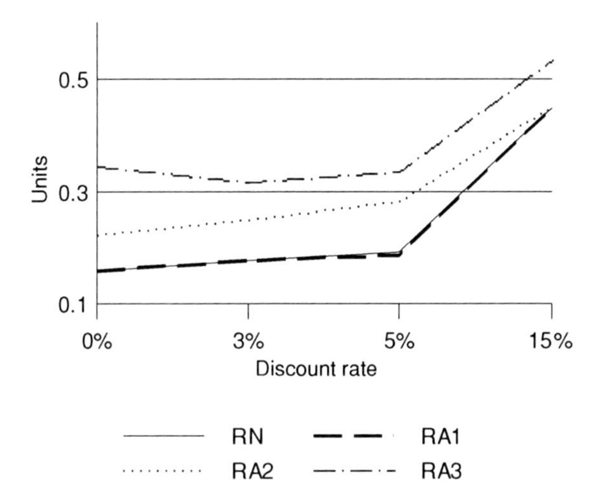


(a) Mean annual probability of invasion

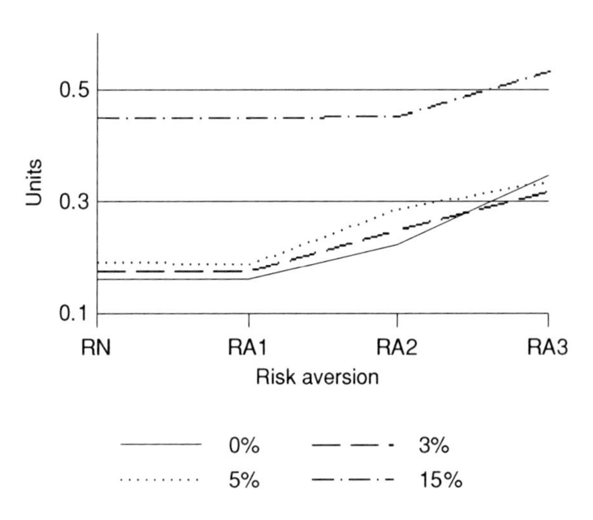


(b) Mean annual probability of invasion

Figure 6.7. The influence of risk aversion and discounting on the probability of invasion

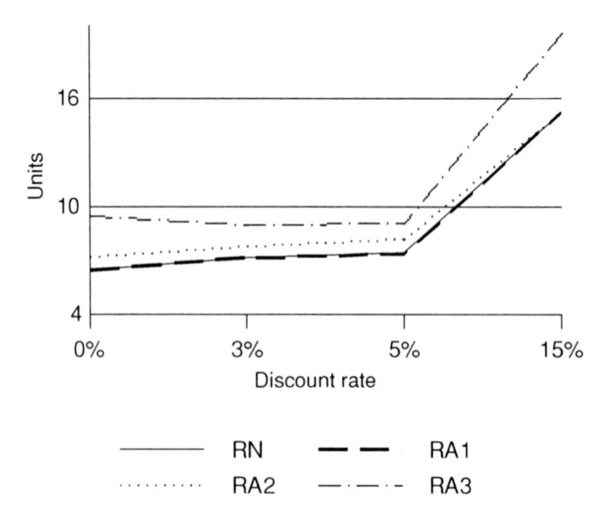


(a) Mean annual invader abundance

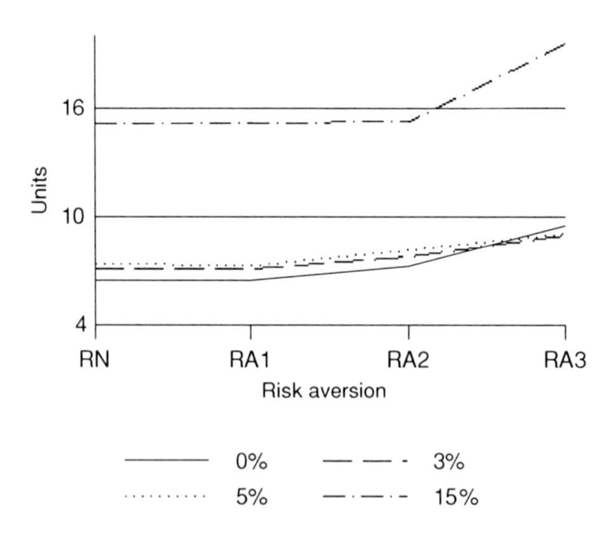


(b) Mean annual invader abundance

Figure 6.8. The influence of risk aversion and discounting on invader abundance

rises faster for changes in the interest rate and not as fast for changes in the degree of risk aversion. This in turn is due to prevention (control) decreasing (increasing) faster for changes in the interest rate than for changes in the degree of risk aversion. Overall, the consequences of changes in risk aversion and discounting on the optimal mix of prevention, control and adaptation can be readily observed by the impacts on the invader population and probability of invasion (cf. Figures 6.7 and 6.8). Increasing either the rate of time preference or the degree of risk aversion increases both the probability of invasion (which may cause damage through established populations) and invader populations (which cause damage).

**Result 4**. *First, a more risk-averse manager implies (holding the discount rate constant) a lower level of overall welfare – but the relative magnitude depends on the level of discounting. Second, if the initial discount rate is non-zero, further increases in the discount rate (holding risk aversion constant) also cause welfare to fall. On the flip side, these suggest with a more farsighted and less risk-averse manager, the greater the welfare, as he or she substitutes into prevention and away from control so firms need not adapt by as much.*

*Support*: As the probability of invasion and growth (from invader populations) rises, the consequences of damages outpace any reduced prevention costs and mean annual welfare falls (see Figure 6.9). Table 6.6 presents the changes in *cumulative* welfare from the baseline of risk-neutral risk preferences and no discounting. For comparability, the measures are not discounted and found as the sums of expected annual welfare over each fifty-year time horizon.[25]

For each level of risk aversion apart from the highest, increasing the discount rate lowers cumulative welfare. For the extreme level of risk aversion, low levels of discounting actually increase cumulative welfare in comparison to a zero discount rate. This is because discounting smoothes control and prevention over time, reducing fluctuations in invasion probabilities and invader populations. Large increases in the discount rate make the manager so shortsighted that invasion probabilities and populations are allowed to rise to high levels requiring immediate control and adaptation, lowering cumulative welfare.

The impacts on cumulative welfare for changes in risk aversion depend on the magnitude of the discount rate. At low discount rates, there is no change in cumulative welfare for a small increase in risk aversion from risk

[25] The change in cumulative welfare from the baseline to a policy of no action is –$2,166,494,205.

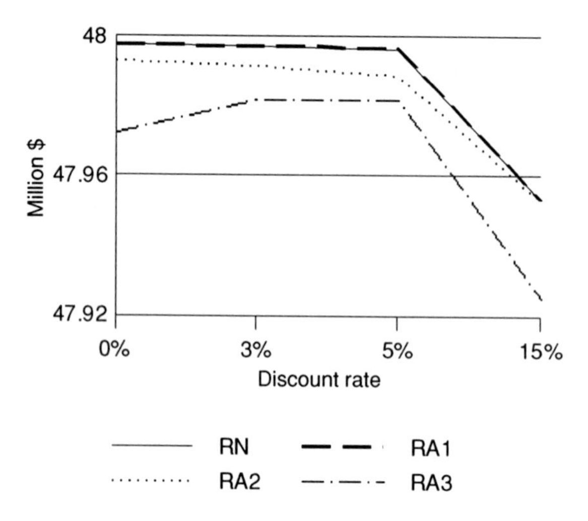


(a) Mean annual welfare

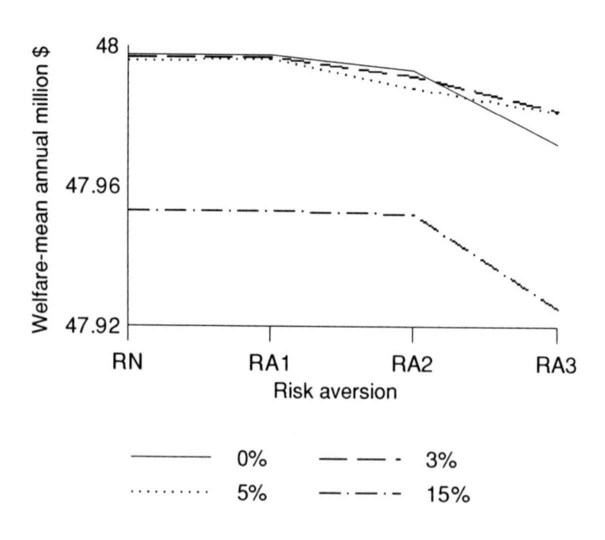


(b) Mean annual welfare

Figure 6.9. The influence of risk aversion and discounting on welfare

Table 6.6 *Changes in cumulative welfare from baseline ($)*

| | Discount rates | | | |
|---|---|---|---|---|
| Risk aversion | 0 | 3 % | 5 % | 15 % |
| RN | Baseline | −$13,580 | −$62,360 | −$2,219,330 |
| RA1 | $0 | −$13,580 | −$45,180 | −$2,214,360 |
| RA2 | −$228,100 | −$302,140 | −$453,280 | −$2,250,120 |
| RA3 | −$1,252,360 | −$780,840 | −$795,080 | −$3,609,050 |

neutrality (RN → RA1). As the degree of risk aversion is increased to RA2 and RA3, cumulative welfare falls. In contrast, at high discount rates, the first increment of risk aversion increases cumulative welfare over that for RN (due to a slight increase in control near the end of the time horizon), while further increases in risk aversion all lower cumulative welfare. Welfare falls because increasing discount rates and risk aversion induces a manager to reduce prevention and increases control, which causes firms to adapt more. Firms are forced to adapt to satisfy exogenous demand. The increased expenditures on control and adaptation dominated the reductions in prevention and therefore welfare falls.

## 5 Conclusions

This chapter presents a dynamic theory of endogenous risk to frame the question on how to manage the prevention and control of non-indigenous species (e.g. zebra mussels in the great lakes). Our model accounts for both biological and economic circumstances of invasions and the feedbacks between the two systems. How differences in manager preferences over time and for risk bearing influence the optimal mix of public prevention and public control were explored, and how that affects private adaptation. In general, the impacts are species specific, resting on whether direct effects on prevention and control dominate the other through indirect effects. The model was then implemented using stochastic dynamic programming to consider how preference changes affect the mix of prevention and control, the probability of invasion and the overall welfare of the system.

While ambiguities still exist, as expected, less risk-averse managers who are farsighted invest more in prevention, less in control and require less private adaptation by firms than more risk-averse and myopic managers. Reduced risk aversion on the part of the manager yields lower probabilities of invasion, lower populations and increased welfare. More farsighted and less risk-averse managers achieve greater welfare as they switch to prevention from control such that firms can reduce their investments in adaptation. This raises a confusing issue for invasive species management. Ex ante one might expect discounting and risk aversion to have similar effects on a manager's mix of risk-reduction technology – prevention should increase with both a lower discount rate and greater risk aversion. The opposite effect is found, however, for risk aversion. While a more farsighted manager does invest in more prevention, a more risk-averse manager does not. Risk aversion cuts two ways. Risk aversion induces a manager to want to avoid risk – both from the invader and from the input used. Since prevention is a riskier input relative to control, a more risk-averse manager goes with the safer bet – control. A better understanding of how such effects might influence the actual implementation of invasive species policy suggests more exploration into the underlying preferences of managers would be worthwhile.

## REFERENCES


Bellman, R. 1961. *Adaptive Control Processes: A Guided Tour.* Princeton, NJ: Princeton University Press.

Berndt, E. R., and Hesse, D. M. 1986. Measuring and assessing capacity utilization in the manufacturing sectors of nine OECD countries. *European Economic Review*. **30**. 961–989.

Bossenbroek, J. M., Nekola, J. C. and Kraft, C. 2001. Prediction of long-distance dispersal using gravity models: zebra mussel invasion of inland lakes. *Ecological Applications*. **11**. 1778–1788.

Briys, E. and Schlesinger, H. 1990. Risk aversion and the propensities for self insurance and self protection. *Southern Economic Journal*. **57**. 458–467.

Brown, R. S. and Christensen, L. R. 1981. Estimating elasticities of substitution in a model of partial static equilibrium: An application to U.S. agriculture, 1947–74. In E. R. Berndt and B. C. Fields (eds.). *Modeling Natural Resource Substitution*. Cambridge, MA: MIT Press. 209–229.

Caves, D. W., Christensen, L. R. and Swanson, J. A. 1981. Productivity growth, scale economies, and capacity utilization in U.S. railroads, 1955–74. *American Economic Review*. **71**. 994–1002.

Christensen, L. R. and Green, W. H. 1976. Economies of scale in U. S. electric power generation. *Journal of Political Economy*. **84**. 655–676.

Crocker, T. D. and Tschirhart, J. 1992. Ecosystems, externalities and economies. *Environmental and Resource Economics*. **2**. 551–567.

Dionne, G. and Eeckhoudt, L. 1985. Self-insurance, self-protection and increased risk aversion. *Economics Letters*. **17**. 39–42.

Eeckhoudt, L. and Hammitt, J. K. 2004. Does risk aversion increase the value of mortality risk? *Journal of Environmental Economics and Management*. **47**. 13–29.

Ehrlich, I. and Becker, G. 1972. Market insurance, self-insurance, and self-protection. *Journal of Political Economy*. **80**. 623–648.
Feder, G. 1979. Pesticides, information, and pest management under uncertainty. *American Journal of Agricultural Economics*. **61**. 97–103.
Finnoff, D., Shogren, J., Leung, B. and Lodge, D. 2005. The importance of bioeconomic feedback in invasive species management. *Ecological Economics*. **52**. 367–381.
Holt, C. A., and Laury, S. K. 2002. Risk aversion and incentive effects. *American Economic Review*. **92**. 1644–1655.
Lee, K. 1998. Risk aversion and self-insurance-cum-protection. *Journal of Risk and Uncertainty*. **17**. 139–150.
Leung, B., Lodge, D. M., Finnoff, D., Shogren, J. F., Lewis, M. A. and Lamberti, G. 2002. An ounce of prevention or a pound of cure: bioeconomic risk analysis of invasive species. *Proceedings: Biological Sciences*. **269**. 2407–2413.
Lichtenberg, E. and Zilberman, D. 1986. The econometrics of damage control: why specification matters. *American Journal of Agricultural Economics*. **68**. 261–273.
Lodge, D. M. 2001. Responses of lake biodiversity to global changes. In F. S. Chapin III, O. E. Sala and E. Huber-Sannwald (eds.). *Future Scenarios of Global Biodiversity*. Heideberg: Springer. 277–312.
Mack, R. N., Simberloff, D., Lonsdale, W. M., Evans, H., Clout, M. and Bazzaz, F. A. 2000. Biotic invasions: causes, epidemiology, global consequences, and control. *Ecological Applications*. **10**. 689–710.
Pimentel, D., Lach, L., Zuniga, R. and Morrison, D. 1999. Environmental and economic costs of nonindigenous species in the United States. *Bioscience*. **50**. 53–65.
Ricciardi, A., Neves, R. and Rasmussen, J. 1998. Impending extinctions of North American freshwater mussels (*Unionoida*) following the zebra mussel (*Dreissena polymorpha*) invasion. *Journal of Animal Ecology*. **67**. 613–619.
Settle, C., Crocker, T. D. and Shogren, J. F. 2002. On the joint determination of biological and economic systems. *Ecological Economics*. **42**. 301–311.
Shogren, J. F. 2000. Risk reductions strategies against the 'explosive invade'. In C. Perrings, M. Williamson and S. Dalmazzone (eds.). *The Economics of Biological Invasions*. Northampton, MA: Edward Elgar.
Sickles, R. C. and Streitwieser, M. L. 1998. An analysis of technology, productivity, and regulatory distortion in the interstate natural gas transmission industry: 1977–1985. *Journal of Applied Econometrics*. **13**. 377–395.
von Neumann, J. and Morgenstern, O. 1944. *Theory of Games and Economic Behavior*. Princeton, NJ: Princeton University Press.
Weitzman, M. 1994. On the 'environmental' discount rate. *Journal of Environmental. Economics and Management*. **26**. 200–209.
Williamson, M. 1996. *Biological Invasions*. London: Chapman and Hall.

*Section C*

# International trade

# 7 Trade and renewable resources in a second-best world: an overview[1]

*Erwin H. Bulte and Edward B. Barbier*

## 1 Introduction

The past decade has witnessed a proliferation of texts on trade and pollution. Compared with this rapidly expanding literature, there are relatively few contributions on trade and renewable resource management.[2] This imbalance in the economics literature is not readily explained by lack of popular interest. The impact of trade liberalisation on renewable resource management and conservation is a highly contentious issue, fiercely debated outside academia by international bodies (e.g. the Convention on International Trade in Endangered Species, the International Tropical Timber Organization, the World Trade Organization and the World Bank), non-governmental organizations (e.g. TRAFFIC, the World Wildlife Fund, Greenpeace and Friends of the Earth) and the popular media (e.g. *The Wall Street Journal, The Economist, The Environment* and *New Scientist*). Mass demonstrations against globalisation and the freeing of world trade in recent years in Genoa, Copenhagen, Seattle and other cities hosting meetings of international policy-makers dominated the news worldwide, and the alleged negative impact of free trade on environmental resources was a major theme during these demonstrations.

While the topic 'trade and renewable resources' might be capable of arousing strong emotions in the public, it is a fair question to ask whether it is sufficiently different from other fields in economics to warrant attention as a separate and emerging academic field (albeit obviously an applied one). We argue that this is indeed the case. Compared with the literature on trade and agriculture, environment or exhaustible resources, the economics literature on trade and renewable resources stands apart

[1] We would like to thank the Council of the European Association of Environmental and Resource Economists (EAERE) for the opportunity to deliver the speech on which this paper is based at the 12th Annual Conference in Bilbao, Spain. We would also like to thank three anonymous referees for helpful suggestions and comments on an earlier draft.

[2] For instance, two recent surveys on renewable resource management in the *Journal of Economic Literature* and the *Journal of Environmental Economics and Management* do not mention this topic (Brown 2000; Wilen 2000).

for at least three reasons: i) the key role played by the institutional context as reflected in the resource management regime (i.e. optimal management vs. open access); ii) the inherently dynamic nature of resource management, with stock size adjusting over time to the opposing forces of replenishment and harvesting; and iii) the associated complex environmental issues beyond concern with just resource extraction (e.g. habitat conversion, non-use values, bio-invasions, biodiversity, etc.). Many resource stocks are not simply a production factor for the traded commodity at hand; they may also contribute to the stability and productivity of ecological systems that provide invaluable services to mankind and affect the welfare of individuals directly.

Imperfectly defined or enforced property rights and failure to internalise all external effects in extraction implies that natural resource management typically takes place in a 'second-best world'. Ever since pioneering work by Lipsey and Lancaster (1956), economists know that trade liberalisation in the presence of pre-existing distortions might yield ambiguous welfare results. The second-best nature of resource management makes it a particularly interesting topic for economic research on the effect of trade liberalisation.

We suggest that there are two distinct, albeit not necessarily conflicting, views on the relationship between trade and renewable resources. These two perspectives reflect the focal points of ecology and economics as scientific disciplines. Ecologists are typically interested in maintaining the integrity of ecosystems and ecological functions, whereas economists are often assumed to care predominantly about human welfare. In recent times, however, these two perspectives are increasingly converging. For example, for many environmental management problems, economists frequently consider ecological functions, or 'services', to be important arguments in welfare functions. Equally, ecologists are realizing that protecting and enhancing ecosystems requires understanding and controlling of the way in which humans exploit these systems to enhance their welfare. As will become apparent in our review of the economics literature on trade and renewable resources, recent advances in this field are increasingly adopting an integrated economic-ecological perspective to analysing this topic. Unfortunately, when it comes to the impact of international trade on the environment, the popular perception is that the views of some economists and some ecologists are still at polar extremes.

Consider these typical positions taken by antagonists and protagonists of free trade. Environmentalists often espouse the 'anti-free trade' view, which centres around concerns about economic scale relative to ecological limits, distribution, the balance of power between multinational enterprises and national governments, and the implied effects of globalisation on incentives for domestic governments to regulate resource use. The WWF, for example, argues that the World Trade Organization (WTO) threatens the environment and believes it is no coincidence that 'the Earth has lost 30 percent of its natural wealth' at the same time as 'the volume of world trade is 14 times greater than it was in 1950, and is growing at twice the pace of other economic activities'. Trade boosts production, consumption and transport – all to the detriment of resource systems. Trade liberalisation undermines important environmental treaties and might set the stage for a regulatory 'race to the bottom' in response to concerns about the private sector's competitiveness on international markets. In addition, trade liberalisation is claimed to potentially affect the social fabric of rural communities managing common property resources, undermining local institutions geared towards sustainable resource management. Causal observation of the devastating impact of the ivory trade on elephant populations, or the effect of the tropical timber trade on forest management in the Philippines or Ivory Coast, lends some credibility to the concerns of environmentalists.

The response of the caricature economist, in contrast, is that trade is unambiguously 'good'. By exploiting scale economies or differences in technologies, factor endowments or preferences, trade essentially relaxes a binding constraint and enhances welfare. While the first-order effects of trade liberalisation on resource management and stock conservation are likely to be ambiguous, all participating countries will experience Pareto-improving gains when conditions for a Walrasian economy are satisfied. Thus any 'losers' from the resource impacts arising from trade liberalisation could be potentially compensated by the 'winners' in the rest of the economy. Of course, economists acknowledge that actual economies do not satisfy Walrasian conditions. But the general perception is that, if trade does trigger substantial damage to the environment, such outcomes are typically associated with the presence of domestic distortions. Various case studies on trade and renewable resources seem to support this view. For example, the ivory trade resulted in excessive elephant slaughtering only because property rights to elephants (and ivory) were not enforced and range states had no incentive to protect and harvest this resource in a sustainable manner (Barbier *et al.* 1990). Likewise, the timber trade ravaged Philippine forests only because corrupt policy-makers had easy access to tempting rents (Ross 2001). Economists argue that the right response in such cases is to address the underlying problem through domestic regulation or environmental treaties, and not to restrict trade. Quite to the contrary, economists often argue that trade is good for conservation through the 'environmental Kuznets curve' (EKC) argument – trade stimulates economic growth and richer people demand more conservation of stocks which higher-income countries can now afford to protect. Similarly, trade can provide the incentives for regulators to manage their resource base more carefully, because higher prices for the resource may imply a 'payoff' for such a strategy (Swanson 1994). For example, although excessive timber-related deforestation is a major problem in many tropical forest countries, Barbier (2001, p. 147) argues that 'producer countries that take a long-term perspective on the development of their forest industries and through sustainable management of their forests are likely to gain substantially from the expanding international trade in timber products'.

In light of such opposing views it is no surprise that there are repeated calls for incorporating trade rules for improved resource management into the WTO. However, as Copeland and Taylor (2003, p2) point out, 'there is a rather large gap between what we know about the relationship between international trade and renewable resource management, and what we would need to know in order to evaluate policy proposals, design new international treaties, or amend WTO obligations'. Similarly, while multilaterial conventions like CITES and CBD have recently embraced the use of economic incentives to promote sustainable and efficient use of resources and wildlife, there is a lack of insight into how to make progress on this front.

The main objective of the chapter is to provide a survey of the current literature and to discuss state-of-the-art knowledge about the impact of trade liberalisation on (i) the incentive to invest in stock conservation, and (ii) welfare in resource-dependent economies. When stepping back from the theoretical Walrasian economy to allow for the existence of policy and market imperfections, we find that trade liberalisation generally has ambiguous effects in terms of both impacts. Opening up for trade can increase welfare, but when institutions are imperfect it might just as likely have the opposite effect. Similarly, environmentalists are sometimes right to fret about the consequences of trade for resource management, but there exist circumstances where trade promotes conservation and where banning trade could be detrimental and puts species at risk. Another objective of the current chapter, therefore, is to demonstrate the conditions which produce outcomes that conform to either the typical economist's or the typical environmentalist's views of trade and the environment, and to search for common ground between these two positions.

The importance of gaining a better understanding of the impact of trade on management of renewable resources is underscored by the simple observation that exports of key renewable resources continue to increase. Bourke and Leitch (2000) note that forestry has become a global activity (with ownership of forests and plans, concession rights increasingly held by foreign companies) and that exports of most forest products, as well as their value, have expanded considerably over the past twenty-five years.[3] Similarly, Vannuccini (2003) writes that some 40 per cent of world fish production enters international trade and that net exports of developing countries grew from $4 billion in 1981 to $17.7 billion in 2001. Within global trends there have been changes in the importance of different countries as exporters – for example, as the pattern of the timber trade shifts to value-added processed products such as wood pulp and paper, wood-based panels and furniture, developing countries such as Indonesia, Malaysia, Brazil, Chile and the Asian newly industrialising countries are emerging as leading exporters (Barbier 2001). Partly this may be due to changing markets (both domestic and international) and other factors. However, this also represents changes in resource conditions because of exactly the processes discussed later in this chapter. For example, Brander and Taylor (1997a, 1997b) argue it is no coincidence that countries like the Philippines and Cote d'Ivoire, with imperfect property rights to their valuable forest resources, have turned from net exporter to net importer of wood products.

We have organised the chapter such that our presentation of the results loosely follows historical developments in the literature. In the 1970s and 1980s, following the rapid spread of optimal control methods throughout the resource economics field, most of the work assumed the perspective of a benevolent planner who either harvests the resource or controls the harvesting by private agents at zero cost.[4] The main results of this literature are discussed in section 2. The perspective drastically changed in the 1990s and there was growing attention to the *polar opposite case* of open access resource extraction. This firmly places us in a second-best world and salient features of these models are addressed in section 3. North–South models are discussed in section 4. Currently the pendulum swings back again, moving from the case where there are virtually no institutions (open access) to the socially optimal case. But rather than taking the existence of a planner for granted, this new literature treats the institutional context as endogenous, following from the incentives and

[3] Globally, exports of industrial roundwood have increased by 22 per cent since 1970 to 120 million cubic meters (cum) in 1997; sawnwood and wood pulp have almost doubled to 113 million cum and 35 million metric tons, respectively. Wood-based panels have increased fivefold to 50 million cum, and paper and paperboard have quadrupled to 87 million metric tons.

[4] Of course, there were a few very influential papers in the 1950s dealing with open access (Scott, Gordon). But these papers did not deal with trade explicitly and their main insights were not used in trade models until the 1990s.

constraints that agents have at the micro level. This literature, as well as some other work on intermediate cases between the polar extreme cases, is discussed in section 5. The conclusions and recommendations for future research ensue.

## 2 Benchmark 1: optimal management and perfect property rights

The simplest analysis of trade and renewable resources is not about trade at all. Rather, it is about the consequences of changing the relative price of resource commodities. In this section we take the simple case of optimal resource management in a partial equilibrium setting as a starting point and then gradually complicate the analysis by adding general equilibrium considerations and trade with exogenous and endogenous prices. Throughout the chapter we assume that the country in question is 'resource abundant', which implies it faces a higher resource price after liberalisation and becomes a net exporter (at least in the short run). Of course, some countries are resource scarce so that the opposite holds and many of the effects discussed below will be reversed.

Regardless of whether a country is resource abundant or scarce, the welfare effects of trade liberalisation under optimal management with perfect property rights must be beneficial.[5] Trade relaxes a binding constraint and makes society as a whole better off. As usual, there are distributional issues as well – the resource industry and consumers may gain or lose, depending on whether the country is resource abundant or scarce. The impact on producer surplus is readily assessed by comparing the rents or profits associated with harvesting at low and high prices. For an optimally managed resource it always holds that raising the price of the resource commodity is consistent with increasing the net present value (NPV) of harvesting. The effect on steady state rent is more complex because of the backward-bending supply curve that is implied by hump-backed growth functions such as the logistic. While higher prices will unambiguously lower the optimal stock in the simple model, the associated equilibrium harvest level may go up or down.

[5] Maintaining the renewable resource stock is also likely to generate wider social values, or 'stock externalities', such as biodiversity values, watershed protection, carbon sequestration and non-use values. These values can be incorporated in the model by including the resource stock, $S$, as a direct argument in the welfare function, typically increasing the optimal resource stock in equilibrium. Failing to account for externalities (such that *suboptimal management* is taking place) implies that welfare effects of a change in the terms of trade are generally ambiguous (Anderson and Blackhurst 1992). This is demonstrated formally in a model of renewable resource management and trade by Barbier and Schulz (1997).

In the following section we consider in more detail the impact of trade liberalisation on stock conservation in a partial equilibrium context.

### 2.1 *Single-market bio-economic model*

Consider the most basic bio-economic model where a planner maximises the net present value of welfare from harvesting a resource stock and where the resource commodity is initially traded domestically only. Assume that harvesting ($H$) is defined by the well-known Schaefer production function (time arguments are omitted for convenience):

$$H = qES \tag{1}$$

where $q$ is a parameter (the so-called catchability coefficient), $E$ is aggregate extraction effort (a control variable for the planner) and $S$ is the extant resource stock, growing over time according to a logistic function. This defines the equation of motion for the stock

$$dS/dt = G(S) - H = \gamma S(1 - S/K) - qES \tag{2}$$

where $\gamma$ and $K$ are parameters (for now), representing the intrinsic growth rate and carrying capacity, respectively. Under autarky, the price received per unit of the resource varies according to the inverse demand function $p = D(H)$ with $D' < 0$. The planner maximises the discounted sum of consumer surplus and resource rent associated with this sector and his current value Hamiltonian ($\mathrm{H_c}$) reads as

$$\mathrm{H}_c = \int_0^H D(z)dz - \frac{cH}{qS} + \lambda[G(S) - H] \tag{3}$$

where $\lambda$ is the shadow price of the resource stock, $c$ represents the per unit cost of harvesting effort and where we have used $E = H/qS$ (from (1)). The necessary conditions for an interior steady state (with constant resource stock and shadow price) are

$$H^* = G(S^*) \text{ and} \tag{4}$$

$$r = G'(S^*) + \frac{cH^*}{S^*[qSD(H^*) - c]} \tag{5}$$

where $r$ is the interest rate and (*) denotes an optimal value. According to equation (4), any stock growth should always be harvested and equation (5) defines that, at the margin, the return to a unit of the resource in situ should equal the exogenous return on investments elsewhere in the economy. Condition (5) implies that the rate of return of the renewable resource may be broken up in two parts: the impact of a change in the stock on growth and the impact of a change in stock on harvesting costs (see Clark 1990 for details).

What happens to extraction in the resource sector when this economy is opened up for trade? Assume that we are talking about a small economy that faces an infinitely elastic demand function, or an exogenous world price $p$ for the traded resource commodity. With net benefits of extraction redefined to account for the fixed price as $B = pH - cE$ (i.e. the government now considers only resource rents or profits for producers), the equilibrium condition that defines the optimal stock, denoted by (^), becomes

$$r = G'(\hat{S}) + \frac{c\hat{H}}{\hat{S}[q\hat{S}p - c]} \tag{5.5'}$$

The implications of opening for trade for domestic stock conservation in equilibrium are now straightforward. In this simple context they boil down to the question of whether the new price $p$ is greater or smaller than the one under autarky defined by $D(H^*)$. Since the growth function is strictly concave ($G'' < 0$), it follows that in equilibrium $dS/dp < 0$. Therefore it holds that the domestic resource stock will (i) be unaffected if by accident $p = D(H^*)$ holds, (ii) be augmented whenever $p < D(H^*)$, and (iii) be smaller whenever $p > D(H^*)$ holds. For a resource-abundant country, therefore, the stock will fall after opening up for trade.

This unambiguous result has been used to interpret the effect of trade measures on resource conservation. For example, if restricting international trade in sea horses, ivory, exotic pets or tropical timber lowers the net price received by exporters (through a tariff), equation (5.5′) predicts that as a result the stock will increase. Trade sanctions appear to 'work' in this case.

However, it is clear from this simple model that this outcome depends on some rather restrictive assumptions. For example, Barbier and Rauscher (1994) demonstrate that a more realistic model that allows for a positive resource stock externality (e.g. biodiversity benefits) as well as extraction for both export and domestic consumption will lead to an ambiguous trade policy outcome. That is, one can no longer be certain that any trade intervention such as a ban or a trade tariff that lowers the terms of trade of the resource-exporting economy will always increase the long-run equilibrium resource stock of the economy. As we show next, other important extensions to the single-market model, such as incorporating the opportunity cost of habitat conservation, also indicate that trade sanctions are unlikely to 'work' unambiguously in terms of enhancing long-run resource stocks.

### 2.2 *Adding opportunity costs of habitat*

The basic model in 2.1 assumes that the only alternative to not harvesting a species is to leave it unexploited. This implies that the model is more suitable to analyse the case of managing marine systems (mainly fisheries) and trade, but less applicable to the case of terrestrial systems where alternative uses to land than just nature conservation exist. If the social planner in the economy does consider the opportunity cost of setting aside land as habitat in order to maintain the resource stock, foregoing potential returns from, say, agriculture or plantation forestry, then resource conservation in the long run will be affected.

To capture such effects, the basic model can be extended to include two sectors, e.g. agriculture and forestry, which are dependent on the same resource base, e.g. land or habitat area. In essence, such an extension now creates a $2 \times 1$ (two sectors and one factor) model of the small open economy, as land or habitat conversion by the non-traded sector (agriculture) affects production from the traded sector dependent on renewable resource exploitation (forestry). Such a model is especially important for analysing a small open economy dependent on exploitation of terrestrial-based renewable resources for export (e.g. timber or wildlife products), which is threatened by widespread land conversion to another economic activity (e.g. agriculture). In developing economies, such a model is relevant to analysing agricultural conversion of wildlife and biodiversity habitat (Swanson 1994; Barbier and Burgess 1997) and forestland (Barbier and Burgess 1997), or aquaculture conversion of mangroves (Barbier 2003).

If the economy takes into account this opportunity cost, then in the long run it will conserve less of the resource stock (i.e., $S^*$ is lower). As demonstrated by Barbier and Schulz (1997), including the opportunity cost of conserving land to maintain the resource stock implies that the comparative static effects of increasing the relative price of the resource commodity are now ambiguous. This leads to two opposite effects. On the one hand, the result will be increased exploitation of the resource stock as exports become more profitable; on the other, there is now less pressure to increase habitat conversion, as the value of wild lands (supporting replenishment of the valuable resource) increases. If the latter effect is strong, then banning or restricting trade is counterproductive as it triggers habitat conversion, undermining the system's ability to support the key resource in the long run.[6] Barbier *et al.* (1990) use this as an important

[6] Adding to the ambiguity, when there are non-use values associated with the stock, income effects may play a role as well – extra revenues from resource sales may lower the marginal utility of consumption and increase demand of 'nature'.

argument against banning the trade in ivory to protect elephants.[7] A similar argument suggests that trade sanctions on tropical timber products could actually further deforestation in exporting countries, as trade bans and punitive tariffs would increase the returns to agricultural conversion, which is the major cause of much tropical deforestation globally (Barbier *et al.* 1994; Barbier 2001).

### 2.3 *The 2 × 2 small open economy*

The small open economy model with renewable resources can be extended to a two-sector, two-factor (2 × 2) model. Such a model was developed by Kemp and Long (1984) to examine the conditions under which a small open economy may choose to specialise in production and exports of a relatively resource-intensive good or in a relatively labour-intensive good.[8] The resource harvesting process they consider is somewhat simpler than the one discussed above as the authors consider only the aggregate amount of resource extracted and thus harvesting costs are assumed independent of stock size. For the optimal stock in equilibrium, then, it must hold that $G'(S) = r$, or marginal growth of the wild stock must equal the discount rate.

For a small open economy the task is to choose the rate of harvest, $H$, so as to maximise

$$W = \int_0^\infty Y(p, H, L)e^{-rt}dt \tag{6}$$

subject to $dS/dt = G(S) - H$. In (6), $Y$ represents the aggregate output of the two-sector economy, which is a function of the terms of trade, harvest and the total endowment of the Ricardian labour factor, $L$. Both goods are produced with constant returns to scale technology. The terms of trade, $p$, represents the relative price of good 1 in terms of good 2, i.e. $p = P_1/P_2$, with the second good being the relative resource-intensive good. For given $p$ constant, Kemp and Long consider how aggregate

[7] Bulte and van Kooten (1999) analyse the ivory trade ban case in more detail, solving a Stackelberg game between regulator and poachers. There are damages associated with elephant conservation (akin to the opportunity cost of habitat). It is shown that the trade ban lowers elephant numbers when discount rates applied by African range states are sufficiently low (below 5 per cent) and that the reverse holds when discount rates are 'high'. The impact of the trade ban on the regulator's incentive to enforce anti-poaching regulation is further discussed in section 5.

[8] Unlike the two-sector, two-factor (2 × 2) model of open access management (pioneered by Brander and Taylor 1997 and discussed below), it is assumed that both factors are combined to produce the two goods.

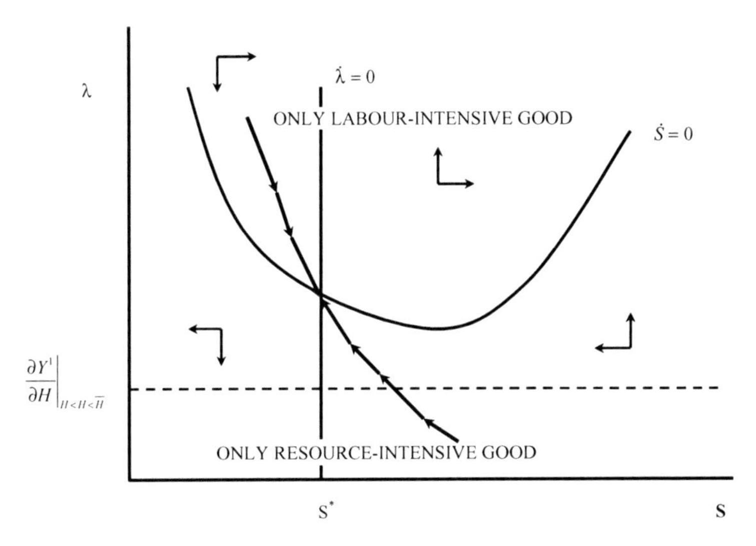


Figure 7.1. An equilibrium for the 2×2 small open economy
Source: Kemp and Long (1984)

output, $Y$, changes with increases in the harvest rate. Over a defined interval $(\underline{H}, \bar{H})$ the economy produces both the labour- and resource-intensive goods as the value marginal product of labour in production is the same. Over this interval, the aggregate output function is linear with respect to $H$ (i.e. $\partial Y/\partial H$ is constant) and this interval defines the 'threshold' for the economy to switch from specialising in one type of good to the other. Outside this interval, the economy specialises in one of the goods and $Y$ has the normal concavity properties with respect to harvest.

Kemp and Long demonstrate that in the steady state, which is a saddle point, it may be optimal for the economy to produce and export only the labour-intensive good, only the resource-intensive good, or produce both goods. However, as the economy passes along one or other of the stable arms towards the saddle, production may switch from one pattern of specialisation to another during the approach dynamics.

For example, Figure 7.1 illustrates the case where the economy specialises in the labour-intensive good (which does not imply that harvests are zero). The saddle path and equilibrium are depicted in $(\lambda, S)$ space, where as before $\lambda$ is the shadow price of the resource stock. As shown in Figure 7.1, the switching threshold occurs well below the steady state $(\lambda^*, S^*)$. It follows that, if the economy begins with relatively low resource stocks, $S < S^*$, then it will follow the left-hand arm of the saddle path and always specialize in the labour-intensive good. In contrast, if the economy is relatively resource abundant initially, $S > S^*$, then the economy will specialize first in producing and exporting the resource-intensive good but will switch eventually to specializing in the labour-intensive good. Note that, along this right-hand arm of the saddle path, the economy continues to overexploit the resource stock, which eventually declines to the steady state level $S^*$.

Kemp and Long also consider the effects of an increase in the terms of trade, $p$, i.e. an increase in the price of the relatively resource-intensive good. The result is an increase in the interval of time during which the relatively resource-intensive good is produced and exported, and a decrease in the time in which the economy specializes in the labour-intensive good. In Figure 7.1, this is represented by a shifting up of the line defined by $\frac{\partial Y}{\partial H}\big|_{H<H<\bar{H}}$, and if this effect is sufficiently large so that this line now exceeds the steady state $(\lambda^*, S^*)$, it is possible that a small open economy will specialize in the resource-intensive good or produce both goods.

The Kemp and Long model is clearly a highly simplified $2 \times 2$ model of a small open economy. No consideration is made of the opportunity cost of habitat, stock externalities or even the cost of resource harvesting. Nevertheless, the conditions under which a small open economy may choose to specialize in production and exports of a relatively resource-intensive good or in a relatively labour-intensive good prove an important contrast to other models of resource-trade relationships.

For example, Matsuyama (1992) showed how trade liberalisation may lower welfare in a model with external benefits in the non-resource (or non-agricultural) sector – a departure from the first best world discussed thus far. If there are increasing returns at the sector level due to the spillover benefits of firms, then reallocation of labour from manufacturing to resource extraction will lower the returns of firms that remain in manufacturing (an external cost that will be ignored by individuals). If opening up for trade induces such a reallocation of labour, then total welfare may fall. This effect has been postulated as one potential explanation of the so-called resource curse effect – an empirical regularity suggesting that countries well endowed with resources tend to grow more slowly than their resource-poor counterparts.[9]

[9] This effect is more likely to eventuate when countries are richer in point resources (like oil fields and mines) than in diffuse resources such as agricultural land (see Leite and Weidmann 1999; Isham *et al.* 2003). Other possible explanations for the resource curse include Dutch disease (Sachs and Warner 1997) and rent seeking (Torvik 2002) and also the potential adverse effect of resource wealth on institutional development, indirectly impacting on economic performance (Isham *et al.* 2003).

## 3 The polar opposite benchmark: open access

One of the most important institutional frameworks to consider in a trade-renewable resource model is the situation of unregulated common property or open access. There is a long tradition in bio-economic models of the fishery in analysing open access problems.

In recent years, there has been increased recognition that the open access resource management problem extends beyond fisheries to other renewable resources, notably forests, wildlife, mangroves and other terrestrial-based resources. Only recently, however, have the implications of open access for the impacts of trade on renewable resource management been explored.

However, not all trade-renewable resource models that incorporate open access resource exploitation necessarily arrive at the same conclusions. The results for resource conservation and welfare differ depending on whether models are constructed in a partial or general equilibrium setting, and depending on assumptions regarding market structure, technology and scale economies. To illustrate these models, we adopt a similar approach as in discussing trade-renewable resource models under optimal resource management. First we consider the simplest case of a single-market bio-economic model. We then examine the case of the $2 \times 2$ model of a small open economy. The discussion of North–South models that incorporate various assumptions concerning open access resource exploitation will be discussed in section 4.

### 3.1 *Single-market bio-economic model*

As a counterpart to the basic single-market optimal management model developed in section 2.1, consider now the same model under open access exploitation. The key feature of the latter model is that aggregate extraction effort, $E$, is determined by the profits generated by resource extraction. That is, an increase in profits will lead to greater extraction in the open access industry, whereas a decline in profits will reduce extraction. This suggests that the dynamics of effort in open access resource extraction, both for autarky and free trade, can be represented by the following equation motion of

$$\dot{E} = \phi[D(H)qS - c]E \tag{7}$$

This differential condition implies that expectations are constantly adapting in response to observations in the field – some kind of 'backward-looking' behaviour (for a model of rational expectations, see Berck and Perloff 1984).

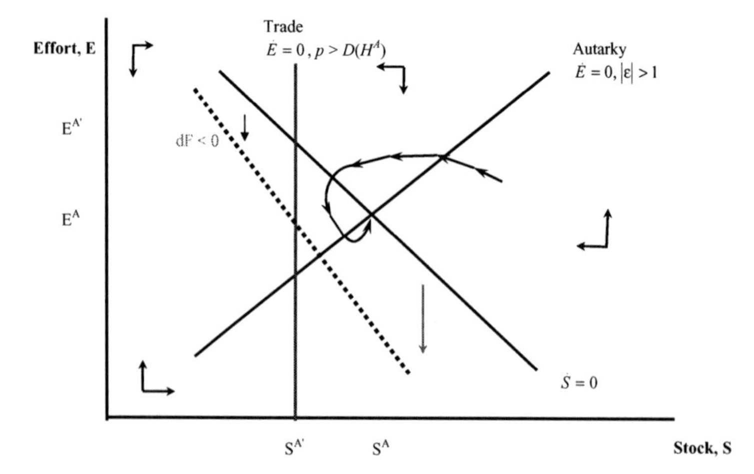


Figure 7.2. Open access equilibrium in the single-market bio-economic model
Source: Adapted from Barbier (2003)

We need to combine (7) with an equation that defines the dynamics of the resource stock. As before, we can assume both a logistic stock growth function for $G(S)$, and Schaefer harvest, $H = qSE$. This implies that stock dynamics are represented by equation (2) above.

Under autarky, the price received per unit of the resource varies according to the inverse demand function, $p = D(H)$ with $D' < 0$. Thus equation (7) indicates how exploitation effort adjusts, at some rate $\phi$, to the profits from harvesting. In the long run, under open access, profits from exploitation are driven to zero and the renewable resource stock is constant. The long-run steady state is therefore

$$E^A = \frac{G(S^A)}{qS^A} = \frac{\gamma(1 - S^A/K)}{q} \text{ and} \tag{8}$$

$$S^A = \frac{c}{D(H^A)q} \tag{9}$$

where the superscript '$A$' is used to denote open access equilibrium values. The long-run steady state and the dynamic path corresponding to this state are depicted in Figure 7.2 for the case where the price elasticity of demand, $\varepsilon = D(H)/D'(H)H$, is elastic, i.e. $|\varepsilon| > 1$.

Once again, it is insightful to see what happens to extraction in the resource sector when the economy is opened for trade. If the economy is small, it faces an exogenous world price, $p$, for the resource commodity it produces. Long-run condition (9) becomes simply

$$S^A = \frac{c}{pq} \tag{9'}$$

which is a vertical line in Figure 7.2. The implications of opening for trade for resource conservation are straightforward and similar to the optimal management case, in that they once again depend on whether the new price $p$ is greater or smaller than the price received under autarky defined by $D(H^A)$. That is, the domestic resource stock will be (i) unaffected if by accident $p = D(H^A)$ holds, (ii) augmented whenever $p < D(H^A)$, and (iii) smaller whenever $p > D(H^A)$ holds.

Assuming that prices rise as a consequence of opening up for trade, it follows that the resource stock will fall.[10] The latter effect is depicted in Figure 7.2, where the new open access equilibrium for the resource stock is indicated as $S^{A''}$. As also shown in Figure 7.2, however, equilibrium effort will rise, to $E^{A'}$.[11] Thus in the simple bio-economic model, the stock and harvest effects of opening up of trade are the same in the open case as in optimal management. However, these two cases do differ in one important respect. Under open access, all rents from resource extraction are dissipated in the long run. Opening up the economy for trade does not affect this outcome; rent in the steady state is zero and remains so. In other words, there are no welfare effects (in terms of changes in producer surplus) in the resource sector.

The open access model has been extended to include the impacts of habitat conversion (e.g. Bulte and Horan 2003). For example, Barbier and Strand (1998) consider a model of an export-oriented shrimp fishery in Campeche, Mexico that produces shrimp for export. According to ecological evidence, the fishery is supported by coastal mangrove systems that serve as breeding and nursery habitat for the shrimp fry. Thus, by assuming that increasing mangrove area $F$ effectively 'extends' the carrying capacity of the fishery, Barbier and Strand modify the net growth equation (2) to

$$\dot{S} = G(S) - H = \gamma S\left(1 - \frac{S}{K(F)}\right) - qSE, \quad K'(F) > 0 \tag{10}$$

Starting with an open access equilibrium under trade conditions (i.e. point $(S^{A'}, E^{A'})$ in Figure 7.2), Barbier and Strand show that the effect of threats to mangrove habitat, through coastal developments in Campeche,

[10] It is easy to see from (9′) that $\frac{dS}{dp} = -\frac{S}{p} < 0$. This also implies a unitary price-stock elasticity response.

[11] From (8), $\frac{dE}{dS} = -\frac{\gamma}{qK}$ and therefore $\frac{dE}{dp} = \frac{\gamma S}{pqK} > 0$.

is to reduce equilibrium effort in the fishery, but leave the resource stocks unchanged. This is shown in Figure 7.2 by the rotating down of the $\dot{S} = 0$ curve. The destruction of mangrove habitat creates a temporary disequilibrium in which stocks fall and so does harvest. Because export prices remain unchanged, a loss is made in the fishery causing some fishers to leave so that effort declines. This will mean less exploitation so stocks will recover and in the long run zero profits prevail once again. Thus in the new equilibrium, stocks return to their steady state level, $S^{A'}$, but effort has fallen (i.e. $E^A < E^{A'}$). The result must be lower levels of harvest and gross revenues in the fishery.[12]

In general, if a small open economy is exploiting and exporting a renewable resource under open access conditions, we would expect habitat conversion to have similar effects. This is readily apparent from (9′) and (10). If a reduction in natural habitat (i.e. a decline in $F$) shifts down the growth function, either by reducing $K$ or $\gamma$, then steady-state effort and thus harvest will fall but not steady-state resource stocks.

### 3.2 *The 2 × 2 small open economy*

Brander and Taylor (1997a) construct a 2 × 2 model of a small open economy by combining an open access renewable resource model with a standard Ricardian model of international trade.[13] One of the two goods is a resource good produced using labour and the resource stock. The other good is a generic 'manufactures' good produced using labour. Brander and Taylor consider such a model to be applicable to understanding the effects of international trade on open access renewable resource exploitation and believe that the insights of such a model are particularly relevant to small open developing economies. Overexploitation of many renewable natural resources – particularly the conversion of forests to agricultural land – often occurs in developing countries because property rights over a resource stock are hard to define, difficult to enforce, or costly to administer.

[12] Barbier and Strand employ this model, under the assumption that the fishery is a price-taker in export markets, to estimate the long-run losses from mangrove deforestation on the open access shrimp fishery in Campeche. See Barbier (2003) for the example of applying the model with a finite elasticity of demand to a case study in Thailand of the impacts of the expansion of aquaculture shrimp, a leading export product, on mangrove conversion and off-shore fisheries. The analysis demonstrates why Thailand has chosen to expand coastal aquaculture to increase export earnings from shrimp production, despite the economic consequences of the accompanying mangrove deforestation for coastal fisheries.

[13] Qualitatively similar results are obtained for a 2 × 2 × 2 model (see Brander and Taylor 1998).

The country has a fixed labour force and produces and consumes two goods. The 'manufactures' good, $M$, is treated as the numeraire whose price is normalised to one. In this constant returns to scale sector (but see below), one unit of labour is used to produce one unit of $M$ and, hence, labour's value marginal product in manufacturing is also one. It follows that, given competitive labour markets, the wage rate in the economy is one if both goods are produced. The second good is harvest, $H$, from a renewable resource stock, which is subject to the standard net biological growth relationship. Harvest is produced with the Schaefer production function $H = qES$ and the labour constraint implies $L = E + M$. The effect of open access exploitation in the resource sector is to ensure that the price of the resource good must equal its unit cost of production in equilibrium. That is, as all rents from using the resource stock are dissipated and only labour costs are incurred in harvesting, the equilibrium open access harvesting condition is always

$$p = w\frac{E}{H} = \frac{w}{qS} = \frac{1}{qS} \tag{11}$$

where $p$ is the (relative) price of the resource good. Equation (11) states that, under open access, the price of the resource good must equal its unit cost of production. Since the wage rate, $w$, in the economy is one (with diversified production), the unit labour requirements, and thus costs, of the resource sector are inversely related to the size of the stock.

To complete their model, Brander and Taylor assume a representative consumer endowed with one unit of labour, who has Cobb-Douglas preferences for both goods ($u = H^{\beta} M^{1-\beta}$). As this implies that both goods are essential, in autarky manufactures, $M$ must also be produced. The authors show that the ratio of the resource's intrinsic growth rate ($\gamma$) to total labour in the economy, $\gamma/L$, determines autarky relative prices. This ratio defines Ricardian 'comparative advantage'. Thus for some sufficiently high ratio of $\gamma/L$ a country would have an autarky price of the resource good less than the world price and can be considered relatively 'resource abundant'. For $\gamma/L > q$, the small economy may specialize in manufactures or the resource good (depending on relative prices) or be characterised by diversified production. For $\gamma/L < q$ it is impossible for the country to specialize in the resource sector.

Figure 7.3 illustrates the effects of opening of trade in a resource-abundant economy – a country where prices under autarky were below the world market level $p^*$ (or a country where $\gamma/L$ is sufficiently large). Figure 7.3a compares the initial post-trade impacts and the transition to the steady state, whereas Figure 7.3b contrasts steady state utility under autarky with various trade scenarios. Denoting $p^A$ and $S_A$ as the

(a) Temporary equilibrium and transition to a steady state

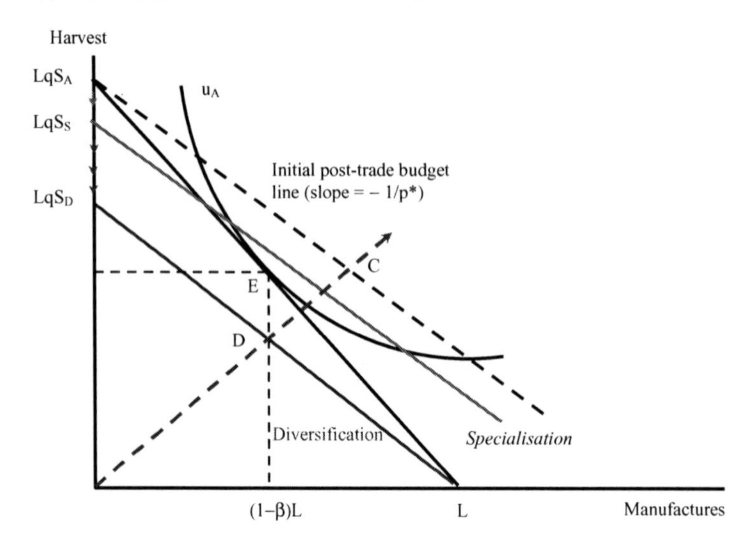


(b) Steady-state utility and the terms of trade

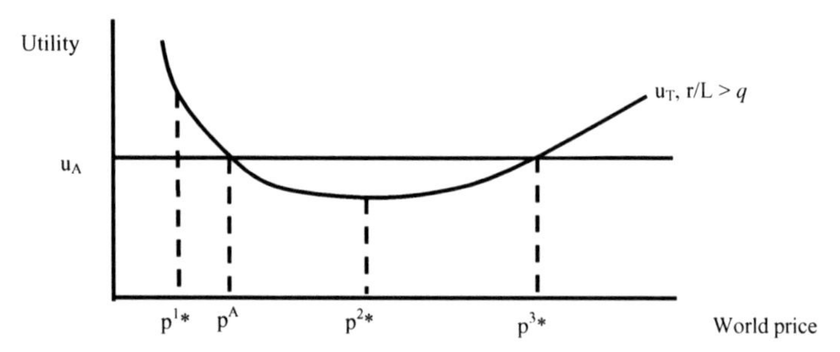


Figure 7.3. Open access resource exploitation and trade in a resource-abundant economy
Source: Brander and Taylor (1997)

autarky equilibrium resource price and stock respectively, we have from (11) $p^A = 1/q\,S_A$ as the initial condition describing this equilibrium. In Figure 7.3a, the initial autarky production and consumption point is given by *E*, with $\beta$ and $(1-\beta)$ representing the share of labour employed in the resource and manufacturing sectors respectively. The production possibility frontier under autarky is the steep line with intercept $Lq\,S_A$ that goes through point *E*. If $p^* > p^A$ when trade opens (world market prices exceeding prior domestic prices under autarky), then the economy immediately specializes in the resource good. The condition $p^* q S_A > 1$ implies that the value marginal product of labour in harvesting exceeds the prevailing domestic wage in the resource sector. The temporary equilibrium production point moves to the vertical axis at $Lq S_A$ and the economy's initial post-trade budget line has a slope $-1/p^*$ (represented by the dotted line in Figure 7.3) and it lies outside the autarky production possibility frontier. This implies that the economy exports initially the resource good, imports manufactures and experiences temporary gains from trade as the new consumption point is now $C$.

However, the initial trading equilibrium cannot be sustained. All labour has entered the resource sector, which will result in the temporary harvest rate rising above the steady-state autarky level. The harvest rate will exceed resource growth and $S$ will decline. As the resource stock falls, Schaefer production implies that harvests will also decline and as shown in Figure 7.3a, the vertical intercept of the production possibility frontier shifts down as indicated by the arrows. Two possible steady-state outcomes may result.

First, if the resource stock stabilises at a level that can sustain the entire labour force at a wage rate exceeding one, then the economy can specialize in production and export of the resource good in the long run. This is indicated by one line in Figure 7.3a, which is the small country's free-trade budget line that has a vertical intercept, and production level, of $Lq S_S$ and an intercept on the horizontal axis beyond $L$. As depicted in the figure, the specialized steady state would allow the country to gain from trade. However, this need not be the case. Steady-state consumption levels under complete trade specialization may not necessarily be higher than in autarky, and depending on the relationship between the terms of trade and steady-state utility, the economy may or may not have gained from trade.

Figure 7.3a also illustrates the case of steady-state diversification. In this case, the resource stock falls to a level, $S_D$, so that not all the labour is allocated to harvesting and its value marginal product equals one. The economy will consume at point $D$, and in comparison to autarky, international trade reduces the small country's steady-state utility unambiguously. While nominal income is unaffected by the opening up for trade, real income has fallen as the consumer price of the resource is now higher than before.

Figure 7.3b illustrates the consequences of trade for a resource-abundant economy, $\gamma/L > q$. The flat line labelled $u^A$ represents the country's steady-state per capita utility under autarky, whereas $u^T$ represents the country's steady-state utility under trade, which is a function of different world prices, $p^*$, for the resource good. The standard gains from trade are a convex function of the difference between the world and autarky price and are minimised if the world price equals the autarky price ($p^A = p^*$). In that case, trading and autarky utility are also equal. At all prices below $p^A$ the economy would export manufactures and experience steady-state gains from trade. There is some possible world price $p^1*$, such that at world prices below this level the economy would stop being diversified and specialize completely in manufactures. At world prices just above $p^A$ the economy would be an exporter of the resource good but diversified in production. In this range, steady-state utility under trade would be less than it would be under autarky. However, if world prices rise to $p^{2*}$, the economy would specialize in the production of the resource good. This price level minimises steady-state utility under trade. Above $p^{2*}$ additional increases in the world price are beneficial to the economy and there is some price, $p^{3*}$, beyond which steady-state gains from trade would occur relative to autarky.

The problem highlighted by Brander and Taylor is that too much harvesting takes place in autarky because there are no secure property rights to the resource. Opening up for trade makes matters worse for resource-abundant countries. Conversely, resource-scarce countries are better off as they now start importing the resource good. Brander and Taylor conclude that, as the problem lies with the open access nature of exploitation in the resource-abundant economy, the first-best policy would be for the small open country to switch to more efficient resource management policy through simply establishing property rights.[14] However, as they acknowledge, there are many policy and institutional distortions that work against such solutions, particularly in developing countries and other resource-abundant small open economies. Consequently, Brander and Taylor (1997a, p. 550) argue in favour of 'second-best approaches'. In a dynamic context where alternative assets are available, the exporting country could impose 'a modified Hartwick's rule' and reinvest temporary gains from selling a resource good on world markets.[15]

In an extension to the analysis by Brander and Taylor, Hannesson (2000) demonstrates that their results may depend critically on the assumption that the manufactures good sector is constant returns to scale.

[14] But see Emami and Johnston (2000), who demonstrate that, in the context of imperfect property rights, resource management by only one country may benefit one or both trading partners, but may also reduce welfare for both, when compared with the case in which neither manages its resource sector.

[15] The Brander and Taylor model has been extended by Smulders *et al.* (2004) to include habitat (as emphasised in section 2) and a third sector that demands land (say agriculture). While open access to resource stocks gives rise to within-industry externalities, it is shown that habitat destruction may create across-industry externalities.

For example, in Brander and Taylor's model, the steady-state national income in terms of manufactures does not change, as long as the country does not specialize fully in open access resource extraction. However, Hannesson argues that it is not at all unlikely for economies heavily dependent on extractive industries and with a locational disadvantage in manufacturing to have diminishing returns in the latter sector. As a consequence, the equilibrium national income of a small open economy in terms of manufactures is likely to rise from trade, even if harvested exports are exploited under open access, as the country is now able to import manufactures at a constant world price rather than having to acquire these goods through reallocating resources with diminishing returns.[16] By shifting labour from manufacturing to harvesting, the marginal and average return to labour in manufacturing increases (see also de Meza and Gould 1992). Note that an opposite result obtains when we assume increasing returns to scale in manufacturing instead (e.g. Matsuyama 1992). Increasing returns may be plausible, for example, because of spillover benefits in manufacturing – a key assumption in the endogenous growth literature.

Hannesson (2000) also demonstrates that, with diminishing returns to manufacturing, moving from an open access regime to optimal management may or may not lead to an improvement in welfare. Such an 'immiserising effect' of a transition from open access to optimal management will occur if the demand for the resource good is inelastic so that the value of harvested output is less than under open access and more labour is withdrawn from the resource sector. The imperfection that drives this result is that insiders in the manufacturing sector cannot prevent outsiders (formerly harvesting the resource) from spilling into 'their sector', adversely affecting the return to their labour.

## 4 Trade between North and South

After discussing the extreme cases of the perfect planner and open access, we now turn to a series of more complex intermediate cases. In this section we will present results from models with trade between different types of countries: North and South. North and South are assumed to be nearly identical (except perhaps in terms of initial resource stocks) but differ in terms of the institutional framework that shapes resource extraction.

[16] When the two goods are substitutes, and thus the indifference curves are linear, these gains from trade always dominate. However, with non-linear indifference curves, such as the case with a Cobb-Douglas utility function, the gains from trade are more ambiguous and it is possible to obtain the same results as Brander and Taylor, even with diminishing returns in manufacturing. See Hannesson (2000).

In the next section we present the static model by Chichilnisky (1994) where the North is assumed to have perfect property rights whereas the South has none – section 2 meets section 3.[17] In section 4.2 we present a model by Karp *et al.* (2001) where institutions are imperfect in both the North and South, but where the common pool externalities are worse in the South because there are more people using the pool. The Karp model is based on some different assumptions with respect to utility and production than most of the papers in this literature. This nicely complements the other work and demonstrates that some of the conventional key assumptions are far from innocuous.

### *4.1 North-South: trade between polar opposite extremes*

Extending the $2 \times 2$ optimal resource management model to a two-country, or North-South, $2 \times 2 \times 2$ model does not in itself yield further insights beyond the basic Kemp and Long model discussed previously. As the authors demonstrate through rigorous proofs, their $2 \times 2$ optimal resource management model, whether applied to exhaustible or renewable resources, is fully consistent with standard Hecksher-Ohlin theorems and is therefore applicable to any number of identical countries or regions with identical factor endowments, technologies, preferences and institutional arrangements, and which face competitive world factor and goods markets. However, a more complex and interesting case arises when we allow for the possibility that there are institutional differences between the trading partners. This is exactly what Chichilnisky (1994) does. She analyses a North-South model of trade and resource management when the two countries are identical except that they differ in the pattern of ownership of an environmental resource used as an input to production. Specifically, resource owners in North have perfect property rights whereas management in South is characterised by unregulated common property.[18]

The harvested resource does not appear in the utility function directly as a consumption good. Instead, the resource flow serves as an input in the production of two goods, A and B. Both goods are produced using Leontief or fixed proportions production technologies with the harvested

[17] Technically speaking, this is not quite true. Brander and Taylor consider the case of complete open access where individual harvesters ignore the external costs of their harvesting. Chichilnisky considers an unregulated common property model where individuals take into account a share of their external costs. The latter collapses to the former whenever the number of individuals approaches infinity.

[18] Strictly speaking, Chichilnisky does not consider an H-O model as she considers the case where the supply of inputs is not given but is determined by (relative) prices.

resource and another factor (labour or capital) as inputs. It is assumed that good B is more resource intensive than good A (which is more intensive in the other input). A key element in the model is the following question: how much of the resource is harvested as an intermediate input for the production of goods A and B? For a given resource stock and any given price, harvesting will be greater when property rights are weak. The common property resource supply curve lies below the optimal supply curve, implying that resources appear to be more abundant.[19] One major result is that while no trade is necessary for efficiency when the two countries are identical,[20] trade will nevertheless occur when property-rights regimes differ. Despite the fact that neither country has a real comparative advantage in producing the resource-intensive good, the lack of property rights for a common-property resource in South leads it to produce and export resource-intensive goods in the steady state. In other words, the country with weak property rights gains an apparent comparative advantage. Distortions, rather than true economic advantages, trigger and drive trade flows.

Karp *et al.* (2001, 2003) define the apparent resource stock as $\delta S$, where $\delta$ is a parameter that measures the degree of the distortion (in an unregulated common property, this distortion is increasing in the number of people harvesting the resource, hence $\delta(n)$ with $\delta'(n) > 0$). In Figure 7.4 the direction of trade is illustrated for North and South – two identical economies but possibly with different stocks ($S^{\mathrm{N}}$ and $S^{\mathrm{S}}$, respectively) and South has imperfect property rights. The 45° line (i.e. identical stocks in North and South) defines zero real comparative advantage. In contrast, the 'no-trade line' defines the case where $S^{\mathrm{N}} = \delta S^{\mathrm{S}}$, or the condition with zero apparent advantage. If the ratio $S^{\mathrm{N}}/S^{\mathrm{S}}$ is such that the system is above the 45° line, South exports the resource-intensive good and has a true advantage in doing this. For $S^{\mathrm{N}}/S^{\mathrm{S}}$ below the 'no-trade line', North exports the resource-intensive good. For $S^{\mathrm{N}}/S^{\mathrm{S}}$ between the 45° line and the 'no-trade line,' the direction of trade will be inefficient as South has an apparent advantage and is an exporter of the resource-intensive good, while North has a true advantage.

If South exports the resource-intensive good, the good will be traded at a price below social costs, even if factor prices are equal across the world, all markets are competitive. This implies that the country with well-defined resource property rights (the North) ends up overconsuming

[19] Chichilnisky considers a static model where the harvest function is strictly concave in effort. In a dynamic setting where the resource stock adjusts to harvesting pressures, it can be shown that equilibrium supply is backward bending (e.g. Clark 1990).

[20] Note there are no scale economies in production that could make specialization and trade beneficial.

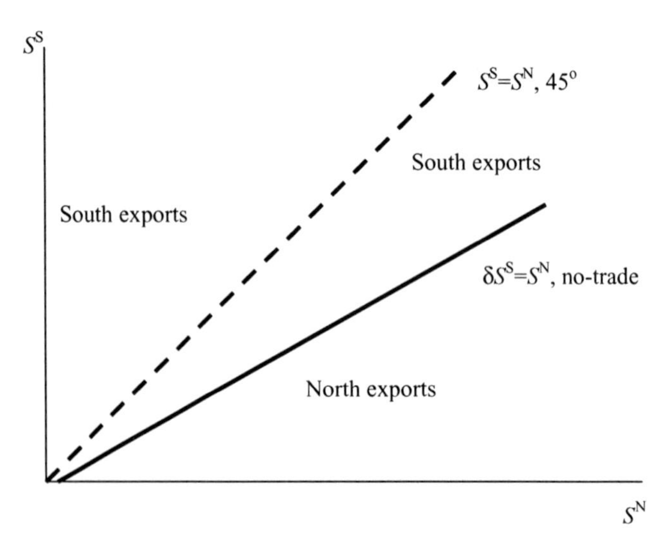


Figure 7.4. Trade patterns between North and South
Source: Karp *et al.* (2001)

the resource-intensive good. Trade exacerbates a pre-existing distortion, which could make the South worse off than under autarky.

What would happen in a dynamic framework? Brander and Taylor (1997b) examine the case of a two-country model where a country governed by a planner meets a country characterised by open access management and solve for steady states. While the country without property rights may export the resource good (and suffer a welfare loss as discussed in section 3) in case of 'mild overuse,' a trade reversal is obtained when relative prices give rise to 'severe overuse' in the open access country. The open access country will first exploit its apparent advantage and export the resource good, but following stock depletion will become a net importer. Efficient management in the short run therefore may result in a comparative advantage over time. If this is the case, both countries experience gains from trade.

### 4.2 *Imperfect property rights in North and South*

Building on Chichilnisky, the model by Karp, Sacheti and Zhao (2001) analyses a more complex case of North–South trade. The main assumptions differ from those of Chichilnisky in the following respects:

i) Rather than contrasting private property in the North versus unregulated common property in the South, the Karp *et al.* model assumes imperfect property rights in *both* countries. While there is no monitoring and enforcement in the North, the common property problem is assumed to be more serious in the South because there are more people utilising the Southern pool, shifting the supply curve of harvested resources out as it inflates the apparent stock (given an actual certain resource stock $S$).

ii) There are two arguments in the utility function: consumption of a subsistence good $A$ with unit price and the resource-intensive consumption good (and intermediate input in the harvesting process of the resource good) $B$. Subsistence good A is assumed to have an income elasticity of demand equal to unity for consumption levels below income threshold $A^*$ and an income elasticity equal to zero for income levels greater than $A^*$. That is, any (extra) income below the threshold will be spent on the subsistence good $A$ and any income in excess of the constraint will be allocated to consumption of $B$.

iii) There are two production factors: the harvested resource flow and some other input (which may be labour or capital). Unlike Chichilnisky, who assumes this factor can be supplied at a certain cost, Karp *et al.* assume its availability is exogenously given. Consistent with earlier discussions, let us assume this additional factor is labour.

The combination of zero substitution in production and consumption, combined with a given availability of labour, implies that the model may have multiple stable steady states. Three cases are depicted in Figure 7.5, graphing the harvesting rate $H/S$ and three realizations of the resource growth rate $G(S)/S$. The kink in the piecewise linear harvest rate path occurs because of the assumption of fixed proportions in production between the inputs 'harvested resource' and labour. For resource stocks below the kink ($S < Sc$), the resource flow is completely employed whereas labour is not (a feature not uncommon in some resource-dependent communities). As the resource stock increases, so does the associated harvest level. Imperfect property rights imply that, for any given resource stock $S < S^c$, there will be excessive extraction – too much $B$ will be allocated to harvesting and not enough will be consumed. At threshold level $S = S^c$, labour is fully employed and the distortion does not affect supply under autarky. Further increasing the stock does not trigger more harvesting as there is no labour to match the resource in production – aggregate harvesting is constant for larger resource stocks, kinking the harvest schedule. The kinked path enables (but does not guarantee) three intersections between harvest and growth. For slow (fast) growing resources, the system settles in a unique 'low' ('high') steady state. For intermediate growth rates, low and high steady states occur simultaneously and initial conditions determine the long-run outcome of the system. By shifting the harvest curve up or down, the stable steady states

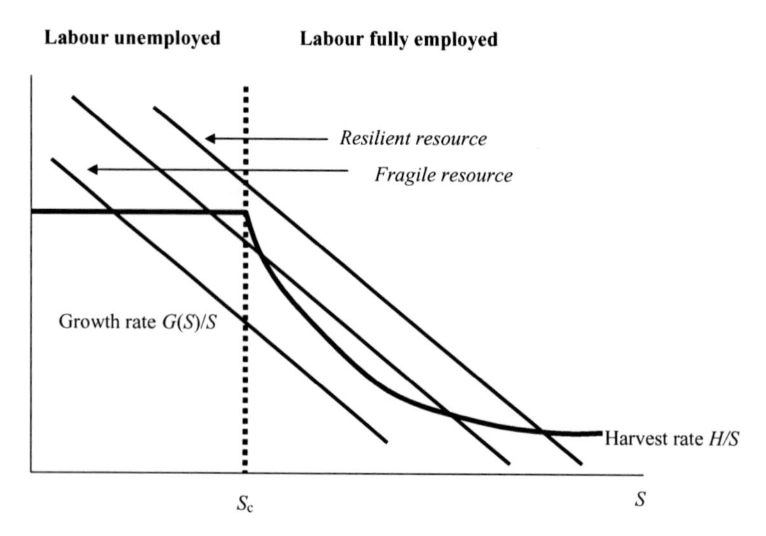


Figure 7.5. Harvesting and growth in the absence of substitution possibilities
Source: Karp *et al.* (2001)

shift accordingly – quantitative changes to the system. However, more dramatic outcomes are also possible as equilibria might appear or disappear altogether.[21] Cases where the system 'jumps' from one stable steady state to another (say from 'low' to 'high' or vice versa) are interpreted as qualitative changes.

What happens when the North and the South move from autarky to trade? When labour is fully employed in both countries, aggregate harvesting is unaffected by trade liberalisation,[22] but the share of extraction is affected – some harvesting will shift from North to South where property rights are weaker. When some labour is unemployed in one of the two trading partners under autarky, it is possible that trade increases the aggregate level of extraction as well.[23] This happens because trade enables countries to reallocate their production and specialize in those

[21] For example, the curves will shift in response to property rights reform. Karp *et al.* (2003) explore the consequences of harmonisation of environmental regulation when trade is driven by distortions rather than true underlying advantages. Among other things they show that, while upward harmonisation (increasing property rights in the South towards the level in the North) is preferable in the long run as it increases the odds of ending up in a high steady state, even *downward* harmonisation can be good for welfare.

[22] This is an artifact of the model, following from (i) the assumption that income elasticity of good $A$ is zero, (ii) fixed proportions in production, and (iii) a fixed total labour stock in both countries.

[23] When some labour is unemployed after trade in both countries, trade does not affect production or consumption and is irrelevant.

products in which they have a comparative advantage. In other words, the (apparently) resource-scarce country will specialize in the production of resource-extensive good $A$. This might enable it to reach a situation where eventually all labour is employed such that aggregate production increases.

The common theme throughout this survey has been to consider the effects of trade liberalisation on welfare and stock conservation. In light of the argument above about multiple steady states in autarky, it is perhaps no surprise that there are also multiple stable candidates under trade for 'intermediate' values of the resource growth, as in Figure 7.5. Initial conditions then determine where the system 'settles down'. For resources that grow sufficiently slow (fast), there exists a unique equilibrium where some (none) of the labour is unemployed. In equilibria where all labour is employed in both countries, the South with its weak property rights will export the resource-intensive good and as a consequence Southern stocks will be lower than under autarky. The reverse holds for Northern stocks.

In the short run, where the apparent resource stocks are fixed, trade may affect production and consumption and, hence, welfare. Trade may trigger an inefficient direction of trade where the South exports the resource-intensive good because its lax property rights regime provides it with a larger *apparent* stock, even though its actual stock is lower. If the direction of the trade flow is efficient, the volume might still be excessive. If these effects are sufficiently strong to outweigh the usual benefits of trade associated with comparative advantages, trading might lower welfare in the South. These effects have also been identified by Chichilnisky. But there are additional long-run effects on the resource stocks to consider as well. The trade pattern identified above might result in a collapse of the local stock and the South may possibly become an importer rather than an exporter of the resource-intensive good – a reversal in apparent comparative advantage.[24] This allows the Southern stock to recover, but could also set the stage for a phase of resource degradation in the North. Trade links the dynamics of resource stocks in North and South and the topologies of the general equilibrium may change qualitatively as a result. That is, when resource stocks adjust in the long run, equilibria may appear or disappear.

Karp, Sacheti and Zhao show that the long-run welfare effects of trade are complex and ambiguous – there are cases (parameter combinations) where the North pulls up the South so that freeing trade eventually makes both countries better off – the economist's dream scenario. Alternatively, there are circumstances where the South drags the North down, such

[24] Brander and Taylor (1997b) also obtain this 'reversal'.

that both countries are worse off – the environmentalist's nightmare. The latter result is not feasible in a model with a rational planner and perfect property rights in the North; one needs a model with imperfect property rights in the North. Of course, mixed outcomes are also feasible. Karp *et al.* interpret these results as common ground between free traders and environmentalists; both perceptions on the welfare and conservation effects of trade liberalisation may be correct and the fixed proportions model allows analysts to identify which perspective of the world is likely to prevail in specific conditions. The model suggests a key role for the intrinsic growth rate in this respect – for sufficiently low rates (or fragile stocks) the pessimistic outcome is more likely to occur.[25]

## 5 Endogenous institutions and property rights regimes

Thus far we have considered the effects of trade liberalisation in polar opposite institutional settings: the benevolent dictator and the open access or unregulated common property case.[26] Or, put differently, the cases of perfect property rights and no property rights, respectively. These all-or-nothing cases are stylised extremes and 'real world harvesting' usually involves some intermediate property regime instead. While resource stocks may be formally owned by governments, it is typically the case that private agents do the harvesting. Firms have an incentive to harvest in excess of their quota or under-report their catches when these are taxed. The government must devote scarce resources to monitoring and enforcement to ensure that the resulting harvest schedule is efficient. Alternatively, the stock may be owned by a group of users who collectively decide on the management of the co-called common pool, but each individual has an incentive to cheat and free-ride on the other's conservation efforts.

We would like to make the following three observations. First, the degree to which the *de jure* stock owner is willing to monitor and enforce its property rights is determined by relative prices and, hence, by the trading regime. Schulz (1996) has argued that, by virtually eliminating the legal value of harvesting, trade bans might result in the cessation of all monitoring. In that case open access 'poaching' for an illegal trade or domestic markets will ensue such that, from an environmentalist's perspective, the trade ban might be counterproductive. Second, the incentive

[25] When the rate is very low, however, trade does not harm the economy as the autarky outcomes would also be dismal.

[26] With unregulated common property management the number of firms exploiting the resource is fixed. Market failure in such a setting is less severe (e.g. Baland and Platteau 1996).

of individuals and firms to cheat and harvest in excess of agreements (or poach when access is denied) is also endogenously determined. Finally, the effectiveness of regulation may be affected by trade as well. Trade liberalisation may expand the set of 'outside options' for individuals in a common pool, thereby making ostracism a less effective deterrent to enforce cooperative extraction from the common pool.

In other words, trade both impacts on the incentives to cooperate and the incentive to regulate – the net effect on the institutional context is unclear *a priori*. Because of the many possible linkages, there is no escape but to build on formal models to learn what mechanisms might exist. The economics discipline is increasingly attempting to cope with this issue and in this section we will discuss three different approaches.

### 5.1 *Enforcement, profits and welfare*

Suppose a resource stock is formally 'owned' by a firm (say, a forest concessionaire or landowner extensively managing his land), but that no-encroachment enforcement is costly. The owner must decide whether to invest in securing his property rights and if so, how much exactly? Alternatively, he may allow others to use it under open access conditions and dissipate any rents. de Meza and Gould (1992) provide an early and general analysis of the fact that the resource must be sufficiently valuable to expend funds on exclusion, monitoring and enforcement. Hotte *et al.* (2000) related this insight to the case of renewable resources and trade. Households are again endowed with a unit of labour that is allocated to either manufacturing or extraction (legal or illegal). In contrast to the Brander and Taylor studies in section 3 (but consistent with Hannesson 2000), it is assumed that there are decreasing returns with respect to labour in manufacturing.

Two types of labour may be extracting the same stock – legal and illegal (poaching) labour ($L_H$ and $E$, respectively). The parameter $\phi$ measures the strength of enforcement and with more enforcement the return from poaching labour is smaller. This occurs because poachers have to avoid detection by the owner. It is assumed that only a share $(1 - \phi_i)$, where $0 \leq \phi \leq 1$, of the poaching labour $E_i$ is effectively geared towards extraction on plot $i$ (owned by firm $i$). The firm (resource owner) can manipulate parameter $\phi$, by choosing enforcement intensity as this affects the precautionary efforts that poachers must incur. The costs associated with achieving enforcement level $\phi$ are $c(\phi)$ where $c' > 0$ and $c'' > 0$.[27]

[27] For additional work along these lines, see Bulte and van Kooten (1999).

Households compare the return to their labour in both activities and either work in manufacturing or legal extraction (with the wage rate determined by the marginal product of labour), or in poaching (where labour earns its average product, driven by the sum of legal and effective illegal effort). The firm chooses legal harvesting and enforcement effort. Both variables have in common that they drive out illegal harvesting effort: legal harvests depress the average return to labour and enforcement increases the costs of avoiding detection. The firm first optimally sets its level of enforcement, $\phi$. Depending on parameters, $\phi^* = 0$ (true open access) or $\phi^* > 0$.[28] Next, the firm decides about the hiring of legal labour. For the case where $\phi^* = 0$, firms do not bother to employ anybody as there are no rents to be gained from (legal) harvesting. In contrast, when $\phi^* > 0$ legal labour is employed up to the level where it exactly crowds out poaching – the entry-deterrence employment level involving rent dissipation for poachers but positive profits for the firm (whose labourers are more productive as they do not worry about detection). While the latter outcome generates positive profits for the firm, it is shown that enforcing property rights may involve a welfare loss at the level of society as a whole. Enforcement of private property rights implies that labour switches to manufacturing (depressing its marginal return which implies a fall in labour income) and also involves enforcement costs.[29] Since the effect of enforcement on labour income in manufacturing is external to the firm, it does not affect its optimal enforcement stringency. The firm only compares gross profits and private enforcement costs.

Under autarky, firms may set $\phi^* = 0$ or $\phi^* > 0$, depending on key parameters. Assume that firms do not bother to enforce their property rights such that *de facto* extraction takes place under conditions of open access. How does opening up for trade affect the firm's enforcement decision? The authors find that the firm's optimal level of enforcement is increasing in the resource price and decreasing in the prevailing wage rate. Hence, when resource prices increase as a result of trade liberalisation, the firm may suddenly find enforcement privately profitable and respond by switching from $\phi^* = 0$ to $\phi^* > 0$. The firm will hire some poachers and others will be forced into manufacturing. Profits of the firm increase but, as outlined above, welfare in society as a whole may fall – another example of immiserising trade. Hotte and co-authors also analyse a more involved dynamic model where the resource stock contracts or grows in

[28] The owner will set $\phi^* = 0$ if enforcement combined with the hiring of legal labour to deter entry by poachers earns negative profits. In contrast, if enforcement plus entry deterrence employment earns positive profits, the owner will set $\phi^* > 0$.

[29] For more information on comparing income under open access and private property, see Weitzman (1974) and de Meza and Gould (1987, 1992).

response to natural growth and harvesting. They show that the increase in enforcement brought about by higher price raises equilibrium stock levels. Endogenising property rights thus opens the possibility that trade might be good for conservation and bad for welfare – reversing some of the insights of section 2.

### 5.2 *Trade and common pool management*

The analysis in 5.1 was based on the assumption that an owner can raise the costs of illegal extraction by raising the effort that poachers *must* incur in order to avoid detection. Nobody ever gets caught and punished because either there is no enforcement and all harvesting is 'illegal' ($\phi^* = 0,\ L_H = 0$), or all illegal harvesting is crowded out ($\phi^* > 0,\ E = 0$). Copeland and Taylor choose another perspective. They assume imperfect enforcement by the resource owner and recognise that individuals may have an incentive to harvest illegally (or harvest in excess of an agreed-upon quota), but must balance the gains from 'cheating' against the expected penalties if caught.

Consider the familiar case where there is a single production factor (labour) that is allocated between CRS manufacturing or resource extraction from a common pool. Unlike in the earlier work of Brander and Taylor (1997), Copeland and Taylor assume the presence of a benevolent resource manager or village elder who sets rules on resource use to maximise steady state utility of the group. The manager decides how much labour households can spend in the common resource (say $l^*$) and attempts to enforce its rule by monitoring the behaviour of the group members (which translates into a certain probability of detection, $\rho$). Next, individuals have to decide whether to behave in accordance with the rule ($l = l^*$) or allocate extra labour ($l > l^*$) to harvesting. If too much time is spent in the commons, households run the risk of being detected (probability $\rho$, independent of the extent of cheating). If this happens, the household is ostracised and is denied access to the common again. Cheating households that are caught should support themselves by working in manufacturing henceforth.[30] Assume the resource stock is large enough to make extraction more profitable than manufacturing.

The tradeoff that households make is relatively straightforward. Adhere to the rules and earn some income in extraction and some in

[30] Note the difference with the model of Hotte *et al.* (2002) where the resource owner tried to drive out illegal effort by reducing the return to poaching (by increasing legal harvesting or enforcement). Copeland and Taylor instead focus on the disciplining effect of punishment, which depends on the forgone profits of working in the commons relative to the wage in manufacturing.

manufacturing. Alternatively, it is possible to take a risk and cheat by spending all the time in the common (recall the detection probability is assumed independent of the degree of violation, hence a maximum violation is optimal when cheating). The optimal choice depends on comparing the expected present value of income from these options. Formally, the discounted benefits of not cheating in the current period are defined as

$$V^{N}(t) = [ph^{*} + (1 - l^{*})w]\mathrm{d}t + (1 - r\mathrm{d}t)\, V^{R}(t + \mathrm{d}t) \qquad (12)$$

where $h^{*} = ql^{*}S$ is the harvest associated with the allowed time in the common, $w$ is the wage rate in manufacturing and $V^{R}$ is the (unrestricted) continuation value. The latter might differ from the benefits of cooperation since playing by the rules in the current period leaves all options open for the future (including cheating). Future benefits are discounted. The benefits of cheating in the current period are

$$\begin{aligned} V^{C}(t) = (pqS)\mathrm{d}t + (1 - r\mathrm{d}t)\, &[\rho \mathrm{d}t\, V^{M}(t + \mathrm{d}t) \\ &+ (1 - \rho \mathrm{d}t)\, V^{R}(t + \mathrm{d}t)] \end{aligned} \qquad (13)$$

where $V^{M}$ is the present value of working in manufacturing henceforth. Note from (12) that cheaters who are not caught have the same set of options for the future as those who abided by the rules. From the above, these options are defined as $V^{R}(t) = \max[V^{N}(t), V^{C}(t)]$. Comparing the benefits and losses from cheating (or the costs associated with ostracism) defines a forward-looking incentive constraint. The manager must choose $l^{*}$ such that this incentive constraint is satisfied, otherwise people will rationally choose to ignore the rules. After some manipulation, Copeland and Taylor demonstrate that in steady state the incentive constraint boils down to

$$l^{*}(pqS - w) \geq \frac{r}{r + \rho}(pqS - w) \qquad (14)$$

When resource rents are positive ($pqS - w > 0$), this condition simply implies $l^{*} > r/(r + \rho)$ or the allowable harvest should be sufficiently large to deter cheating. The threshold level is composed of parameters reflecting impatience and the probability of detection. However, there is no guarantee that the rents associated with such a policy will be positive (i.e. $pqS > w$). When the returns of harvesting according to the threshold fall below those in manufacturing, people will switch their occupation. If we define the open access allocation of labour to the common as $L^{O}$, the incentive constraint is written as

$$L^{*} = \min\left[L^{O}, \frac{r}{r + \rho}N\right] \qquad (15)$$

where $N$ is the number of households in the village. The planner takes into account the incentive constraint when formulating a policy that will not be violated. There are three possibilities. First, the incentive constraint is not binding and no individual is tempted to increase his harvest effort when aggregate harvesting is at the optimal level. The outcome is conventional first best management as outlined in section 2. Second, the incentive constraint binds even as the system approaches open access harvesting. This may happen when the resource good's price is very low so that $L^O$ is small and lower than the other term in brackets in (15). The planner is not able to redirect the allocation of labour from harvesting to manufacturing and the outcome is as modelled in section 3. Depending on key parameters, notably the growth rate, catchability coefficient, detection and death rates, and the size of the human population, such an outcome may be temporary (that is, depending on relative prices) or permanent. Finally, and most interestingly, it may be the case that the incentive constraint binds in steady state while there are positive profits associated with current harvesting. This yields an intermediate allocation of labour and intermediate level of stock depletion (between first best and open access) and defines a 'constrained optimum'. The institutional context thus defined therefore measures whether it is feasible and necessary for an imperfectly informed planner to constrain agents' harvesting.

How does trade fit into this analysis? Copeland and Taylor distinguish between three different categories of countries: (i) countries which are never able to move beyond open access harvesting, (ii) countries which might secure some form of a 'constrained optimal regime' if the conditions are favourable, and (iii) countries which achieve the fully efficient outcome given the right set of parameters. If moving from autarky to trade increases the price of the resource good, countries in categories (ii) and (iii) can move from open access to limited property rights. If prices rise further, category (iii) countries can even make the final step to the full cooperative outcome (and countries from the other categories will not become worse off). The reason for this transition is that rising prices trigger a flow of labour entering the resource sector and a fall of the stock. Eventually, for category (ii) and (iii) countries, both arguments in (4) are of equal size and the incentive constraint begins to bind. Depending on whether the first best allocation of labour to the common pool $L^*$ is greater or smaller than $N[\mathrm{r}/(\rho + \mathrm{r})]$ for any arbitrarily high value of $p$, the country can reach the first best (or not). For $L^* > N[\mathrm{r}/(\rho + \mathrm{r})]$, first best harvesting can be sustained while meeting the incentive constraint.

Copeland and Taylor also find that a country can have open access harvesting or (limited) control over harvesting, depending on the price of the resource good. Raising the value of the stock provides the incentive to generate rents from harvesting. Transition towards controlled harvesting is facilitated by rapid replenishment of the stock, good detection possibilities $\rho$ and a small size of the group $N$. The key insight of these models is that observing conditions of *de facto* open access under autarky need not imply equally wasteful management under trade. Trade may be a pre-condition for management reform and arguing the other way around might be counterproductive. Yet they also warn that some countries (category (i) countries) will not benefit from trade-induced higher prices. These countries will 'lose', as spelled out in section 3.

### 5.3 *Trade and corruption*

In the previous two sub-sections we explored the possibility that higher prices for resource commodities might result in more efficient management. Implicitly it was assumed that harvesters do not change their behaviour, other than intensifying their extraction effort in response to higher prices. But typically harvesters have additional instruments at their disposal other than harvest effort. For example, harvesters might organise themselves and lobby for more lenient regulation. Alternatively, they may bribe planners for special favours in their pursuit of resource rents. Corruption is increasingly recognised as a major issue in 'real world' management of resources like oil fields (Karl 1997) and forests (Ross 2001), and analysing the consequences of trade liberalisation on the incentive to bribe therefore is important.

For economists, the static common agency model by Grossman and Helpman (1994) provides a logical starting point to consider this issue. In this model, a self-interested planner maximises a linear objective function that includes social welfare ($W$) and bribes from interest groups ($T$) as arguments, and a linear welfare weight to quantify tradeoffs between them:

$$\text{Max}\,\Pi = [\alpha W + (1 - \theta) T] \tag{16}$$

where $\Pi$ represents the objective for the planner and $\theta$ is the weight of welfare in the planner's objective function. In the sections with a planner thus far we implicitly assumed that $\alpha = 1$ and that the planner has eye only for the social good. This is clearly unrealistic, certainly for many resource-rich countries, as is readily gleaned from the various corruption indices that are available (e.g. World Bank governance indicators or Transparency International data).

There are multiple interest groups in society – say resource firms versus environmentalists – 'bidding' on a menu of possible policies announced by the planner. In the first stage of the game the lobby groups offer the government a so-called bribe schedule that links bribes to the policies implemented. In the second stage, the government chooses the 'optimal' policy, taking objective function (16) and bribe schedules as given, and collects the bribes. In the third stage production and consumption take place, taking environmental policies as given. The model should be solved through backward induction.

In other words, there exists a political market where policies are exchanged for bribes and where the sum of bribes is balanced against welfare. The welfare weight $\theta$ is typically assumed constant and may be interpreted as a measure of corruption in the economy – the lower $\theta$, the more corrupt is the planner. Interestingly, empirical work by Leite and Weidmann (1999) suggests that the extent of corruption is increasing in natural resource abundance. Given positive values of $\theta$, more extensive lobbying translates into more distorted policies, steering the economy further away from the first best allocation of factors. In the context of resource economics, firms may have incentives to bribe the planner to pay lower taxes or receive more generous harvest quotas.

Assume resource firms lobby for extending the harvest quota beyond the socially optimal quota. Privately optimal quota might diverge from socially optimal ones because firms and planners apply a different discount rate, or because there are external effects in harvesting or conservation. As demonstrated by Bernheim and Whinston (1986), the Nash equilibrium outcome is characterised by two conditions: (1) maximisation of the government objective function, and (2) maximisation of the joint utilities of lobby groups and planner. This implies that outcomes on the political market are *locally truthful*, or that the firm's willingness to pay (bribe) is increasing in the policy's benefits. Outcomes are on the bargaining frontier and are Pareto efficient in the sense that no actor (lobby group or otherwise) can be made better off without making someone else worse off.

Since the firm's benefits of securing a larger quota are increasing in the resource price, the equilibrium transfer from firm to planner is determined by this price too. When trade liberalisation changes relative prices it affects political pressures and, hence, the balance on the political market. Following this reasoning, a resource-abundant country that opens up for trade will experience an increase in lobbying and, as a result, larger quotas (in the short run) at the detriment of welfare and stock conservation – another example of endogenous institutions.[31] While this effect

[31] Leite and Weidmann (1999) also consider the effect on corruption of opening up for trade and find a reverse effect – more 'open' economies are typically characterised by less extensive corruption. Trade regulations (as occurring at intermediate levels of 'trade openness') are a source of rents and thereby trigger further bribing (see also Baland and Francois 2000). Removing such regulations lowers the potential for rent seeking and corruption and provides a force that works in an opposite direction from the 'price effect' of trade liberalisation discussed in the main text.

has yet to be analysed fully in the context of resource management, such an approach could yield similar insights as studies of the relation between pollution, trade and corruption.[32]

However, the interaction effects between corruption, trade and resource conversion can also be complex. For example, a recent cross-country analysis of the economic factors underlying tropical deforestation and agricultural land expansion indicates that the influence of corruption on land conversion may depend on what happens to a country's terms of trade as well as its degree of resource dependency (Barbier 2004). The presence of significant interaction effects between the terms of trade and corruption and primary product export dependency suggests caution in assuming that an important policy mechanism by which the rest of the world can reduce land conversion in developing economies is through sanctions, taxation and other trade interventions that reduce the terms of trade of these economies.

Such a finding is consistent with the theoretical models reviewed here. Throughout this chapter we have noted that the impact of higher prices on the institutional setting is ambiguous: while the 'optimal enforcement models' in sections 5.1 and 5.2 implied that higher prices may result in more efficient resource management, corruption is a force that may pull in an opposite direction. By enhancing the incentive to bribe, greater corruption may cause deviations from optimal management. In such a sub-optimal world, it is difficult to predict how trade interventions and other policies will affect renewable resource management in the exporting country.

## 6 Conclusions

This chapter has provided only a brief overview of the literature on trade and renewable resource management. The key motivation for the chapter is the lack of consensus between economists and environmentalists about the desirability of international trade and that these opposing

[32] While we cannot do justice to the rapidly growing field of corruption and environmental regulation, we can highlight a few examples. Evidently, corruption enables firms to evade stringent regulation by paying a bribe rather than abating emissions. Lopez and Mitra (2000) consider the effect of corruption on the relation between income and pollution levels and find that both the level and 'turning point' of the EKC are affected by the degree of corruptability. Frederiksson and Svensson (2003) analyse the relation between corruption, political instability and environmental policy. Damania *et al.* (2003) find a negative direct effect of corruption on the standard for lead content in gasoline (as well as an interaction effect between trade openness and corruption). Frederiksson *et al.* (2004) analyse the effects of corruption and industry size on energy policy outcomes (finding that policy stringency is inversely related to corruptability and positively related to lobbying costs and that capitalists' and workers' lobbying efforts are negatively related).

views offer different recommendations for reform of WTO policies. While many economists argue that trade is not the issue and that domestic and international institutions for environmental management should be strengthened if resource conservation is a goal, environmentalists often take the opposite stand and argue the case for severely restricting and regulating trade to protect biodiversity and critical renewable resource stocks.

We organised the chapter such that it loosely followed the main developments in the literature. The guiding principle in structuring the literature is the changing perspective on the role of institutions in resource management. In the 1970s and 1980s, following standard models in resource economics, most of the work assumed the perspective of a benevolent planner or sole owner with full property rights. In the 1990s, with the increasing recognition of the implications for resource management of open access resource extraction, more attention was given to the second-best setting where property rights are either *de facto* or *de jure* absent. More recent contributions treat the institutional context as endogenous and model the incentives and constraints that agents have at the micro level to influence resource management outcomes.

Our summary of the literature suggests a very mixed overall picture, both in terms of welfare and effects on stock conservation. While generally it is possible to obtain relatively clean results for the polar extreme cases of perfect management and open access, it is evident that more realistic assessments generally imply ambiguous outcomes. The interplay of economic, ecological and institutional variables therefore determines whether trade is overall 'good' or 'bad', which provides quite a bit of common ground between economists and environmentalists. This has important implications for international policy-making concerned with trade and renewable resource management issues. Neither the 'conventional' view of economists that trade impacts on resource management can safely be ignored, nor the equally 'simplistic' view of environmentalists that trade is the source of resource losses and must therefore be curbed, is a good starting point for recommending specific trade policies and reforms for most of the pressing biodiversity and renewable resource management problems facing the world today. While trade restrictions and impediments lower welfare in a first-best world, it is evident that export and import measures may promote welfare in exporting countries when, say, enforcement of property rights is imperfect. The presence of international non-use values associated with resource conservation, of spillover regulatory benefits, may perhaps also justify trade interventions on certain occasions (albeit clearly not a first-best approach to maximise global welfare).

Space limitations do not permit a thorough treatment of trade policy (but see Anderson and Blackhurst 1992; Barbier and Rauscher 1994; Barbier *et al.* 1990 and 1994; Schulz 1996, Brander and Taylor 1998, and others). While tariffs for most resource commodities have declined, there is evidence that non-tariff measures continue to play a large role in shaping trade patterns (Bourke and Leitch 2000; Barbier 2001). Tariff escalation – higher rates on higher levels of procession – is also still exerting an influence on the structure of resource sectors in developing countries.[33] However, whether the net welfare effects of such distortions are positive or negative is hard to determine in general. It appears as if each specific management problem, whether it be control of ivory poaching, tropical forest conservation, fisheries management, protection of endangered species or preservation of biodiversity 'hot spots', must be analysed on a case-by-case basis in order to determine the linkages between the key economic, ecological and institutional factors that are driving the problem. Only through such careful analysis can the impacts of trade and resource management and economic welfare be identified and only then can possible policy recommendations be identified.

In conducting this summary of the economics literature on trade and renewable resources, we are aware that new developments are already occurring in this rapidly evolving literature. Several additional research topics have a direct bearing on the focus of the current chapter, but cannot be addressed here for reasons of space. One example is the risk and welfare impacts posed by biological invasions, or 'bio-invasions'. Trade typically involves transport, which implies the risk of introducing non-native species. Managing this complex and unpredictable issue has recently received ample attention (e.g. Horan *et al.* 2002; Barbier and Shogren 2004; McAusland and Costello 2004) and will likely continue to be a topic of interest for some time. Another example is the effect of trade on habitat and biodiversity. If trade triggers specialization and if this in turn impacts on land use (e.g. forestry versus agricultural land use options), then trade liberalisation can be linked to biodiversity loss through the so-called species-area curve (Polasky *et al.* 2004).

[33] For example, consider the case of forest commodities. Rates for most forest products are 5 per cent or lower, but rates for processed goods like plywood are typically 10–15 per cent. Tariff escalation discourages local processing, negatively affecting the scope for investment and industrialisation in exporting countries, and the ability of exporters to capture a larger share of the value added associated with processing resource commodities. Depending on linkages and scale economies (as well as alternative employment and investment opportunities in the economy and the institutional context within which harvesting takes place), this may or may not seriously affect the scope for modernisation. There are cases where tariff escalation has been countered by export bans of unprocessed commodities. Often such efforts to promote industrialisation have spurred inefficient and uncompetitive industries.

Recent research also suggests that eco-labelling is relevant in the context of trade and resource management (e.g. Swallow and Sedjo 2002; Nunes and Riyanto 2001). In the context of an information problem, eco-labelling allows consumers to identify the environmentally friendly (e.g. 'sustainably managed') alternatives and express their preference for such commodities. It also enables complying producers to earn a green premium, although evidence of such premiums for sustainably produced timber and biodiversity-friendly shade-grown coffee is modest (Barbier 2001; Nunes and Riyanto 2001). Another issue that has recently been explored is the interaction between wild species, stockpiled commodities from the wilds and the impact of legalised trade from *ex situ* stocks on the incentive to poach (Kremer and Morcom 2000) and coordinate on extinction scenarios (Bulte *et al.* 2003).

Finally, we will briefly mention a few issues that have received little attention in the literature thus far and that could possibly be of interest to consider in the future. First, since trade liberalisation affects factor flows it is relevant to know how such flows affect factor income in the non-resource sector (e.g. de Meza and Gould 1992). The literature is silent on increasing returns to scale at firm or sector level, but exploring this possibility would link the literature with that on trade and pollution (Neary 1999) and the so-called 'resource curse' (Matsuyama 1992; Sachs and Warner 1997). As discussed above in the context of Matsuyama's model, when labour flows out of manufacturing, as will be the case for a resource-abundant country that opens for trade, scale economies will lower income in manufacturing. If external linkages are strong enough, this effect can dominate any gains from trade. Conversely, when trade triggers an inflow of factors, for example because of enhanced enforcement in the resource sector, this brings a positive welfare effect.

Other overlooked issues thus far include the simple observation that most models do not consider more than one production factor (labour) in addition to the resource stock or flow and that key results can change when we distinguish between multiple factors. The distinction between mobile and immobile factors (capital versus labour) also seems apt – a questionable simplification in light of evidence that, for example, international logging firms are 'footloose' (Marchak 1995). In a similar vein, the effect of trade on investment, technology diffusion and capital accumulation is relevant. Trade may also impact on preferences for nature conservation through the impact on income, as demand for conservation has always been considered income elastic. Imperfect competition and strategic interaction between jurisdictions sharing access to common stocks or output markets have yet to be analysed (see Ruseski (1998) for such a model that does not involve trade). Finally, empirical work seems to lag far behind theory. It appears as if the literature on trade and resource management has only just begun.

## REFERENCES


Anderson, K. and Blackhurst, R. (eds.). 1992. *The Greening of World Trade Issues*. Hemel Hempstead: Harvester Wheatsheaf.

Baland, J. and Francois, P. 2000. Rent seeking and resource booms. *Journal of Development Economics*. **61**. 527–542.

Baland, J. and Platteau, J. 1996. *Halting Degradation of Natural Resources*. Oxford: Clarendon Press.

Barbier, E. B. 2001. International Trade and Sustainable Forestry. In Günther G. Schulze and Heinrich W. Ursprung (eds.). *International Environmental Economics: A Survey of the Issues*. Oxford: Oxford University Press. 114–147.

Barbier, E. B. 2003. Habitat fishery linkages and mangrove loss in Thailand. *Contemporary Economic Policy*. **21** (1). 59–77.

Barbier, E. B. 2004. Explaining agricultural land expansion and deforestation in developing countries. *American Journal of Agricultural Economics*. **86** (5). 1347–1353.

Barbier, E. B. and Burgess, J. C. 1997. The economics of tropical forest land use options. *Land Economics*. **73** (2). 174–195.

Barbier, E. B. and Rauscher, M. 1994. Trade, tropical deforestation and policy interventions. *Environmental and Resource Economics*. **4**. 75–90.

Barbier, E. B. and Schulz, C. E. 1997. Wildlife, biodiversity and trade. *Environment and Development Economics*. **2** (2). 33–60.

Barbier, E. B. and Shogren, J. R. 2004. Economic growth and the endogenous risk of biological invasion. *Economic Inquiry*. **42** (4). 587–601.

Barbier, E. B. and Strand, I. 1998. Valuing mangrove–fishery linkages – a case study of Campeche, Mexico. *Environmental and Resource Economics*. **12**. 151–166.

Barbier, E. Burgess, J. C. Swanson, T. M. and Pearce, D. W. 1990. *Elephants, Economics and Ivory*. London: Earthscan.

Barbier, E. Burgess, J. C. Bishop, J. T. and Aylward, B. A. 1994. *The Economics of the Tropical Timber Trade*. London: Earthscan.

Berck, P. and Perloff, J. M. 1984. An open access fishery with rational expectations. *Econometrica*. **52**. 489–506.

Bernheim, B. D. and Whinston, M. 1986. Menu actions, resource allocation and economic influence. *Quarterly Journal of Economics*. **101**. 1–31.

Bourke, I. J. and Leitch, J. 2000. *Trade Restrictions and Their Impact on International Trade in Forest Products*. Rome: Food and Agricultural Organization of the United Nations.

Brander, J. and Taylor, M. S. 1997a. International trade and open access renewable resources: the small open economy case. *Canadian Journal of Economics*. **30**. 526–525.

Brander, J. and Taylor, M. S. 1997b. International trade between consumer and conservationist countries. *Resource and Energy Economics*. **19**. 267–279.

Brander, J. and Taylor, M. S. 1998. Open access renewable resources: trade and trade policy in a two country model. *Journal of International Economics*. **44**. 181–209.

Brown, G. 2000. Renewable natural resource management and use without markets. *Journal of Economic Literature*. **38**. 875–914.

Bulte, E. H. and van Kooten, G. C. 1999. The ivory trade ban and elephant numbers: theory and application to Zambia. *American Journal of Agricultural Economics*. **81**. 453–466.

Bulte, E. H and Horan, R. D. 2003. Habitat conservation, wildlife extraction and agricultural expansion. *Journal of Environmental Economics and Management*. **45**. 109–127.

Bulte, E. H., Horan, R. D. and Shogren, J. F. 2003. Elephants: comment. *American Economic Review*. **93**. 1437–1445.

Chichilnisky, G. 1994. North–south trade and the global environment. *American Economic Review*. **84**. 851–874.

Clark, C. W. 1990. *Mathematical Bioeconomics*. New York: Wiley Interscience.

Copeland, B. R. and Taylor, M. S. 2003. *Trade, tragedy and the commons*. Department of Economics, University of British Columbia, Working Paper.

Damania, R., Frederiksson, P. G. and List, J. A. 2003. Trade liberalization, corruption, and environmental policy formation: theory and evidence. *Journal of Environmental Economics and Management*. **46**. 490–512.

de Meza, D. and Gould, J. R. 1987. Free access versus private property in a resource: income distributions compared. *Journal of Political Economy*. **95**. 1317–1325.

de Meza, D. and Gould, J. R. 1992. The social efficiency of private decisions to enforce property rights. *Journal of Political Economy*. **95**. 561–580.

Emami, A. and Johnston, R. 2000. Unilateral resource management in a two-country general equilibrium model of trade in a renewable fishery resource. *American Journal of Agricultural Economics*. **82**. 161–172.

Frederiksson, P. G., Vollebergh, H. and Dijkgraaf, E. 2004. Corruption and energy efficiency in OECD countries: theory and evidence. *Journal of Environmental Economics and Management*. **47**. 207–231.

Frederiksson, P. G. and Svensson, J. 2003. Political instability, corruption and policy formation: the case of environmental policy. *Journal of Public Economics*. **87**. 1383–1405.

Grossman, G. and Helpman, E. 1994. Protection for Sale. *American Economic Review*. **84**. 833–850.

Hannesson, R. 2000. Renewable resources and the gain from trade. *Canadian Journal of Economics*. **33**. 122–132.

Horan, R. D., Lupi, F., Perrings, C. and Bulte, E. H. 2002. Management of invasive species under ignorance. *American Journal of Agricultural Economics*. Principal Paper Session. **84**. 1303–1310.

Hotte, L., van Long, N. and Tian, H. 2000. International trade with endogenous enforcement of property rights. *Journal of Development Economics*. **62**. 25–54.

Isham, J., Woodcock, M., Pritchett, L. and Busby, G. 2003. The varieties of resource experience: how natural resource export structures affect the political economy of economic growth. Economics Discussion Paper 03–08. Vermont: Middlebury College.

Karl, T. L. 1997. *The Paradox of Plenty: Oil Booms and Petro-States*. Berkeley, CA: University of California Press.

Karp, L., Sacheti, S. and Zhao, J. 2001. Common ground between free traders and environmentalists. *International Economic Review*. **42**. 617–647.

Karp, L., Zhao, J. and Sacheti, S. 2003. The long run effects of environmental reform in open economies. *Journal of Environmental Economics and Management*. **45**. 246–264.

Kemp, M. C. and Long, N. V. 1984. The role of natural resources in trade models. In Jones, R. and P. Kenen (eds.). *Handbook of International Economics* vol I. Amsterdam: North Holland.

Kremer, M. and Morcom, C. 2000. Elephants. *American Economic Review*. **90**. 212–234.

Leite, C. and Weidmann, J. 1999. *Does mother nature corrupt? Natural resources*, corruption and economic growth. IMF Working Paper WP/99/85.

Lipsey, R and Lancaster, K. 1956. The general law of second best. *Review of Economic Studies*. **24**. 11–32.

Lopez, R. and Mitra, S. 2000. Corruption, pollution and the environmental Kuznets curve. *Journal of Environmental Economics and Management*. **40**. 137–150.

Marchak, M. P. 1995. *Logging the Globe*. Montreal, Quebec: McGill-Queen's University Press.

McAusland, C. and Costello, C. 2004. Avoiding invasives: trade-related policies for controlling unintentional exotic species introductions. *Journal of Environmental Economics and Management*. **48**. 954–977.

Matsuyama, K. 1992. Agricultural productivity, comparative advantage and economic growth. *Journal of Economic Theory*. **58**. 317–34.

Neary, J. 1999. *International Trade and the Environment: Theoretical and Policy Linkages*. Invited lecture. Annual Conference of the European Association of Environmental and Resource Economists. Oslo. 25–27 June.

Nunes, P. and Riyanto, Y. 2001. *Policy instruments for creating markets for biodiversity*: certification and ecolabeling. FEEM Working Paper 72. Rome: Italy.

Polasky, S. Costello, C. and McAusland, C. 2004. On trade, land-use, and biodiversity. *Journal of Environmental Economics and Management*. **48**. 911–925.

Ross, M. L. 2001. *Timber Booms and Institutional Breakdown in Southeast Asia*. Cambridge: Cambridge University Press.

Ruseski, G. 1998. International fish wars: the strategic role for fleet licensing and effort subsidies. *Journal of Environmental Economics and Management*. **36**. 70–88.

Sachs, J. D. and Warner, A. M. 1997. *Natural Resource Abundance and Economic Growth*. NBER Working Paper Series, WP 5398. Cambridge, MA: National Bureau of Economic Research.

Schulz, C. 1996. Trade policy and ecology. *Environmental and Resource Economics*. **8**. 15–38.

Smulders, J. A., van Soest, D. P. and Withagen, C. 2004. International trade, species diversity and habitat conservation. *Journal of Environmental Economics and Management*. **48**. 891–910.

Swallow, S. and Sedjo, R. 2002. Voluntary eco-labeling and the price premium. *Land Economics*. **87**. 272–284.

Swanson, T. 1994. The economics of extinction revisited and revised: a generalised framework for the analysis of the problems of endangered species and biodiversity losses. *Oxford Economic Papers*. **46**. 800–821.

Torvik, R. 2002. Natural resources, rent seeking and welfare. *Journal of Development Economics*. **67**. 455–470.

Vannuccini, S. 2003. *Overview of Fish Production, Utilization, Consumption and Trade*. Rome: Fishery Information, Data and Statistics Unit, Food and Agricultural Organization of the United Nations.

Weitzman, M. 1974. Free acces vs private ownership as alternative systems for managing common property. *Journal of Economic Theory*. **8**. 225–234.

Wilen, J. 2000. Renewable resource economists and policy: what differences have we made? *Journal of Environmental Economics and Management*. **39**. 306–327.

# 8 International trade and its impact on biological diversity

*Rafat Alam and Nguyen Van Quyen*

## 1 Introduction

For the past twenty-five years, the world has been moving towards a free trade regime. At the same time the concern for the impact of free trade on natural resources is increasing. There is debate among the environmentalists and the economists on the impact of trade on welfare and biodiversity (see Chapter 18 of this volume). Environmentalists 'worry that trade will expand the scope of market failures, put added strain on the environment and lead to degradation of natural resource stocks in the long run' (Karp *et al.* 2001, p. 617), which in turn will decrease the welfare of both import and export countries. Many economists argue, however, that free trade will improve social welfare and rectify environmental externalities provided markets function efficiently, property rights over biodiversity resources are well defined and non-market values of natural resources are accounted for in the production process.

The fact that most of the world's biodiversity-rich land lies in the populated and poor South makes the situation even worse. The biodiversity-rich South is already overburdened to meet the demand of its own population for biodiversity-derived goods (such as agricultural products, timber and non-timber forest products), while free trade, it is argued, adds further pressure to overuse and overexploit biodiversity resources.

Yet as incomes grow in Northern countries, their consumers are displaying an increasingly stronger preference for so-called 'green products'. Eco-labelled and certified fair-trade products are gaining wider acceptance and increasing their market share as a significant proportion of northern consumers are willing to pay a premium price for such products. Similar to the quality-differentiated goods in the manufacturing sector, many quality-differentiated goods are emerging from the agricultural sector. The quality-differentiating characteristic of such goods is not in their taste or appearance but rather in the 'environment-friendly' manner with which they were produced.

Over the past decade several papers have explained the complex relationship between trade and renewable natural resources (see Chapter 18 of this volume). A significant part of this literature has shown that institutional and market failures in resource-intensive countries lead to overexploitation of these resources and a decrease in social welfare. For example, North–South trade models developed by Chichilinisky (1994) and Karp *et al.* (2001) show how a resource-intensive country may overexploit its biodiversity resources when it fails to define and enforce property rights over these resources. Habitat destruction by land conversions and agriculture is indicated as another main cause for the loss of biodiversity around the world (e.g. Reid *et al.* 1989; Southgate *et al.* 1991; Barbier *et al.* 1994; Smulders *et al.* 2004). Several other papers (e.g. Swanson 1994; Barbier and Schulz 1997 develop models that describe the impact of trade and land conversion on a country's natural resource base and its exports. In these models, the property rights in the South are weakly defined while the resource sector produces an exporting product and competes with the agricultural sector. Recently, Polasky *et al.* (2004) constructed a model where consumers of both import and export countries are identical and equally concerned about biodiversity loss but the relative endowments of biodiversity vary between countries. The model again shows how trade liberalisation may lead to overexploitation of biodiversity resources and decrease social welfare. Smulders *et al.* (2004) constructed a model with three sectors: manufacturing, agriculture and resource extraction. In their model, agriculture and resource extraction compete for the same habitat area, while land has poorly defined property rights. Their model shows how free trade leads to overexploitation of the resource base and a short-term welfare gain due to reduced search cost for the resource good.

The current chapter complements and contributes to this growing literature on trade and renewable natural resources. In particular we develop a model in a static general equilibrium context that shows that even under growing 'green consumerism', free trade when combined with agricultural and population growth can lead to the depletion of biodiversity resources. The model uses the concept of international trade in vertically differentiated products which is in line with the work of Dixit (1979), Dixit and Norman (1978), Flam and Helpman (1982) and Copeland and Kotwal (1995) on vertical differentiation of product quality. Our framework considers the case where there are two types of agricultural products – one produced in the South that requires the conversion of biodiversity-rich land and is, hence, perceived as a lower-quality product by green consumers and the other produced in the North that does not require such conversion and is assumed to be of higher quality. Under conditions of free trade these quality-differentiated goods are traded in the international market. By including in our model a discount factor for 'biodiversity-depleting' products we incorporate in the analysis the Northern green consumer's current trend of differentiating between certified and non-certified goods and their readiness to pay a premium for the better-quality good that destroys less biodiversity. But this type of discounting does not show that green consumers value biodiversity *per se* or receive utility from the conservation of biodiversity in the South and are ready to pay for it. In this model, we elaborate the impact of this green discounting on land conversion.

The chapter derives some very important results that may contribute towards designing appropriate policy instruments and mechanisms for biodiversity conservation. First, it finds that free trade increases the clearing of undeveloped land in the South and in that way may lead to further destruction of biodiversity. But at the same time it is shown that free trade also increases the welfare of the Northern and Southern consumers. Second, the terms of trade of the South rise with increasing land development even when its agricultural goods are discounted by Northern green consumers. Yet in the case of large-scale land development, the terms of trade may reverse and the South may become a net importer of agricultural goods.

Third, if the Southern consumers also become sensitive towards biodiversity loss, their utility decreases with trade liberalisation. Fourth, when considering a more realistic situation of two types of consumers in the North – 'green' consumers who are sensitive to biodiversity loss and 'grey' consumers who are indifferent (similar to Southern consumers) – free trade may destroy further levels of biodiversity. Fifth, we show that as the income share of 'green' consumers increases in the North, biodiversity loss may decrease. Sixth, the model suggests that in the absence of free trade, only Southern population growth can decrease the stock of biodiversity. Free trade puts an added pressure on this trend. Further, under free trade population increases in the North may also lead to enhanced rates of biodiversity loss. Yet the population increase in the South affects biodiversity to a greater extent than does the population increase in the North.

Lastly, we show that if there is some technology that may decrease the use of land in the agricultural sector in the South and if the North subsidises this technology, then biodiversity loss may be decreased. Yet though this transfer may also increase the utility of Southern consumers, it decreases the welfare of Northern consumers.

The structure of the chapter is as follows. Section 2 describes the model, while section 3 explains the impact of free trade on biodiversity.

Section 4 discusses the impact of trade on biodiversity when there are two types of consumers in the North – green and grey – while section 5 explains the impact of population growth on biodiversity. Section 6 discusses the impact of new technology and North to South subsidisation of this technology on reversing the trend of land conversion-induced biodiversity loss. Finally, section 7 concludes with the policy insights derived from the model.

## 2 The model

### *2.1 Production technologies*

Consider a world in which there are two types of countries – country 1 (the North) and country 2 (the South). Each of these countries produces two goods – one manufacturing good, good 1, and one agricultural good, good 2 – using labour and land inputs. In what follows the countries will be indexed by $i = 1, 2$ and the goods by $j = 1, 2$. In each country, the industries producing the two goods are assumed to be perfectly competitive. Let the production function for producing good $j$ in country $i$ be given by

$$Y_{ij} = \min\left[\frac{L_{ij}}{\ell_{ij}}, \frac{A_{ij}}{a_{ij}}\right] \tag{1}$$

In (1), $Y_{ij}$, $L_{ij}$, and $A_{ij}$ denote, respectively, the output of good $j$ produced in country $i$ and the labour and land inputs used to produce this output. Also, $\ell_{ij}$ and $a_{ij}$ are two positive constants representing, respectively, the labour input and the land input required to produce one unit of the good in question.

The labour and land endowments of country $i$ are denoted by $\bar{L}_i$ and $\bar{A}_i$ respectively. In each country, part of the land endowment has already been developed and is ready for use as input in the production process; the remaining part is still in a state of wilderness and must be cleared or converted before being used as a factor of production. Let $A_i$ be the amount of land in country $i$ that has already been cleared and is currently available for use as input in the production process. Further, we shall assume that all the land in the North has been developed, i.e. $A_1 = \bar{A}_1$. We also assume that in the South, the wilderness region, with area $\bar{A}_2 - A_2$, is rich in biological diversity and part or all of this region can either be cleared and used for food production or can be conserved. We assume that the value of logging through land clearing is minimal and so it is not included in our model. In addition, we assume that the undeveloped land in the South has clearly defined property rights that can be either vested in the state or private agents (as the form of ownership does not alter the results of the model qualitatively). The proxy variable used for biodiversity is the amount of undeveloped land in the South. Such a spatial proxy for biodiversity has certain advantages. First, in light of the definitional ambiguity of the term biodiversity as well as difficulties in measuring and capturing all aspects of the value of biodiversity, a spatial measure such as 'undeveloped land' is much more inclusive than, say, using 'number of species'. Second, as biodiversity entails irreversibilities, using a spatial scale may prove to be a better safeguard from a conservation point of view. Moreover, we assume that the biodiversity-rich 'undeveloped land' in the model does not produce any products but can yield agricultural products only after it is developed. This implies that there is no competition for natural resources from the production of any other products. Finally, the production side of the model also assumes that the total amount of labour needed to clear $\Delta A_2$ units of wilderness land in the South is given by

$$C(\Delta A_2) = \gamma_0 + \gamma_1 \Delta A_2 + \frac{1}{2}\gamma_2 (\Delta A_2)^2 \tag{2}$$

where $\gamma_0, \gamma_1, \gamma_2$ are positive parameters. Thus, if $\Delta A_2$ unit of wilderness land is cleared, the total amount of land offered for rent in the South will be $A_2 + \Delta A_2$.

## 2.2 *Preferences*

In what follows, we shall assume that the land input used in the manufacturing sector is derived from the stock of developed land, while land input used in the agricultural sector comes from what is left of the stock of developed land and, possibly, from part of the undeveloped land after it has been converted. Consumers in the North are taken to receive disutility from this biodiversity loss and will accordingly discount the benefits of the agricultural goods when produced by converting wilderness into agricultural land. Therefore, the utility of the Northern consumers is affected not only by the amount of these goods but also by the damage inflicted on the environment from their production. We further assume that the consumers have perfect information about the origin of production of the agricultural goods they purchase. To capture these ideas, we shall assume that the preferences of a typical consumer in the North are represented by the following utility function:

$$u_1\left(x_{11}, x_{12}^1, x_{12}^2\right) = x_{11}^{1-\alpha_1}\left[x_{12}^1 + \left(1 - \frac{\varepsilon_1 \Delta A_2}{\bar{A}_2 - A_2}\right) x_{12}^2\right]^{\alpha_1} \tag{3}$$

In (3), $x_{11}$ represents the amount of the manufacturing good consumed by a consumer in the North, while $x_{12}^1$ and $x_{12}^2$ represent the amounts that such a consumer purchases of agricultural goods originating from the North and from the South respectively. Also, $\varepsilon_1$ and $\alpha_1$ are two parameters in the range of (0, 1). As can be seen from (3), the expression $[1 - \frac{\varepsilon_1 \Delta A_2}{\bar{A}_2 - A_2}]$ represents the weight assigned to one unit of the agricultural good produced in the South. Note that when $\Delta A_2 = 0$, there is no further biodiversity loss and the agricultural goods produced in the South are considered to be of the same quality as those produced in the North. This weight declines linearly as more and more new land is brought into production and is equal to zero when all the wilderness land is converted into agricultural land. This discount factor may be viewed as being derived from a type of 'guilt feeling' of Northern consumers as they may perceive that by consuming 'biodiversity-intensive' Southern agricultural good they are indirectly providing incentives for biodiversity destruction. Inclusion of such a discount factor captures the Northern green consumer's current trend for demanding differentiated eco-labelled and fair-trade products. Yet this type of discounting does not necessarily suggest that such green consumers value biodiversity *per se* or receive utility simply by preserving biodiversity in the South.

Likewise, the preferences of the consumers in the South are represented by the following utility function:

$$u_2\left(x_{21}, x_{22}^1, x_{22}^2\right) = x_{21}^{1-\alpha_2}\left[x_{22}^1 + \left(1 - \frac{\varepsilon_2 \Delta A_2}{\bar{A}_2 - A_2}\right) x_{22}^2\right]^{\alpha_2} \quad (4)$$

where $x_{21}$, $x_{22}^1$ and $x_{22}^2$ represent the consumption of manufacturing goods and the consumption of agricultural goods produced in the North and in the South respectively. Also, $\varepsilon_2$ and $\alpha_2$ are two positive parameters strictly less than 1. As specified by (4), the preferences of consumers from the South are allowed to be different from those of the North. First, consumers from the North and from the South might have different preferences over biodiversity loss. Due to the income disparity between the two regions, we shall assume that $\varepsilon_2 < \varepsilon_1$, i.e. consumers in the South (under current income levels) have weaker preferences over biodiversity conservation. Second, preferences between the two regions over manufacturing and agricultural goods are also allowed to be different when $\alpha_1 \neq \alpha_2$. We also assume that the partial elasticity of the agricultural good (i.e. its contribution to the utility) is lower than the manufacturing good in both regions (i.e. $\alpha_2 < 0.5$). Yet we assume that the partial elasticity of agricultural goods is greater in the South than in the North (i.e. $\alpha_1 < \alpha_2$).

## 2.3 *Profit maximisation*

Let $p^i_j, i = 1, 2, j = 1, 2$ denote the price of good $j$ originating in country $i$. Because the manufacturing sector in the South is assumed not to be more detrimental to the environment than that in the North, the same price must apply to the manufacturing goods produced in both regions, that is, $p^1_1 = p^2_1 = p_1$. However, as some of the wilderness area in the South may be converted into agricultural land, we could have $p^1_2 \neq p^2_2$. Also, let $\omega_i$ and $r_i$ denote, respectively, the wage rate and the rental rate of land in each country $i = 1, 2$.

The representative firm in each sector $j$ in each country $i$ solves the following profit maximisation problem:

$$\max_{(L_{ij}, A_{ij})} \left[ p^i_j Y_{ij} - \omega_i L_{ij} - r_i A_{ij} \right] \tag{5}$$

As for the firms in the South that hire labour to convert wilderness land into agricultural land, they solve the following profit maximisation problem:

$$\max_{\Delta A_2} \left[ r_2 \Delta A_2 - \omega_2 C(\Delta A_2) \right] \tag{6}$$

where, as noted above, $\Delta A_2$ represents the portion of wilderness used as input in the production of the agricultural goods.

## 2.4 *Utility maximisation*

The representative consumer in the North solves the following utility maximisation problem:

$$\max_{(x_{11}, x^1_{12}, x^2_{12})} x_{11}^{1-\alpha_1} \left[ x^1_{12} + \left( 1 - \frac{\varepsilon_1 \Delta A_2}{\bar{A}_2 - A_2} \right) x^2_{12} \right]^{\alpha_1} \tag{7}$$

subject to the following budget constraint:

$$p_1 x_{11} + p^1_2 x^1_{12} + p^2_2 x^2_{12} - m_1 = 0 \tag{8}$$

where $m_1$ represents individual income. The solution of this utility maximisation problem is straightforward. First, we can observe that a fraction of income equal to $\alpha_1$ will be spent on agricultural goods and the remaining fraction on manufacturing goods. More precisely, the demand for manufacturing goods is given by

$$x_{11} = \frac{(1 - \alpha_1) m_1}{p_1} \tag{9}$$

while the demand for agricultural goods from the North and South are given respectively by

$$x_{12}^1 = \frac{\alpha_1 m_1}{p_2^1}, x_{12}^2 = 0 \text{ if } p_2^1\left(1 - \frac{\varepsilon_1 \Delta A_2}{\bar{A}_2 - A_2}\right) < p_2^2 \tag{10}$$

$$x_{12}^1 = 0, x_{12}^2 = \frac{\alpha_1 m_1}{p_2^2} \text{ if } p_2^1\left(1 - \frac{\varepsilon_1 \Delta A_2}{\bar{A}_2 - A_2}\right) > p_2^2 \tag{11}$$

$$x_{12}^1 + \left(1 - \frac{\varepsilon_1 \Delta A_2}{\bar{A}_2 - A_2}\right) x_{12}^2 = \frac{\alpha_1 m_1}{p_2^1} \text{ if } p_2^1\left(1 - \frac{\varepsilon_1 \Delta A_2}{\bar{A}_2 - A_2}\right) = p_2^2 \tag{12}$$

Furthermore, the representative consumer in the South solves the following utility maximisation problem:

$$\max_{(x_{21}, x_{22}^1, x_{22}^2)} x_{21}^{1-\alpha_2}\left[x_{22}^1 + \left(1 - \frac{\varepsilon_2 \Delta A_2}{\bar{A}_2 - A_2}\right) x_{22}^2\right]^{\alpha_2} \tag{13}$$

subject to the following budget constraint:

$$p_1 x_{21} + p_2^1 x_{22}^1 + p_2^2 x_{22}^2 - m_2 = 0 \tag{14}$$

where $m_2$ is this consumer's income. The solution of this utility maximisation problem is given by:

$$x_{21} = \frac{(1-\alpha_2)m_2}{p_1} \tag{15}$$

$$x_{22}^1 = \frac{\alpha_2 m_2}{p_2^1}, \quad x_{22}^2 = 0 \text{ if } p_2^1\left(1 - \frac{\varepsilon_2 \Delta A_2}{\bar{A}_2 - A_2}\right) < p_2^2 \tag{16}$$

$$x_{22}^1 = 0, \quad x_{22}^2 = \frac{\alpha_2 m_2}{p_2^2} \text{ if } p_2^1\left(1 - \frac{\varepsilon_2 \Delta A_2}{\bar{A}_2 - A_2}\right) > p_2^2 \tag{17}$$

$$x_{22}^1 + \left(1 - \frac{\varepsilon_2 \Delta A_2}{\bar{A}_2 - A_2}\right) x_{22}^2 = \frac{\alpha_2 m_2}{p_2^1} \text{ if } p_2^1\left(1 - \frac{\varepsilon_2 \Delta A_2}{\bar{A}_2 - A_2}\right) = p_2^2 \tag{18}$$

### 2.5 *Autarkic equilibrium*

First we explore the autarkic equilibrium under which the economies of the two regimes are closed and they consume what they produce domestically. The price and allocation system will be given by the lists $\mathcal{P} = ((p_j^i)_{j=1}^2, \omega_i, r_i)_{i=1}^2$ and $\mathcal{A} = (((Y_{ij}, L_{ij}, A_{ij})_{j=1}^2)_{i=1}^2, (X_{i1}, X_{i2})_{i=1}^2)$. For the South, the term $\Delta A_2$ is added in this list. A pair $(\mathcal{P}, \mathcal{A})$ is said to constitute an equilibrium if the following conditions are satisfied.

First, for each $i$ and each $j$, the production plan $(Y_{ij}, L_{ij}, A_{ij})$ maximises the profit of the representative firm in sector $j$ of country $i$ when the price system $\mathcal{P}$ prevails. Second, $\Delta A_2$ is the part of the wilderness area converted to agricultural land in the South when the price system $\mathcal{P}$ prevails. Third, $(X_{i1}, X_{i2})$ is the consumption bundle that maximises the utility of the representative consumer in each country subject to the aggregate budget constraint. More precisely, $(X_{11}, X_{12})$ is the solution of the problem constituted by (7) and (8), where the income of the representative consumer in country 1 is given by

$$m_1 = GDP_1 = p_1^1 Y_{11} + p_2^1 Y_{12} \tag{19}$$

For simplicity and without loss of generality, we assume that $\varepsilon_2 = 0$, i.e. consumers from the South are totally insensitive to biodiversity loss. Then the solution that emerges is

$$x_{11} = \frac{(1 - \alpha_1)m_1}{p_1^1} \tag{20}$$

$$x_{12} = \frac{\alpha_1 m_1}{p_2^1} \tag{21}$$

where $(X_{21}, X_{22})$ is the solution of the problem constituted by (13) and (14), where the income of the representative consumer in country 2 is given by

$$m_2 = GDP_2 = p_1^2 Y_{21} + p_2^2 Y_{22} \tag{22}$$

The solutions are then given by:

$$x_{21} = \frac{(1 - \alpha_2)m_2}{p_1^2} \tag{23}$$

$$x_{22} = \frac{\alpha_2 m_2}{p_2^2} \tag{24}$$

Fourth, the following market-clearing conditions must hold for the North and South:

$$Y_{i1} = \frac{(1 - \alpha_i)\left(p_1^i Y_{i1} + p_2^i Y_{i2}\right)}{p_1^i} \tag{25}$$

$$Y_{i2} = \frac{\alpha_i \left(p_1^i Y_{i1} + p_2^i Y_{i2}\right)}{p_2^i} \tag{26}$$

$$\bar{L}_1 = Y_{11}\ell_{11} + Y_{12}\ell_{12} \tag{27}$$

$$\bar{L}_2 = Y_{21}\ell_{21} + Y_{22}\ell_{22} + C(\Delta A_2) \tag{28}$$

$$A_1 = a_{11} Y_{11} + a_{12} Y_{12} \tag{29}$$

$$A_2 + \Delta A_2 = a_{21} Y_{21} + a_{22} Y_{22} \tag{30}$$

In what follows, we consider an equilibrium under which the output of each sector in each country is positive, i.e. $Y_{ij} > 0, i = 1, 2, j = 1, 2$. Because the technology in each sector is linear, profit maximisation implies that the representative firm in each sector makes zero profits. The zero-profit conditions are expressed by the following equations:

$$p_j^i = \omega_i \ell_{ij} + r_i a_{ij}, i = 1, 2, j = 1, 2 \tag{31}$$

Furthermore, in the South, wilderness land will be cleared until the marginal cost of land clearing is equal to the rental rate of land, i.e.

$$r_2 = \omega_2(\gamma_1 + \gamma_2 \Delta A_2), \tag{32}$$

and

$$r_2 = \omega_2(\gamma_2 \Delta A_2), \tag{32A}$$

if $\gamma_1 = 0$.

For the South, equations (22), (23), (24), (25), (26), (28), (30), (31), (32) constitute a system of nine equations in nine unknowns:

$$p_j^2, j = 1, 2, \omega_2, r_2, \Delta A_2, Y_{2j}, j = 1, 2, X_{21}, X_{22}.$$

Due to Walras's law, only eight of these equations are independent. Choosing the wage rate in the South as the numeraire, we will have eight independent equations in eight unknowns, which can be solved to find the autarkic equilibrium in the South.

For the North, equations (19), (20), (21), (25), (26), (27), (29), (31) constitute a system of eight equations in eight unknowns: $p_j^1, j = 1, 2, \omega_1, r_1, Y_{1j}, j = 1, 2, X_{11}, X_{12}$. Due to Walras's law, only seven of these equations are independent. Choosing the wage rate in the North as the numeraire, we will have seven independent equations in seven unknowns, which can be solved to find the autarkic equilibrium in the North.

Although the model can be solved algebraically, the large number of parameters makes interpretation of the results cumbersome. Therefore, we proceeded by solving the model numerically by assuming some reasonable values for the parameters that characterise the model.[1] Numerical solutions are useful because they reveal a number of important results of the model. For simplicity and without loss of generality, we assume that $\varepsilon_2 = 0$, i.e. consumers from the South are totally insensitive to biodiversity loss and that $\gamma_0 = \gamma_1 = 0$. Also, we assume that the technology used for the production of the agricultural good is land intensive while that used for the derivation of the manufacturing good is labour intensive. It thus would cost more to produce agricultural goods in the North than

[1] The values used for the parameters are listed in Table A1 in the Appendix.

Table 8.1 *Autarky prices of the agricultural and manufacturing goods*

| | Price of agricultural goods | Price of manufacturing goods |
|---|---|---|
| North | 0.55 | 0.40 |
| South | 0.14 | 0.43 |

in the South. By the same reasoning, we assume that it costs more to produce manufacturing goods in the South.

The solutions of the autarkic equilibrium give us the following results. First (in accordance with our assumptions) we find that the South is more land abundant and it has a technological and cost advantage in the production of the land-intensive agricultural product. The North, meanwhile, is more labour abundant and it has a technological and cost advantage in the production of the manufacturing good. This result is also reflected through the prices of the two goods in the North and South under autarky. Table 8.1 illustrates the results and clearly shows that under autarky, the price of the agricultural (manufacturing) good is lower (higher) in the South than in the North. This reflects our initial assumptions that the South has a comparative advantage in the production of the agricultural good and the North in the production of the manufacturing good such that when free trade is allowed the South will export agricultural goods and the North will export manufacturing goods.

### 2.6 *General equilibrium under free trade*

When the two economies are open for free trade, the world markets will clear and the solution procedure will be similar to the one highlighted above except for the addition of a few auxiliary variables and equations in the system of general equilibrium.

We now assume that two types of 'good 2' are produced. One is derived in the South and is biodiversity intensive and the other is produced in the North. These are denoted by $X_{i2}^1$ and $X_{i2}^2$, respectively. The list $(X_{i1}, X_{i2}^1, X_{i2}^2)$ denotes the consumption bundle that maximises the utility of the representative consumer in each subject to the aggregate budget constraint. More precisely, $(X_{11}, X_{12}^1, X_{12}^2)$ is the solution of the problem constituted by (7) and (8), where the income of the representative consumer in country 1 is given by

$$m_1 = GDP_1 = p_1^1 Y_{11} + p_2^1 Y_{12} \tag{33}$$

Here the solutions are

$$x_{11} = \frac{(1-\alpha_1)m_1}{p_1^1} \tag{34}$$

$$x_{12}^1 + \left(1 - \frac{\varepsilon_1 \Delta A_2}{\bar{A}_2 - A_2}\right) x_{12}^2 = \frac{\alpha_1 m_1}{p_2^1} \tag{35}$$

and $(X_{21}, X_{22}^1, X_{22}^2)$ is the solution of the problem constituted by (13) and (14) where the income of the representative consumer in country 2 is given by

$$m_2 = GDP_2 = p_1^2 Y_{21} + p_2^2 Y_{22} \tag{36}$$

The solutions are then

$$x_{21} = \frac{(1-\alpha_2)m_2}{p_1^2} \tag{37}$$

$$x_{22}^1 + \left(1 - \frac{\varepsilon_2 \Delta A_2}{\bar{A}_2 - A_2}\right) x_{22}^2 = \frac{\alpha_2 m_2}{p_2^2} \tag{38}$$

Also, the following world market-clearing conditions must hold:

$$Y_{11} + Y_{21} = \frac{(1-\alpha_1)\left(p_1^1 Y_{11} + p_2^1 Y_{12}\right)}{p_1^1} + \frac{(1-\alpha_2)\left(p_1^2 Y_{21} + p_2^2 Y_{22}\right)}{p_1^2} \tag{39}$$

$$Y_{12} = X_{12}^1 + X_{22}^1 \tag{40}$$

$$Y_{22} = X_{12}^2 + X_{22}^2 \tag{41}$$

Equations (27) to (32) must also hold for this free trade case. In addition, we consider an equilibrium under which the output of each sector in each country is positive, i.e. $Y_{ij} > 0, i = 1, 2, j = 1, 2$.

As the manufacturing good produced in the South is assumed not to be harmful to the environment, its price must be identical to the price of the manufacturing goods produced in the North and hence we will have:

$$p_1^1 = p_1^2 \tag{42}$$

In general equilibrium, if the North imports the agricultural goods produced in the South, we must have the condition

$$p_2^1 \left(1 - \frac{\varepsilon_1 \Delta A_2}{\bar{A}_2 - A_2}\right) = p_2^2 \tag{43}$$

When (43) holds, (12) applies for the representative consumer in the North and we will have

$$X_{12}^{1} + \left(1 - \frac{\varepsilon_1 \Delta A_2}{\bar{A}_2 - A_2}\right) X_{12}^{2} = \frac{\alpha_1 \left(p_1^1 Y_{11} + p_2^1 Y_{12}\right)}{p_2^1} \tag{44}$$

Also, when (43) holds, (17) will apply for the representative consumer in the South due to the assumption that $\varepsilon_2 < \varepsilon_1$ and we will thus have

$$X_{22}^{1} = 0 \tag{45}$$

$$X_{22}^{2} = \frac{\alpha_2 \left(p_1^2 Y_{21} + p_2^2 Y_{22}\right)}{p_2^2} \tag{46}$$

Together, equations (33), (34), (36), (37), (39) to (41), (27) to (32), (42) to (44) and (46) constitute a system of seventeen equations in seventeen unknowns:

$$p_j^i, i = 1, 2, j = 1, 2, \omega_i, i = 1, 2, r_i, i = 1, 2,$$
$$\Delta A_2, Y_{ij}, i = 1, 2, j = 1, 2, X_{12}^1, X_{12}^2, X_{22}^1, X_{22}^2$$

Once again, by evoking Walras's law, only sixteen of these equations are independent and by choosing the wage rate in the South as the numeraire, we will have sixteen independent equations in sixteen unknowns, which can be solved to find the equilibrium of this two-country world.

In the same fashion as the autarky solution and following the same assumptions, we proceeded with solving the model numerically by assuming certain reasonable values for the parameters that characterise the model. The results of this analysis are presented in the following section.

## 3 Impact of trade on biodiversity

### *3.1 From autarky to free trade*

Our model simulations can be used to examine the common view found in the literature that free trade degrades the environment and accelerates biodiversity loss. In the numerical solutions, we compute the amount of new agricultural land cleared in the South. Table 8.2 shows these results for different levels of population or labour force.

The results from Table 8.2 clearly show that the amount of cleared land increases as the South moves from autarky to free trade for different population sizes. The finding thus supports the view that free trade increases biodiversity loss. The result is quite intuitive. As the South has a competitive advantage in agricultural goods, it will export these goods as the economy opens up for trade. In autarky, the South was producing

Table 8.2 *Change in the amount of cleared land under autarky and free trade for different population sizes*

| | $L_2 = 1.5$ | | $L_2 = 2$ | |
|---|---|---|---|---|
| Variables | Autarky | Free trade | Autarky | Free trade |
| $\Delta A_2$ | 0.1729 | 1.3370 | 0.6394 | 1.6486 |
| $U_2$ | 3.3217 | 3.6021 | 4.3153 | 4.5112 |
| $U_2/L_2$ | 2.2145 | 2.4014 | 2.1577 | 2.2556 |
| $U_1$ | 5.5509 | 6.7349 | 5.5509 | 6.5491 |
| $U_1/L_1$ | 1.8503 | 2.2449 | 1.8503 | 2.1830 |
| $p_2$ | 0.11 | 0.18 | 0.14 | 0.19 |

Notes: $L_2$ = the size of the labour force in the South
$\Delta A_2$ = the change in agricultural land or the amount of cleared land
$U_2$ = the total utility of the Southern consumers
$U_2/L_2$ = the per capita utility of the Southern consumers
$p_2$ = the price of the agricultural good

food only to feed its own people, but now it will also produce food for exports. This increases the pressure on land resources. To produce more agricultural goods, the South clears the untouched biodiversity-rich land and brings it into cultivation. Also, as the utility of Southern consumers is not sensitive to biodiversity loss, moving from autarky to free trade increases both total and per capita utility in spite of the increase in biodiversity loss. This increase in utility provides an incentive for the South to push for higher exports by clearing more land and in the process causing further biodiversity loss. We can also observe that the price of the agricultural good rises when the economy moves from autarky to free trade.

For Northern consumers who can now consume more agricultural goods at a lower price, both their gross and per capita utility increase as it moves from autarky to free trade. But it is also important to note that both the gross and per capita utility of the Northern consumers decrease under free trade as the population of the South increases. This is due to the biodiversity loss that emerges from population growth in the South.

### 3.2 *The impact of positive environmental sensitiveness of Southern consumers*

In the model so far we have assumed that the Southern consumers are insensitive to the loss of biodiversity, i.e. $\varepsilon_2 = 0$. If we relax this constraint (i.e. $\varepsilon_2 > 0$), biodiversity loss will now decrease both the gross and per capita utility of the Southern consumers. This is shown in Table 8.3.

Table 8.3 *Impact on Southern consumer utility of positive environmental sensitivity to biodiversity loss*

| $\varepsilon_2$ | $U_2$ *[autarky]* | $U_2$ *[free trade]* |
|---|---|---|
| 0.00 | 4.3153 | 4.5112 |
| 0.15 | 4.1775 | 4.1394 |
| 0.25 | 4.0854 | 3.8915 |

It is interesting to note in Table 8.3 that under both autarky and free trade, the gross utility of Southern consumers decreases as their sensitivity towards biodiversity loss increases. Moreover, when the South moves from autarky to free trade, the gross utility of the Southern consumers decreases. This implies that free trade may decrease both the utility of the Southern consumers and their stock of biodiversity when people in the South are sensitive to biodiversity loss. Further, if the sensitivity of the Southern consumers becomes highly positive and $\varepsilon_2 > \varepsilon_1$, then the South may lose the comparative advantage in agricultural goods and start importing them from the North.

### 3.3 *The impact of biodiversity loss on the terms of trade*

The model assumes that Northern consumers are sensitive to biodiversity loss and these preferences are reflected in their utility function. The consumption of the agricultural good produced in the South, which destroys biodiversity, is discounted by Northern consumers through the discount factor $(1 - \frac{\varepsilon_1 \Delta A_2}{\bar{A}_2 - A_2})$. Also, when Northern consumers purchase both the agricultural goods produced at home and the agricultural good produced in the South, we have: $p_2^1(1 - \frac{\varepsilon_1 \Delta A_2}{\bar{A}_2 - A_2}) = p_2^2$, according to (43). Now the terms of trade for the South, given by the ratio of the price of the exported good over the price of the imported good, will be given by

$$p^* = \frac{p_2^2}{p_1^2} = \frac{l_{22} + a_{22}\gamma_2 \Delta A_2}{l_{21} + a_{21}\gamma_2 \Delta A_2} \tag{47}$$

Differentiating the above terms of trade with respect to $\Delta A_2$ provides the following expression:

$$\frac{\partial p^*}{\partial \Delta A_2} = \frac{\gamma_2(a_{22}l_{21} - a_{21}l_{22})}{(l_{21} + a_{21}\gamma_2 \Delta A_2)^2} \tag{47A}$$

By assumptions $a_{22} > a_{21}$ and $l_{21} > l_{22}$ we will have $\frac{\partial p^*}{\partial \Delta A_2} > 0$ which implies that the terms of trade increase as $\Delta A_2$ rises. So, even with the green consumer's discounting of the Southern agricultural product, the Southern terms of trade increase with increases in land clearing. This works as an incentive to clear more land in the South. But with large-scale land clearing, the price of the Southern agricultural product may increase enough to reverse the pattern of trade in agricultural goods, possibly resulting in the South becoming a net importer of such goods.

## 4 Consumer heterogeneity in the North

We now extend the model to the case in which there are two types of consumers in the North, a green type that is sensitive to biodiversity loss and a grey type that is indifferent to such loss. Furthermore, the latter type of consumer is assumed to behave similarly to Southern consumers who have been assumed to be insensitive to the degradation of their environment. We also assume that $\alpha_2 = \alpha_{2grey}$. The green consumers are assumed to constitute a fraction $\theta$ of the Northern population such that $0 < \theta < 1$ and the size of the grey consumer equal to 1-$\theta$. We also assume that all Northern consumers have the same income level and that the utility functions of the two types of consumers are given by:

$$u_{1green}\left(x_{1green,1}, x^1_{1green,2}, x^2_{1green,2}\right) = [x_{1green,1}]^{1-\alpha_{1green}} \times \left[x^1_{1green,2} + \left(1 - \frac{\varepsilon_1 \Delta A_2}{\bar{A}_2 - A_2}\right) x^2_{1green,2}\right]^{\alpha_{1green}} \tag{48}$$

$$u_{1grey}\left(x_{1grey,1}, x^1_{1grey,2}, x^2_{1grey,2}\right) = [x_{1grey,1}]^{1-\alpha_{1grey}} \times \left[x^1_{1grey,2} + \left(1 - \frac{\varepsilon_1 \Delta A_2}{\bar{A}_2 - A_2}\right) x^2_{1grey,2}\right]^{\alpha_{1grey}} \tag{49}$$

The budget constraints of the green and grey Northern consumers are given, respectively, by

$$p_1 x_{1green,1} + p^1_2 x^1_{1green,2} + p^2_2 x^2_{1green,2} - \theta m_1 = 0 \tag{50}$$

$$p_1 x_{1grey,1} + p^1_2 x^1_{1grey,2} + p^2_2 x^2_{1grey,2} - (1-\theta) m_1 = 0 \tag{51}$$

The world market clearing conditions provide that: $p^1_2(1 - \frac{\varepsilon_1 \Delta A_2}{\bar{A}_2 - A_2}) = p^2_2$. The green consumers demand for manufacturing and agricultural goods will be given, respectively, by

$$x_{1green,1} = \frac{(1-\alpha_{1green})\theta m_1}{p_1} \tag{52}$$

$$x^1_{1green,2} + \left(1 - \frac{\varepsilon_1 \Delta A_2}{\bar{A}_2 - A_2}\right) x^2_{1green,2} = \frac{\alpha_{1green}\theta m_1}{p^1_2} \tag{53}$$

Table 8.4 *Change in the amount of cleared land with two types of Northern consumers*

| | $L_2 = 2$ | | |
|---|---|---|---|
| Variables | Autarky | Free trade | Free trade with two groups |
| $\Delta A_2$ | 0.6394 | 1.6486 | 1.7894 |
| $U_2$ | 4.3153 | 4.5112 | 4.5623 |
| $U_2/L_2$ | 2.1577 | 2.2556 | 2.2812 |
| $p_2^2$ | 0.14 | 0.19 | 0.21 |
| $p_2^2/p_1^2$ | – | 0.43 | 0.44 |

From the world market clearing condition we get $p_2^1 > p_2^2$ and as the grey consumers in the North are insensitive to biodiversity loss, they will consume agricultural goods exported from the South only when they are relatively cheaper. Accordingly, their utility maximisation will provide the following demand functions for manufacturing and agricultural goods:

$$x_{1grey,1} = \frac{(1 - \alpha_{1grey})(1 - \theta)m_1}{p_1} \tag{54}$$

$$x^2_{1grey,2} = \frac{\alpha_{1grey}(1 - \theta)m_1}{p_2^2} \tag{55}$$

With these utility maximisation results we can obtain a different set of solutions for the two consumer groups under investigation. The results that we obtain through the numerical solutions for this extension of the model are interesting and support our previous claims. Table 8.4 shows the comparative picture of the key variables under autarky, free trade and free trade with two types of Northern consumers. It clearly establishes our previous claim that biodiversity loss through clearing of land increases if a country moves from autarky to free trade. In fact, it even increases further if a portion of the Northern consumers is insensitive to biodiversity loss. Both the total and per capita utility in the South increase as we move from autarky to free trade to free trade with two groups of Northern consumers. Also, the price of good 2 and the terms of trade for the South increase as we move from free trade to free trade with two types of Northern consumers. Both of the above two outcomes emerge as there are now incentives to Southern economies to clear more land and increase their exports of agricultural goods.

Lastly, Table 8.5 shows that as the income share of green consumers in the North becomes larger, the conversion of biodiversity-rich land

Table 8.5 *Change in the amount of cleared land under free trade with two types of Northern consumers and different levels of income shares for the 'green consumers' ($L_2 = 2$)*

| Different values of $\theta$ | $\Delta A_2$ |
|---|---|
| 0.5 | 1.8239 |
| 0.6 | 1.7894 |
| 0.7 | 1.7546 |

decreases. This implies that only via preference-related discounting of the Northern consumers can the rate of biodiversity loss in the South be reduced. Yet this result holds only if there is no population growth in the south.

## 5 Impact of population growth on biodiversity

We can also derive important insight from the above models over the impact of population growth on biodiversity loss. Our results support the claim that rapid population growth in the South is a major determinant of global biodiversity decline. Table 8.6 shows that in all three cases that we have examined (autarky, free trade and free trade with two groups of Northern consumers), as population size increases, the amount of converted biodiversity-rich land also increases. Population growth coupled with free trade makes the situation worse, while population growth coupled with free trade with two types of Northern consumers further aggravates the situation. Also, under both free trade scenarios, the total utility of the South and its terms of trade improve with population increase. These two factors work as an incentive to increase the export of agricultural goods and cause more land clearing and associated biodiversity loss. But at the same time the per capita utility declines with population increase in the South. The latter two trends may in the long run deteriorate overall national welfare.

Furthermore, beyond the population increase in the South, the population increase in the North can also create pressures for biodiversity decline. Although the Northern countries may be sensitive to biodiversity loss, with increases in their population their demand for agricultural goods will increase which will raise their imports of Southern agricultural goods. This will induce a further expansion of the agricultural sector in

Table 8.6 *The impact on biodiversity of population increase in the South*

| | Autarky | | Free trade | | Free trade with two groups | |
|---|---|---|---|---|---|---|
| Variables | $L_2 = 1.5$ | $L_2 = 2$ | $L_2 = 1.5$ | $L_2 = 2$ | $L_2 = 1.5$ | $L_2 = 2$ |
| $\Delta A_2$ | 0.1729 | 0.6394 | 1.3370 | 1.6486 | 1.4864 | 1.7894 |
| $U_2$ | 3.3217 | 4.3153 | 3.6021 | 4.5112 | 3.6666 | 4.5623 |
| $U_2/L_2$ | 2.2145 | 2.1577 | 2.4014 | 2.2556 | 2.4444 | 2.2812 |
| $P_2^2$ | 0.11 | 0.14 | 0.18 | 0.19 | 0.19 | 0.21 |
| $p_2^2/p_1^2$ | – | – | 0.397 | 0.43 | 0.41 | 0.44 |

Table 8.7 *The impact on biodiversity levels of population increase in the North under free trade*

| | Free trade | | |
|---|---|---|---|
| Variables | $L_1 = 3$ | $L_1 = 3.5$ | $L_1 = 4$ |
| $\Delta A_2$ | 1.6486 | 1.9131 | 2.1729 |
| $U_2$ | 4.5112 | 4.6108 | 4.7226 |
| $U_2/L_2$ | 2.2556 | 2.3054 | 2.3613 |
| $p_2^2/p_1^2$ | 0.43 | 0.45 | 0.47 |

the South and will lead to more conversion of biodiversity-rich land (see Table 8.7).

We can also note that with population increase in the North, the gross and per capita utility of Southern consumers and the terms of trade of the South increase as well. This will again act as an incentive for further expansion of the Southern agricultural sector and increase biodiversity loss.

Biodiversity loss will escalate further when we have population increases in both the North and South. Comparing columns 2 to 3 in Table 8.8 we see that when only the Northern population increases, the amount of cleared land increases from 1.6486 to 1.9131. Comparing columns 2 and 4, we see that an increase in the population in the South leads to an increase in land conversion from only 1.6486 to 1.9457. Finally, if we compare columns 2 and 5, we see that when population increases in both the South and North the amount of cleared land is at its highest level compared with the previous cases. This highlights population increase as one of the main drivers of biodiversity loss. Moreover, under free trade, we have higher levels of biodiversity decline in the case of a population increase in the South than in the North.

Table 8.8 *The impact on biodiversity of population increase under free trade when population increases both in the North and the South*

| Variables | $L_1 = 3$ $L_2 = 2$ | $L_1 = 3.5$ $L_2 = 2$ | $L_1 = 3$ $L_2 = 2.5$ | $L_1 = 3.5$ $L_2 = 2.5$ |
|---|---|---|---|---|
| $\Delta A_2$ | 1.6486 | 1.9131 | 1.9457 | 2.2019 |
| $U_2$ | 4.5112 | 4.6108 | 5.3565 | 5.4395 |
| $U_2/L_2$ | 2.2556 | 2.3054 | 2.1426 | 2.1758 |
| $p_2^2/p_1^2$ | 0.43 | 0.4507 | 0.4536 | 0.48 |

Table 8.9 *The impact of a technology subsidy on the 'cleared land' and the utility of Northern and Southern consumers*

| | Amount of subsidy | | |
|---|---|---|---|
| Variables | T = 0.05 | T = 0.08 | T = 0.11 |
| $\Delta A_2$ | 1.2811 | 1.2445 | 1.2053 |
| $U_2$ | 3.6937 | 3.7488 | 3.8040 |
| $U_1$ | 6.7296 | 6.7275 | 6.7261 |

## 6 Impact of technology subsidies on biodiversity loss

We make a final extension of the model in order to examine the impact of technology subsidies on biodiversity and on the utilities of the consumers in both regions. We assume that the North provides a lump-sum subsidy of the amount 'T' to the South. This subsidy 'T' is subtracted from the North's GDP and added to the South's. We further assume that this subsidy is used by the South to improve technological efficiency in the agricultural sector, so that less amount of per unit land is used to produce the same amount of agricultural goods. In this way the subsidy decreases land clearing and biodiversity loss in the South. This impact is reflected through the following equation added to the model:

$$a_{22} = \delta - \sigma.\ell_0 \tag{56}$$

where $\ell_0 = T/\omega_2$ (i.e. the amount of labour used to invent new technology that uses a lower amount of land in the agricultural sector) and where $\delta$ and $\sigma$ are two parameters. Table 8.9 displays the simulated results derived from the solution of the model with these extra assumptions.

The results in Table 8.9 clearly show that under free trade, if a subsidy is provided from the North to the South that improves the land-use efficiency in the South, then we will observe decreases in the amount of biodiversity-rich land converted in the South. The subsidy also increases the gross utility of Southern consumers while that of the consumers in the North will decrease at a lower rate.

## 7 Policy implications and conclusion

This chapter has lent support to some of the concerns of anti-globalisation activists that when there is no market for the biodiversity resources themselves, free trade may deplete the world's biodiversity. Biodiversity resources provide different goods and services that often do not have any market. For example, ecosystem services that are produced by biodiversity-rich land increase the productivity of the agricultural sector and yet that sector does not pay for them. There are other non-use values (e.g. carbon sequestration) associated with 'undeveloped land' which also have no market. In fact, the market for many of these non-use values is global in nature and scope, which implies that the South alone could not correct for the distortion generated from the missing market problem. Due to absence of markets for these services and products derived from biodiversity-rich land, the opportunity cost of their alternative use in agriculture becomes significant and causes land clearing. Further, when developing countries cannot monetise the value of the conserved biodiversity, they have no alternative but to deplete their rich biodiversity resource stock in order to meet pressing subsistence needs for the current population, even though this may be welfare decreasing for future generations in these countries as well as in the world at large.

Nevertheless, the chapter has also provided support to the concerns of environmental and resource economists that even without trade, unsustainable population growth alone can deplete biodiversity resources. Further, the 'ecological footprint' concept developed by Dally and Goodland is also supported by this model. Free trade coupled with Southern rapid population growth and agricultural expansion can augment the 'ecological footprint' of the South, leading to rapid biodiversity loss.

Demand-side mechanisms like discounting of biodiversity-depleting products and supply-side mechanisms such as eco-friendly agricultural technologies can have a positive impact to decrease biodiversity loss. But merely discounting Southern agricultural products and not being concerned about the biodiversity resources directly may not provide sufficient incentives for biodiversity conservation in the long run. If Northern 'green' consumers really care about biodiversity loss in the South, then biodiversity should be an argument in their utility function. In such a case, their valuation of Southern biodiversity can be measured and they should be willing to pay for the conservation of biodiversity in the South equal to that value. Not valuing biodiversity directly and only discounting differentiated products as shown in this model can only slow down the depletion of biodiversity stock, but cannot provide a more sustainable long-term solution.

The above discussion leads to the conclusion that policy planners should try to create markets for the different goods and services provided by biodiversity-rich lands (see Chapter 1 in this volume). For example, they may further explore alternative markets such as bio-prospecting or eco-tourism in order to appropriate economic values for the conservation of the biodiversity stock. Also, the biodiversity-rich developing countries have to diversify with respect to domestic production and trade. They need to decrease their dependence on export of natural resources and agricultural products and diversify their export towards manufacturing products. Above all, if consumers in all regions of the world really care about the existence of biodiversity, they should be prepared to pay a premium for the conservation and provision of this global public good. Hence, institutions such as the Global Environmental Fund should become the principal mechanisms for the conservation of biodiversity.

## REFERENCES


Barbier, E. B. and Schulz, C. E. 1997. Wildlife, biodiversity and trade. *Environment and Development Economics*. **2**. 145–172.

Barbier, E. B., Burgess, J. C., Bishop, J. T. and Aylward, B. A. 1994. *The Economics of Tropical Timber Trade*. London: Earthscan.

Bulte, E. H. and Barbier, E. B. 2003. Trade and renewable resources in a second best world: an overview. *Unpublished Draft Keynote Address in the 12th Annual Conference of the European Association of Environmental and Resource Economics*. Spain. 28–30 June.

Chichilinisky, G. 1993. North south trade and the dynamics of renewable resources. *Structural Change and Economic Dynamics*. **4**. 219–248.

Chichilinisky, G. 1994. North-south trade and the global environment. *American Economic Review*. **84**. 851–874.

Copeland, B. R. and Kotwal, A. 1995. Product quality and the theory of comparative advantage. *European Economic Review*. **40**. 1745–1760.

Dixit, A. 1979. Quality and quantity competition. *Review of Economic Studies*. **46**. 587–599.

Dixit, A. and Norman, V. 1978. Advertising and welfare. *Bell Journal of Economics*. **9**. 1–17.

Flam, H. and Helpman, E. 1982. Vertical product differentiation and north-south trade. *The American Economic Review*. **77**. 810–822.

Karp, L., Sacheti, S. and Zhao, J. 2001. Common ground between free-traders and environmentalists. *International Economic Review*. **42**. 617–647.

Polasky, S., Costello, C. and McAusland, C. 2004. On trade, land use and biodiversity. *Journal of Environmental Economics and Management*. **48**. 911–925.

Reid, W. V. and Miller, K. R. 1989. *Keeping Options Alive: The Scientific Basis for Conserving Biodiversity. Washington: World Resource Institute.*

Smulders, S., Van Soest, D. and Withagen, C. 2004. International trade, species diversity and habitat conservation. *Journal of Environmental Economics and Management*. **48**. 891–910.

Southgate, D., Sierra, R. and Brown, L. 1991. The causes of tropical deforestation in Ecuador: a statistical analysis. *World Development*. **19**. 1145–1151.

Swallow, S. K. 1990. Depletion of the environmental basis for renewable resources: the economics of interdependent renewable and nonrenewable resources. *Journal of Environmental Economics and Management*. **19**. 281–296.

Swanson, T. M. 1994. The economics of extinction revisited and revised: a generalised framework for the analysis of the problems of endangered species and biodiversity losses. *Oxford Economic Papers*. **46**. 800–821.

World Bank. 1998. *World Resources 1998–99: A Guide to the Global Environment.* Washington, DC: World Bank Publications.

World Development Report. 1992. New York: Oxford University Press.

WTO. Special Study, "Trade and Environment". 2000. New York: Oxford University Press.

## APPENDIX

Table 8.A1 *Parameters values used for numerical calculations*

| Parameter and values |
|---|
| $l_{11} = 0.3$ |
| $a_{11} = 0.1$ |
| $l_{12} = 0.15$ |
| $a_{12} = 0.4$ |
| $l_{21} = 0.4$ |
| $a_{21} = 0.2$ |
| $l_{22} = 0.1$ |
| $a_{22} = 0.3$ |
| $\alpha_1 = 0.3$ |
| $\alpha_2 = 0.4$ |
| $\alpha_{1green} = 0.3$ |
| $\alpha_{1grey} = 0.4$ |
| $\theta = 0.6$ |
| $\varepsilon_1 = 0.4$ |
| $\bar{A}_2 = 5$ |
| $A_1 = 1.5$ |
| $L_1 = 3$ |
| $A_2 = 2$ |
| $L_2 = 2$ |

*Part II*

# The value of biodiversity

*Section A*

# Concepts

# 9 Designing the legacy library of genetic resources: approaches, methods and results

*Timo Goeschl and Timothy Swanson*

## 1 Introduction

Of the many ways in which biodiversity might be conceptualised, one of the most important is as the diversity of the set of genetic resources (see also Chapters 21 and 23 in this volume). The diversity of the set of genetic resources refers to the amount of information contained within biological systems, commonly assessed at the level of genes. The 'resource' aspect arises out of the fact that biologically sourced information is a key input into research and development (R&D) processes that are used to address problems important to society. The life science sector, for instance, uses this information in order to conduct research on problems in both the agricultural and the health sectors.[1]

The idea that biodiversity may contain information on how to find new sources of pharmaceuticals, crops, etc. has been present in the genetic resources valuation literature from its very beginnings (Oldfield 1989). Equally present has been the recognition that the R&D options available in genetic resources are lost as the genetic base is narrowed (Swanson 1995, 1993). There is a long-recognised importance placed by economists on the retention of genetic resources for the performance of useful R&D and the issues surrounding the management of this resource for this function (Brown and Swierzbinski 1988).

The question of how to manage the informational values inherent within genetic resources may be asked in various ways. One particularly instructive analogy is the maintenance of the collection of all previously published written works within a library (Weitzman 1998). Consider the important issues that are raised when deciding how to manage the information that has been deposited in books throughout human

[1] Once found, these solutions are incorporated within bio-technological innovations such as new crops and new pharmaceuticals. Given the importance of bio-technologies, researchers have long suspected that viewing biodiversity in regard to its inherent genetic resources might give rise to important economic reasons for biodiversity conservation.

history.[2] An argument can be made that even though the preservation of all of these books is expensive, it is worthwhile to do so because – among other things – these books act as a repository of valuable information for which a need might arise at some future point in time. However, it may not be worthwhile to preserve *all* books given that there is a cost in doing so (shelf space). There may be obvious inefficiencies from doing so, such as the retention of two books that contain most of the same ideas or the provision of scarce shelf space to a volume that has never been checked out. There may also be non-obvious benefits from maintaining those volumes that are currently under-appreciated, such as the arrival of the future reader who is able to ascertain what the author really intended or the reprioritisation of the problem to which the volume was addressed. The need for some type of cost-benefit criterion in the design of the library is apparent, but it is very difficult to discern how a cost-benefit analysis undertaken now might take into account the needs and preferences for information over a potentially infinite time horizon.

In the literature on managing biological information inherent in biodiversity, several different approaches have been taken to the subject of designing this 'legacy library'. One approach has been to think about how to maximise the amount of information contained within a given-sized library. This would be, for instance, the interest of a librarian whose aim it is to afford the widest possible array of different books within an allotted shelf space constraint. The proper implementation of this objective is then the one that optimises the stock of distinct pieces of information preserved in the legacy library.

Another view of the same design problem considers the demand side of the equation. Consider in this instance a library user who comes to the library with a specific question and thus requires a particular piece of information. This piece of information may be available in any book with a certain probability and the objective of the designer is to maintain the optimum-sized collection of books given the impact of the marginal retention on the search for that information within the library. In this context the valuation of the legacy library is based on the user's demand for specific pieces of information, but the impact of the size of the library

[2] It may not seem obvious to the reader that biological resources would hold as much potentially useful information as a published work, but there are good reasons to expect that many biological characteristics present in living resources are representative of strategies successful within a contested environment and hence potentially useful in addressing other biological problems of that nature. This is the fundamental reason for reliance upon biological resources for R&D in the life science industries (Swanson 1995).

on that information is a two-edged sword: it might contain the necessary information but also provides the size of the set which must be searched in order to acquire that information. It is the problem of looking for a particular solution within a library that might come to resemble Citizen Kane's warehouse.

A related but distinct depiction of the demand side for biological information is given by extending this approach across the largest possible group of users and the longest possible period of time. In this approach the question concerns the needs of something more like a public research institution that is designing its research library with the future in mind. Its management objective then is to provide for a flow of useful information to a succession of unspecified users over a possibly infinite time horizon. This flow of information is generated by repeated searches in the library for answers to unpredictable questions that arise predictably over time. In this view, the design of the legacy library will be determined by its capacity to provide a flow of answers to new questions in perpetuity.

In summary, three main questions have been asked regarding the design of the legacy library of genetic resources: How to construct the library richest in information given the costs? How to design the library to optimise the search for a given piece of information? How to design the library to provide the optimal stock of information to meet the demands arising from an endless stream of unpredictable problems? These are very different questions that will obviously lead to very different answers regarding the design of the legacy library; however, taken together they provide insights on all the myriad issues raised in this introduction.

In this chapter, we present the distinct methods associated with the questions formulated above and report on a number of key results generated in these studies. The method associated with the first question uses the genetic diversity inherent in a collection of conservation candidate species as a measure for the inherent future R&D values. Based on this information, the most diverse 'library' is constructed. The second question has been tackled with the use of three different methods. Their common element is that all have at their core a specific variant of a search model. We discuss the different implementations employed and report on the ongoing debate about what constitutes efficient search in a genetic library. The last question has been approached through two models, with one using a real-options approach, the other developing a model of endogenous dynamic search in which biodiversity is a productive R&D input. We then conclude with a discussion on how these approaches can jointly inform us about the value of genetic resources.

## 2 Maximising the diversity of a library

### 2.1 *The approach*

Starting with Oldfield (1989), the preservation of biological diversity has repeatedly been cast as a problem of supplying the broadest portfolio of biological options. What makes this problem hard to solve is the difficulty of defining a meaningful measure of the stock of diversity. One solution to this problem can be found in the hugely influential literature on diversity measures pioneered by Solow *et al.* (1993) and Weitzman (1992). Weitzman's work focuses on the question of how to measure diversity in a way that could serve as an objective in a traditional cost-benefit analysis framework of maximising net benefits under some budget constraint.

The building block of Weitzman's measure of diversity is a concept of pair-wise distance between two objects that contains information about the degree of dissimilarity. This measure of dissimilarity between an object $j$ and a collection of objects $S$ is

$$d(j, S) = \min_{i \in S}(i, j) \tag{1}$$

i.e. the distance to the set $S$ is the distance to the 'closest relative' in the set. This difference should be analogous to a first difference or derivative of a 'diversity function' (to be defined) such that the marginal contribution of object $j$ is

$$V(S \cup j) - V(S) = d(j, S) \tag{2}$$

where $V(\bullet)$ denotes a function uniquely quantifying diversity. The problem is to construct $V(\bullet)$ such that it holds for all possible $j$ and $S$. If the pair-wise distance measures between objects are known, then these are sufficient to construct a measure of diversity ('collective dissimilarity') of any set containing more than two objects. In fact, it permits the definition of a rigorous but universal measure of 'diversity' as a scalar measure rooted in concepts of relatedness.

What makes this result relevant to researchers interested in genetic resources is that phylogenetic information about species and the systematic taxonomies that build on this information contain exactly such a measure of pair-wise distances. This allows Weitzman to demonstrate that maximising biological diversity (in a taxonomic sense) is equivalent to maximising the information content of the stock of candidate species. More precisely yet, diversity is 'the first derivative of information content with respect to uncertainty' (Weitzman 1998). To arrive at the optimal stock of biological diversity, society should then solve the following problem:

$$\max \Phi(X) = BX + U(X) \quad \text{subject to } CX \leq M \tag{3}$$

where $X$: vector of independent survival probability of each species
$B$: vector of species' individual direct net benefits
$U$: value of diversity function $V(\bullet)$ generated by $X$
$C$: individual costs of diversity maintenance
$M$: budget constraint.

In subsequent papers, Weitzman demonstrates the applicability of the concept of diversity functions for developing robust rules for conservation programmes. Weitzman (1993) applies the framework to crane conservation under simplifying assumptions about direct use values. In Weitzman (1998), diversity functions are used to develop a ranking criterion for species competing for scarce conservation resources. In the light of the setting in which the ranking criterion is developed, it might be referred to as 'Noah's rule'. The ranking criterion incorporates not only information on the distinctiveness value of a species to be included in a programme under a budget constraint but also the direct utility of a species, the increase in the probability of survival as a result of being conserved and the marginal cost of survivability gains. In this, Noah's rule delivers a first-order approximation to the optimal conservation strategy for the policy-maker. Metrick and Weitzman (1998) contrast the implications of this ranking criterion for endangered species protection in the USA with empirical evidence on public support for the listing of a species, the actual listing decision and the per-species spending on listed species.

There have been a number of refinements and generalisations of the diversity measures pioneered by Weitzman and Solow and Polasky. Nehring and Puppe (2002) generalise Weitzman (1992), which is based on a single attribute (genetic distance), to a multi-attribute setting. Weikard (2002) demonstrates that Weitzman's approach is not limited to examining issues of biodiversity at the genetic level. The same logic can be used to examine – and hence rank in terms of conservation priority – entire ecosystems.

The development of a scalar measure based on shared and distinct evolutionary histories between species has been hailed as a remarkable contribution to the literature on optimal biodiversity management. However, both the taxonomic implementation of the diversity function and its assumptions have also been critically assessed in the light of their usefulness in designing actual conservation policies, their ability to deliver an optimal portfolio of R&D options and their focus on information supply.

The criticism of the approach for the purpose of designing conservation policies centres on the information requirement of its algorithm. For successful implementation, the approach assumes perfect information regarding diversity in nature. This is perfect information regarding not only the existing range of diversity in forms of species known to exist but also the genetic make-up of all species identified within that range. In the real world, both assumptions are not even approximately met (Mainwaring 2001). The actual number of extant life forms is unknown and only estimates are available, and the present and foreseeable state of genomic analysis of all species is insufficient both for developing diversity functions and for solving the associated optimisation problem without considerable cost. If genetic-level information is costly to produce, the question arises as to whether the extra information acquired delivers sufficient benefits to warrant its acquisition. In a comparison of conservation site rankings based on phylogenetic criteria and on traditional species richness criteria, Polasky *et al.* (2001) find that there is little difference. The additional information required on the genetics may therefore not deliver significant welfare gains that would justify its collection.

Apart from the practical considerations of conservation site rankings, the second and rather more fundamental question concerns the usefulness of a phylogenetic approach to managing biological information with a view to R&D. The fundamental contribution of diversity measures is that they develop a rigorous unit of analysis that is rooted in an objective standard, namely a systematic taxonomy based on shared and distinct evolutionary histories of species. However, there are serious doubts over whether genetic distance maps smoothly into other goods or services that people may have preferences over (Mainwaring 2001). For example, it is not clear that a stock optimised with respect to phylogenetic diversity is optimal with respect to R&D options. This is only the case if the process by which traits useful to humans are selected, retained and/or discarded in species is the same as the process of evolution of these species. More generally yet, the genetic diversity concept is not grounded in preferences or in a mechanism linking genetic distance with 'some well defined concept of usefulness or desirability' (Brock and Xepapadeas 2003).

The third fundamental question to be raised is whether a supply-side approach is the most meaningful way of thinking about the problem of designing a legacy library of genetic information. While the library's diversity is a reasonable objective, it may not be the only or not even the most important one. Demand-side and cost considerations may be critical for the decision: even a highly diverse library is not very valuable if it costs too much to find the information required or if there are too few books on key areas of interest. The emphasis on demand and cost issues is at the core of the following approaches.

## 3 How to search in a library

Searching is costly. The value attributed to a library therefore depends on how productive is the search for a desirable piece of information that is possibly available in the library. The longer the expected duration of the search (in terms of number of trials for instance), the less valuable is the collection of books. The approaches to valuing genetic resources presented in this section emphasise this point with force. They make clear that each genetic resource represents a cost as well as a potential benefit in its role as a member of the set of searchable objects. We start with an empirical study on the current value of a collection of genetic resources before studying the papers that develop an explicit bio-prospecting framework.

### *3.1 The production function approach: the value of past searches*

A significant share of genetic resources available for use today is held *ex situ* in gene banks rather than *in situ* in nature or managed environments on land. In some instances, these genetic resources deposited as accessions in gene banks have a well-documented use history that can shed light on the R&D value of biological diversity, in particular in the context of crop breeding. Gollin and Evenson (1998) develop such a model that allows a monetary estimate of these R&D values.

The starting point for models of this kind is the search-theoretic literature originating with Evenson and Kislev (1976). This paper develops a stochastic model of (productivity-enhancing) information production based on a search over a given distribution. The context in which the model is developed is that of applied agricultural R&D aimed at yield enhancement through crop breeding. Gollin and Evenson (1998) extend and apply this information production model in order to estimate the contribution of genetic resources to plant breeding. The objective of the exercise is to ascertain the value of diverse germplasm in agricultural R&D and to convert the R&D values of genetic resources so derived into money equivalents. The contribution of this approach is that it delivers a robust and empirically measurable estimate of the historical contribution of genetic resources to plant breeding. This is done by specifying an 'R&D production function' in terms of new cultivars released and then estimating the extent to which its various component parts have contributed to the past production of new information. An R&D production function in the context of plant breeding, for example, would have to consist of at least: i) the scientific input (human capital); ii) the technological input (physical capital); iii) the genetic resource input (natural capital). The theory of a production function states that increases in these various inputs would result in increases in the desired output: new modern plant varieties (Gollin and Evenson 1998).

Gollin and Evenson (1998) apply this theoretical framework to conduct an empirical study which attempts to estimate the relative contribution of genetic resources in the R&D process in plant breeding. Here the R&D production function of new plant varieties $N$ is specified as

$$N = f(L, K, G) \tag{4}$$

where $L$: level of input from human capital (scientists)
$K$: level of input from physical capital (technology, machinery)
$G$: level of input from genetic capital (biological diversity).

The empirical study is based upon the record of plant breeding at the International Rice Research Institute since 1960 and estimates the extent to which new varieties of rice were attributable to the various forms of investments. This study estimated that approximately 35 per cent of the production of modern new rice varieties has been attributable to the genetic resource input into the R&D function. This implies that the inputs supplied by plant breeders in rice breeding (human and technological) generated no more than 65 per cent of the useful information within modern plant varieties. The imputed present value of a single landrace accession according to this study was $86–272 million. The imputed present value of 1,000 accessions with no known history of use was $100–350 million. Given that the initial stock of rice germplasm (in 1960) was 20,000 accessions, the added stock of germplasm since that time (about three times as many accessions) has been estimated to be responsible for fully 20 per cent of the green revolution in rice production.[3]

This study gives an indication of the scale of the property rights failure outlined in the previous section. In the context of rice production, diverse germplasm contributes 35 per cent of the 'total input' required for the production of a new plant variety. Since the existing commercial varieties lose their resistance rapidly in the context of large-scale

[3] The studies conducted by Gollin and Evenson (1998) used as a measure of 'genetic resource inputs' the number of plant varieties held within a public gene bank. Of course, it is crucial that – for additional varieties to provide additional value – the varieties be dissimilar from those already held and be inclusive of proven resistance strategies (evolved in a natural system).

monocultural production, this implies that a large proportion of rice production is attributable to this one factor. The loss of this factor through inadequate investment would not constitute a small-scale inefficiency.[4]

This literature has been expanded recently, among others by Zohrabian *et al.* (2003), who show that even when information on desirable production traits is incomplete or fuzzy, estimates on the lower bound of the value of the marginal accession lie above the upper bound of the expected cost of conservation.

The derivation of the value of a germplasm collection for R&D is of course first and foremost a measure of its current contribution to productivity. The main interest in genetic resources is instead their future contribution to R&D through bio-prospecting activities. This forward-looking perspective is taken up in the search-theoretic approaches presented in the following section.

### 3.2 Search-theoretic approach

In an influential article on the value of bio-prospecting, Simpson *et al.* (1996) develop a search-theoretic perspective on the problem that is inspired by Brown and Swierzbinski (1988). They ground the value of biodiversity in the activity of 'biodiversity prospecting' by an R&D-intensive industry and deduce the marginal willingness to pay for an additional sample to be prospected when screening of samples is costly. The aim of their work is to quantify the willingness of private firms to invest in the conservation of biodiversity when the value of each sample is the outcome of a Bernoulli trial (the screen). In other words, they evaluate genetic resources from the vantage point of expected private profits from research.

The typical model features a fixed probability $p$ of identifying a valuable trait in a sample where valuable traits give rise to a product with fixed revenue $R$ through a process of further R&D. The cost of screening a sample is fixed at level $c$. The expected value of a search over $n$ samples can then be expressed as $V(n)$ which is

$$V(n) = pR - c + (1 - p)(pR - c) + (1 - p)^2(pR - c) + \ldots\ldots \tag{5}$$

The marginal value of the $n$th sample is then

$$v(n) = (pR - c)(1 - p)^n \tag{6}$$

[4] The inefficiency described here is that which would result if too many lands were converted to monoculture, leaving too little area for the generation of newly evolved genetic varieties in response to changes in the biological environment.

The empirical problem with the formulation in equation (6) is that the probability of a 'hit', $p$, is the most important parameter for estimating $v(n)$, but that data on $p$ are notoriously difficult to obtain. Simpson *et al.* solve this dilemma by evaluating the expected value of the marginal species under the most optimistic conditions. One interesting finding is that the function mapping the probability of success in any single trial to the value of the marginal species is single-peaked and strongly skewed to the right. This means that once the probability of a successful trial is such that the expected marginal value of a trial exceeds the cost of the trial, the value will rise very rapidly to its maximum value and then decrease again rapidly. This observation is crucial as it shows several points. Sampling costs are an essential determinant of the marginal value and studies that do not take into account these costs (e.g. Pearce and Puroshothoman 1995; Farnsworth and Soejarto 1985) are bound to overestimate the marginal value significantly. Second, the fact that the marginal value of the species is not a monotonously increasing function of the probability of success in the Bernoulli trial brings an issue to the fore that had previously been overlooked by many researchers, namely the presence of substitutability between species.

The degree of relative scarcity of 'successful' traits is one of the key elements in the search-theoretic perspective: more than one sample can be a 'success' in the Bernoulli trial, such that once a trial has been successful, there is no further need for sampling.[5] If substitutability is very scarce, i.e. the probability of success is very low, then the marginal value is depressed since the expected revenue from the marginal trial is too low to warrant a high volume of trials. If substitutability is not scarce, then the expected revenue from the marginal trial is too low to warrant a high volume since it is very likely that a success has occurred already. In other words, if there is much redundancy within the stock of samples, a significant proportion of the samples can be discarded prior to screening with little loss of expected revenue since it is very likely that a success will be found within the remaining portion.

Based on a number of reasonable assumptions regarding the market value of a product and other parameters, Simpson *et al.* (1996) derive an upper bound for the willingness to pay for the marginal sample and translate this into a per-area WTP for conservation using the common MacArthur-Wilson approach of relating habitat size to the extant stock of biodiversity. Based on this computations, the maximal WTP for a hectare

[5] The biological equivalent is that there may be abundance of species with very similar genetic make-up and that the same bio-active compound (that results in a 'success' in the screen) can be produced by species of completely different genetic structure.

of biodiverse lands in Western Ecuador, one of the 'biodiversity hot spots', is $20.63. The rainforests of the Amazon elicit only $2.59 per hectare. This implies that most areas with even extraordinary biodiversity do not justify significant payments from the pharmaceutical industry for their preservation. The conclusion of Simpson *et al.* is that there is little reason to expect that the industrial use of genetic resources will result in their preservation by private investors.[6]

Simpson and Craft (2001) expand on this paper by considering the social value of biodiversity for R&D in terms of producer and consumer surplus as opposed to the private industry valuation alone. In this paper, they combine the search model with a product differentiation model in the spirit of Salop (1980). Even though values may lie in a somewhat higher region, the fundamental logic of redundancy still holds: to the extent that genetic resources give rise to products that are near substitutes, the marginal value of an additional species declines rapidly.

### 3.3 *Limitations and problems of the search-theoretic approach*

Even though Simpson *et al.*'s (1996) work has been highly influential in the literature, a number of shortcomings have been pointed out and the problems with this approach to valuation of biodiversity as an R&D input are by now well studied. Rausser and Small (2000) challenge the results of the paper on the grounds that if there is prior information about which areas are more likely to produce information on which problems, search can be conducted 'efficiently' rather than by 'brute force' as modelled by Simpson *et al.* (1996) (SSR henceforth). As a result, the values of marginal biodiversity should be altered significantly. As proof, Rausser and Small (2000) (RS henceforth) offer a generalised version of the search model proposed by Simpson *et al.* (1996) that differs in two respects meant to represent an 'organising scientific framework'. The first is that the probability of a 'hit' is no longer identical across leads. This is meant to reflect the availability of prior knowledge on the economic potential of different leads. The second is that the sequence of searches across leads is informed by these differences. Testing leads in an order of declining hit probability $p_i$, the marginal value of the $n$th lead is then

$$v_n = a_n\,[p_n(R - V_{n+1}) - c] \tag{7}$$

[6] This search-theoretic approach has been considerably refined in order to include differences in the value of individual hits (Gollin *et al.* 2000) or situations in which the assumption of independence between the probability distribution of individual traits is violated (Simpson and Sedjo 1998).

with $a_n = \prod_{i=1}^{n} (1 - p_i)$ being the probability that the search is carried to the $n$th stage given probability $p_i$ that the $i$th lead will generate a success and with $V_{n+1}$ denoting the value of continuing the search.

Rausser and Small (2000) perform a numerical comparison with Simpson *et al.* (1996) by translating their framework into an area-based measure in which the heterogeneity of leads is mimicked through heterogeneity of species density per area while the hit probability of each species is constant. Their calculations indicate, for example, that the biodiversity hotspot of Western Ecuador should be more properly valued at $9,117 per hectare rather than $21 as calculated by SSR. For all but one of the biodiversity hotspots considered by SSR, RS find that SSR's methodology under-estimates the marginal values of biodiverse lands by a factor of 300 or more.

In a comment, Costello and Ward (2004) cast doubt on the validity of the comparison between RS's and SSR's numerical results. They demonstrate in detail the contribution of differences in parameters and their use between RS and SSR that gives rise to the factor of deviation. The factors contributing most to the deviation arise from the number of species included in the search (a factor of 12.5) and the ecological model parameters (a factor of 4). The efficient organisation of the search, the key departure of RS, has a negligible impact on the marginal value. Overall, organising search efficiently improves the marginal value of the marginal hectare of biodiverse land by about 4 per cent. Much more important than the efficient organisation of the search is therefore the ability to truncate the search. This is equivalent to saying that information on what collection to conduct the search is likely to be much more important than the decision on how to search a given library.

A second, and perhaps more fundamental, problem with the search-theoretic literature is that it casts the issue of the valuation of biodiversity for R&D as a process of sequential search for a given target. The value of genetic resources thus reduces to the chance that a given species might provide a solution to a given problem, multiplied by the reward that the innovator might expect to receive by virtue of such a discovery. An important implication of this approach is that the process of sequential search used by these bio-prospectors implies that a solution to a problem may be found before the set of all genetic resources is exhausted. Hence there may be some redundancy within the set of conserved genetic resources and the marginal value of maintaining a larger set would then be diminishing. However, if a solution today is not a solution tomorrow, problems continue to emerge over time in a dynamic fashion. In this case, the values derived on the basis of a static representation of the search problem may be a misleading proxy for the true R&D value of biodiversity. This idea is taken up in the last section.

## 4 Managing the flow of biological information: how frequently will the library be consulted?

In a world in which solutions, once found, retain their value in perpetuity, it will be worthwhile to find solutions to existing problems early in the planning period or not at all. In other words, for the purposes of R&D, the value of the library is exhausted after all existing problems that can be solved with an expected non-negative net benefit have been addressed. The two remaining approaches question the permanence of solutions found in the library and point to the need to return to the library time and time again in order to find new answers to new problems. If this is the case, then the objective of designing the legacy library is its capacity to satisfy not only the information demand of current users with known problems but also future users with new problems that are yet unknown.

Where do these new problems come from? The two approaches presented here point to two different sources: biological evolution and uncertainty. The engine driving the 'renewability of problems' in Goeschl and Swanson (2002) is the fundamental biological adaptation and selection mechanisms that constitute natural evolution. As systems become more large-scale and uniform, they become more inherently unstable and problems arise predictably (see also Weitzman 1998). In Kassar and Lasserre (2004), it is the evolving uncertainty over needs, tastes and the reliability of the current state of knowledge that gives rise to values of species becoming stochastic processes. In both cases, the faster these dynamic processes, the more valuable biodiversity becomes.

### *4.1 Evolutionary processes of natural selection*

Goeschl and Swanson (2002) start with a simple observation from the plant-breeding industry. In that sector, it is a stylised fact that a widely used modern plant variety experiences declining yields in each of its years of usage, resulting in commercial obsolescence and replacement in a period of approximately five to ten years. One reason is that in plant breeding, as in many other R&D-intensive sectors, innovators (plant breeders) are engaged individually in a contest of innovation against one another; the breakthrough of one can make the other's products commercially obsolete. However, these innovators are also engaged in a contest against an entirely different competitor, also functioning in a manner to render their prior innovations obsolete. That source is the ongoing adaptation of the pests and pathogens to innovations applied widely within a biological context. Plant breeders' R&D efforts are increasingly addressed to the ongoing problems of pest adaptation and resistance.[7] The widespread application of the single plant variety results in the selection of that part of the pest/pathogen population that succeeds within that environment and generates the forces that result in its own demise. Thus, the solution provided by R&D must necessarily be impermanent, and a flow of such problems will arrive predictably, although the precise nature of each specific problem is unpredictable. In order to provide for a stock of information to address this predictable flow of problems, an important input into plant breeder R&D is the diversity of genetic resources (see Evenson *et al.* 1998).

Based on this observation, Goeschl and Swanson (2002) (GS henceforth) develop a more generalised framework in which genetic resources are used in R&D at the base of an industry that addresses recurring problems of resistance, as in the pharmaceutical or plant-breeding industries. The R&D process is one in which firms are engaging in a continuing contest of innovation against a background of both creative destruction (Schumpeterian competition) and adaptive destruction (natural selection and adaptation). This framework demonstrates that the search-theoretic model is a problematic description of the wider context in which genetic resources play a role because it incorporates only a single search rather than the more permanent dynamic characteristics of biological phenomena.

Taking into account the dynamic context of new problems arising out of the application of solutions to previous problems, GS propose a model in which the problem of conservation reduces to choosing between proportion of land used for conservation purposes ($v$) versus the proportion used for intensive production $F(.)$ of a final agricultural output. The purpose of holding land for conservation is two-fold: it both increases the stock of information available for search for innovations (thus increasing the arrival rate of innovations $\phi i(v)$) and it decreases the scale of application of the production technology (and thus decreases the rate of adaptation to the previous innovation $a(v)$). The design of genetic resource conservation policies balances the value of retaining and restraining future innovation capacities against the costs of current forgone consumption.

[7] Pests and disease now account for average annual crop losses of 28.9 per cent, increasing with each year of the use of a given plant variety (Evans 1993; Oerke *et al.* 1994; Scheffer 1997). A recent survey found that plant breeders cited pest resistance as the primary focus of their activities (Swanson and Luxmoore 1998).

In this framework the marginal value of land held in biodiversity, *MVLB*, is

$$MVLB = \frac{[\phi i'(v) - \lambda a'(v)\gamma^{-1}](\gamma - 1)F(\bullet)}{r - [\phi i(v) - \lambda a(v)\gamma^{-1}](\gamma - 1)} \tag{8}$$

which is the net present value of the net increase in productivity in terms of final output F from two sources. The first is $\phi i'(v) > 0$, the marginal increase in the rate of arrival of innovations by virtue of a greater amount of biodiverse land preserved. The second is $\lambda a'(v)\gamma^{-1}$, the marginal reduction in the rate of biological innovation on account of reduced selection pressure.[8] The own discount rate applied in the denominator is the composite rate used by the social planner which takes into account the rate of technological and biological innovation. All of these costs and benefits are measured in terms of the step-size of innovation ($\gamma$), which represents the gain or loss from one more or less innovation within the system. The MVLB is then the integral across an infinite time horizon of the benefits from innovations lost if the genetic resource base is narrowed; its opportunity cost is the marginal value of one more unit of homogenous production (net of the adaptation costs that this implies).

The analysis then shows that the valuation of the demand side of the library from the perspective of one-time library users is fundamentally different from the valuation of the same problem by a planner attempting to ensure the capacity of the library for the long term. The question then really is: do we now know the precise number and nature of the problems we will need to address across all of time? If so, then the simple search model is appropriate; if not, it is important to consider the informational value of currently unvalued information.

## 4.2 *Real options*

Kassar and Lasserre (2004) develop a real-options framework to arrive at a characterisation of the insurance value of maintaining these uncertain biodiversity values in the face of uncertainty about the future. As an illustration, they consider a two-species framework with species values $v_1$ and $v_2$. The species are perfect substitutes. In a static world of certainty, there would be no reason to retain the less valuable species if its conservation incurs a positive cost. However, if species values evolve stochastically, for example in the manner of Brownian motion such that

$$\frac{dv_i}{dt} = \alpha(v, t)dt + \sigma(v, t)dz, \quad i = 1, 2 \tag{9}$$

[8] Recall that $a' < 0$.

then the decision maker may find it worthwhile *not* to exercise the option of letting the species that is less valuable at time $t$ go extinct. The reason is that the most preferred species today may not be the most preferred species tomorrow on account of stochastic fluctuations in their respective values. The decision maker therefore has to decide whether to disinvest in the currently less preferred species (thus saving on conservation costs) or to retain it for the future in the expectation that it will provide an alternative to the currently preferred species. Whether the option is exercised is dictated by a trigger value ratio between the species $V_i\left(\frac{v_j}{v_i}\right)^*$. These considerations lead to the derivation of the marginal value of biodiversity as the difference of retaining the option (at the cost of preservation) or disinvesting in the marginal species. Assume species 2 is marginal, but currently retained (i.e. $v_2 > v_2^*(v_1)$). This means that species 1 is in use. Then the marginal value of species 2 is

$$V_{m2}(v_1, v_2) = F_1(v_1, v_2) - V(v_1), \quad v_2^*(v_1) < v_2 \leq v_1 \tag{10}$$

with $F_1(v_1, v_2)$ denoting the value of holding on to species 2 and $V(v_1)$ denoting the value of exercising the disinvestment option. The value of holding on to species 2, $F_1(v_1, v_2)$, has three components: the continuation payoff associated with the status quo, the expected payoff from exercising the option later, and the expected payoff from substituting species 2 for species 1 at some later point in time.

Generalised to more than two species, Kassar and Lasserre (2004) show that the marginal value of biodiversity increases with uncertainty (defined as the variability in the path that the species values are following). This is because the insurance value of having additional marginal species that can substitute for species that have undergone a loss of value increases – since the conditions for future substitution are more likely with greater variability. Likewise, negative correlation in the stochasticity of the evolution of species values increases the marginal value of biodiversity; this implies a greater likelihood that one species will be able to substitute for the other if and when its value declines.

One key conclusion from Kassar and Lasserre (2004) is that in a world in which conservation projects are carried out under uncertainty, substitutability in use works in exactly the opposite fashion from its description in Simpson *et al.* (1996). The closer the substitutability today, the more likely it is that the marginal species today will be called upon to replace the currently most preferred species. Substitutability is a reason for disinvestment only with respect to the probabilistic properties of the species. The higher the correlation of the stochastic processes, the lower the value of the optional species.

## 5 Conclusion

We are now able to return to the question we set out in the introduction to this chapter: how should the legacy library be designed in order to meet the needs of society? We would say that this depends most of all on the priorities that societies place on the various problems we have identified. These problems might be categorised by reference to the various approaches developed in the papers reviewed here, as follows: i) concerns regarding the maximum supply of distinctiveness given the costs of storage; ii) concerns regarding the search costs for solutions (the time and resources absorbed in the process of identifying relevant information from within the mass available); and iii) concerns regarding the future path of problems and the implied changes in those costs (the path of opportunity and research costs).

In terms of biodiversity, persons who place high priorities on maximising the supply of distinctiveness in the legacy library given current opportunity costs of forgoing investing resources into production would focus on the approach advanced by Weitzman. Here, the issue is one of taking a given budget and using it most effectively (and irrespective of the values being generated). The problem devolves to the question of securing the largest number of distinct books (or the greatest quantity of distinctiveness) within a given library space. However, distinctiveness in this instance is defined without reference to any use or usefulness to humans, but merely as a measure of the base genetic variability. While the approach can be extended to encompass value dimensions other than distinctiveness, there are two cogent criticisms of this approach. The first is that the supply-side orientation overlooks that additional search investments have to be carried out to utilise the resources inherent in the legacy library. The other is that it is essentially static and thus places little emphasis on the potential values from biodiversity conservation (as a means of solving problems important to human societies).

If biodiversity does generate important values to human societies, it would seem that these values would be important to the issues concerning optimal management of that resource. The latter two approaches incorporate this perspective, but differ in how the value of diversity is generated. In the case of SSR, the search framework indicates that the problem is one of identifying the useful information as quickly as possible – the emphasis is upon the time and resources expended in the process of search for a solution to a specific problem. In this context, the existence of the biodiversity resource generates costs and benefits in equal measure, by being both information and obstacle. The problem devolves to that of Citizen Kane's warehouse: how to find the one jewel in the mass of meaningless objects? The most cogent criticism of this approach is that, although rooted in the idea of diversity as useful information, it also casts diversity as the major obstacle to its own usefulness. In a world in which the global storehouse of genetic information is (we would argue) becoming ever smaller, it would seem that this is an interesting but potentially inadequate manner in which to cast the problem of biodiversity management. Although it is clear that the objective of biodiversity management should be the maximisation of the informational value of this resource, it would not appear that the most important tradeoffs concern the loss of these solution concepts in a mass of diversity.

The final framework to consider is that within the real options literature. Here the objective remains the same as in SSR (i.e. the maximisation of the informational value of biodiversity) but the core problem concerns: how do we make such a permanent decision on the size and content of the library today (given irreversibility of the decline of biodiversity) when the information being saved must be able to address the problems of tomorrow? Here the problem is to decide which volumes to retain in the context of pressures for an ever-shrinking library when the subjects they must address are not yet knowable. Are the popular volumes of today a good predictor of the important information for tomorrow? Should we maintain a library of a size adequate to address the number of problems that exist today, or will there be a greater or smaller number of problems in the future? The answers from the options-based literature indicate that today's decision should use current information, but also provide an additional hedge against future uncertainties. This implies the retention of larger libraries with a greater diversity of information than current circumstances would imply. The most cogent criticism of this literature is that it implicitly assumes stagnant (or even regressive) technological progress and a lack of substitutability for biodiversity-sourced information. It is, of course, possible that technological change might greatly reduce, or even eliminate, the need to source solutions from within biodiversity, and such technological change would result in all of the uncertainties being resolved against the conservation of resource. This would be the case of the displacement of the legacy library with internet-based search engines.

So, how should society invest in its legacy library? Different readers might respond to this question differently, depending on: i) whether or not they consider human needs to be the basis of decision making concerning biodiversity; ii) whether they consider the present search for solutions to be the important decision-making criterion or the longer-term search for solutions to be the important criterion; and iii) whether or not they are technologically optimistic concerning the displacement of biodiversity as a fundamental source of solutions. We would argue that in the presence of irreversibilities, the current generation has the responsibility to make this decision by reference to the longer-term welfare of future generations and that it may be dangerous to assume that technological change will solve this problem if we do not.

## REFERENCES

Brock, W. A. and Xepapadeas, A. 2003. Valuing biodiversity from an economic perspective: a unified economic, ecological and genetic approach. *American Economic Review*. **93** (5). 1597–1614.

Brown, G. and Swierzbinski, J. 1988. Optimal genetic resources in the context of asymmetric public goods. In V. K. Smith (ed.). *Environmental Resources and Applied Welfare Economics*. Washington: RFF. 293–312.

Costello, C. and Ward, M. 2004. Search, bioprospecting, and biodiversity conservation: comment. Working Paper. Santa Barbara, CA: University of California.

Evans, L. T. 1993. *Crop Evolution, Adaptation and Yield*. Cambridge: Cambridge University Press.

Evenson, R. E. and Kislev, Y. 1976. A stochastic model of applied research. *Journal of Political Economy*. **84**. 265–281.

Evenson, R. E., Gollin, D. and Santaniello, V. (eds.). 1998. *Agricultural Values of Genetic Resources*. London: CABI.

Farnsworth, N. and Soejarto, D. 1985. Potential consequences of plant extinction in the United States on the current and future availability of prescription drugs. *Economic Botany*. **39** (2). 231–240.

Goeschl, T. and Swanson, T. 2002. The social value of biodiversity for R&D. *Environmental and Resource Economics*. **22**. 477–504.

Gollin, D. and Evenson, R. E. 1998. Breeding values of rice genetic resources. In R. E. Evenson, D. Gollin, and V. Santaniello (eds.). *Agricultural Values of Plant Genetic Resources*. Wallingford: CABI. 179–193.

Gollin, D., Smale, M and Skovmand, B. 2000. Searching an ex situ collection of wheat genetic resources. *American Journal of Agricultural Economics*. **82** (4). 812–827.

Kassar, I. and Lasserre, P. 2004. Species preservation and biodiversity value: a real options approach. *Journal of Environmental Economics and Management*. **48**. 857–879.

Mainwaring, L. 2001. Biodiversity, biocomplexity, and the economics of genetic dissimilarity. *Land Economics*. 77 (1). 79–83.

Metrick, A. and Weitzman, M. 1998. Conflicts and choices in biodiversity preservation. *Journal of Economic Perspectives*. **12** (3). 21–34.

Nehring, K. and Puppe, C. 2002. A theory of diversity. *Econometrica*. **70** (3). 1155–1198.

Oerke, E-C., Dehne, H-W., Schönbeck, F. and Weber, A. 1994. *Crop Production and Crop Protection. Estimated Losses in Major Food and Cash Crops*. Amsterdam: Elsevier.

Oldfield, M. L. 1989. *The Value of Conserving Genetic Resources*. Sunderland, MA: Sinauer.

Pearce, D. and Puroshothoman, P. 1995. The economic value of plant-based pharmaceuticals. In T. Swanson (ed.). *Intellectual Property Rights and Biodiversity Conservation*. Cambridge: Cambridge University Press.

Polasky, S., Csuti, B., Vossler, C. A. and Meyers, S. M. 2001. A comparison of taxonomic distinctness versus richness as criteria for setting conservation priorities for North American birds. *Biological Conservation*. **97** (1). 99–105.

Rausser, G. C. and Small, A. A. 2000. Valuing research leads: bioprospecting and the conservation of genetic resources. *Journal of Political Economy*. **108** (1). 173–206.

Salop, S. 1980. Monopolistic competition with outside goods. *Bell Journal of Economics*. **10**. 141–156.

Scheffer, R. 1997. *The Nature of Disease in Plants*. Cambridge: Cambridge University Press.

Simpson, R. D. and Craft, A. 2001. The value of biodiversity in pharmaceutical research with differentiated products. *Environmental and Resource Economics*. **18** (1). 1–17.

Simpson, R. D. and Sedjo, R. A. 1998. The value of genetic resources for use in agricultural improvement. In R. E. Evenson, D. Gollin, V. Santaniello (eds.). *Agricultural Values of Plant Genetic Resources*. Wallingford: CABI. 55–67.

Simpson, R. D., Sedjo, R. A. and Reid, J. W. 1996. Valuing biodiversity for use in pharmaceutical research. *Journal of Political Economy*. **104** (1). 163–185.

Solow, A., Polasky, S. and Broadus, J. 1993. On the measurement of biological diversity. *Journal of Environmental Economics and Management*. **24** (1). 60–68.

Swanson, T. 1993. *The International Regulation of Extinction*. London: Macmillan and New York: New York University Press.

Swanson, T. (ed.). 1995. *Intellectual Property Rights and Biodiversity Conservation*. Cambridge: Cambridge University Press.

Swanson, T. and Luxmoore, R. 1998. *Industrial Reliance on Biodiversity*. Cambridge: WCMC.

Weikard, H.-P. 2002. Diversity functions and the value of biodiversity. *Land Economics*. **78** (1). 20–27.

Weitzman, M. 1992. On diversity. *Quarterly Journal of Economics*. **107** (2). 363–405.

Weitzman, M. 1993. What to preserve? an application of diversity theory to crane preservation. *Quarterly Journal of Economics*. **108** (1). 157–183.

Weitzman, M. 1998. The Noah's Ark problem. *Econometrica*. **66** (6). 1279–1298.

Zohrabian, A., Traxler, G., Caudill, S. and Smale, M. 2003. Valuing pre-commercial genetic resources: a maximum entropy approach. *American Journal of Agricultural Economics*. **85** (2). 429–436.

# 10 Why the measurement of species diversity requires prior value judgements

*Stefan Baumgärtner*

## 1 Introduction

In the discussion about biodiversity loss and conservation (cf. Wilson 1988; McNeely *et al.* 1990; Watson *et al.* 1995; Millennium Ecosystem Assessment 2005), the two issues of (i) quantitative measurement of biodiversity and (ii) its economic valuation play a major role. Concerning the first issue, there exists an extensive ecological literature (see, for example, Pielou 1975; Magurran 1988; Purvis and Hector 2000) to which, lately, economists have made important contributions (e.g. Solow *et al.* 1993; Weitzman 1992, 1993, 1998; Weikard 1998, 1999, 2002; Nehring and Puppe 2002, 2004). Concerning the second issue, there also exists an extensive and still growing literature (e.g. Hanley and Spash 1993; Pearce and Moran 1994; Smith 1996; Goulder and Kennedy 1997; Nunes and van den Bergh 2001; Freeman 2003).

The conventional wisdom on the relation between these two issues seems to be that the quantitative measurement of biodiversity is a value-free task which precedes the valuation of biodiversity. In contrast to this view, I shall argue in this chapter that the measurement of biodiversity requires prior value judgements about biodiversity and its role in ecological-economic systems. The argument proceeds as follows. Section 2 briefly discusses why and how biodiversity can be thought of as an economic good, so as to provide a background for the discussion of the measurement of biodiversity. Section 3 then surveys different ecological and economic measures of species diversity. A conceptual comparison of those reveals systematic differences between the two. In Section 4, I shall critically discuss these differences and argue that they are related to a fundamental difference in the philosophical perspective on biodiversity between ecologists and economists. This difference in basic value judgements about biodiversity between ecologists and economists leads to different measures of species diversity. I therefore conclude in section 5 that the measurement of species diversity requires prior value judgements as to what purpose species diversity serves in ecological-economic systems.[1]

## 2 Biodiversity as an economic good

According to a classic definition, 'economics is the science which studies human behaviour as a relationship between ends and scarce means which have alternative uses' (Robbins 1932, p. 15). In this sense, biological diversity can be thought of as an economic good (Baumgärtner 2006, Chapter 7; Heal 2000). It is obviously scarce; and it satisfies human needs and allows people to achieve certain ends. This pertains to a multitude of different roles and functions of biodiversity, e.g. in the provision of food, fuel, fibre, industrial resources, pharmaceutical substances, bio-indicators for science; in the regulation of ecosystem functioning and stability, nutrient cycling, water run-off, soil fertility, pollination, cleansing of water and air, local climates; in the control of pests and disease; or in its aesthetical, recreational and educational function (Watson *et al.* 1995; Daily 1997; Millennium Ecosystem Assessment 2005).

Corresponding to the very different human needs which are being satisfied by this natural resource, one can also attribute *economic value* to it. For goods which are being traded on markets one can (under certain conditions) take the market price as expressing their economic value. For biological diversity, however, there is the problem that the resource is not, or only partially, being traded on markets. In order to determine its total economic value, or individual components thereof, one therefore needs to employ (direct or indirect) methods for non-market valuation.[2] These methods can, in principle, also be used to determine the total economic value of biodiversity (Watson *et al.* 1995, pp. 844–858).[3] Examples include the replacement cost method, the averting expenditure/avoiding costs method, the production function method, the hedonic pricing method, the travel cost method, or the contingent valuation method (e.g. Hanley and Spash 1993; Smith 1996; Bateman *et al.* 2002; Freeman 2003). All these methods operate based on the assumption that the object of valuation – biodiversity, or individual components thereof – should be objectively described and quantitatively specified before it can

[1] This chapter builds on material that I have developed in detail elsewhere (Baumgärtner 2005, 2006, Chapters 7 and 8).

[2] For an introduction to the concept of *total economic value* see, for example, Pearce and Turner (1990, p. 129), Pearce (1993) and Turner (1999). This concept can also be applied to biodiversity (McNeely 1988, p. 14ff; Watson *et al.* 1995, p. 830ff).

[3] Nunes and van den Bergh (2001) as well as Pearce and Moran (1994, p. 48) stress the considerable difficulties which occur when using these methods for the valuation of biodiversity.

be valued. Obviously, the conventional wisdom seems to be that the quantitative measurement of biodiversity should precede its valuation. However, the quantitative measurement of biodiversity already requires prior value judgements, as I shall demonstrate in the following.

## 3 The measurement of biodiversity

There exist a multitude of different biodiversity measures and indices (see surveys by Pielou 1975; Magurran 1988; Purvis and Hector 2000; Baumgärtner 2005), which may be roughly classified as *ecological* or *economic* biodiversity indices. Among the ecological indices are concepts traditionally used by ecologists such as species richness, Shannon-Wiener-entropy, Simpson's index, or the Berger-Parker-index (Pielou 1975; Magurran 1988; Begon *et al.* 1998; Ricklefs and Miller 2000). These indices have recently been complemented by indices that were – after pioneering contributions of ecologists (May 1990; Erwin 1991; Vane-Wright *et al.* 1991; Crozier 1992) – put on a rigorous axiomatic basis by economists in economic journals (Weitzman 1992, 1993, 1998; Solow *et al.* 1993; Weikard 1998, 1999, 2002; Nehring and Puppe 2002, 2004). One may therefore refer to them as economic biodiversity indices. In the following, a conceptual comparison of the ecological and economic biodiversity indices is provided.

### *3.1 Focus: species and ecosystems*

Biological diversity can be considered on different hierarchical levels of life: gene, population, species, genus, family, order, phylum, ecosystem, etc. (Groombridge 1992). This chapter is concerned with the level of species, as this is the level of organisation which is currently being given most attention in the discussion of biodiversity conservation policies.[4] That is, biodiversity is here considered in the sense of species diversity.

In order to describe the species diversity of an ecosystem and to compare two systems in terms of their diversity, one can build on different structural characteristics of the system(s) under study. These include the following:

- the *number* of different species in the system
- the characteristic *features* of the different species, and
- the relative *abundances* with which individuals are distributed over different species.

[4] Ceballos and Ehrlich (2002) have pointed out that the loss of populations is a more accurate indicator for the loss of 'biological capital' than the extinction of species.

Intuitively, it seems plausible to say that a system is more diverse than another one if it comprises a higher number of different species, if the species in the system are more dissimilar from each other and if individuals are more evenly distributed over the different species. A simple example can illustrate this idea. Consider two systems, A and B, which both consist of eight individuals of insect species: system A comprises six monarch butterflies, one dragonfly and one ladybug; system B comprises four swallowtail butterflies and four ants. Obviously, according to the first criterion (species number), system A has a higher diversity (three different species) than system B (two different species). But according to the third criterion (evenness of relative abundance) one may as well say that system B has a higher diversity than system A, because there is less chance in system B that two randomly chosen individuals will be of the same species. And as far as the second criterion goes (characteristic species features), one would have to start by saying what the characteristic species features actually are, which can then be used to assess the aggregate dissimilarity of both systems.

Before discussing these ideas in detail, let's first introduce a formal and abstract description of the ecosystem whose species diversity is of interest. Let $n$ be the total *number* of different species existent in the system and let $S = \{s_1, \ldots, s_n\}$ be the set of these species. Each $s_i$ (with $i = 1, \ldots, n$) represents one distinct species. In the following, $n \geq 2$ is always assumed. Let $m$ be the total number of different relevant features, according to which one can distinguish between species, and let $F = \{f_1, \ldots, f_m\}$ be the list of these *features*. Each $f_j$ (with $j = 1, \ldots, m$) represents one distinct feature. For example, possible features could include the following: being a mammal/bird/fish, being a herbivore/carnivore/omnivore, unit biomass consumption/production, being a 'cute little animal', etc. Then one can characterise each species $s_i$ (with $i = 1, \ldots, n$) in terms of all features $f_j$ (with $j = 1, \ldots, m$). Let $x_{ij}$ be the description of species $s_i$ in terms of feature $f_j$, so that $x = \{x_{ij}\}_{i=1,\ldots,n\ j=1,\ldots,m}$ is the complete characterisation of all species in terms of all relevant features.

The *abundance* of different species in the ecosystem is described by the distribution of absolute abundances of individuals over different species. Let $a_i$ be the *absolute abundance* of individuals of species $s_i$ (with $i = 1, \ldots, n$), which may be measured either as the number of individuals of that species or as the total bio-mass stored in all individuals of that species. The *relative abundance* of species $s_i$ is then given as $p_i = a_i / \sum_{i=1}^{n} a_i$. Let $p = (p_1, \ldots, p_n)$ be the vector of relative abundances. By construction of $p_i$, one has $\sum_{i=1}^{n} p_i = 1$ and $0 \leq p_i \leq 1$, where $p_i = 0$ means that species $i$ is absent from the system and $p_i = 1$ (implying $p_j = 0$ for all $j \neq i$) means that species $i$ is the only species in the system. If species abundances are measured by counting individuals of that species, the relative abundance $p_i$ indicates the probability of obtaining an individual of species $s_i$ in a random draw from all individuals in the system. When abundances are measured in biomass, the relative abundance $p_i$ indicates the relative share of the ecosystem's biomass stored in individuals of species $s_i$. Without loss of generality, assume that $p_1 \geq \ldots \geq p_n$, i.e. species are numbered in the sequence of decreasing relative abundance, such that $s_1$ denotes the most common species in the system whereas $s_n$ denotes the rarest species.

Altogether, the formal description of an actual or potential ecosystem state $\Omega$ comprises the specification of $n$, $S$, $m$, $F$, $p$ and $x$, which completely describes the composition of the ecosystem from different species as well as all species in terms of their characteristic features. In the following, a *biodiversity measure* of the ecosystem $\Omega$ means a mapping $D$ of all these data on a real number:

$$D : \Omega \rightarrow IR \text{ with } \Omega = \{n, S, m, F, p, x\} \tag{1}$$

That is, I consider only biodiversity measures which characterise the species diversity of an ecosystem by a single number ('biodiversity index'). The various measures differ in what information about the ecosystem state, $\Omega$, they take into account and how they aggregate this information to an index.

### 3.2 *The basic index: species richness*

The simplest measure of biodiversity of an ecosystem $\Omega$ is just the total number $n$ of different species found in that system. This is often referred to as

$$D^R(\Omega) = n \tag{2}$$

Species richness is widely used in ecology as a measure of species diversity. One example is the long-standing and recently revitalised diversity-stability debate, i.e. the question of whether more diverse ecosystems are more stable and productive than less diverse systems (cf. McCann 2000). Another example are the so-called species-area relationships, which are important for the present biodiversity conservation debate because they are virtually the only tool to estimate the number of species that go extinct due to large-scale habitat destruction (MacArthur and Wilson 1967; Whitmore and Sayer 1992; May *et al.* 1995; Rosenzweig 1995; Gaston 2000; Kinzig and Harte 2000).[5] Species richness is also the biodiversity

[5] The well-established species-area relationships state that species richness $n$ increases with the area $l$ of land as $n \sim l^z$, where $z$ (with $0 < z < 1$) is a characteristic constant for the type of ecosystem.

indicator implicitly used in the public discussion, which often reduces biodiversity loss to species extinction.

In the species richness index (2), all species that exist in an ecosystem count equally. However, one might argue that not all species should contribute equally to an index of species diversity. Two different strands have evolved in the literature, both of which develop indices in which different species are given different weight. The first strand, which has evolved mainly in ecology, weighs different species according to their relative abundance in the system. This is vindicated by the observation that the functional role of species may vary with their abundance in the system. These biodiversity indices are discussed in section 3.3 below. The other strand, which has been contributed to the discussion of biodiversity mainly by economists, stresses that different species should be given different weight in the index due to the characteristic features they possess. These biodiversity indices are discussed in section 3.4 below.

### *3.3 Indices based on relative abundances*

Ecologists have tackled the problem of incorporating the functional role of species in a measure of species diversity by formulating diversity indices in which the contribution of each species is weighted by its relative abundance in the ecosystem (Pielou 1975; Magurran 1988; Begon *et al.* 1998; Ricklefs and Miller 2000). Intuitively, rare species should contribute less than common species to the biodiversity – in the sense of 'effective species richness' – of an ecosystem. A general measure for the effective number $\nu$ of species, which uses the information about pure species number $n$ and the distribution of relative abundances $p = (p_1, \ldots, p_n)$ to build on this intuition, is the following:

$$V_\alpha(\mathrm{n}, \mathrm{p}) = \left( \sum_{\mathrm{i}=1}^{\mathrm{n}} \mathrm{p}_\mathrm{i}^\alpha \right)^{1/(1-\alpha)} \quad \text{with } \alpha \geq 0 \tag{3}$$

This measure has a number of desired properties, which have made it the foundation for various biodiversity indices in ecology:

1. The measure (3) – more exactly: its logarithm $H_\alpha = \log \nu_\alpha$ – is well known from information theory where it has been introduced by Rényi (1961) as a generalised entropy. Its properties are well studied and understood (Aczél and Daróczy 1975).
2. The maximal value of $\nu_\alpha(n, p)$ increases with the number $n$ of different species.
3. For given $n$, the measure $\nu_\alpha$ takes on values between 1 and $n$, depending on $p$. Hence, it can be interpreted as an effective species number.

4. For given $n$, the measure $\nu(n, p)$ assumes its maximal value – pure species richness $n$ – when all species have equal relative abundance, i.e. $p_i = 1/n$ for all $i = 1, \ldots, n$. In this case of an absolutely even distribution of relative abundances, the effective number $\nu$ of different species equals the total number $n$ of different species in the system. Further, the measure $\nu(n, p)$ decreases with increasing unevenness of the distribution of relative abundances $p$. This means dominance of a few species or, more generally, an uneven distribution of relative abundances brings down the index of effective species number $\nu$ from its maximal value which is given by pure species richness $n$. It also follows that the index assumes its minimal value when a system is dominated by one single species, with all other species having negligible relative abundances, i.e. $p_i \approx 0$ for all $i = 1, \ldots, n$ except $i = i^*$, where $i^*$ denotes the dominant species and $p_{i^*} \approx 1$. In this case, $\nu_\alpha(n, p)$, which means that the effective number of different species is approximately one.
5. The parameter $\alpha \geq 0$ weighs the influence of evenness of the distribution of relative abundances $p$ against the influence of pure species number $n$ when calculating the effective species number $\nu$. For $\alpha \geq 0$, the evenness of the distribution of relative abundances $p$ is completely irrelevant and the effective species number $\nu$ is simply given by the pure species number $n$. The larger $\alpha$, the higher is the weight of the evenness in the calculation of the effective species number (3). For $\alpha \to \infty$, the pure species number $n$ is completely irrelevant and the effective species number $\nu$ is exclusively determined by how (un)evenly the relative abundances of species are distributed.
6. For different values of the parameter $\alpha$ one can recover from expression (3) different species diversity indices that are well established in ecology (Hill 1973). They can thus be considered as special cases of the general measure (3):
   - With $\alpha = 0$ one obtains the *species richness index* already introduced in section 3.2 above:

     $$D^{\mathrm{R}}(\Omega) = \nu_0(n, p) = n \tag{4}$$

     That is, to zeroth order effective species number is just pure species richness.
   - With $\alpha = 1$ one obtains the *Shannon Wiener-index:*

     $$D^{SW}(\Omega) = \nu_1(n, p) = \exp\ H \text{with } H = -\sum_{i=1}^{n} p_i \log p_i \tag{5}$$

where $H$ is well known from statistics and information theory as the Shannon-Wiener expression for entropy (Shannon 1948; Wiener 1961).

- With $\alpha = 2$ one obtains *Simpson's index* (Simpson 1949):

$$D^S(\Omega) = \nu_2(n, p) = 1 \Big/ \sum_{i=1}^{n} p_i^2 \quad (6)$$

It is based on the underlying idea that the probability of any two individuals drawn at random from an infinitely large ecosystem belonging to different species is given by $\sum_{i=1}^{n} p_i^2$. The inverse of this expression is taken to form the biodiversity index, so that $D^S$ increases with the evenness of the distribution of relative abundances.

- With $\alpha \to +\infty$ one obtains the *Berger-Parker index* (Berger and Parker 1970; May 1975) as

$$D^{BP}(\Omega) = \nu_{+\infty}(n, p) = 1/p_1 \quad (7)$$

that is, the inverse relative abundance of the most common species. It can be interpreted as an effective species number in the sense that $1/p_1$ gives the equivalent number of equally abundant (hypothetical) species with the same relative abundance as the most abundant species in the system. Obviously, the Berger-Parker index considers only the relative dominance of the most common species in the system, neglecting all other species.

One of the properties of the biodiversity measure (3) is that for given $n$ and $p$ the value of $\nu_\alpha(n, p)$ decreases with $\alpha$. As the most widely used diversity indices can all be expressed as special cases of Equation (3) for different values of $a$, it becomes evident that the results for the effective species number in a given system yielded by these indices are related in the following way:

$$n = D^R \geq D^{SW} \geq D^S \geq D^{BP} > 1 \quad (8)$$

### 3.4 *Indices based on characteristic features*

The biodiversity indices discussed in section 3.3 all take the species richness of an ecosystem, properly adjusted by the distribution of relative abundances so that rare species are given less weight than common species, to be a measure of diversity. According to these indices, systems with more, and more evenly distributed, species are found to have a higher biodiversity than systems with less, or less evenly distributed, species. This procedure has been criticised for not taking into account the (dis)similarity between species. For example, a system with 100 individuals of some plant species, 80 individuals of a different plant species and 50 individuals of yet another plant species will be found to have exactly the same biodiversity, according to these indices, as a system with 100 individuals of some plant species, 80 individuals of a mammal species and 50 individuals of some insect species. Yet intuitively one would say that the latter has a higher biodiversity. This intuition is based on the (dis)similarity between the various species.

In order to account for the (dis)similarity of species when measuring biodiversity, one needs a formal representation of the characteristic features of species. Based on these characteristic features, the (dis)similarity of species can be measured and taken into account when constructing a biodiversity index. Two different approaches exist so far. One has been initiated by ecologists (see May 1990; Erwin 1991; Vane-Wright *et al.* 1991; Crozier 1992) and put on a rigorous axiomatic basis, enhanced and popularised by Weitzman (1992, 1993, 1998). Here it shall therefore be called the Weitzman approach.[6] It builds on the concept of a distance function to measure the pairwise dissimilarity between species. The diversity of a set of species, in this approach, is then taken to be an aggregate measure of the dissimilarity between species. This approach is most appealing when applied to phylogenetic diversity. The other approach, developed by Nehring and Puppe (2002, 2004), generalises the Weitzman approach. It builds directly on the characteristic features of species and their relative weights. Both approaches will now be discussed in detail.

### 3.4.1 *The Weitzman index*

Weitzman (1992) defines a diversity measure, $D(S)$, of a set $S$ of species based on the fundamental idea that the diversity of a set of species should be an aggregate measure of the pairwise dissimilarity between species. The dissimilarity between two species, $s_i$ and $s_j$, is conceptualised by their distance $d(s_i, s_j)$ in feature space. The pairwise distances of all species are the elementary data upon which the diversity measure builds. Weitzman (1992, 1993) suggests the use of taxonomic or phylogenetic information to determine the pairwise distances between species, but also states that any other quantifiable trait of species could be used for that purpose as well, e.g. morphological or functional features. A distance function can, of course, also be meaningfully defined when species differ in more than one feature.

[6] Solow *et al.* (1993) and Weikard (1998, 1999, 2002) have developed biodiversity indices that follow a very similar logic.

Weitzman's (1992) diversity index $D(S)$ of a set $S$ of species is then defined recursively by

$$D^{W}(Q \cup \{s_i\}) = D^{W}(Q) + \delta(s_i, Q) \text{ for all } s_i \in S/Q \quad (9)$$

where $D^{W}(\{s_j\}) = D_0 \in \mathrm{IR}_+$ for all $s_j \in S$ and $\delta(s_i, Q) = \min_{s_j \in Q} d(s_i, s_j)$ for all $s_i \in S \setminus Q$.

This means that the calculation of the index starts from an arbitrarily chosen start value $D_0 \in \mathrm{IR}_+$ assigned to the set that contains only one species, irrespective of what species $s_j$ that is. Depending on the particular application, $D_0$ may be chosen to be zero or a very large number. One then calculates the biodiversity index of an enlarged subset $Q'$ of $S(0 \subset Q' \subseteq S)$, that one obtains when adding species $s_i \in S \setminus Q$ to the set $Q$, $Q' = Q + \{s_i\}$, by adding the increase in diversity $\delta(s_i, Q)$, which species $s_i$ adds to the diversity of the subset $Q$. This increase in diversity is calculated as the minimal distance between the added species $s_i$ and any of the species $s_j$ in the subset $Q$. So, the recursive algorithm (9) allows one to calculate the diversity of a set $S$ of species, starting from the arbitrarily chosen diversity value of a single species set, $D_0$, and then adding one species after the other until the whole set $S$ is complete.

One problem with the recursive definition (9) is that, in general, its outcome is path dependent, i.e. the value calculated depends on the particular sequence in which species are added when constructing the full set $S$. Therefore, the diversity function as defined by Equation (9) is, in general, not unique. The Weitzman approach is most appealing, however, when applied to the special case when the feature space is ultrametric.[7] In this special case the recursive definition (9) is not path dependent but uniquely defines a diversity index. Ultrametric distances have an interesting geometric property which is also ecologically relevant. A set $S$ of species characterised by ultrametric distances can be represented graphically by a hierarchical or phylogenetic tree and any phylogenetic tree can be represented by ultrametric distances. In a phylogenetic tree the distance $d(s_i, s_j)$, which indicates the dissimilarity between species $s_i$ and $s_j$ is given by the vertical distance to the last common ancestor of $s_i$ and $s_j$, and the diversity $D^{W}(S)$ of the set $S$ of all species is given by the summed vertical length of all branches of the tree.

### 3.4.2 *The Nehring-Puppe index*

Even more general than Weitzman's distance-function approach is the so-called 'multi-attribute approach' proposed by Nehring and Puppe (2002,

[7] A space is called *ultrametric* if the pairwise distances between any three points in space have the property that the two greatest distances are equal:

$$max\{d(s_i, s_j), d(s_j, s_k), d(s_i, s_k)\} = mid\{d(s_i, s_j), d(s_j, s_k), d(s_i, s_k)\} \quad \text{for all} \quad s_i, s_j, s_k \in S.$$

2004). Like Weitzman, they base a measure of species diversity on the characteristic features of species. In contrast to Weitzman, the elementary data are not the pairwise dissimilarities between species but the characteristic features $f$ themselves. From the different features $f$ and their relative weights $\lambda_f \geq 0$, which may be derived from the individuals' or society's preferences for the different features, Nehring and Puppe construct a diversity index as follows:

$$D^{NP}(\Omega) = \sum_{f \in F: \exists s_i \in s \textit{ with ``} s_i \textit{ has feature } f\textit{''}} \lambda_f. \tag{10}$$

In words, the diversity index for a set $S$ of species is the sum of weights $\lambda_f$ of all features $f$ that are represented by at least one species $s_i$ in the ecosystem. Each feature shows up in the sum at most once. In particular, each species $s_i$ contributes to the diversity of the set $S$ exactly the relative weight of all those features which are possessed by $s_i$ and not already possessed by any other species in the set.

Nehring and Puppe also show that under certain conditions the characterisation of an ecosystem by its diversity $D^{NP}$ uniquely determines the relative weights $\lambda_f$ of the different features. This means that in assigning a certain diversity to an ecosystem one automatically reveals an (implicit) value judgement about the relevant features according to which one distinguishes between species and one describes an ecosystem as more or less diverse.

## 4 Critical assessment of ecological and economic biodiversity indices

### *4.1 Conceptual comparison*

Comparing the ecological and economic biodiversity indices reviewed in section 3 above at the conceptual level, it is obvious that the two classes are distinct by the information they use for constructing a diversity index (Figure 10.1). While the ecological measures (section 3.3) use the number $n$ of different species in a system as well as their relative abundances $p$, the economic ones (section 3.4) use the number $n$ of different species as well as their characteristic features $f$. In a sense, the indices discussed in section 3.3 are 'heterogeneity indices' rather than 'diversity' indices (Peet 1974), as they are based on richness and evenness but completely miss out features. The indices discussed in section 3.4 are 'dissimilarity indices' rather than 'diversity' indices, as they are based on richness and dissimilarity but completely miss out abundances. Both kinds of indices contain pure species richness as a special case.

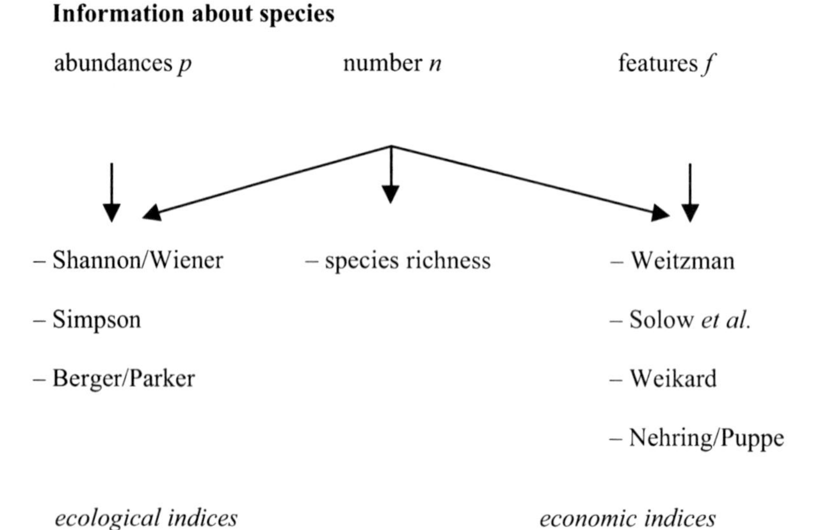


Figure 10.1. Biodiversity indices differ by the information on species and ecosystem composition they use.

Up to now, there do not exist any encompassing diversity indices based on all ecological information considered here – species richness $n$, abundances $p$ and features $f$. A logical next step at this point could be to construct a general diversity index based on species richness, abundances and features, which contains the existing indices as special cases. However, one should not jump to this conclusion too quickly. It is important to note that the ecological and economic diversity indices have come out of very different modes of thinking. They have been developed for different purposes and are based on fundamentally different value systems. Therefore, they may not even be compatible. This point is addressed next.

### *4.2 The relevance of abundances and features*

From an economic point of view, relative abundances are usually considered irrelevant for the measurement of diversity. The reason is that in economics the diversity issue is usually framed as a choice problem. Diversity is then a property of the choice set, i.e. the set of feasible alternatives to choose from. Individuals facing a situation of choice should consider only the list of possible alternatives (say, the menu in a restaurant) rather than the actual allocation which has been realised as the result of other people's earlier choices (say, the dishes on the other tables in a restaurant). Furthermore, when economists talk about product diversity, relative abundances are irrelevant since there is the possibility of production.[8] If all people in a restaurant order the same dish from the menu, this dish will be produced in the quantity demanded; if all people order different dishes, different dishes are produced. In any case, the diversity of the choice set is determined by the diversity of the order list (the menu) and not by the actual allocation of products (the dishes on the tables).

This argument has influenced economists' view on biodiversity as well. Economists consider biological diversity as a form of product diversity, i.e. a diverse resource pool from which one can choose the most preferred option(s). And this diversity is essentially determined by the choice set, i.e. the list $S$ of species existent in an ecosystem (e.g. Weitzman 1992, 1993, 1998). The actual abundances of individuals of different species, in that view, do not matter.

Ecologists, in contrast, often argue that biological species living in natural ecosystems – even when considered merely as a resource pool to choose from – are different from normal economic goods for a number of reasons (e.g. Begon *et al.* 1998; Ricklefs and Miller 2000). First, individuals of a particular species cannot simply be produced, at least not so easily, not for any species and not in any given number. Second, there are direct interactions between individuals and species within ecosystems, which heavily influence survival probabilities and dynamics in an ecosystem. For that sake, relative abundances matter. Third, while some potential ecosystems (in the sense of relative abundance distributions) are viable in situ, others are not.

Hence, it becomes apparent that the two types of biodiversity measures – the ecological ones and the economic ones – aim at characterising two very different aspects of the ecosystem. While the ecological measures describe the actual and potentially unevenly distributed allocation $\Omega$ of species, the economic measures characterise the abstract list $S$ of species existent in the system.

### 4.3 *Different philosophical perspectives on diversity*

The underlying reason for this difference between the ecological and economic measures of biodiversity can be found in the philosophically distinct perspective on diversity between ecologists and economists. Ecologists traditionally view diversity more or less in what may be called a 'conservative' perspective, while economists predominantly have what

[8] While the scarcity of production factors may limit the *absolute* abundances of the produced products, all possible *relative* abundances can be produced without restriction.

may be called a 'liberal' perspective on diversity (Kirchhoff and Trepl 2001).

In the conservative view, which goes back to Gottfried Wilhelm Leibniz (1646–1716) and Immanuel Kant (1724–1804), diversity is an expression of unity. By viewing a system as diverse, one stresses the integrity and functioning of the entire system. The ultimate concern is with the system at large. In this view, diversity may have an indirect value in that it contributes to certain overall system properties, such as stability, productivity or resilience at the system level. In contrast, in the liberal view, which goes back to René Descartes (1596–1650), John Locke (1632–1704) and David Hume (1711–1776), diversity enables the freedom of choice for autonomous individuals who choose from a set of diverse alternatives. The ultimate concern is with the well-being of individuals. In this view, diversity of a choice set has a direct value in that it allows individuals to make a choice that better satisfies their individual subjective preferences. Once one alternative has been chosen, the other alternatives, and the diversity of the choice set, are no longer relevant.

Of course, the integrity and functioning of the entire system will also be important for the well-being of autonomous individuals who simply want to choose from a set of diverse alternatives. For example, today's choice may impede the system's ability to work properly in the future and, therefore, to provide diversity to choose from in the future. This is an intertemporal argument, which combines (i) an argument about diversity's importance at a given point in time for individuals, who want to make an optimal choice at this point in time, and (ii) an argument about diversity's role for system functioning and evolution over time. From an analytical point of view, one should distinguish these two arguments. This underlies the distinction between the conservative and the liberal perspective, which is analytical to start with.

These two distinct perspectives on diversity – the conservative one and the liberal one – correspond to some extent with the two types of biodiversity measures considered here (section 3): the ecological measures that take into account relative abundances and the economic measures that deliberately do not take into account relative abundances. The ecological measures are based on a conservative perspective in that their main interest is to represent biodiversity as an indicator of ecosystem integrity and functioning. With that concern, the distribution of relative abundances is an essential ingredient in constructing a biodiversity index. In contrast, the economic measures are based on a liberal perspective in that their main interest is to represent biodiversity as a property of the choice set from which economic agents – individuals, firms or society – can choose to best satisfy their preferences. With that concern, it seems plausible that the actual distribution of relative abundances is not taken into account when constructing a biodiversity index.

## 5 Conclusion

The question of how to measure species diversity is intimately linked to the question of what species diversity is good for. This is not a purely descriptive question, but also a normative one. There are many possible answers, but in any case an answer requires value judgements. Do we consider species diversity as valuable because it contributes to overall ecosystem functioning – either out of a concern for conserving the working basis of natural evolution, or out of a concern for conserving certain essential and life-supporting ecosystem services, such as oxygen production, climate stabilisation, soil regeneration and nutrient cycling (Perrings *et al.* 1995; Daily 1997; Millennium Ecosystem Assessment 2005)? Or do we consider species diversity as valuable because it allows individuals to make an optimal choice from a diverse resource base, e.g. when choosing certain desired genetic properties in plants for developing pharmaceutical substances (Polasky *et al.* 1993; Polasky and Solow 1995; Simpson *et al.* 1996; Rausser and Small 2000), or breeding or genetically engineering new food plants (Myers 1983, 1989; Plotkin 1988)?

These are examples for different value statements about biodiversity which are made on the basis of different fundamental value judgements: in the former case dominates the conservative perspective, in the latter the liberal one. As I have shown here, these two perspectives lead to different measures of species diversity, the ecological measures and the economic measures. Of course, there is a continuous spectrum in between these two extreme views on why species diversity is valuable and how to measure it. But in any case, one is led to conclude, the measurement of species diversity requires prior value judgements as to what purpose species diversity serves in ecological-economic systems.

### REFERENCES

Aczél, J. and Daróczy, Z. 1975. *On Measures of Information and their Characterizations*. New York: Academic Press.

Bateman, I. J., Carson, R. T., Day, D., Hanemann, M., Hanley, N., Hett, T., Jones-Lee, M. W., Loomes, G., Mourato, S., Özdemiroglu, E., Pearce, D. W., Sugden, R. and Swanson, J. 2002. *Economic Valuation with Stated Preference Techniques – A Manual*. Cheltenham: Edward Elgar.

Baumgärtner, S. 2005. Measuring the diversity of what? and for what purpose? a conceptual comparison of ecological and economic biodiversity indices. *Working Paper*, available at http://ssrn.com/abstract=894782

Baumgärtner, S. 2006. *Natural Science Constraints in Environmental and Resource Economics. Problem and Method.* University of Heidelberg, available at http://www.ub.uni.heidelberg.de/archiv/6593

Begon, M., Harper, J. L. and Townsend, C. R. 1998. *Ecology – Individuals, Populations, and Communities.* 3rd ed. Sunderland: Sinauer.

Berger, W. H. and Parker, F. L. 1970. Diversity of planktonic foraminifera in deep sea sediments. *Science.* **168**. 1345–1347.

Ceballos, G. and Ehrlich, P. R. 2002. Mammal population losses and the extinction crisis. *Science.* **296**. 904–907.

Crozier, R. H. 1992. Genetic diversity and the agony of choice. *Biological Conservation.* **61**. 11–15.

Daily, G. C. (ed.). 1997. *Nature's Services: Societal Dependence on Natural Ecosystems.* Washington, DC: Island Press.

Erwin, T. L. 1991. An evolutionary basis for conservation strategies. *Science.* **253**. 750–752.

Freeman, A. M. 2003. *The Measurement of Environmental and Resource Values. Theory and Methods.* 2nd ed. Washington, DC: Resources for the Future.

Gaston, K. J. 2000. Global patterns in biodiversity. *Nature.* **405**. 220–227.

Goulder, L. H. and Kennedy, D. 1997. Valuing ecosystem services: philosophical bases and empirical methods. In Daily, G. C. (ed.). *Nature's Services: Societal Dependence on Natural Ecosystems.* Washington, DC: Island Press. 23–47.

Groombridge, B. (ed.). 1992. *Global Biodiversity: Status of the World's Living Resources. A Report Compiled by the World Conservation Monitoring Centre.* London: Chapman & Hall.

Hanley, N. and Spash, C. L. 1993. *Cost-Benefit Analysis and the Environment.* Aldershot: Edward Elgar.

Heal, G. 2000. Biodiversity as a commodity. In S. A. Levin (ed.). *Encyclopedia of Biodiversity,* vol. 1. New York: Academic Press. 359–376.

Hill, M. O. 1973. Diversity and evenness: a unifying notation and its consequences. *Ecology.* **54**. 427–431.

Kinzig, A. P. and Harte, J. 2000. Implications of endemics-area relationships for estimates of species extinctions. *Ecology.* **81** (12). 3305–3311.

Kirchhoff, T. and Trepl, L. 2001. Vom Wert der Biodiversität – über konkurrierende politische Theorien in der Diskussion um Biodiversität. *Zeitschrift für angewandte Umweltforschung – Journal of Environmental Research.* Special Issue **13**. 27–44.

MacArthur, R. H. and Wilson, E. O. 1967. *The Theory of Island Biogeography.* Princeton, NJ: Princeton University Press.

Magurran, A. E. 1988. *Ecological Diversity and its Measurement.* Princeton, NJ: Princeton University Press.

May, R. M. 1975. Patterns of species abundance and diversity. In M. L. Cody and J. M. Diamond (eds.). *Ecology and Evolution of Communities.* Cambridge, MA: Harvard University Press. 81–120.

May, R. M. 1990. Taxonomy as destiny. *Nature.* **347**. 129–130.

May, R. M., Lawton, J. H. and Stork, N. E. 1995. Assessing extinction rates. In R. M. May and J. H. Lawton (eds.). *Extinction Rates.* Oxford: Oxford University Press. 1–24.

McCann, K. S. 2000. The diversity-stability debate. *Nature.* **405**. 228–233.

McNeely, J. A. 1988. *Economics and Biological Diversity: Developing and Using Economic Incentives to Conserve Biological Resources*. Commissioned by the International Union for Conservation of Nature and Natural Resources (IUCN), Gland, Switzerland.

McNeely, J. A., Miller, K. R., Reid, W. V., Mittermeir, R. A. and Werner, T. B. 1990. *Conserving the World's Biological Diversity*. Commissioned by the IUCN, World Resources Institute, Conservation International, World Wildlife Fund-US and the World Bank. Washington, DC: World Bank.

Millennium Ecosystem Assessment 2005. *Ecosystems and Human Well-Being: Synthesis Report*. Washington, DC: Island Press.

Myers, N. 1983. *A Wealth of Wild Species: Storehouse for Human Welfare*. Boulder, CO: Westview Press.

Myers, N. 1989. Loss of biological diversity and its potential impact on agriculture and food production. In D. Pimentel and C. W. Hall (eds.). *Food and Natural Resources*. San Diego, CA: Academic Press. 49–68.

Nehring, K. and Puppe, C. 2002. A theory of diversity. *Econometrica*. **70**. 1155–1198.

Nehring, K. and Puppe, C. 2004. Modelling phylogenetic diversity. *Resource and Energy Economics*. **26**. 205–235.

Nunes, P. A. L. D. and van den Bergh, J. C. J. M. 2001. Economic valuation of biodiversity: sense or nonsense? *Ecological Economics*. **39**. 203–222.

Pearce, D. W. 1993. *Economic Values and the Natural World*. London: Earthscan.

Pearce, D. W. and Moran, D. 1994. *The Economic Value of Biodiversity*. London: Earthscan.

Pearce, D. W. and Turner, R. K. 1990. *Economics of Natural Resources and the Natural Environment*. New York: Harvester Wheatsheaf.

Peet, R. K. 1974. The measurement of species diversity. *Annual Review of Ecological Systems*. **5**. 285–307.

Perrings, C., Mäler, K.-G., Folke, C., Holling, C. S. and Jansson, B.-O. (eds.). 1995. *Biodiversity Loss: Economic and Ecological Issues*. Cambridge: Cambridge University Press.

Pielou, E. C. 1975. *Ecological Diversity*. New York: Wiley.

Plotkin, M. J. 1988. The outlook for new agricultural and industrial products from the tropics. In E. O. Wilson (ed.). *BioDiversity*. Washington, DC: National Academy Press. 106–116.

Polasky, S. and Solow, A. R. 1995. On the value of a collection of species. *Journal of Environmental Economics and Management*. **29**. 298–303.

Polasky, S., Solow, A. and Broadus, J. 1993. Searching for uncertain benefits and the conservation of biological diversity. *Environmental and Resource Economics*. **3**. 171–181.

Purvis, A. and Hector, A. 2000. Getting the measure of biodiversity. *Nature*. **405**. 212–219.

Rausser, G. C. and Small, A. A. 2000. Valuing research leads: bioprospecting and the conservation of genetic resources. *Journal of Political Economy*. **108**. 173–206.

Rényi, A. 1961. On measures of entropy and information. In J. Neyman (ed.). *Proceedings of the Fourth Berkeley Symposium on Mathematical Statistics and Probability, Vol. I*. Berkeley: University of California Press. 547–561.

Ricklefs, R. E. and Miller, G. L. 2000. *Ecology*. 4th ed. New York: W. H. Freeman.

Robbins, L. 1932. *An Essay on the Nature and Significance of Economic Science*. London: Macmillan.

Rosenzweig, M. L. 1995. *Species Diversity in Space and Time*. Cambridge: Cambridge University Press.

Shannon, C. E. 1948. A mathematical theory of communication. *Bell System Technical Journal*. **27**. 379–423 and 623–656.

Simpson, E. H. 1949. Measurement of diversity. *Nature*. **163**. 688.

Simpson, R. D., Sedjo, R. A. and Reid, J. W. 1996. Valuing biodiversity for use in pharmaceutical research. *Journal of Political Economy*. **104**. 163–185.

Smith, K. V. 1996. *Estimating Economic Values for Nature: Methods for Non-Market Valuation*. Cheltenham: Edward Elgar.

Solow, A., Polasky, S. and Broadus, J. 1993. On the measurement of biological diversity. *Journal of Environmental Economics and Management*. **24**. 60–68.

Turner, R. K. 1999. The place of economic values in environmental valuation. In I. J. Bateman and K. G. Willis (eds.). *Valuing Environmental Preferences – Theory and Practice of the Contingent Valuation Method in the US, EU, and Developing Countries*. Oxford, UK: Oxford University Press. 17–41.

Vane-Wright, R. I., Humphries, C. J. and Williams, P. H. 1991. What to preserve? systematics and the agony of choice. *Biological Conservation*. **55**. 235–254.

Watson, R. T., Heywood, V. H., Baste, I., Dias, B., Gamez, R., Janetos, T., Reid, W. and Ruark, R. (eds.). 1995. *Global Biodiversity Assessment* (published for the United Nations Environment Programme). Cambridge: Cambridge University Press.

Weikard, H.-P. 1998. On the measurement of diversity. Working Paper No. 9801. University of Graz, Austria: Institute of Public Economics.

Weikard, H.-P. 1999. *Wahlfreiheit für zukünftige Generationen*. Marburg: Metropolis.

Weikard, H.-P. 2002. Diversity functions and the value of biodiversity. *Land Economics*. **78**. 20–27.

Weitzman, M. 1992. On diversity. *Quarterly Journal of Economics*. **107**. 363–405.

Weitzman, M. 1993. What to preserve? an application of diversity theory to crane conservation. *Quarterly Journal of Economics*. **108**. 155–183.

Weitzman, M. 1998. The Noah's Ark problem. *Econometrica*. **66**. 1279–1298.

Whitmore, T. C. and Sayer, J. A. (eds.). 1992. *Tropical Deforestation and Species Extinction*. London: Chapman & Hall.

Wiener, N. 1961. *Cybernetics*. Cambridge, MA: MIT Press.

Wilson, E. O. (ed.). 1988. *BioDiversity*. Washington, DC: National Academy Press.

*Section B*

# Techniques

# 11 Combining TCM and CVM of endangered species conservation programme: estimation of the marginal value of vultures (*Gyps fulvus*) in the presence of species–visitors interaction[1]

*Nir Becker, Yael Choresh, Moshe Inbar and Ofer Bahat*

## 1 Introduction

Using different valuation techniques in order to estimate the value of endangered species is well documented in the literature. Those benefits can be contrasted against the protection cost or against alternative uses of the habitat that might risk their existence. However, performing a cost-benefit analysis (CBA) should take into account issues such as the value of the marginal individual, tradeoff analysis among competing goals and a feedback interaction between the size of a species' population versus number of visitors allowed in a particular wildlife park.

The central aim of this chapter is to examine how employing the travel cost method (TCM) in conjunction with the contingent valuation method (CVM) can provide insights as to whether the protecting measures and associated allocated budget for the conservation of a particular wildlife species are in accordance with public priorities. This question is examined in a case study assessing the values associated with the protection of the Griffon vulture (*Gyps fulvus)* via the development of one particular protection method, namely feeding stations. We also performed a simple CBA of the total conservation efforts at the national level and compared it with the total benefit derived from the increased number of vultures over a given period of a national protecting plan. Our second aim is to show how valuation techniques can be used for wildlife policy analysis in

[1] We thank the Israeli Ministry of the Environment for their financial support. Jeff Bennet and Andreas Kontoleon provided helpful comments, as did participants in the departmental seminar at the Department of Agricultural Economics and Management at the Hebrew University. This paper is dedicated to the memory of Orna Eshed, the Gamla nature reserve biologist, who inspired our research but did not live to see it through.

two other respects: entrance-fee policy and allocation of efforts to protect species among competing sites.

We see two contributions in the study discussed in this chapter. First, it measures the value of the marginal individual of an endangered species in order to conduct a cost-benefit analysis. Since policy should be judged at the margin, calculating the average values of extinction species provides the wrong signal for benefit estimation (Bulte and Van Kooten 1999; Kontoleon and Swanson 2003). We used a multiple-scenario questionnaire format which enabled us to trace a demand function for the species and by that receive the value of the marginal individual. The second contribution lies in the policy implications of using TCM and CVM beyond determining the cost-benefit question of 'should we conserve or not?'.

The study addressed two policy issues: the debate between revenue vs. efficiency and the issue of site interaction. In times of budget constraints, it is a necessity to consider other options for financing nature conservation beyond the general budget (Van Sickle and Eagles 1998). Hence alternative sources of revenues (such as from recreation) should also be considered even though they may entail loss in welfare. Using the demand functions derived from the TCM analysis we can reveal the impact of the tradeoff between revenue increases and welfare decreases.

With respect to the issue of site interaction, there is a need to combine in the analysis the number of individual vultures and number of visitors. There is a positive feedback between these two parameters which policy makers should take into account as a higher number of vultures bring more visitors. We estimate the value visitors place on the marginal vulture through the CV study while the TC analysis allows us to assess the number of additional visitors. Combining results from both methods could reveal the optimal number of vultures that should be targeted at each site as well as how many more visitors they would attract (González-Cabán *et al.* 2003).

The Israeli Red List of Threatened Animals classifies the Eurasian Griffon vulture, *Gyps fulvus*, as Vulnerable (Dolev and Perevolotsky 2002). The population of the Griffon vulture, once numerous and abundant throughout its breeding range, has suffered from a severe decline during the last century. In Israel, although protected by law, the population has declined from over 1,000 breeding pairs in the second half of the nineteenth century (Mendelssohn and Leshem 1983; Tristram 1885) to a present number of about 140 breeding pairs (Bahat, personal communication). This worrisome decline is a result of hunting, excessive usage of pesticides (Mendelssohn 1972; Shlosberg 2001), electrocutions (Lehman *et al.* 1999) and improved pastoral hygiene which has resulted in a reduction in the available food for vultures (Cramp and Simmons 1980; Wilbur 1983).

The Griffon vulture is an obligatory carrion feeder which forages over extensive areas (Cramp and Simmons 1980; Mundy *et al.* 1992). In Israel, its main food source is cattle (Bahat 1995). Since this food source is not always available, and since natural food sources have become scarce due to changes in land use in recent years, there is a need for supplementary feeding in the form of a feeding station. Such feeding stations have proven to be successful in sustaining stable vulture populations in Israel and elsewhere (Bahat *et al.* 2002; Mundy *et al.* 1992).

Since 1996, the Israel Electric Company has joined forces with the Israel Nature and National Parks Authority and the Society for the Protection of Nature in Israel in a mutual effort to protect the Griffon vulture and other endangered raptors. The budget for this project is allocated to various protecting measures, including the operation of feeding stations (Bahat *et al.* 2002).

The benefit of protecting vultures is both ecological and social: as scavengers, vultures are crucial to the well-being of the environment, by releasing it from dead animals that could otherwise be hazardous both to wildlife and to humans. Yet vultures also have significant recreational interest. While vultures are not a common sight in other parts of the country, the Gamla Nature Reserve in Northern Israel attracts significant crowds who are able to observe the soaring flight patterns and breeding efforts of the vultures.

In the next section we will review the operation of feeding stations as a management tool in an overall scheme of protecting vultures. We will also present studies valuing use and non-use values of endangered species and the methods used for such valuation. This review will include also CVM-TCM interaction, entrance-fee analysis and estimation of the value of the marginal individual. Section 3 will describe the study sites. Section 4 presents the TCM and CVM results from samples derived from the Gamla and Hai-Bar Nature Reserves, as well as from the general population. Section 5 will provide a break-even point analysis for feeding stations as a means of preserving the population of vultures at the two specific sites as well as a cost-benefit analysis of the total investment in vulture conservation in Israel. Furthermore, we will examine two additional policy issues: entrance-fee analysis under different policy goals, as well as an analysis of the optimal allocating of conservation efforts between two competing sites. The last section concludes.

## 2 Literature review

### *2.1 Feeding stations*

One of the conservation efforts of protecting vultures and preventing their decline is the operating of feeding stations where food quality can be assured and the availability of food can attract vultures to areas where they once were abundant (Mundy *et al.* 1992).

A well-operated feeding station can also serve as a source of bone fragments to the breeding parents, to compensate for the absence of the main bone-crushers in their foraging areas (Richardson *et al.* 1986). In the absence of bone fragments, the parents search for various substitutes which are equally hard, such as pieces of metal, swallow them after filling their crop and regurgitate them at the nest (Mundy *et al.* 1992). These pieces are useless as a supply of calcium for the nestling's skeleton and can also be harmful, even lethal. As a result, the nestling suffers from rickets and may not reach the fledgling stage (Houston 1978).

This management tool was first used in South Africa in 1966, where a feeding scheme for the Bearded vulture, *Gypaetus barbatus*, was conducted (Butchart 1988). In France, the use of feeding stations started in the Pyrenees in 1969 (Terrasse 1985). Right after that, feeding stations were established in other places in Europe and in the USA, as part of the reintroduction programme of the endangered California condor, *Gymnogyps californianus* (Wilbur *et al.* 1974). The population of the Black vulture, *Aegypius monachus*, in Greece is also recovering as the result of operating feeding stations (Vlachos *et al.* 1999).

In order to deal with rickets, the South African Vulture Study Group has begun to provide bone fragments in feeding stations. Rickets declined from 16.9 per cent in 1976 to 3.7 per cent in 1983 (Richardson *et al.* 1986).

In Israel, the Nature Reserves and Parks Authority started the operating of feeding stations in 1972. A network of sixteen feeding stations was spread all over the country. These feeding stations were placed in vultures' foraging zones and are located in areas where carcasses can be provided on a regular basis (Bahat *et al.* 2002).

A routine supplement of bone fragments at the local feeding station at Gamla Nature Reserve began in 1998. As a result, the number of bone fragments found in the nests increased and the number of nests containing artifacts decreased. Furthermore, the number of nestlings suffering from rickets dropped (Ben-noon *et al.* 2003).

### 2.2 *The value of wildlife viewing and protecting*

Two of the most commonly used economic valuation methods are the travel cost and contingent valuation methods.

One of the major difficulties in determining the optimal amount of wildlife protection is that the economic benefits associated with species conservation are often non-marketed. The 'extra' value gained over costs is an estimate of the net economic benefits, or consumer surplus derived from wildlife viewing. Brown (1993) estimated the value of viewing elephants in Kenya through the application of both TCM and CVM. The total net economic value per foreign visitor on a wildlife-viewing safari was calculated. A portion of that total value, 12.6 per cent, was allocated to viewing elephants specifically, a value which was translated to $23–27 million. As a result of such studies, the Kenya Nature Protecting Authority has realised the economic implications of declining elephant populations due to poaching.

Similarly, using TCM and CVM techniques the annual recreational value of wildlife viewing in Lake Nakuru National Park in Kenya was found to be $7.5–15 million (Navrud and Mungatana 1994). The flamingos, *Phoenicopterus minor*, in the lake accounted for more than one third of this value. Considering that Lake Nakuru is just one of several parks in Kenya, and that wildlife viewing is becoming an important part of the global trend of increasing ecotourism, the results of this study suggest that sustainable management of wildlife resources could provide a significant and much-needed revenue source for the country in the future. This economic potential can also secure the preservation of wildlife and hence provide the possibility for a 'win-win' outcome.

Another application of the CVM to the conservation of endangered species was undertaken in the State of Victoria, Australia (Jakobsson and Dragun 2001). Two CVM questionnaires were used. The results show a higher value for protecting the Leadbeaters' possum (A$40–84 million) than for competing activities in the area (such as timber cut from the region). Furthermore, the value people place on conservation of all endangered species (A$160–340 million) was larger in at least one order of magnitude than the direct expenditure on conserving flora and fauna in the area (about A$10 million per year).

Most of the wildlife valuation studies deal with the value of its entire habitat (or nature reserve) or the value of a representative individual of a given population of an endangered species. However, policy decisions are often made on the margin. Bulte and Van-Kooten (1999, 2000) argue that for species such as the ancient temperate rainforests and minke whales (*Balaenoptera acutorostrata)*, the likelihood of going below the critical species threshold level should dictate whether policy decisions should be led by teleological (utilitarian) or deontological (Kantian) reasoning. Their conclusion is that in many cases it is necessary to depart from the utilitarian approach. Yet their argument can be reversed in our case. As long as the species has gone beyond the critical level and as long as there is an increasing marginal cost of protecting the species, a marginal analysis should be applied.

Other researchers who dealt with the value of the marginal species include Kontoleon and Swanson (2003), González-Cabán *et al.* (2003) and Paulrod (2004). Paulrod estimated the value of the marginal benefit of angling in Sweden for sport fishing. He found that the marginal value of catch can vary from a few Swedish Krone to a few hundred Swedish Krone depending on site location and type of fish. Therefore, it is important to take marginal valuation into account when undertaking resource allocation decisions.

Kontoleon and Swanson (2003) and González-Cabán *et al.* (2003) dealt with land-based species. Both tried to use marginal valuation in order to estimate the value of habitat, though Kontoleon and Swanson (2003) dealt with an endangered species (the panda, *Ailuropoda melanoleuca)* while González-Cabán (2003) dealt with deer hunting. The benefits assessed in the first study are derived from preservation of the species itself (non-use values), while in the latter study benefits are derived from hunting (use values). Kontoleon and Swanson used a CV study in which respondents were asked to relate to three panda conservation scenarios – cages, pens and free in the wild – with each scenario entailing a different amount of land allocation. They found strong evidence for decreasing marginal values per hectare which ranges from $0.72/hectare to $0.000054/hectare depending on the scenario. González-Cabán *et al.* (2003) assessed the added recreational benefits from increasing deer population by a programme of prescribed burning. The study was undertaken in the San Bernardino National Forest in Southern California. Prescribed burning improves deer habitat and as such attracts more hunters. The value of the marginal hunting trip was translated to the value of a marginal deer and from there on to the marginal value of land. It was found that the value of land decreases from more than US$7920/acre down to US$1200/acre as one goes from the first to the 8,500th acre. Again, contrasting this finding with the cost of prescribing fire can be of great assistance to decision makers as to how many acres to devote to that activity. In our study, we estimated the value of the marginal species, the Griffon vulture, but did not translate it into hectares but into conservation efforts, namely number of feeding stations. As in Kontoleon and

Swanson (2003), respondents were presented with three scenarios which differed in the number of soaring vultures in the sky that a respondent sees.

One way of dealing with obtaining marginal values is to combine CVM and TCM studies. González-Cabán *et al.* (2003) used TCM results of additional trips as a response to added population in order to achieve the marginal value of a deer. Yet there are other reasons for combining TCM–CVM studies such as calibrating CVM hypothetical responses. If one can decompose the CVM results to its use and non-use values and if one can compare the use value derived from CVM to the use value derived from TCM, it might be of interest to include the non-use value also as a reliable estimate. Carson *et al.* (1996) provides a comprehensive literature review of such studies. It is interesting to note that while one would expect CVM estimates to be larger (as they include both use and non-use values), most studies found evidence of the opposite. The CVM/TCM value ranges from 0.3 to 0.5. Carson *et al.* (1996) provides some reasons for this finding while we will also refer to this issue further on in the chapter. In our case study we combine CVM and TCM in order to decide about the optimal effort allocation between two competing sites. Once an additional vulture is added to the site, there is an added value to existing visitors. However, in a dynamic setting, their numbers would increase and would boost the added value to the site even further. The added benefit depends on the elasticity of the demand function derived from the TCM.

### 2.3 *Entrance fee analysis*

Entrance fees for natural sites or parks can be an important contribution to preserving the site and raising revenues for managing it through the best available means. Fees can also be a rationing tool to prevent peak-season congestion of visitors which, in return, has an adverse affect on the site. In times of increasing budget limits, natural sites which raise their own sources of funds can partly alleviate the lack of resources usually provided by central government.

However, entrance fees are controversial on two grounds. First, they have negative distributional implications. Some argue that nature-based recreation is a community necessity and should be provided to its inhabitants. The other argument relates to the efficiency provision of a public good. Once a site such as a museum or nature reserve is built and established, there is no extra cost from an additional visitor (at least for low levels of congestion). Thus, there is no justification for charging an entrance fee, at least from a marginal cost perspective. This argument assumes that the museum or the nature reserve is a pure public good.

Charging a positive price for such goods may deter some users which will lead to a social deadweight loss. This reasoning holds as long as parks are considered to be pure public goods. However, in most cases there are marginal costs associated with added visitors, for example extra roads, more parking spaces, more hiking trails, etc. Hence, the need for optimal park pricing emerges.

The problem with pricing natural resources is that it tries to achieve too many goals with only one tool, namely the entrance fee. Without having a clear definition of society's welfare function and the tradeoffs between various policy objectives, all that remains to be done is to compare different prices and see how they achieve these different goals.

Optimal pricing of public goods is a persistent and common problem observed in both developed (Van Sickle and Eagles 1998; Mendes 2000; Herath and Kennedy 2004) and developing nations (Mercer *et al.* 1995). Not only natural parks face this dilemma. Willis (2003) analysed and compared three pricing options for the public park of Bosco di Capodimonte in Naples, Italy: efficient pricing which means zero entrance fees, maximising revenue and covering operation costs. Each pricing mechanism dictates revenues, cost recovery as well as the magnitude of any welfare losses. In our study we employ the same criteria, namely cost recovery and maximum revenue, but also trace the revenue function in order to identify the relevant location on the curve in which tradeoffs exist between revenue increases and welfare decreases.[2]

## 3 The study sites

Gamla Nature Reserve is located at the centre of the Golan Heights in Northern Israel. The reserve contains the highest waterfall in Israel, archaeological sites and the largest Griffon vulture nesting colony in the country. An average of around 100,000 tourists visit the area each year.

Hai-Bar Nature Reserve is located on the Southeastern outskirts of Haifa, in the heart of Mount Carmel, and aims to provide the means for the breeding and rehabilitation of animals that were once common in the Mediterranean area, in order to eventually release them into the wild. An average of about 45,000 people per year visit this reserve.

[2] One way of dealing with the undesired distributional implications of a uniform price was suggested by Mendes (2000), where differential price is assessed based on willingness to pay. In this study we did not employ this option as to make it applicable, a mechanism for differentiating people at the gate is required. Beyond its potentially politically incorrect nature, this pricing approach also entails some degree of paternalism since the government decides what low-income people can do with 'vouchers' distributed to them. Another suggested pricing mechanism is to decrease congestion. We have excluded this pricing option as well because of lack of data. This remains, however, an issue for future research.

Table 11.1 *Travel cost – regression Hai-Bar*

| Parameter (variable) | Coefficient | t-stat |
|---|---|---|
| Constant | 0.0359 | 5.454 |
| No. of children | −0.0008 | −2.513 |
| Green organisation membership | −0.0035 | −1.476 |
| Education | 0.0025 | 1.515 |
| Income | 3.6038E-07 | 2.047 |
| Travel cost | −0.0001 | −12.639 |
| | | $R^2 = 0.41$ |
| | | $F = 36.34$ |

Dependent variable: travel frequency

## 4 The valuation process

TCM and CVM questionnaires were distributed among visitors at Gamla and Hai-Bar Nature Reserves. CVM questionnaires were also distributed within a representative sample of the general population.

### *4.1 TCM*

TCM was conducted in order to estimate the use values of the sites, reflected in the travel costs incurred by the visitors. The TC function would capture the negative relationship between demand for visits and travel costs. In addition, we controlled for membership in a green organisation, education and income levels by adding them as additional explanatory variables.

In total, 170 questionnaires were distributed at Gamla Nature Reserve (NR) from January to June 2002; 143 were usable (85 per cent). At Hai-Bar NR, 270 out of the 296 questionnaires were usable (91 per cent). The questionnaires were distributed from November 2002 to April 2003.

#### *4.1.1 Calculating TCM*

Travel cost was calculated based on the abovementioned socio-economic variables as well as the cost of travel, the opportunity cost of time and the entrance fee to the site. The regression results are given in Tables 11.1 and 11.2 for Hai-Bar and Gamla respectively.

As can be seen from the tables, at Hai-Bar, the coefficients on travel cost, number of children and income are significant at the 99 per cent level. At Gamla, only the travel cost coefficient is significant. This is theoretically consistent as people would purchase fewer trips if they live

Table 11.2 *Travel cost – regression Gamla*

| Parameter (variable) | Coefficient | t-stat |
|---|---|---|
| Constant | 0.155 | 1.115 |
| Travel cost | −0.0002 | −4.63 |
| Income | −1.35578E-06 | −0.090 |
| Length of trip | −0.002 | −0.520 |
| No. of children | 5.36222E-05 | 0.083 |
| Education | −0.0186 | −0.420 |
| | | $R^2 = 0.40$ |
| | | F = 35.45 |

Dependent variable: travel frequency

further from the site. The result of the income coefficient at Hai-Bar is also expected, while the sign for 'number of children' is somewhat surprising as it would be expected to be positive on account of the significant educational role of the Hai-Bar reserve. A possible explanation might be that families with children are attracted to other types of recreational sites.

We used a zonal TCM approach which results in the following demand functions for Gamla and Hai-Bar NRs as given in equations (1) and (2) respectively.[3] These demand functions are estimated after holding all significant variables at their mean level.

$$P = TC_G = 293 - 0.0037\,(VI_G) \qquad R^2 = 0.913 \tag{1}$$

$$P = TC_H = 506 - 0.0133(VI_H) \qquad R^2 = 0.823 \tag{2}$$

where: $TC_G$ = visiting price for Gamla NR
$TC_H$ = visiting price for Hai-Bar NR
$VI_G$ = number of visitors to Gamla NR
$VI_H$ = number of visitors to Hai-Bar NR

### *4.1.2 Calculating the value of viewing vultures*

The value of viewing vultures was calculated according to the relative importance visitors attributed to this experience and was found to be between 85 per cent and 92 per cent for Gamla and Hai-Bar NRs. Based

[3] Other functional forms were tested as well: log-linear, linear-log (exponential), log-log and reciprocal. Because we use linear approximation in section 5, we keep error consistency by reporting only the linear case.

on that, we can extract the total value of viewing vultures at both sites, as given in equations (3) and (4) for Gamla and Hai-Bar respectively.

$$TBenefit_G = \sum TC_G = 293\,(VI_G) - 0.00185\,(VI_G)^2 \qquad (3)$$

$$TBenefit_H = \sum TC_H = 506\,(VI_H) - 0.00665\,(VI_H)^2 \qquad (4)$$

The values of the sites as related to viewing vultures are 11.76 M. NIS and 9.84 M. NIS for Gamla and Hai-Bar NRs respectively.[4,5]

### *4.2 CVM*

We used CVM questionnaires to estimate the total value of viewing and protecting vultures. We undertook in person interviews as recommended by the NOAA panel (Arrow *et al.* 1993). CVM questionnaires were distributed to a sample of 150 and 151 visitors at Gamla and Hai-Bar NRs respectively, with the assumption that they too attribute non-use values to viewing and protecting vultures, even if they were users while completing the questionnaire (Shechter *et al.* 1998). Furthermore, this is indeed the relevant population which is able to value the site because of their affinity and familiarity with it (Carson 2000). The questionnaire was also distributed among a sample of 150 individuals from the general population. It was assumed that the willingness to pay (WTP) of this sample would be lower than the WTP of the other two, but greater than zero.

Prior to the final form of the questionnaire, it was handed to four focus groups, which gave their feedback on the clarity and length of the questionnaire. Their distribution of WTP bids was used to formulate the payment card used in the final questionnaire (Arrow *et al.* 1993).[6] A general description of the samples is given in Table 11.3.

The WTP question was presented in three scenarios, adopting the method used by Loomis (1987) at Mono Lake and by Kontoleon and Swanson (2003) on the giant panda. In these studies respondents were presented with three levels of the environmental attribute they were asked to value. We showed people three levels of vulture population density (Figure 11.1). Since at the Gamla NR seeing vultures in the sky is a common sight, respondents were asked about their WTP to prevent their decline, whereas at Hai-Bar NR and the sample of the general population,

[4] 1$ = 4.4 NIS.

[5] In order to keep the consistency with the calculation process, this number of visitors for the value of the sites was derived from the functional form of the demand curve and not from actual data. However, there is a difference of about 15 per cent which is in the acceptable range (Bateman *et al.* 2002).

[6] A full version of the questionnaire can be obtained from the authors upon request.

Table 11.3 *CVM questionnaire – socio-economic characteristics of the three samples*

| | Gamla | Hai-Bar | General pop. |
|---|---|---|---|
| Gender | Men | Men | Men |
| Age | 26–35 | 36–45 | 46–55 |
| Origin | Israeli | Israeli | Israeli |
| Marital status | Married | Married | Married |
| No. of children | 2–3 | 2–3 | 2–3 |
| Residence | Urbanites | Urbanites | Urbanites |
| Green organisation membership | Non-members | Non-members | Non-members |
| Source of knowledge | | | from the media |
| Education | Academic | Academic | Academic |
| Income | Average | Average | Average |

Note: the table provides a description of the most frequent characteristics of the respondents in the sample

**Gamla questionnaire:**

**Stage 1** **Stage 2** **Stage 3**

**Hai-Bar questionnaire:**

**Stage 1** **Stage 2** **Stage 3**

Figure 11.1. Three-stage scenarios

where vultures are not a common sight, respondents were asked about their WTP to increase their number in the sky.

Population density was demonstrated by presenting the respondents with different numbers of soaring vultures. At the Gamla NR, the two scenarios were as follows:

*Scenario 1*: How much are you willing to pay *to prevent a move* from picture 1 to picture 2 (WTP 1)?

*Scenario 2*: How much are you willing to pay *to prevent a move* from picture 2 to picture 3 (WTP 2)?

The scenarios presented at Hai-Bar NR were:

*Scenario 1*: How much are you willing to pay *to enable a move* from picture 1 to picture 2 (WTP 1)?

*Scenario 2*: How much are you willing to pay *to enable a move* from picture 2 to picture 3 (WTP 2)?

The number of soaring vultures represents the actual population density at the site. At Gamla NR, five soaring vultures in the picture represent ninety-five vultures on site (the current situation), two depicted vultures represent thirty-eight actual vultures. At Hai-Bar NR, two soaring vultures in the picture represent five vultures on site (the current situation), while seven vultures in the picture represent eighteen vultures on site.

The purpose of using this method is twofold. First, it allows us to check whether reported values are consistent with declining marginal benefits. Second, it enables us to derive a demand function of marginal WTP with respect to number of vultures.

In addition, a regression was fitted to the WTP where the explanatory variables were income, education, age, gender, marital status and membership of a 'green' organisation. The details results are not presented here for reasons of brevity, but all coefficients had the expected sign which provides an indication of the internal consistency of our results.

#### *4.2.1 CVM results*

Summary statistics of the distribution of the reported WTP amounts for each scenario at each site are presented in Table 11.4.

As the table shows, there is a large difference between the mean and the median WTP figures. This is probably due to the non-normal distribution of the reported birds and the large number of relatively extreme results on the right-hand side of the distribution tail. The issue of mean versus median is important to public decision making, especially in democratic societies in which the outcome is based on majority rather than the mean voting.

Table 11.4 *CVM questionnaire – WTP in the three samples (in NIS)*

| | Gamla | Hai-Bar | General pop. |
|---|---|---|---|
| WTP1 | 50.72 | 41.61 | 36.77 |
| Median | 50.0 | 20.0 | 20.0 |
| Mode | 50.0 | 20.0 | 20.0 |
| Maximum | 200.0 | 150.0 | 200.0 |
| Minimum | 0.0 | 0.00 | 0.0 |
| Standard deviation | 48.17 | 39.82 | 36.7 |
| WTP2 | 66.93 | 45.19 | 37.48 |
| Median | 50.0 | 20.0 | 20.0 |
| Mode | 50.0 | 20.0 | 20.0 |
| Maximum | 300.0 | 200.0 | 150.0 |
| Minimum | 0.0 | 0.0 | 0.0 |
| Standard deviation | 61.45 | 42.79 | 36.5 |
| Total WTP | 117.65 | 86.8 | 74.25 |

Table 11.5 *CVM questionnaire – use and non-use values in the three samples (in New Israeli Sheqels, NIS)*

| | | Gamla | Hai-Bar | General pop. |
|---|---|---|---|---|
| Use value per visitor NIS | | 22.00 | 20.22 | 18.26 |
| Use value for site mil. NIS | | 2.68 | 1.0 | 40.01 |
| Non use value per visitor NIS | Existing value | 54.12 | 32.98 | 26.28 |
| | Option value | 3.41 | 6.94 | 8.53 |
| | Bequest value | 31.30 | 24.82 | 17.15 |
| Total non use value per visitor NIS | | (75.5 %) 88.83 | (74.6 %) 64.74 | (70 %) 51.96 |
| Total non use value for site mil. NIS | | 8.26 | 2.91 | 93.55 |
| Total WTP per visitor NIS* | | 117.65 | 86.7 | 74.25 |
| Total no. of visitors | | 92,700 | 45000 | 1.8 million households |
| Total value of watching vultures | | 10.94 mil. NIS | 3.91 mil. NIS | 133.6 mil. NIS |
| Didn't state vulture viewing as important ( %) | | 1.65 % | 1.76 % | 3.14 % |

The breaking down of the total value into its use and non-use components is shown in Table 11.5.

One of the most striking results from this table is that non-use value consists of about 75 per cent of the total WTP. There are a few plausible explanations for this result. One is that people at the site already exercised their use value (by being there). Hence, their immediate reaction is to declare a non-use value. However, people surveyed off site might want to ensure a possible visit there so their immediate reaction is to declare a use value.

Results of the regression analysis for Gamla, Hai-Bar and the general population samples are given in Tables 11.6, 11.7 and 11.8, while a comparison of the general impact of key variables on WTP is given in Table 11.9.

The total, average and marginal values of the vultures at the sites are given in Table 11.10. As the table shows, the total WTP for the site is 118 NIS and 87 NIS for an average visitor at Gamla and Hai-Bar NRs respectively. If we multiply this value by the number of visitors at the sites, we can derive the total value which is 10.94 M. NIS and 3.91 M. NIS for Gamla and Hai-Bar NRs respectively.

Furthermore, the marginal benefit function can be derived from these results. This was done by plotting a line through the two mean points of the change in number of vultures in the two scenarios. For example, at Gamla NR, an average respondent is willing to pay 51 NIS to prevent a decline of 60 per cent in the number of vultures (i.e. three in the picture or fifty-seven in reality). This entails that 0.894 NIS represents the value for the mean vulture between the fifty-seventh and the ninty-fifth vulture. After completing this analysis, straight-line equations that pass through these mean points were calculated. These equations represent the marginal benefit function and are presented in equations (5) and (6) for Gamla and Hai-Bar NRs respectively.

$$MB_G = 2.11 - 0.0183\,(VU_G) \tag{5}$$

$$MB_H = 9.767 - 0.547\,(VU_U) \tag{6}$$

As can be seen from both equations, the marginal benefit decreases with the number of vultures. In order to calculate the total benefit of the site, we can integrate (5) and (6) to get (7) and (8) as follows:

$$TB_G = 2.11\,(VU_G) - 0.00915\,(VU_G)^2 \tag{7}$$

$$TB_H = 9.76\,(VU_H) - 0.2735\,(VU_H)^2 \tag{8}$$

The above analysis results in a total value of 10.84 M. NIS and 3.16 M. NIS for Gamla and Hai-Bar NRs respectively. This amount is very similar to the mean WTP times the number of relevant population (10.94 M. NIS and 3.91 M. NIS respectively).

In contrast to what one would expect, the value derived from the CVM is smaller than that derived from the TC analysis. One way of explaining

Table 11.6 *CVM questionnaire – regression results of Gamla*

| | WTP1 | | | | WTP2 | | | |
|---|---|---|---|---|---|---|---|---|
| | Coefficient | St. error | t-value | Sig. | Coefficient | St. error | t-value | Sig. |
| Constant | 38.860 | 35.768 | 1.086 | .280 | 25.975 | 49.663 | .523 | .602 |
| Gender | 3.435 | 8.540 | .402 | .688 | 7.059 | 11.789 | .599 | .551 |
| Age | 17.356 | 4.949 | 3.507 | .001 | 16.924 | 7.016 | 2.412 | .017 |
| Origin | 8.991 | 10.566 | .851 | .397 | −8.795 | 14.696 | −.598 | .551 |
| Marital status | −35.764 | 12.211 | −2.929 | .004 | −27.023 | 16.864 | −1.602 | .112 |
| No. of children | −6.031 | 4.097 | −1.472 | .144 | −4.560 | 5.658 | −.806 | .422 |
| Residence | 5.313 | 10.917 | .487 | .627 | 11.812 | 15.042 | .785 | .434 |
| Green organisation | −5.141 | 10.702 | −.480 | .632 | 10.633 | 15.014 | .708 | .480 |
| Education | −1.939 | 4.949 | −.392 | .696 | −1.266 | 6.951 | −.182 | .856 |
| | | $R^2 = 0.215$ | | | | $R^2 = 0.90$ | | |
| | | F = 3.595 | | | | F = 1.272 | | |
| | | Sig. = 0.001 | | | | Sig. = 0.260 | | |

Dependent variable: WTP

Table 11.7 *CVM questionnaire – regression results of Hai-Bar*

| | WTP1 | | | | WTP2 | | | |
|---|---|---|---|---|---|---|---|---|
| | Coefficient | St. error | t-value | Sig. | Coefficient | St. error | t-value | Sig. |
| Constant | 19.039 | 45.242 | .421 | .675 | 69.887 | 47.629 | 1.467 | .145 |
| Gender | −7.488 | 7.690 | −.974 | .332 | −7.639 | 8.034 | −.951 | .344 |
| Age | 11.963 | 4.553 | 2.627 | .010 | 13.373 | 4.812 | 2.779 | .006 |
| Origin | −14.879 | 9.404 | −1.582 | .117 | −17.635 | 9.892 | −1.783 | .077 |
| Marital status | 34.231 | 15.432 | 2.218 | .029 | 37.357 | 16.277 | 2.295 | .024 |
| No. of children | −2.254 | 4.960 | −.454 | .650 | 1.743 | 4.954 | .352 | .726 |
| Residence | −15.078 | 11.105 | −1.358 | .177 | −26.257 | 11.834 | −2.219 | .029 |
| Green organisation | −3.134 | 9.395 | −.334 | .739 | −12.399 | 9.836 | −1.261 | .210 |
| Education | −2.979 | 5.433 | −.548 | .585 | −5.531 | 5.738 | −.964 | .337 |
| Income | 6.353 | 5.729 | 1.109 | .270 | −3.443 | 5.896 | −.584 | .560 |
| | | $R^2 = 0.22$ | | | | $R^2 = 0.28$ | | |
| | | F = 14.57 | | | | F = 20.10 | | |
| | | Sig. = 0.017 | | | | Sig. = 0.045 | | |

Dependent variable: WTP

Table 11.8 *CVM questionnaire – regression results of general population*

| | WTP1 | | | | WTP2 | | | |
|---|---|---|---|---|---|---|---|---|
| | Coefficient | St. error | t-value | Sig. | Coefficient | St. error | t-value | Sig. |
| Constant | −64.839 | 43.370 | −1.495 | .138 | −57.676 | 48.814 | −1.182 | .241 |
| Gender | −2.555 | 7.258 | −.352 | .726 | −8.661 | 8.254 | −1.049 | .297 |
| Age | 5.012 | 3.613 | 1.387 | .169 | 4.848 | 4.352 | 1.114 | .268 |
| Origin | −10.029 | 9.115 | −1.100 | .274 | −5.971 | 10.300 | −.580 | .564 |
| Marital status | 6.140 | 9.005 | .682 | .497 | 12.493 | 10.074 | 1.240 | .218 |
| No. of children | −5.304 | 3.954 | −1.341 | .183 | −2.159 | 4.535 | −.476 | .635 |
| Residence | 12.928 | 8.986 | 1.439 | .154 | 5.525 | 10.205 | .541 | .590 |
| Green organisation | 17.535 | 15.051 | 1.165 | .247 | 21.814 | 16.891 | 1.291 | .200 |
| Source of knowledge | 21.852 | 6.166 | 3.544 | .001 | 21.578 | 7.029 | 3.070 | .003 |
| Education | 2.854 | 4.575 | .624 | .534 | .402 | 5.097 | .079 | .937 |
| Income | 2.287 | 5.490 | .417 | .678 | .869 | 6.190 | .140 | .889 |
| | | $R^2 = 0.35$ | | | | $R^2 = 0.29$ | | |
| | | F = 3.788 | | | | F = 4.554 | | |
| | | Sig. = 0.028 | | | | Sig. = 0.049 | | |

Dependent variable: WTP

Table 11.9 *CVM questionnaire – comparing socio-economic variables of the three samples and their relation to WTP*

| | Gamla | | Hai-Bar | | General pop. | |
|---|---|---|---|---|---|---|
| | Significance | Relation to WTP | Significance | Relation to WTP | Significance | Relation to WTP |
| Gender | | Women have a higher WTP | | Men have a higher WTP | | Men have a higher WTP |
| Age | significant | Older people have a higher WTP | significant | Older people have a higher WTP | | Older people have a higher WTP |
| Origin | | Non-Israelis have a higher WTP | significant | Israelis have a higher WTP | | Israelis have a higher WTP |
| Marital status | significant | Not married have a higher WTP | significant | Not married have a higher WTP | | Not married have a higher WTP |
| No. of children | | Few children have a higher WTP | | Few children have a higher WTP | | Few children have a higher WTP |
| Residence | | Rural have a higher WTP | significant | Urbanites have a higher WTP | significant | Rural have a higher WTP |
| Green organisation | | Non-green have a higher WTP | | Non-green have a higher WTP | significant | Non-green have a higher WTP |
| Source of knowledge | | | | | | Previous knowledge associated with higher WTP |
| Education | | Lower WTP | | Lower WTP | | Lower WTP |
| Income | significant | Higher WTP | | Higher WTP | | Higher WTP |

Table 11.10 *The value of the marginal vulture at each site (in NIS)*

| | Gamla | Hai-Bar | General population |
|---|---|---|---|
| Scenario 1 | From 5 to 2 | From 0 to 2 | From 0 to 2 |
| WTP 1 | 51 | 42 | 37 |
| Scenario 2 | From 2 to 0 | From 2 to 7 | From 2 to 5 |
| WTP 2 | 67 | 45 | 38 |
| Total WTP | 118 | 87 | 75 |
| No. of vultures on site | 95 | 5 | 350 |
| Annual no. of visitors | 92,700 | 45,000 | 1.8 M. households |
| Total value of site | 10.94 M. | 3.91 M. | 135 M. |
| Value of average vulture | 115,000 | 783,000 | 385,714 |
| WTP per vulture | 0.37 | 7.03 | 0.13 |
| Value of marginal vulture | 34,438 | 316,440 | 244,800 |

this is the fact that respondents were approached at the site after actually having incurred the costs associated with visiting the site, a fact that might have influenced their responses.

## 5 Break-even point, cost-benefit and other policy implications

### 5.1 *Costs of feeding stations*

The information on the costs of establishing and operating feeding stations was taken from the financial reports of the Israel Nature and National Parks Authorities (Hatzofe 2003). The annual operating cost of a feeding station (amortised fixed costs plus variable costs) is estimated to be 73,000 NIS.

In order to perform a cost-benefit analysis, we should know how many vultures can be added to the population as a result of operating a feeding station in the area. However, this information is not available at the moment.[7] Therefore, we can instead focus on the number of vultures that the feeding station should add to the population in order for it to cover its costs. As seen from Table 11.10, we can estimate the value of the marginal vulture. Dividing the annual cost of feeding stations by this value gives us the break-even point. Table 11.11 presents the break-even point under four scenarios. Total and use values are calculated only for the mean and median willingness to pay. As can be seen from the table at Gamla, the break-even point ranges from 2–9 vultures annually. At Hai-Bar it ranges from 0.2–2 vultures.

[7] This is a topic for further biological research.

Table 11.11 *Break-even point under different scenarios (number of vultures)*

| | Mean | | Median | |
|---|---|---|---|---|
| Payment site | Total value | Use value only | Total value | Use value only |
| Gamla | 2.12 | 8.65 | 2.49 | 10.17 |
| Hai-Bar | 0.23 | 0.91 | 0.50 | 1.97 |

Table 11.12 *Cost-benefit ratios (CBR) under the different scenarios*

| | Mean | | Median | |
|---|---|---|---|---|
| CBR site | Total value | Use value only | Total value | Use value only |
| Gamla | 1.32 | 0.34 | 1.13 | 0.28 |
| Hai-Bar | 12.21 | 3.09 | 5.62 | 1.43 |
| General population | 9.42 | 2.82 | 5.07 | 1.52 |

### 5.2 *Cost-benefit*

The benefit of the entire vulture population in Israel could be estimated through a cost-benefit analysis of a national project to increase their number. This project is called 'Porsim Kanaf', a conservation project on the major birds of prey in Israel which has been operating since 1996. During the first five years of the project, the number of breeding couples increased from 70 to 140 (Bahat *et al.* 2002). The total budget of the project was estimated to be 3.7 M. NIS. This is equivalent to a cost of 26,000 NIS per additional vulture. Based on that, we can extrapolate on the marginal benefits received by the four scenarios in Table 11.11 to get a cost-benefit ratio for each of them. These results are presented in Table 11.12.

As the table illustrates, only at Gamla and only under the use value criterion is the ratio smaller than 1.

### 5.3 *Some policy implications of valuation techniques*

The most evident policy implication of valuing endangered species is the cost-effectiveness of preserving them, as shown earlier. However, combining TC and CV techniques can serve as a basis for further policy decision making such as (i) optimal revenue consideration and (ii) allocating conservation efforts among competing sites.

### *5.3.1 Revenue estimation*

Revenue or profit consideration is a vital part of nature organisations. In times of budgetary cutbacks, the government will give higher priorities to more pressing issues even though nature conservation can produce positive net benefit results. In this case, the alternative is to manage the given nature reserve as a commercial entity, at least to complement declining government funds. Sometimes the situation is even more complicated. In Israel, for example, some of the sites are open to the public free of charge while others charge a fixed price without taking into consideration the demand for the site. In such situations, a cross subsidy is the only (or main) source of financing the free sites.

TC studies can be of use if the nature conservation authorities would want to consider some flexibility with respect to their pricing policies. We illustrate this kind of analysis for the two vulture colonies in Gamla and Hai-Bar NRs.

There are two extreme options the nature conservation authorities can implement. They can either allow visitors free of charge or charge an entrance fee in order to maximise revenues. There are various in-between pricing possibilities such as covering operation costs or pricing in a way that the number of visitors will not pass some critical ecological threshold. In this study we considered the following options:

- current situation
- maximise revenues
- charge entrance fee such that total operating costs would be covered.

Based on equations (1)–(4), we can calculate the characteristics of each of the four scenarios. Results are presented in Table 11.13. The current entrance fee in all nature sites which charge a price is 18 NIS. Table 11.13 shows the number of visitors as well as revenues and benefit associated with scenario 1 (current situation).[8] Since we are dealing with linear functions, revenues are maximised at the point where the marginal revenue is equal to zero. With respect to scenario 2 (revenue maximisation), entrance fees should be raised by a significant amount (814 per cent at Gamla and 1,406 per cent at Hai-Bar) so that total revenues (across both sites) will go up from 2.01 M. NIS to 10.67 M. NIS (an increase of

[8] All calculations are done on the estimated demand function so there is a slight change between the reported results and the actual ones. It was decided to keep to the functional form even in the current situation in order to be consistent with the estimation error in the TCM demand functions.

Table 11.13 *Pricing, revenues and welfare*

| Site scenario | Gamla | Hai-Bar |
|---|---|---|
| 1. Current situation: | | |
| Visitors (000) | 74 | 38 |
| Price (NIS) | 18 | 18 |
| Revenue (M. NIS) | 1.33 | 0.68 |
| Welfare (M. NIS) | 11.76 | 9.84 |
| 2. Maximum revenue: | | |
| Visitors (000) | 40 | 19 |
| Price (NIS) | 146.5 | 253 |
| Revenue (M. NIS) | 5.86 | 4.81 |
| Welfare (M. NIS) | 8.69 | 7.27 |
| 3. Zero profit: | | |
| Visitors (000) | 76 | 37 |
| Price (NIS) | 13 | 20 |
| Revenue (M. NIS) | 1 | 0.74 |
| Welfare (M. NIS) | 12.12 | 8.96 |

531 per cent). However, welfare measures (aggregated across both sites) go down from 21.6 M. NIS to 15.96 M. NIS at the most (a decrease of 26 per cent). Finally, if the goal is to cover costs, we can see in scenario 3 that entrance fees should be lowered slightly at Gamla and modestly raised at Hai-Bar.

The tradeoff between maximum revenue and maximum welfare can be shown graphically (Figure 11.2) where the two tradeoff functions for Gamla and Hai-Bar NRs are shown. The horizontal axis represents welfare while the vertical axis shows revenues (both in M. NIS). The first two scenarios are also shown. Note that the relevant area in which policy-makers need to make a decision is at the region where the function is downward sloping. Though raising the admission price to its revenue-maximising level may be impractical, choosing a price between the current low level and the revenue-maximising level may prove a viable policy compromise.

### *5.3.2 Species–visitors interaction*

Deriving the demand function for the marginal vulture can allow us to address further policy questions. Some of these questions include: what is the optimal investment policy in two competing sites with and without budget limitations? How should we take into account the interaction

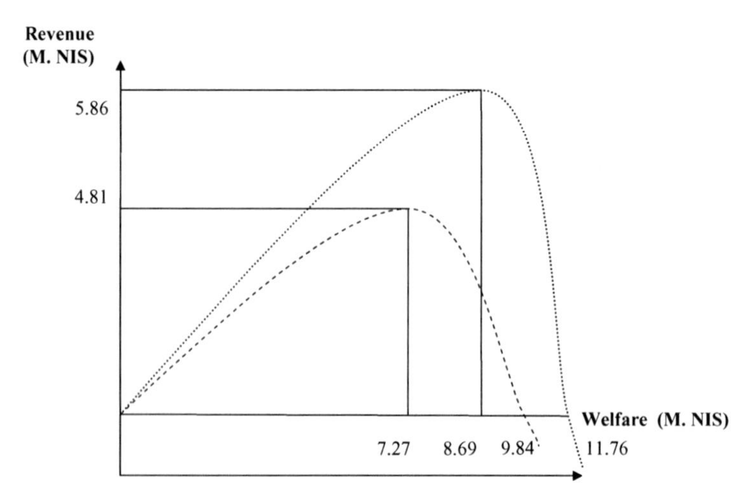


Figure 11.2. Revenues versus welfare in two alternatives

between the quantity of vultures at a given site and the number of visitors associated with that quantity?

Vultures fly over long distances in search of food, so a well-managed regional effort allocation could shift vultures to more desirable sites.[9] Visitors are also affected by the number of Griffon vultures in a given colony apart from the pricing approach implemented in the nature reserve. However, in this section we assume that revenue maximisation is not a viable option and the sites should be open to the public and financed from the central government like other public goods.

Based on the marginal and total function for number of Griffon vultures at each site presented in equations (5)–(8), we can calculate the benefits of the four scenarios presented in Table 11.14. All scenarios are based on the CV equations (and not TC functions) in order to keep the same size of error. Furthermore, CV functions include number of vultures while TC functions include number of visitors. In the last scenario, however, we will combine the two functions.

Scenario 1 refers to the current situation. The total value of vultures at each site is 10.84 M. NIS and 3.16 M. NIS for Gamla and Hai-Bar NRs respectively. Scenario 2 presents the 'no restriction case'. That is, by how much should we increase the population at the two sites if the only

[9] It should be stressed that we still have a long way to go in order to understand these dynamics so this kind of interaction can serve only as a method of thinking. Furthermore, we are still far away from knowing the marginal costs of preserving vultures at each site and in particular whether they are equal or not.

Table 11.14 *Summary of scenarios for species–visitors interaction*

| Site scenario | Gamla | Hai-Bar |
|---|---|---|
| 1. Current situation: | | |
| Visitors (000) | 79 | 38 |
| Vultures | 95 | 10 |
| Benefit (M. NIS) | 10.844 | 3.164 |
| 2. No restriction scenario: | | |
| Visitors (000) | 79 | 38 |
| Vultures | 115 | 18 |
| Benefit (M. NIS) | 11.19 | 3.92 |
| 3. Max. welfare subject to same number of vultures for total number of visitors: | | |
| Visitors (000) | 79 | 38 |
| Vultures | 89 | 16 |
| Benefit (M. NIS) | 10.60 | 3.91 |
| 4. Max. welfare with change in number of vultures *and* number of visitors: | | |
| Visitors (000) | 71 | 52 |
| Vultures | 94 | 11 |
| Benefit (M. NIS) | 10.79 | 8.37 |

criterion is to equate marginal benefits to zero (assuming a negligible marginal cost)? This entails an increase of ten (11 per cent) and eight (80 per cent) vultures at Gamla and Hai-Bar NRs respectively. It should be noted that the number of visitors in this scenario remains constant relative to the baseline current situation. The increase in the total benefit of the two sites is 3.2 per cent and 24.4 per cent respectively. This is due to the already low marginal benefit of the marginal vulture. There are, of course, other reasons for further increasing the number of vultures (e.g. insurance policy) but these are ignored here.

Scenario 3 describes the case where the number of visitors is fixed at the base-line level and for a representative visitor we equate the marginal benefit of vultures at the two sites and then multiply the result by the current number of visitors. This was accomplished by equating the marginal benefit between the two sites subject to the constraint of an aggregate number of 105 vultures across both sites. The main result here is that more effort should be allocated towards the Hai-Bar NR rather than the Gamla NR. By reshifting six vultures we would be able to increase the total welfare of both sites from 14 M. NIS to 14.51 M. NIS (about 4 per cent).[10] This scenario can be used also if we want to analyse the

[10] The underlying assumption here is that the marginal cost of preserving one vulture is equal between the two sites. If not, we should equate each marginal benefit function to its

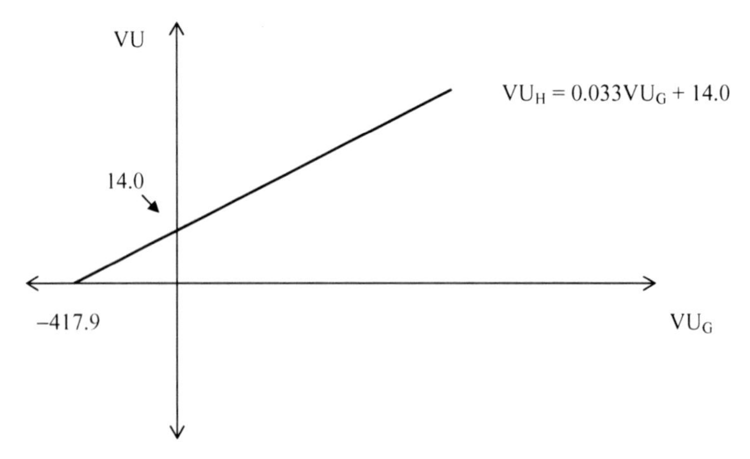


Figure 11.3. Expansion path

optimal investment path in nature conservation by not limiting ourselves to a given number of vultures. Deriving the expansion path can prove useful both in cases where policy-makers are concerned with increasing the population and in cases where there is a decline in their number and decisions have to be made as to how to prioritise conservation funds from a limited budget. Solving for the expansion path by equating the marginal benefits at both sites without a constraint on the number of vultures results in equation (9):

$$VU_H = 14.0 + 0.033\,(VU_G) \tag{9}$$

This equation is also illustrated in Figure 11.3.

As we can see, there is an absolute priority to invest in Hai-Bar NR until we reach a stable population of fourteen vultures. From there on, there is a linear relationship of 1:30 for vulture conservation at Gamla NR vs. Hai-Bar NR.

Finally, the last scenario (scenario 4) is the most flexible one. Here we allocate conservation efforts while taking into account the impact on the number of visitors. There is an interaction between TCM and CVM functions of each site which should be utilised. The linkage is given by the fact that if we invest in vulture conservation, we increase the welfare of a representative visitor. This was made known from equations (2) and (6). If we substitute the benefit difference in the travel cost equation we can get the impact on the number of visitors. This number can now serve

own marginal cost. In fact, the marginal cost at Hai-Bar is greater than the one at Gamla, which points out that allocating six more vultures to Hai-Bar NR is an overestimation had we known the true marginal cost function at each site.

as the basis for a new summation when calculating the marginal benefit of an additional vulture and multiplying by the (new) number of visitors.

For a representative site, the total benefit difference between the current and optimal number of vultures is given by:

$$\Delta TB_i = [a(VU_i^*) - b(VU_i^*)^2] - [a(\bar{VU}_i) - b(\bar{VU}_i)^2] \tag{10}$$

where: $VU_i^*$ = optimal number of vultures

$\bar{VU}_i$ = current number of vultures

and $a$ and $b$ are the marginal benefit function parameters.

In order to apply this value to the representative visitor, we insert the result of equation (10) into the demand function derived from the TC analysis and we obtain

$$VI_i = e - f(\Delta TB_i) \tag{11}$$

where $e$ and $f$ are the demand function coefficients of the TC function and the cost difference is the LHS of equation 10 (assuming as before that admission was free).

Applying the model to Gamla and Hai-Bar NRs, we see that at Gamla NR the decline of vultures by 1 per cent reduces the number of visitors by 10 per cent. This, however, is compensated by increasing the number of vultures at Hai-Bar NR by 10 per cent which increases the number of visitors by 27 per cent. It is obvious that since Hai-Bar NR is closer to metropolitan areas, this would be an expected result. The overall benefit from the two sites combined is now 19.16 M. NIS, which is the largest figure obtained from all scenarios considered.

## 6 Policy and conclusions

Valuing endangered species usually requires decisions to be made on the margin. This chapter reports on the results of a study that uses TC and CV techniques in order to estimate the value of the marginal species – in this case the Griffon vulture in Israel. A cost-benefit analysis was also carried out both on the regional level with respect to assessing the welfare implications of one particular conservation activity (namely feeding stations) and on the national level with respect to assessing broader vulture conservation policy options. The former of these analyses was undertaken by comparing the costs of feeding stations to the estimated value of the marginal vulture derived from the two valuation methods. It was found that protecting vultures passes a national cost-benefit test and that feeding stations are economically viable in generating on average 0.23–2.12 vultures annually.

In the latter analysis two additional policy issues were analysed: entrance fee policy and effort allocation. With respect to pricing policy it was shown that by charging the revenue-maximising entry fee level, policy-makers can generate a large increase in revenues compared with the current situation, but at the same time this will bring about a substantial welfare reduction. The region where the tradeoff between the revenues and welfare is relevant to the policy-makers was identified.

Further, effort allocation was shown to be important, especially when a change in population size can bring different numbers of visitors. It was shown how combining CVM and TCM results can provide insights as to the overall optimal allocation of vultures and visitors between the two competing sites analysed in the study.

In sum, the chapter makes the following broader policy contributions. First, there are solid economic arguments to invest in 'charismatic' wildlife species, even if their population size is above its critical survival threshold level. Second, assuming we have good ecological appreciation of the cost-effectiveness of feeding stations, there is a welfare-enhancing rationale for differentiating efforts among different ecotourism sites. Third, pricing mechanisms can be used as a management tool for decision makers in order to achieve different goals. Finally, our research highlights the importance of creating a comprehensive database of critical survival threshold levels of different species which would allow wildlife policy decisions to be made at the margin which will lead to a more efficient allocation of conservation funds and efforts.

## REFERENCES

Arrow, K., Solow, R., Portney, P. R., Leamer, E. E., Radner, R. and Schuman, H. 1993. *Report of the NOAA Panel on Contingent Valuation*. Washington, DC: Resources for the Future.

Bahat, O. 1995. Physiological adaptations and foraging ecology of an obligatory carrion eater – the Griffon vulture (*Gyps fulvus*). Ph.D. dissertation. Tel-Aviv University, Tel-Aviv.

Bahat, O., Hatzofe, O. and Perevolotsky, A. 2002. 'Porsim Kanaf' – protecting the vultures and raptors. A report for the first five years, 1996–2001. Tel Aviv, Israel: Israel Nature and Parks Authority, The Society for the Protection of Nature in Israel and Israel Electric Company (in Hebrew).

Bateman, I. J., Carson, R. T., Day, B., Hanemann, M., Hanley, N., Hett, T., Jones-Lee, M., Loomes, G., Mourato, S., Ozdemiroglu, E., Pearce, D. W., Sugden, R. and Swanson, J. 2002. *Economic Valuation in Stated Preference Techniques: A Manual*. Cheltenham: Edward Elgar and Northampton, MA: USA.

Ben-noon, G., Eshed, O., Court, L., Hatzofe, O. and Bahat, O. 2003. Calcium provision to Griffon vulture chicks as part of a management scheme implemented at Gamla Nature Reserve. In abstract of presentations, 6th World Conference on Birds of Prey and Owls. Budapest. May.

Brown, G. Jr. 1993. The viewing value of elephants. 146–155. In E. B. Barbier (ed.). *Economics and Ecology*. London: Chapman & Hall.

Bulte, E. and Van Kooten, G. C. 1999. Marginal valuation of charismatic species: implications for conservation. *Environmental and Resource Economics*. **14** (1). 119–130.

Bulte, E. and Van Kooten, G. C. 2000. Economic science, endangered species and biological loss. *Conservation Biology*. **14** (1). 113–119.

Butchart, D. 1988. Give a bird a bone. *African Wildlife*. **42**. 316–322.

Carson, R. T. *et al.* 1996. Contingent valuation and revealed preferences methodologies: comparing the estimates for quasi-public good. *Land Economics*. **72**. 80–200.

Carson, R. T. 2000. Contingent valuation: a user's guide. *Environmental Science and Technology*. **34**. 1413–1418.

Cramp, S. and Simmons, K. E. L. (eds.). 1980. *Handbook of the Birds of Europe, the Middle East and North Africa: The Birds of the Western Palearctic. Vol. 2: Hawks to Bustards*. Oxford: Oxford University Press.

Dolev, A. and Perevolotsky, A. (eds.). 2002. *Endangered Species in Israel, Red List of Threatened Animals, Vertebrates*. Tel Aviv, Israel: Nature and Parks Authority and the Society for the Preservation of Nature (in Hebrew).

González-Cabán, A., Loomis, J. B., Griffin, D., Wu, E., McCollum, D., Mckeever, J. and Freeman, D. 2003. Economic value of big game habitat production from natural and prescribed fire. Research Paper PSW – RP – 249. Berkley, CA: USDA Forest Service, Pacific South West Research Station.

Herath, G. and Kennedy, J. 2004. Estimating the economic value of Mount Buffalo National Park with the travel cost and contingent valuation models. *Tourism Economics*. **10** (1). 63–78.

Houston, D. C. 1978. The effect of food quality on breeding strategy in griffon vultures. *Journal of Zoology. London* **186**. 175–184.

Hatzofe, O. 2003. Financial report. Tel Aviv, Israel: Israel Nature and Parks Authority. Internal document (in Hebrew).

Jakobsson, K. M. and Dragun, A. K. 2001. The worth of a possum: valuing species with the contingent valuation method. *Environmental and Resource Economics*. **19**. 211–227.

Kontoleon, A. and Swanson, T. 2003. The WTP for property rights for the giant panda: can a charismatic species be an instrument for conservation of natural habitat? *Land Economics*. **79**. 483–499.

Lehman, R. N., Ansel, A. R., Garret, M. G., Miller, A. D. and Olendorff, R. R. 1999. Suggested practices for raptor protection on power lines: the American story. 125–144. In M. Ferrer and G. F. E. Janss (eds.). *Birds and Power Lines*. Madrid: Quercus.

Loomis, J. B. 1987. Balancing public trust resources of Mono Lake and Los Angeles' water right: an economic approach. *Water Resource Research*. **23**. 1149–1456.

Mendelssohn, H. 1972. The impact of pesticides on bird life in Israel. *ICBP Bull.* **11**. 75–104.

Mendelssohn, H. and Leshem, Y. 1983. The status and conservation of vultures in Israel. 86–98. In S. R. Wilbur and J. A. Jackson (eds.). *Vulture Biology and Management*. Berkeley, CA: University of California Press.

Mendes, I. 2000. Pricing recreation use of national parks for more efficient nature conservation: an application to the Portuguese case. *European Environment*. **13** (5). 288–302.

Mercer, E., Kramer, R. and Sharma, N. 1995. Rain forest tourism – estimating the benefits of tourism development in a new national park in Madagascar. *Journal of Forest Economics*. **1** (2). 239–270.

Mundy, P., Butchart, D., Ledger, J. and Piper, S. 1992. *The Vultures of Africa*. Randburg, South Africa: Accorn Books and Russel Friedman Books.

Navrud, S. and Mungatana, E. D. 1994. Environmental valuation in developing countries: the recreational value of wildlife viewing. *Ecological Economics*. **11**. 135–151.

Paulrod, A. 2004. Economic valuation of sport fishing in Sweden. Ph.D. Thesis submitted to the Swedish University of Agricultural Sciences. Uppsala.

Richardson, P. R. K., Mundy, P. J. and Plug, I. 1986. Bone crushing carnivores and their significance to osteodystrophy in griffon vulture chicks. *Journal of Zoology*. **210** (15). 23–43.

Shechter, M. Reiser, B. and Zaitzev, N. 1998. Measuring passive use value. *Environmental and Resource Economics*. **12**. 457–478.

Shlosberg, A. 2001. Toxicological risks of raptors in Israel. 8–30. In A. Shlosberg and O. Bahat (eds.). Proceedings of an International Workshop on The Risk of Toxicoses from Pesticides and Pollutants in Raptors and Wildlife in Israel – The Present Situation and Recommendations for the Future. Jerusalem Biblical Zoo, Israel. April 2000. Tel-Aviv: SPNI.

Terrasse, J. F. 1985. The effects of artificial feeding on Griffon, bearded and Egyptian vultures in the Pyrenees. 429–430. In I. Newton and R. D. Chancellor (eds.). *Conservation Studies on Raptors*. Norwich: WGBPO, Pastom Press.

Tristram, H. B. 1885. *The Fauna and Flora of Palestine*. London: Palestine Exploration Fund.

Van Sickle, K. and Eagles, P. F. J. 1998. Budgets, pricing policies and user fees in Canadians parks' tourism. *Tourism Management*. **19** (3). 225–235.

Vlachos, C. G., Bakaloudis, D. E. and Holloway, G. J. 1999. Population trends of black vulture, *Aegypius monachus*, in Dadia Forest, North Eastern Greece following the establishment of a feeding station. *Bird Conservation International*. **9**. 113–118.

Wilbur, U. S., Sanford, R., Carrier, W. D. and Borneman, J. C. 1974. Supplemental feeding program for California condor. *Journal of Wildlife Management*. **38**. 43–346.

Wilbur, S. R. 1983. The status of vultures in Europe. 75–80. In S. R. Wilbur and J. A. Jackson (eds.). *Vulture Biology and Management*. Berkeley, CA: University of California Press.

Willis, K. G. 2003. Pricing public parks. *Journal of Environmental Planning and Management*. **46** (1). 3–17.

# 12 Valuing ecological and anthropocentric concepts of biodiversity: a choice experiments application[1]

*Michael Christie, Nick Hanley, John Warren, Tony Hyde, Kevin Murphy and Robert Wright*

## 1 Introduction – the challenge of valuing biodiversity

Society needs to make difficult decisions regarding its use of biological resources, for example in terms of habitat conservation, or changing how we manage farmland through agri-environmental policy (Hanley and Shogren 2002). Environmental valuation techniques can provide useful evidence to support such policies by quantifying the economic value associated with the protection of biological resources. Pearce (2001, p. 29) argues that the measurement of the economic value of biodiversity is a fundamental step towards its conservation since 'the pressures to reduce biodiversity are so large that the chances that we will introduce incentives [for the protection of biodiversity] without demonstrating the economic value of biodiversity are much less than if we do engage in valuation'. Assigning monetary values to biodiversity is thus important since it allows the benefits associated with biodiversity to be directly compared with the economic value of alternative resource use options (Nunes and van den Bergh 2001). OECD (2001) also recognises the importance of measuring the economic value of biodiversity and identifies a wide range of uses for such values, including demonstrating the value of biodiversity, in targeting biodiversity protection within scarce budgets, and in determining damages for loss of biodiversity in liability regimes.

More generally, the role of environmental valuation methodologies in policy formulation is increasingly being recognised by policy-makers. For example, the Convention of Biological Diversity's Conference of the Parties decision IV/10 acknowledges that 'economic valuation of biodiversity and biological resources is an important tool for well-targeted and calibrated economic incentive measures' and encourages parties,

[1] We thank the Department of the Environment, Food and Rural Affairs for funding this research, and members of the steering committee for many useful comments.

governments and relevant organisations to 'take into account economic, social, cultural and ethical valuation in the development of relevant incentive measures'. The EC Environmental Integration Manual (2000) provides guidance on the theory and application of environmental economic valuation for measuring impacts to the environment for decision-making purposes. Within the UK, HM Treasury's 'Green Book' provides guidance for public sector bodies on how to incorporate non-market costs and benefits into policy evaluations.

The idea of placing economic values on the environment has been challenged by many authors on a variety of grounds, from ethical objections to participatory perspectives. However, what concerns us here is not whether one should attempt to place economic values on changes in biodiversity, but rather what are the particular difficulties of doing so. These include incommensurate values or lexicographic preference issues (Spash and Hanley 1995; Rekola 2003) and – the issue we focus on here – people's understanding of complex goods (Hanley *et al.* 1996; Christie 2001; Limburg *et al.* 2002).

Stated preference valuation methods ideally require survey respondents to make informed value judgements on the environmental goods under investigation. This requires information on these goods to be presented to respondents in a meaningful and understandable format, which in turn will enable them to express their preferences consistently and rationally. Herein lies the problem: many studies have found that members of the general public have a low awareness and poor understanding of the term 'biodiversity' (Spash and Hanley 1995; DEFRA 2002). If an individual is unaware of the characteristics of a good, then it is unlikely that he/she has well-developed preferences for it which can then be uncovered in a stated preference survey.

An additional complication is that biodiversity itself is not uniquely defined by ecologists. Scientists are in general agreement that the number of species per unit of area provides a useful starting point (Harper and Hawksworth 1995; Whittaker 1977). Although such a measure appears to be relatively straightforward, issues such as what constitutes a species and what size of area to count species over-complicate this measure. Even if these questions were resolved, ecologists recognise that some species, such as keystone species, may be more important and/or make a greater contribution to biodiversity than others. Many of these concepts are complex and it is questionable as to whether the public is capable of truly understanding these ecological concepts. Indeed, our understanding of the way in which members of the general public think about and value biodiversity is limited. We do not know whether the public understands, or is even aware of, ecological concepts such as species, habitats and ecosystems.

The issues highlighted above indicate that research that attempts to value changes in biodiversity will be challenging, since it requires the analyst to first identify which of the ecological concepts of biodiversity are considered important by the general public, and then find appropriate language in which these biodiversity concepts can be meaningfully conveyed to the public in ways which are consistent with underlying ecological ideas on what biodiversity is.

In this chapter we develop a novel approach to the valuation of biodiversity. In particular, we draw heavily on the ecological literature relating to the definition and measurement of biodiversity. This literature is then used to feed into the design of a choice experiment which examines public values of various attributes of biodiversity. In what follows, we first present an overview of the way that ecologists consider biodiversity. Next, we provide a review of some of the key studies that have attempted to value biodiversity. Finally, we present our case study research on the value of biodiversity, with particular reference to how we overcome issues of presenting complex and often new information in valuation studies.

## 2 An ecologist's perspective of biodiversity

### *2.1 Defining biodiversity*

The concept of biological diversity, originally simply meaning 'number of species present', appears to have been first developed in the sense in which it is used today during the 1970s and early 1980s (Peet 1974; Lovejoy 1980a,b; Norse and McManus 1980), despite attempts to strangle the idea at birth (Hurlbert 1971). A few years later, Norse *et al.* (1986) defined biological diversity at the genetic (within-species), species (species numbers) and ecological (community) level. The contracted term 'biodiversity' came from a 'National Forum on Biodiversity' held in the USA in 1986, the proceedings of which brought the term, and concept, into more general use (Wilson 1988).

Although there are many possible definitions, perhaps the most widely accepted is that provided in Article 2 of the 'Convention on Biological Diversity' (signed by 157 national and supra-national organisations) at the 1992 UN Conference on the Environment and Development:

> 'Biological diversity' means the variability among living organisms from all sources, including, inter alia, terrestrial, marine and other aquatic systems and the ecological complexes of which they are part; this includes diversity within species, between species and of ecosystems.

Biodiversity is frequently divided into a hierarchy of three levels: ecosystems/habitats, species and genes. Ecosystems are defined as communities of co-occurring species of plants and animals plus the physical environment; as such they are difficult to define and delimit. At the other end of the spectrum, genes are currently still difficult to identify and count. Thus, species counting is the obvious tool for measuring biodiversity. Therefore, although biodiversity may be measured at levels from genome to biome (Colwell and Coddington 1995; Roy and Foote 1997; Hawksworth 1995; Lovejoy 1995; Magurran 1988), measurement of the number of species present within a defined target area is generally accepted as one of the simplest measures of biodiversity. This, of course, raises the problem of defining the target area in which to record the inventory of species, which again requires the ability to define and delimit ecosystems. A further complicating factor relates to how to account for taxonomical variation between species (Harper and Hawksworth 1995).

For the practical reasons outlined above, ecologists most frequently describe biodiversity as some function of the number of species per unit area, even when they are interested in defining habitat, ecosystem or regional diversity. Part of the driver for this methodological approach is scientists' desire to quantify variation. However, the general public may not be motivated by the same desire and may value higher levels of diversity (e.g. habitat biodiversity) without reference to species counting. Simply counting the number of species in a given area does not, however, fully define biodiversity. In particular, both ecologists and the public might consider that some species are more important than others.

### 2.2 *Are all species equal?*

An important issue for valuing biodiversity is a reflection of the values which human beings place on the presence or absence of different organisms. People tend to assign values, consciously or unconsciously, to different organisms. The 'cuteness' concept is an obvious issue: furry and feathery organisms, and attractive plants, are preferred by most (though not all) people to poisonous snakes, weeds and the smallpox virus (May 1995). Closely related to the cuteness concept is that of flagship species or charismatic species. These are high-profile, impressive species (such as top predators), or species linked to local identity such as national birds or plants (Noss 1990). Species which possess characteristics which humans value (such as speed) tend to be regarded in higher esteem than species that do not. Although cute and charismatic species are clearly important for biodiversity in terms of human values, there appears to be no scientific indicator or measure of the cuteness or charisma of a species and thus it is difficult to incorporate such attributes into ecological measures of biodiversity.

Rarity (on whatever scale) is a second attribute which contributes to the assigned 'value' of species within the biodiversity of an ecosystem or habitat. This concept is inherent in the wide range of active management measures in place for conservation (e.g. Biodiversity Action Plans (BAPs), Environmentally Sensitive Areas etc.: see Potter 1988; Robinson 1994; Brotherton 1996; Simpson *et al.* 1996) and their driving policy measures worldwide (e.g. Article 19 of the EC Structure Regulation 797/85 for ESAs; EC Species Directive etc.). The issue often reflects basically irreconcilable structures (such as political boundaries versus natural distributions of organisms), and is commonly allied to public pressure for conservation of preferred 'rare' organisms. There are, however, well-recognised difficulties in using rarity as a measure of value in biodiversity assessment (McIntyre 1992). Furthermore, species may be rare for a variety of different reasons, not all necessarily deserving of higher conservation status. For example, newly evolved species are likely to be rare by definition, and many such species are likely to fail to become established, but does this matter? Species with very exacting habitat requirements or those at high trophic levels are unlikely ever to have been abundant, but should they be awarded the same conservation priority as formerly common species that have recently become rare at the hands of mankind?

The principal issue is whether it is desirable to assign weightings to individual species (or groups of species) which reflect their perceived value to people, and, if so, how to do this, preferably in a quantitative manner. There have been several attempts to assign weightings to rare species, usually in the context of assessing the 'conservation value' of a site for practical management terms. Examples include Dony and Denholm (1985) for small woodland sites in southern England and Ali *et al.* (2000) for desert vegetation in the Eastern Sahara. Such schemes usually incorporate some estimate of weighting for the rarer species based on their frequency of occurrence across a defined part of the planet's surface (whether on a local or broader scale).

Another useful and practical approach to account for rarity is to make an assessment of the likely threat that a species will become extinct. Such a hierarchy of threat of extinction is currently used in the IUCN red data list, which identifies five levels of endangerment: extinct, extinct in the wild, critically endangered, endangered and vulnerable.

### 2.3 *Do certain species contribute more than others to the biodiversity of an area?*

Thus far, we have argued that species richness appears to be the most useful practical measure of biodiversity, and that the general public's preferences for individual species may be influenced by charismatic/anthropocentric factors such as cuteness or rarity. Such factors, however, have little meaning in terms of an ecologist's perception of the importance of a species. Ecologists have identified certain 'keystone' species which make significant contribution to enhancing the biodiversity of an area.

The value of certain species in influencing (positively or negatively) habitat or resource provision for other species is a primary consideration here. This can occur directly, as in the case of British oaks, which provide resources assisting the survival of a great range of other organisms, including food for a range of herbivores; nest and feeding sites for Lepidoptera, birds and arboreal mammals; habitat for obligate oak-associates such as gall-wasps; mycorrhizal fungi; bark- and leaf-dwelling fungi; a range of pathogens; and epiphytic bryophytes and lichens, to name but a few (Morris and Perring 1974). Alternatively, keystone species may be less abundant species from higher up in the food chain. The classic example is the North American sea otter. After being hunted to the edge of extinction in the nineteenth century, there was a dramatic increase in the sea urchin population (a major component of the otter's diet) which in turn resulted in the disappearance of kelp forests along the American west coast. Thus, such keystone species are thought to be pivotal species about which the diversity of a large part of the community depends. However, that is not to say the community itself will cease to function and become unrecognisable following the loss of a keystone species. Indeed, the UK National Vegetation Classification system (Rodwell 1991) recognises oak woodland communities even in the absence of oak trees! Allied to the keystone species concept is that of the ecological indicator species (Noss 1990). Such species are easy to monitor and variations in their numbers are used to indicate that an environmental change has occurred that is likely to have produced perturbations in the population of several other species with similar habitat requirements.

While the above classification of species as keystone species are based entirely on the theory of ecosystem functioning, ecologists also classify species in terms of their potential importance in conservation. Such species are therefore identified in part by their perceived ability to attract human interest. For example, the terms umbrella species and flagship species are used to describe two related concepts which describe a species' potential impact in promoting conservation. Umbrella species typically require large areas of habitat for their conservation. These are typically large mammals or birds, which need a variety of habitat types or alternatively require large blocks of a single habitat. Thus promoting the conservation of such species (which almost by definition tend to be charismatic mega fauna) also automatically promotes the conservation of large tracts of habitat plus all the other species that share this resource.

The above provides a brief overview of the way that ecologists think about and measure biodiversity. What is clear from this review is that (i) biodiversity is a highly complex issue and (ii) ecologists themselves are not in full agreement with regard to how best to define and measure biodiversity. The choice experiment valuation study reported later in this chapter aims to value some of the key ecological and anthropocentric concepts of biodiversity outlined above. However, before discussing our research, it is first useful to review existing research on the economic value of biodiversity.

## 3 Existing research on the economic value of biodiversity

A general comment on much of the existing biodiversity valuation literature is that it mostly does not value diversity itself, but focuses rather on individual species and habitats (Pearce 2001). In this section we review a number of key studies that have attempted to measure the economic value of different elements of biodiversity. In particular, we distinguish between studies that have valued biological resources (e.g. a particular species, habitat area or ecosystem function) and those which have valued some of the ecological and anthropocentric concepts of biological diversity (e.g. components of biodiversity such as rarity or charismatic species).

### *3.1 The economic value of biological resources*

Studies that have valued biological resources may generally be categorised into four sub-groups of biological resources: genetic, species, habitat and ecosystem functions.

#### *3.1.1 Genetic diversity*

Studies that have quantified genetic diversity have predominantly measured direct-use benefits of biological resources in terms of inputs to the production of market goods such as new pharmaceutical and agricultural products. The majority of studies have based valuations on market contracts and agreements for bio-prospecting by pharmaceutical industries. ten Kate and Laird (1999) provide an extensive review of such bio-prospecting agreements. Franks (1999) provides a useful contribution on the value of plant genetic resources for food and agriculture in the

Table 12.1 *Summary of studies that have valued biological resources*

| Life diversity level | Biodiversity value type | Value ranges | Method(s) selected |
|---|---|---|---|
| Genetic diversity | Bio-prospecting | From $175 000 to $3.2 million | Market contracts |
| Species diversity | Single species | From $5 to $126 | Contingent valuation |
| | Multiple species | From $18 to $194 | Contingent valuation |
| Ecosystems and natural habitat diversity | Terrestrial habitat (passive use) | From $27 to $101 | Contingent valuation |
| | Coastal habitat (passive use) | From $9 to $51 | Contingent valuation |
| | Wetland habitat (passive use) | From $8 to $96 | Contingent valuation |
| | Natural areas habitat (recreation) | From $23 per trip to $23 million per year) | Travel cost, tourism revenues |
| Ecosystems and functional diversity | Wetland life support | From $0.4 to $1.2 million | Replacement costs |
| | Soil and wind erosion protection | Up to $454 million per year | Replacement costs, hedonic price, production function |
| | Water quality | From $35 to $661 million per year | Replacement costs, averting expenditure |

Source: Adapted from Nunes and van den Bergh (2001)

UK and also the contribution of the UK's agri-environmental schemes to the conservation of these genetic resources.

### 3.1.2 *Species diversity*

There have been a large number of studies that have valued particular species. Most of these studies have been undertaken in the US and utilise stated preference techniques, thus enabling both use and passive-use values to be assessed. Nunes and van den Bergh (2001) provide an extensive review of valuation studies that have addressed both single and multiple species. Valuations for single species range from $5 to $126, and for multiple species range form $18 to $194 (see Table 12.1). In the UK, there have been a limited number of studies that have valued both single and multiple species. For example, MacMillan *et al.* (2003) estimated the value of wild geese conservation in Scotland, while White *et al.* (1997 and 2001) examine the value associated with the conservation of UK mammals including otters, water voles, red squirrels and brown hare. MacMillan *et al.* (2001) also take a slightly different perspective by valuing the reintroduction of two species (the beaver and wolf) into native forests in Scotland.

### 3.1.3 *Habitat diversity*

Biological resources may also be described in terms of the diversity within natural habitats. Studies have addressed the valuation of habitats from two perspectives. One approach is to link the value of biodiversity to the value of protecting natural areas that have high levels of outdoor recreation or tourist demand. A second approach to the valuation of natural areas involves the use of stated preference methods. Table 12.1 summarises the range of passive-use values elicited for terrestrial, coastal and wetland habitats. UK examples of CV studies that have valued habitats include: Garrod and Willis (1994) who examined the willingness to pay of members of the Northumberland Wildlife Trust for a range of UK habitat types, Hanley and Craig (1991) who valued blanket bogs in Scotland's flow country and MacMillan and Duff (1998) who examined the public's WTP to restore native pinewood forests in Scotland.

### 3.1.4 *Ecosystem functions and services*

Ecosystem functions and services describe a wide range of life support systems including waste assimilation, flood control, soil and wind erosion and water quality. Many of these functions and services are complex and it is likely that members of the public will possess a poor understanding of these issues. The consequence of this is that attempts to value ecosystem functions and services will be difficult, particularly in methods (such as the stated preference methods) where respondents are required to make a value judgement based on the description of the good in question. Analysts often use other techniques including averting behaviour, replacement costs and production functions to measure the indirect values of ecosystem functions.

### *3.2 The economic value of biological concepts*

A number of valuation studies have attempted to value biodiversity by explicitly stating to respondents that the implementation of a conservation policy will result in an increase in the biodiversity of an area. For example, Garrod and Willis (1997) estimated passive-use values for biodiversity improvements in remote upland coniferous forests in the UK. The improvements in forest biodiversity were described in relation to a series of forest management standards that increased the proportion of broad-leaved trees planted and the area of open spaces in the forest. The marginal value of increasing biodiversity in these forests was estimated to range between £0.30 to £0.35 per household per year for a 1 per cent increase and between £10 to £11 per household per year for a 30 per cent increase in enhanced biodiversity forest area. Willis *et al.* (2003) extend this work to examine public values for biodiversity across a range of UK woodland types. Other studies have assessed public WTP to prevent a decline in biodiversity. For example, MacMillan *et al.* (1996) measure public WTP to prevent biodiversity loss associated with acid rain; whilst Pouta *et al.* (2000) estimate the value of increasing biodiversity protection in Finland through implementing the Natura 2000 programme.

White *et al.* (1997 and 2001) take a slightly different approach and examine the influence of species characteristics on WTP. They conclude that charismatic and flagship species such as the otter attracted significantly higher WTP values than less charismatic species such as the brown hare. They further suggest that species with a high charisma status are likely to command higher WTP values than less charismatic species that may be under a relatively greater threat or of more biological significance in the ecosystem. In a meta-analysis of WTP for a range of species, Loomis and White (1996) also find that more charismatic species, such as marine mammals and birds, attract higher WTP values than other species.

The above review has demonstrated that from those studies that have claimed to value biodiversity, only a handful have actually examined ecological concepts of biological diversity; most studies have alternatively tended to value biological resources. Furthermore, studies that have valued biological diversity currently only provide limited information on the value of the components of biological diversity. Research effort has yet to provide a comprehensive assessment of the value attached to the components of biological diversity such as anthropocentric measures (e.g. cuteness, charisma and rarity) and ecological measures (e.g. keystone species and flagship species). To address this knowledge gap, we designed a choice experiment study to measure the economic value of the various anthropocentric and ecological measures of biodiversity. Details of this are now presented.

## 4 Valuing biodiversity on UK farmland using choice experiments

The aim of this research was to estimate the economic value of various ecological and anthropocentric aspects of biodiversity. To achieve this, a choice experiment study was devised. The policy setting for this study research is the development of policy regarding biodiversity conservation and enhancement on farmland in England and in particular in Cambridgeshire and Northumberland.

### *4.1 What aspects of biodiversity are important to people?*

A principal challenge for the economic valuation of biodiversity is to identify which aspects of biodiversity people feel are important and then develop ways in which these ideas can effectively be communicated to the general public. For a biodiversity concept to be relevant in this context, it has to have ecological significance, be capable of being explained to ordinary people and be something which they might in principle care about. In addition, the analyst needs to design effective ways of conveying this complex and often new information.

In the case study research, a detailed review of the ecological literature on biodiversity was undertaken – see Christie *et al.* (2004) for more detail. In this review, 21 different concepts that ecologists use to describe and measure biodiversity were identified. Clearly, it would be extremely difficult (if not impossible) to attempt to value all of these concepts in a single valuation exercise. Thus, to simplify matters, a conceptual framework was drawn up to provide a simplified and structured framework in which biodiversity could meaningfully be presented to members of the public (see Figure 12.1). This framework is split into sections according to which perspective we take on the importance and meaning of biodiversity: anthropocentric or ecological. Within each of these headings, we identify different aspects of biodiversity that need to be considered for inclusion.

| BIODIVERSITY CONCEPTS | | | | | | | | | | |
|---|---|---|---|---|---|---|---|---|---|---|
| ANTHROPOCENTRIC CONCEPTS | | | | | | ECOLOGICAL CONCEPTS | | | | |
| Charismatic species | Cuteness | Familiar species | Locally important species | Endangered species | Rare species | Keystone species | Umbrella species | Flagship species | Ecosystem function | Ecosystem health |
| *Familiar species of wildlife* | | | | *Rare, unfamiliar species of wildlife* | | *Habitat quality* | | | *Ecosystem processesa* | |

Figure 12.1. Conceptual framework – biodiversity concepts

The final row of the Figure shows the biodiversity attributes that were eventually selected for the experimental design. We now explain how these were chosen.

A series of focus groups comprising members of the general public were arranged. The discussions held in the focus groups aimed to identify the level of understanding that the public had for each of the elements of the framework in Figure 12.1, and also to identify their views on the importance of each element. The framework was then amended to reflect this input from the focus groups.

The key issues identified in the developmental focus groups included:

- Over half of the participants could not remember having come across the term 'biodiversity' before. Most of those who indicated a familiarity with the term were unable to provide a clear or accurate definition of the concept.
- Participants indicated that they were familiar with related terms including 'species', 'habitat' and 'ecosystem'.
- Participants indicated that they were *not* familiar with the majority of the ecological concepts of biodiversity outlined in the third row of Figure 12.1. On a more positive note, it was also found that most participants of the focus groups appeared to be capable of quickly picking up a basic understanding of most biodiversity concepts if these were explained in layman's terms.

The conclusion from this was that a survey on the value of biodiversity would need to employ alternative, non-scientific terminology to meaningfully describe the ecological concepts associated with biodiversity. Based on this focus group evidence, four attributes were identified as being appropriate to describe the diversity of biodiversity concepts to the public:

- *Familiar species of wildlife.* This attribute includes charismatic, familiar (recognisable) and locally symbolic species, and may be considered in terms of both common and rare familiar species.

- *Rare, unfamiliar species of wildlife.* This attribute focuses on those species that are currently rare or in decline which are unlikely to be familiar to members of the public.
- *Habitat.* The protection of habitats and in particular the mix of species that reside within them was considered to be an important component of biodiversity conservation. Of note in this category was the fact that focus group participants were more concerned about achieving a biodiversity outcome (i.e. protecting the range of species within a habitat), rather than a focus on how this might be achieved (e.g. by targeting policy towards the protection of ecologically significant species such as keystone or umbrella species).
- *Ecosystem processes.* The public were also concerned with preserving the 'health' of the ecosystem processes. It was also considered useful to distinguish between ecosystem processes which have a direct impact on humans and those which do not.

### 4.2 *An innovative approach to presenting complex information in choice experiments*

A key factor affecting the validity of choice experiment studies relates to the success to which the good under investigation can be meaningfully, accurately and consistently presented to survey respondents. Although this can be a challenge in many valuation studies, the very fact that only a small proportion of the public have knowingly heard of the term 'biodiversity' before presents a significant challenge to this research. In this study, the survey instrument was required to present a lot of information on biodiversity which was likely to be complex and new to respondents. The majority of valuation studies tend to describe the environmental good under investigation using verbal descriptions, perhaps supported by some written script and / or pictorial images. Although such an approach to presenting the good can be successful with goods that are familiar to survey respondents, evidence gathered in the focus groups indicated that such a standard approach was unlikely to be suitable for presenting biodiversity which was found to be unfamiliar and considered complex. Feedback from focus groups also indicated that the large volume of new information required to be presented on biodiversity was found to lead to both confusion and respondent fatigue. The adoption of a more visual and interactive approach was therefore considered to be more suitable.

For these reasons, the survey was administered using in-person interviews in people's homes. During the interviews, information on biodiversity was conveyed to respondents using a 20-minute MS PowerPoint audio-visual presentation. This has a number of advantages in terms of

Table 12.2 *Summary of biodiversity attributes and levels used in the choice experiment*

<table>
<tr><th></th><th>POLICY LEVEL 1</th><th>POLICY LEVEL 2</th><th>DO NOTHING (Biodiversity degradation will continue)</th></tr>
<tr><td>Familiar species of wildlife</td><td>Protect rare familiar species from further decline</td><td>Protect both rare and common familiar species from further decline.</td><td>Continued decline in the populations of familiar species</td></tr>
<tr><td>Rare, unfamiliar species of wildlife</td><td>Slow down the rate of decline of rare, unfamiliar species.</td><td>Stop the decline and ensure the recovery of rare unfamiliar species</td><td>Continued decline in the populations of rare, unfamiliar species</td></tr>
<tr><td>Habitat quality</td><td>Habitat restoration, e.g. by better management of existing habitats</td><td>Habitat recreation, e.g. by creating new habitat areas</td><td>Wildlife habitats will continue to be degraded and lost</td></tr>
<tr><td>Ecosystem process</td><td>Only ecosystem services that have a direct impact on humans, e.g. flood defence are restored.</td><td>All ecosystem services are restored</td><td>Continued decline in the functioning of ecosystem processes</td></tr>
<tr><td>Annual tax increase</td><td colspan="2">£10 | £25 | £100 | £260 | £520</td><td>No increase in your tax bill</td></tr>
</table>

using a range of formats (pictures, audio tracks and text), which helps minimise respondent fatigue and maximise the effectiveness with which information is conveyed. The PowerPoint presentation introduced survey respondents to a simple definition of biodiversity; 'biodiversity . . . is the scientific term used to describe the variety of wildlife in the countryside'. The narrative that accompanied these slides provided further elaboration of this definition and provided examples to illustrate various aspects of biodiversity. A number of slides then introduced the four attributes of biodiversity that had been identified in the focus groups: familiar species of wildlife, rare unfamiliar species of wildlife, habitat quality and ecosystem services. Each attribute was defined, and the alternative levels of biodiversity enhancements associated with these attributes were introduced – see Table 12.2 for a summary of these attributes and attribute levels. Within these descriptions, named examples of relevant species, habitats and ecosystem services within the study area were provided and images presented. These were included to help respondents attain a clearer understanding of the various aspects of biodiversity being discussed. Respondents were also made aware of alternative motivations that people may have for protecting the various aspects of biodiversity. For example, respondents were reminded that they 'might recognise an individual mammal, reptile, bird or even plant because it possesses impressive features such as being large or colourful, or alternatively that it has a particular significance in local culture'.

Following the presentation of this information, respondents were provided with an opportunity to discuss and clarify with the interviewer any issues of outstanding confusion. Next, a series of slides introduced the case study area (Cambridgeshire or Northumberland). Details presented included a description of the predominant land uses found within the case study areas, and the current levels of biodiversity that exist in those areas. Respondents were then informed that human activities, such as farming and development, were currently threatening overall levels of biodiversity in the area and the consequences of this on the four biodiversity attributes were outlined. Respondents were then informed that the government could introduce policies to help protect and enhance biodiversity in the respective case study areas. Policies described included agri-environmental schemes and habitat re-creation schemes. Details of how such policies could be introduced to specifically enhance the four aspects of biodiversity identified earlier were presented. In each case, the potential improvements were described in terms of the attribute levels used in the choice experiment. Respondents were then asked to think about which aspects of biodiversity they would like to see being protected and enhanced. Finally, at the end of the presentation, respondents were given a further opportunity to clarify any issues of confusion / uncertainty regarding any aspect of the presentation.

The feedback from respondents of a pilot survey indicated that the majority of respondents understood the concepts presented. Respondents also indicated that the presentation of more information (to try to increase understanding) would likely be detrimental to the study as a whole since this would lead to respondent fatigue. Thus, the inclusion of further opportunities for respondents to discuss issues of confusion with the interviewer was seen as a better option to ensure that respondents fully understood the information presented.

### *4.3 The choice task*

Following the PowerPoint presentation, respondents were asked to complete a choice experiment exercise. The choice experiment was introduced as follows:

> In the presentation you were provided with information on different aspects of biodiversity. You were also informed that biodiversity within Cambridgeshire (Northumberland) is under threat. We as a society have some options over how we respond to the threats to biodiversity. We are therefore interested in your opinions on what action you would most like to see taken.
>
> 'We are now going to show you five alternative sets of policy designs that could be used to enhance Cambridgeshire's (Northumberland's) biodiversity. In each set, you will be asked to choose the design which you prefer.'

An example of a choice task was then presented to respondents and the choice task was explained. Once the respondents had undertaken all five choice tasks, they were asked to indicate the main reason that they had for making the choice that they did. This was to allow protest responses to be identified.

We have already explained how biodiversity attributes were selected for inclusion in the choice experiment (above). Each of these attributes was then defined according to three levels of provision, including the status quo and two levels of improvement/enhancement. Table 12.2 provides a summary of the four biodiversity attributes used in the choice experiment, along with the three levels of provision of each attribute.

In choice experiments it is common practice to include a standard option within all choice tasks. In this study, we choose a 'Do nothing' policy option. The 'Do nothing' option was designed to reflect the situation where no new policies would be implemented to protect and enhance biodiversity on farmland in the case study areas. The consequence for this option in terms of the four attributes of biodiversity was then reported as a continued decline in biodiversity in the study area.

The payment vehicle used in the choice experiment was an increase in general taxation. The reasons for using this payment vehicle include the fact that biodiversity enhancement programmes are generally paid for through taxation and also that participants of the focus groups indicated that taxation was their preferred payment option. Six payment levels of taxation were used in the choice experiment, including the £0 level in the status quo option. The actual levels used were identified following a small open-ended pilot contingent valuation study which identified the likely range of bid levels for biodiversity enhancements. These levels were then tested in a pilot choice experiment. The final tax levels used in the choice experiment were: £10, £25, £100, £260, £520, plus the no tax increase in the 'Do nothing' option. Tax rises were annual increases per household for the next five years. The SPSS software package was used to generate a ($3^4 \times 5^1$) fractional factorial orthogonal experimental design, which created 25 choice options. A blocking procedure was then used to assign the options to ten bundles of five choice sets. Thus each respondent was presented with a bundle of five choice tasks, where each choice task comprises two policy options and a status quo.

## 5 Results

In the study, 741 respondents (343 in Cambridgeshire and 398 in Northumberland) each undertook five choice tasks. The survey data were analysed using a conditional logit model (see Hensher *et al.* 2005 for a detailed explanation of this model and Christie *et al.* 2006 for specific details relating to this research). Table 12.3 shows results from the choice experiment data for both Cambridgeshire and Northumberland, based on a conditional logit model. The pseudo-Rho$^2$ value is higher for the latter sample, and is very close to the 20 per cent level suggested by Louviere *et al.* (2000) as indicating a very good fit in this kind of data. The Cambridgeshire model shows significant estimates for all the attribute parameters. In almost all cases, parameter signs are in accord with *a priori* expectations. As may be seen, improving familiar species from continued decline to either protecting rare species only ('FamRare') or protecting all species ('FamBoth') increases utility; moving habitat quality from continued decline to habitat restoration ('HabRestore') or habitat recreation ('HabRecreate') is positively valued; moving ecosystem services from continued decline to a recovery of either directly relevant services alone ('EcoHuman') or all services ('EcoAll') creates higher utility. The only exception is for rare, unfamiliar species. Here, although a

Table 12.3 *Logit models for Cambridge and Northumberland CE samples*

| Attribute | Parameter estimate | t-value |
|---|---|---|
| *Cambridgeshire* | | |
| FAMRARE | 0.126 | 2.1 |
| FAMBOTH | 0.331 | 5.2 |
| RARESLOW | −0.165 | −3 |
| RARERECOVER | 0.408 | 5.7 |
| HABRESTORE | 0.122 | 2.3 |
| HABCREATE | 0.217 | 3.5 |
| ECOHUMAN | 0.19 | 3.2 |
| ECOALL | 0.15 | 2.2 |
| PRICE | −0.004 | −15.2 |
| Pseudo $Rho^2$ | 14 % | |
| N (individuals) | 343 | |
| *Northumberland* | | |
| FAMRARE | 0.309 | 5.1 |
| FAMBOTH | 0.334 | 5.2 |
| RARESLOW | −0.08 | −1.5 |
| RARERECOVER | 0.645 | 8.1 |
| HABRESTORE | 0.243 | 4.7 |
| HABCREATE | 0.253 | 4.3 |
| ECOHUMAN | 0.359 | 5.9 |
| ECOALL | 0.064 | 1 |
| PRICE | −0.003 | −15.3 |
| Pseudo $Rho^2$ | 19 % | |
| N (individuals) | 398 | |

move from continued decline to stopping decline and ensuring recovery ('RareRecover') increases well-being, a move to slowing decline ('Rare-Slow') is negatively valued. All price ('tax') increases reduce utility, as expected.

For Northumberland, the same pattern is repeated, except that the 'EcoAll' and 'RareSlow' attributes are not significant. This means that any improvement in habitat quality or familiar species is positively and significantly valued, as is an improvement in directly relevant ecosystem services ('EcoHuman') – although not an improvement in all services ('EcoAll'). This implies the Northumberland group only cared about ecosystem services that seemed to directly impact on their well-being. The Northumberland group also had a negative value for 'RareSlow', but since this estimate is insignificant, this is unimportant.

Table 12.4 *Implicit prices for Cambridge and Northumberland CE samples*

| Attribute | Implicit price | SE | 95% lower | 95% upper |
|---|---|---|---|---|
| *Cambridgeshire* | | | | |
| FAMRARE | 35.65 | 17.19 | 1.95 | 69.34 |
| FAMBOTH | 93.49 | 18.03 | 58.15 | 128.82 |
| RARESLOW | −46.68 | 15.88 | −77.80 | −15.55 |
| RARERECOVER | 115.13 | 21.22 | 73.53 | 156.72 |
| HABRESTORE | 34.40 | 15.32 | 4.37 | 64.42 |
| HABCREATE | 61.36 | 17.52 | 27.02 | 95.69 |
| ECOHUMAN | 53.62 | 16.97 | 20.35 | 86.88 |
| ECOALL | 42.21 | 19.23 | 4.51 | 79.90 |
| *Northumberland* | | | | |
| FAMRARE | 90.59 | 19.24 | 52.87 | 128.30 |
| FAMBOTH | 97.71 | 18.47 | 61.50 | 133.91 |
| RARESLOW | n/a | | | |
| RARERECOVER | 189.05 | 25.28 | 139.50 | 238.59 |
| HABRESTORE | 71.15 | 16.29 | 39.22 | 103.07 |
| HABCREATE | 74.00 | 17.51 | 39.68 | 108.31 |
| ECOHUMAN | 105.22 | 17.7 | 70.52 | 139.91 |
| ECOALL | n/a | | | |

The statistical equivalence of the parameter estimates of the two models can be compared using a Likelihood Ratio test. The probability value for this test is <0.01, indicating that the models are different. In other words, the valuation of biodiversity attributes varies significantly between the two samples, so that simple benefits transfer of valuation functions is rejected.

Table 12.4 shows the implicit prices estimated from the logit model. These implicit prices show the marginal WTP on average of moving from one level – the excluded level, which in our case is always the worst-case, do nothing level – to a higher level. For example, the value of £35.65 for 'FamRare' for Cambridgeshire means that people were on average willing to pay £35.65 extra per year in higher taxes to move from continued decline in familiar species to a situation where rare, familiar species are protected from further decline. These are *ceteris paribus* values, so should be treated with care in a cost-benefit context. We can see from Table 12.4 that a scale effect is present in almost all cases for Cambridgeshire, meaning that higher levels of protection and enhancement are valued more highly for each attribute, with the exception of the odd result on 'RareSlow', and in the case of the ecosystem function attribute, where the value of protecting only directly relevant ecosystem services ('EcoHuman') is higher than that of protecting all ('EcoAll'). The highest benefits in per-person terms come from ensuring the recovery of rare, unfamiliar species ('RareRecover'). For Northumberland, the implicit prices for 'RareSlow' and 'EcoAll' are omitted, since the parameter estimates were not significantly different from zero. Furthermore, there was little evidence that the Northumberland sample considered the scale effects between the levels of the familiar species and for habitat quality attribute. Highest WTP is associated with ensuring the recovery of rare, unfamiliar species ('Rare-Recover') – the same result as for Cambridgeshire.

## 6 Discussion

Two key questions which can be asked of this data are: is there evidence that the general public are willing to pay additional taxes to support biodiversity conservation, and if so, then why? Here we are firstly interested in whether respondents chose a biodiversity enhancement policy option (Options A or B) as opposed to the 'do nothing' option. In fact, only 15 per cent of respondents chose the 'Do nothing' option. In other words, these respondents were not willing to pay additional taxes to achieve biodiversity enhancements. Eighty five per cent of the choices made by CE respondents were for choice options A or B. This demonstrates that the majority of respondents were willing to pay some amount of additional taxation to attain biodiversity enhancements. This finding was backed up by a contingent valuation study which was undertaken at the same time as the choice experiment (see Christie *et al.* 2006 for details), where positive WTP values existed for three biodiversity protection and enhancement policies. Approximately one-third of respondents in the contingent valuation were unwilling to pay for biodiversity enhancement, compared with 15 per cent in the choice experiment.

In terms of the reasons given by CE survey respondents for making these choices, over half of the respondents (52.6 per cent) stated that they considered that the biodiversity improvements in policy options A or B were 'good value of my money'. Three per cent of respondents giving a zero bid stated that the biodiversity improvements were not good use of their money, while five per cent stated that they already contribute to environmental causes. Protest votes included 'I do not think that increases in taxation should be used to fund biodiversity improvements' (6.5 per cent) and 'The costs of biodiversity improvement should be paid for by those who degrade biodiversity' (14.2 per cent).

Another question our research enables us to address is: what aspects of biodiversity protection policy do the public value most? Examining the implicit prices in Table 12.4 provides some answers. In the choice experiment, familiar species attained positive and significant implicit prices. In Cambridgeshire, scale effects were evident in that the implicit price for the protection of both rare and common familiar species (£93.49) was significantly higher than the protection of only the rare familiar species (£35.65). This was not, however, the case in the Northumberland sample, where the two levels of protection had similar implicit prices (£90.59 and £97.71 respectively for the protection of rare only and rare and common familiar species). In conclusion, evidence from the choice experiment suggests that the public do support policies that target rare familiar species of wildlife, but the evidence is less clear for common familiar species.

The second attribute addressed in the choice experiment related to rare unfamiliar species of wildlife. Two levels of provision were addressed. 'RareSlow' which aimed to 'slow down the rate of the decline in the populations of rare unfamiliar species. . . . . it is likely that some rare unfamiliar species may still become locally and nationally extinct'. The second level 'RareRecover' aimed to 'stop decline and ensure recovery of rare unfamiliar species'. The findings for the 'RareSlow' attribute level were interesting since it was found to be negative in the Cambridgeshire sample (indicating that negative utility would be gained from a slowdown in the decline of the population of rare unfamiliar species – which was not predicted), while the attribute level was not significant in the Northumberland CE model. The implications of this finding was that it appears that the public are unwilling to support policies that simply delay the time it takes for such species to become extinct. This conclusion was further emphasised by the fact that highest implicit prices were attained from the 'RareRecover' attribute level. Thus, the policy implication of these findings is that the public appear to only support policies that aim to achieve recovery of the populations of rare species, rather than those that simply attempt to slow down decline in population numbers. A further implication of these findings relates to the fact that survey respondents were told that they were unlikely to ever see these rare, unfamiliar species. Thus, these values can be considered to represent passive-use values.

The habitat quality attribute was included to assess whether the public valued the restoration of existing habitats ('HabRestore') or the recreation of new habitats on farmland ('HabRecreate'). Both attribute levels were found to be positive and significant in the two case study locations. In Cambridgeshire, the value for habitat restoration (£34.40) was half that for habitat recreation (£61.36), while the same ordering of values was attained for both levels in Northumberland (£71.15 and £74.01 respectively). The reason for this difference may be similar to those stated above for the familiar species attribute. In other words, the Cambridgeshire respondents may have been more able to distinguish between attribute levels and/or the Cambridgeshire sample may have considered that there were very few existing habitats within Cambridgeshire which would benefit from restoration. Again, evidence was not collected to identify which, if any, of these reasons could be verified. However, there was evidence that the public would support policies that aimed to protect and enhance habitats, although the value of the implicit prices was found to be slightly lower than those found for the two species attributes.

Finally, the ecosystem services attribute was included to assess whether the public valued ecosystems that had only a direct impact on humans ('EcoHuman') and all ecosystem services include those which did not directly affect humans ('EcoAll'). The ecosystems services that had direct impacts on humans were found to be both positive and significant. However, the 'EcoAll' attribute level was not significant in the Northumberland model and was lower than the 'EcoHuman' attribute level in the Cambridge sample. It would thus appear that survey respondents 'cared' about ecosystem functions that affect humans, but were less interested in the other ecosystem services.

A final issue of interest relates to the transferability of the CE results between the two case study areas. A Likelihood Ratio test was used to compare the parameter estimates between the Cambridgeshire and Northumberland models (Table 12.3). The probability value for this test is < 0.01, indicating that the two models were different. Based on this evidence we would reject the transfer of the indirect utility functions between the two areas. Another test for benefits transfer undertaken on the choice experiment data was to test whether the implicit prices for each attribute were significantly different from each other between the Cambridgeshire and Northumberland samples. Evidence from Table 12.4 indicates that the 95 per cent confidence intervals for implicit prices do overlap between the models in two out of six cases – for 'FamBoth' and 'HabRecreate'; however, this is largely due to the large standard errors on the implicit prices. So again, there is little evidence in support of benefits transfer in the choice experiment data.

## 7 Conclusions

Policy-makers may benefit from information on the economic value of biodiversity protection, but also on which aspects of biodiversity are most valued by taxpayers. Stated preference methods such as choice experiments can provide these type of value estimates, but implementing these methods is difficult in this particular case since the general public have a rather low level of understanding of what biodiversity is and why it matters. In this study we make use of a novel way of conveying information to respondents, information which is consistent with ecological understanding of what aspects of biodiversity might be considered. We then use choice experiments to estimate the relative values people place on these attributes.

We believe that this research has made a significant contribution to our understanding of the economic value of biodiversity. We demonstrate that the use of innovative information presentation methods in the valuation study enables these methods to value complex and unfamiliar goods. Second, we believe that our approach has made a significant step towards the integration of ecological ideas into a valuation protocol and this provides us with a greater understanding of how the public perceive and value ecological concepts of biodiversity. How policy-makers might choose to use such information is something we have not addressed here. But economists would argue that, in a world of scarce resources and conflicting demands, some information on the relative values society places on biodiversity conservation is better than no information.

## REFERENCES


Ali, M. M., Dickinson, G. and Murphy, K. J. 2000. Predictors of plant diversity in a hyperarid desert wadi ecosystem. *Journal of Arid Environments*. **45**. 215–230.

Brotherton, I. 1996. Biodiversity, spatial extent and protectability. In I. A. Simpson and P. Dennis. (eds.). *The Spatial Dynamics of Biodiversity: Proceedings 5th Annual IALE Conference*. International Association for Landscape Ecology, Stirling, 149–160.

Christie, M. 2001. A comparison of alternative contingent valuation elicitation treatments for the evaluation of complex environmental policy. *Journal of Environmental Management*. **62**. 255–269.

Christie, M., Warren, J., Hanley, N., Murphy, K. and Wright, R. 2004. *Developing Measures for Valuing Changes in Biodiversity*. Report to DEFRA: London.

Christie, M., Hanley, N., Warren, J., Murphy, K., Wright, R. and Hyde, R. 2006. Valuing the diversity of biodiversity. *Ecological Economics*. **58** (2), 304–317.

Colwell, R. K. and Coddington, J. A. 1995. Estimating terrestrial biodiversity through extrapolation. *Philosophical Transactions of the Royal Society of London*. **345**. 101–118.

DEFRA. 2002. *Survey of Public Attitudes to Quality of Life and to the Environment – 2001*. London: DEFRA.

Dony, J. G. and Denholm, I. 1985. Some quantitative methods of assessing the conservation value of ecologically similar sites. *Journal of Applied Ecology*. **22**. 229–238.

Franks, J. R. 1999. In situ conservation of plant genetic resources for food and agriculture: a UK perspective. *Land Use Policy*. **16** (2). 81–91.

Garrod, G. D. and Willis, N. D. 1994. Valuing biodiversity and nature conservation at a local level. *Biodiversity and Conservation.* **3**. 555–565.

Garrod, G. D. and Willis, K. G. 1997. The non-use benefits of enhancing forest biodiversity: a contingent ranking study. *Ecological Economics.* **21**. 45–61.

Hanley, N. and Craig, S. 1991. Wilderness development decisions and the Krutilla-Fisher model: the case of Scotland's 'flow country'. *Ecological Economics.* **4**. 145–164.

Hanley, N., Spash, S. and Walker, L. 1996. Problems in valuing the benefits of biodiversity conservation. *Environmental and Resource Economics.* **5**. 249–272.

Hanley, N. and Shogren, J. 2002. Economics and nature conservation: awkward choices. In Daniel, W. Bromley and Jouni Paavola (eds.). *Economics, Ethics and Environmental Policy.* Oxford: Blackwell. Contested Choices. 120–130.

Harper, J. L. and Hawksworth, D. L. 1995. Preface. In. D. L. Hawksworth (ed.). *Biodiversity: Measurement and Estimation.* London: Chapman and Hall. 5–12.

Hawksworth, D. L. (ed.). 1995. *Biodiversity: Measurement and Estimation.* London: Chapman and Hall. 140.

Hensher, D. A., Rose, J. M. and Greene, W. H. 2005. *Applied Choice Analysis: A Primer.* Cambridge University Press: Cambridge.

Hurlbert, S. H. 1971. The nonconcept of species diversity: a critique and alternative parameters. *Ecology.* **52**. 577–586.

Limburg, K. E., O'Neill, R. V., Costanza, R. and Farber, S. 2002. Complex systems and valuation. *Ecological Economics.* **41** (3). 409–420.

Loomis, J. B. and White, D. S. 1996. Economic benefits of rare and endangered species: summary and meta-analysis. *Ecological Economics.* **18**. 197–206.

Louviere, J. J., Hensher, D. A. and Swait, J. D. 2000. *Stated Choice Methods: Analysis and Application.* Cambridge University Press: Cambridge.

Lovejoy, T. E. 1980a. Foreword. In M. E. Soulé and B. A. Wilcox (eds.). *Conservation Biology: an Evolutionary-Ecological Perspective.* Sunderland, MA: Sinauer Assoc. v–ix.

Lovejoy, T. E. 1980b. Changes in biological diversity. In: G. O. Barney. (ed.). *The Global 2000 Report to the President, Vol. 2 (The Technical Report).* Harmondsworth: Penguin Books. 327–332.

Lovejoy, T. E. 1995. The quantification of biodiversity: an esoteric quest or a vital component of sustainable development? In D. L. Hawksworth (ed.). *Biodiversity: Measurement and Estimation.* London: Chapman and Hall. 81–87.

MacMillan, D.C, Hanley, N. and Buckland, S. 1996 . Contingent valuation of uncertain environmental gains. *Scottish Journal of Political Economy.* **43** (5). 519–533.

MacMillan, D. C. and Duff, E. I. 1998. The non-market benefits and costs of native woodland restoration. *Forestry.* **71** (3). 247–259.

MacMillan, D. C., Duff, E. I. and Elston, D. 2001. Modelling non-market environmental costs and benefits of biodiversity projects using contingent valuation data. *Environmental and Resource Economics.* **18** (4). 391–340.

MacMillan, D., Philip, L., Hanley, N. and Alvarez-Farizo, B. 2003. Valuing nonmarket benefits of wild goose conservation: a comparison of interview and group-based approaches. *Ecological Economics.* **43**. 49–59.

Magurran, A. E. 1988. *Ecological Diversity and its Measurement*. Princeton, NJ: Princeton University Press.

May, R. M. 1995. Conceptual aspects of the quantification of the extent of biological diversity. In D. L. Hawksworth. (ed.). *Biodiversity: Measurement and Estimation*. London: Chapman and Hall. 13–20.

McIntyre, S. 1992. Risks associated with the setting of conservation priorities from rare plant species lists. *Biological Conservation*. **60**. 31–37.

Morris, M. G. and Perring, F. H. 1974. *The British Oak: Its History and Natural History*. Faringdon: E. W. Classey.

Norse, E. A. and McManus, R. E. 1980. Ecology and living resources biological diversity. In: *Environmental Quality 1980: The 11th Annual Report of the Council on Environmental Quality*. Council on Environmental Quality. Washington, DC. 31–80.

Norse, E. A., Rosenbaum, K. L., Wilcove, D. S., Wilcox, D. A., Romme, W. H., Johnston, D. W. and Stout, M. L. 1986. *Conserving Biological Diversity in our National Forests*. Washington, DC: The Wilderness Soc.

Noss, R. F. 1990. Can we maintain biological and ecological integrity. *Conservation Biology*. **4**. 241–243.

Nunes, P. A. L. D. and van den Bergh, J. C. J. M. 2001. Economic valuation of biodiversity: sense or nonsense? *Ecological Economics*. **39**. 203–222.

OECD. 2001. *Valuation of Biodiversity Benefits: Selected studies*. Paris: OECD.

Pearce, D. 2001. Valuing biological diversity: issues and overview. In OECD. *Valuation of Biodiversity Benefits: Selected Studies*. Paris: OECD.

Peet, R. K. 1974. The measurement of species diversity. *Annual Reviews of Ecology and Systematics*. **5**. 285–307.

Potter, C. 1988. Environmentally sensitive areas in England and Wales: an experiment in countryside management. *Land Use Policy*. **5**. 301–313.

Pouta, E., Rekola, M., Kuuluvainen, J., Tahvonen, O. and Li, C-Z. 2000. Contingent valuation of the Natura 2000 programme in Finland. *Forestry*. **73** (2). 119–128.

Rekola, M. 2003. Lexicographic preferences in contingent valuation: A theoretical framework with illustrations. *Land Economics*. **79** (2). 277–291.

Robinson, G. M. 1994. The greening of agricultural policy: Scotland's environmentally sensitive areas (ESAs). *Journal of Environmental Planning and Management*. **37**. 215–231.

Rodwell, J. S. 1991–1995. *British Plant Communities: Volumes 1–4*. Cambridge: Cambridge University Press.

Roy, K. and Foote, M. 1997. Morphological approaches to measuring biodiversity. *Trends in Ecology and Evolation*. **12**. 277–281.

Simpson, I. A., Hanley, N., Parsisson, D. and Bullock, C. H. 1996. Indicative prediction of botanical diversity change in Environmentally Sensitive Areas. In: I. A. Simpson and P. Dennis. (eds.). *The Spatial Dynamics of Biodiversity: Proceedings 5th Annual IALE Conference*. International Association for Landscape Ecology. Stirling. 71–78.

Spash, C. L. and Hanley, N. 1995. Preferences, information and biodiversity preservation. *Ecological Economics*. **12**. 191–208.

ten Kate and Laird 1999. *The Commercial Use of Biodiversity: Access to Genetic resources and Benefit-Sharing*. London: Earthscan Publications.

White, P. C. L., Gregory, K. W., Lindley, P. J. and Richards, G. 1997. Economic values of threatened mammals in Britain: A case study of the otter *Lutra lutra* and the water vole *Arvicola terrestris*. *Biological Conservation*. **82** (3). 345–354.

White, P. C. L., Bennett, A. C. and Hayes, E. J. V. 2001. The use of willingness-to-pay approaches in mammal conservation. *Mammal Review*. **31**. 2, 151–167.

Whittaker, R. H. 1977. Evolution of species diversity in land communities. *Evolutionary Biology*. **10**. 1–67.

Willis, K., Garrod, G., Scarpa, R., Powe, N., Lovett, A., Bateman, I., Hanley, N. and MacMillan, D. 2003. *The Social and Environmental Benefits of Forests in Great Britain*. Edinburgh: Forestry Commission.

Wilson, E. O. 1988. *Biodiversity*. Washington, DC: National Acad. Press.

# 13 Spatially explicit valuation with choice experiments – a case of multiple-use management of forest recreation sites[1]

*Paula Horne, Peter Boxall and Wiktor Adamowicz*

## 1 Introduction

The recent development of multi-attribute valuation methods has enabled the examination of preferences for environmental goods as defined by their characteristics or attributes. There is a growing number of applications in forest recreation that typically describe the attributes in terms of forest management, congestion levels, evidence of wildlife abundance, the length of travel, etc. (e.g. Boxall *et al.* 1996; Adamowicz *et al.* 1998; Boxall and MacNab 2000). Fewer studies, however, have examined the supply of goods that provide non-use or passive-use values (e.g. Tanguay *et al.* 1995; Adamowicz *et al.* 1998). The spatial dimension in the supply of goods of passive-use value is even less examined. In this chapter, we look at values associated with specific locations and identify spatial preferences for biodiversity conservation. We discuss the use of spatially explicit multi-attribute valuation with an illustration of a case study of forest management in a system of municipal recreation sites. Use of site-specific attributes provided more information and a richer set of policy implications than when only the average measures of attributes over the complex of sites were used.

Municipal recreation forests in the Nordic countries face a variety of demands including recreational use, nature conservation and, sometimes, the generation of revenue from timber harvesting (Hytönen 1995). These different demands require a range of features from the forest

[1] The authors wish to thank Michel Haener, Taina Horne and two anonymous reviewers for their valuable comments. Special thanks are due to Helsinki's Green Area Division of Public Works Department for assisting in data gathering. Financial support from Liikesivistysrahasto and the Ministry of Agriculture and Forestry through the Finnish Biodiversity Research Programme (FIBRE) is gratefully acknowledged. The paper was presented in an IUFRO Division 4 conference in April 2003. The usual disclaimers apply. We would also like to thank the *Journal of Forest Ecology and Management* for kind permission to reprint the original article.

environment. The selected forest management regime as well as the natural forest ecosystem affects the range and abundance of the different features desired by the forest users (e.g. the biodiversity level and the scenery of the site). Since the goal of forest management at outdoor recreation sites is to fulfil the recreational needs of different visitors, forest managers need information on the preferences of forest recreationists. Visitors who are satisfied with the current management conditions seldom state their opinions unless actually asked, while those with a critical view on current management freely let their opinions be known (Sievänen 1992). Thus, use of participatory approaches involving interest groups should be complemented by surveys of representative samples of visitors, since the values and attitudes of interest groups are often found to differ from those of the general public (Wellstead *et al.* 2003).

Earlier studies indicate that Finnish people prefer relatively intensively managed forest landscapes (e.g. Savolainen and Kellomäki 1984; Karjalainen 2001). Despite this, however, awareness of the environmental impacts of intensive forest management has created public pressure to change management practices towards biodiversity-enhancing regimes. Visitors may have seemingly contradictory preferences, on the one hand wishing for easily passable even-aged stands, while on the other hand asking for conservation of biodiversity. This complex set of preferences for forest benefits and opinions on forest management, as well as confusion about the concepts and relationships among different attributes of the forest, has the potential to complicate management planning.

The aim of the study reported in this chapter was to evaluate visitors' preferences for forest management regimes at municipal recreation sites by examining the tradeoffs between different elements of forest management. The city of Helsinki has managed its recreation forests with the goal of providing pleasant recreation environments and conserving the special features of local forests. We focused on the interrelations of use and non-use values, namely scenery at the preferred recreation site as a non-consumptive use value, and species richness as a proxy variable for non-use value of biodiversity. Heterogeneity of the study sites and of visitors' preferences was taken into account in the analysis. Treating the study area as a system of separate recreation sites enabled the examination of preferences for spatial variability in the features of the recreation environment that result from different management practices.

Preferences for forest management at one site may be somewhat different than preferences for forest management over the set of recreation sites in the study area. Thus, forest management strategies could be viewed over the system of spatial units, where the manager faces an option of varying levels of management intensity among the sites. Within this system, the manager could assign different management goals for each site, or integrate all management activities into a management system applied at all sites. Therefore, one management attribute can be the variability in the management regime over the system of recreation sites.

It is often useful for managers to be able to identify and quantify the impact of a change in forest management on visitors' satisfaction. A welfare analysis provides a quantitative measure of the impact of different management scenarios on visitors' satisfaction. Changing the management regime would please some visitors while others might find it disagreeable. Including an indication of heterogeneous preferences in the analysis allows for the examination of who would gain and who would lose because of a management change.

## 2 Methods and data

### 2.1 *The choice experiment method*

In order to evaluate visitor preferences we applied the choice experiment method where respondents are presented with a number of choice sets consisting of two or more alternatives from which one chooses the preferred alternative. Each alternative describes various levels of a set of attributes, influenced by the chosen forest management strategy. Attributes can be quantitative or qualitative in nature, and the ability to combine these two types of data is one of the main benefits of the choice experiment approach. Louviere *et al.* (2000) provide an overview of the choice experiment approach, while Akabua *et al.* (2000) provide examples of the usefulness of the technique for forest managers interested in non-timber values.

Choice experiments are consistent with random utility theory and offer a wide range of information on tradeoffs among the benefits provided by the choices (Adamowicz *et al.* 1994, 1998). The theory is based on probabilistic choice, where individuals are assumed to choose a single alternative which maximises their utility (welfare) from a set of available alternatives. Probabilistic choice models rely on random utility theory which describes the utility of each alternative ($U$) as the sum of systematic and error components. The systematic component, $V$, is a vector of individual and alternative specific attributes that are observable. The presence of an error component, $\varepsilon$, makes the choice random, and the error component includes all the impacts and factors affecting the choice that are not observable by the researcher (Louviere *et al.* 2000). Random utility theory posits that an individual, $n$, chooses the alternative, $i$, from the choice set, $Cn$, if the indirect utility of $i$ is greater than that of any other choice $j$. The following equation identifies this notion:

$$U_{in} > U_{jn} \Rightarrow V_{in} + \varepsilon_{in} > V_{jn} + v_{jn} \qquad \forall j \neq i; i, j \in C_n. \tag{1}$$

The theory describes the probability with which an alternative is chosen given its systematic and error components. The probability of individual $n$ choosing an alternative $i$ is the same as the probability that the utility of alternative $i$ is greater than the utility of any other alternative of the choice set. Thus,

$$P(i) = P(V_{in} + \varepsilon_{in} > V_{jn} + \varepsilon_{jn}) \qquad \forall j \neq i; i, j \in C_n. \tag{2}$$

The conditional logit model is the most commonly used method in the analysis of multi-attribute choices. Assuming that the error components have an IID Gumbel distribution (Ben-Akiva and Lerman 1985, p. 104), the probability of choosing $i$ is

$$P(i) = \frac{\exp^{Vin}}{\sum_j \exp^{Vjn}} \tag{3}$$

The model is estimated using maximum likelihood estimation procedures and assumes a linear-in-parameters functional form for the systematic portion of the conditional indirect utility function (Ben-Akiva and Lerman 1985). The coefficient of an attribute in a linear specification is the marginal utility of that attribute. Utility at various attribute levels can be determined by multiplying the various levels by their marginal utilities.

Observing the choices made and the connections of the different attribute levels to monetary changes can derive measures of economic welfare. The Hicksian compensating variation (CV) for the case we examine can be written as

$$CV = -\frac{V_{jn}^0 - V_{jn}^1}{\alpha} \tag{4}$$

where $\alpha$ is the marginal utility of money and $V_{jn}^0$ and $V_{jn}^1$ are the initial and new states of the resource (Hanemann 1982). The initial state, or status quo, thus provides the basis for economic welfare analysis (Carson *et al.* 1994). Typically the marginal utility of money is derived from the parameters of the choice model on some monetary attribute such as household taxes (e.g. Adamowicz *et al.* 1998) or the costs of travel to sites (e.g. Boxall *et al.* 1996).

### 2.2 *Study area and data collection*

The study area consisted of the five outdoor recreation sites of Luukkaa, Pirttimäki, Vaakkoi, Salmi and Karjakaivo owned by the city of Helsinki. They are all located in the Nuuksio lake uplands (60° 12′N; 24°55′E), about 20–40 km northwest of Helsinki. Nuuksio National Park is located in the vicinity, with pathways connecting it to the recreation sites. All the sites but Vaakkoi have recreation facilities like camping areas, pathways, ski tracks, parking lots and service cabins. There is no fee to access any of the sites. Annual visitor numbers vary from more than 200,000 visits in Luukkaa to about 20,000 visits in Vaakkoi. Miettinen and Horne (1999) report no significant differences in the visitor structure of Luukkaa, Pirttimäki, Salmi and Vaakkoi, while there were more men, young age groups and first-timers among the visitors of Karjakaivo.

The management of these municipal recreation sites is financed through the budget of the city of Helsinki; therefore the municipal taxpayer has some interest in the condition and costs of managing the sites. Forests at the sites are managed to sustain recreation benefits as well as to provide environments that sustain local flora and fauna. Silvicultural methods at the sites have been less intensive than in forests managed strictly for timber and done at a smaller scale with a longer rotation period. Trees are harvested to open the canopy to facilitate regeneration of some species and harvests are planned to provide greater access for recreational users. Harvested trees are collected and sold or used for firewood and construction, but forest-revenue maximisation is not a management goal. The value of harvested trees covers the annual costs of forest operations. There are also patches of forest left for nature conservation within the recreation forest areas.

Data collection was conducted on-site in four of the five recreation sites (Luukkaa, Pirttimäki, Vaakkoi and Salmi) by means of a personal interview after a visit to the site. At the fifth site, Karjakaivo, most visitors come from a nearby physical education centre and therefore, to alleviate concerns that visitors there were not similar to those at the other sites, the survey was not administered to visitors there. There are also features of this site that made interviewing visitors difficult. Visitors were sampled to address both spatial and temporal concerns. Each site had 3–4 sampling points that were selected on the basis of the relative number of visitors and the spatial distribution of access points to the site. Data gathering was conducted from early June to late October 1998 during evenings, weekends and the summer vacation time, as those are the most popular times for visits.

The interviewer selected respondents as they passed the sampling point in accordance with their gender and age. The fraction of each gender and age class in the sample should be proportional to its share of the visitor population. However, the structure of the visitor populations of the recreation sites was unknown. To define the classes, the interviewers observed all visitors passing the sampling point and marked each on a form according to their estimated class of age and gender. Representatives of each class were asked to participate in the survey in proportion to that class's fraction of all visitors. Selection of respondents within each age and gender class was random, generated by the availability of the interviewer, which depended both on the sampling periods and whether occupied with other respondents. Both those who refused and those who answered were marked accordingly, in order to observe the proportion of refusals and the representativeness of the sample. The visitor sample and those who refused to answer were found to be representative of the visitor population observed at the sampling points. About a third of those who were asked to participate refused, usually referring to the short length of their visit or the impatience of their young children.

### 2.3 *Survey instrument*

The survey questionnaire contained the choice experiment questions, as well as questions about attitudes towards forest management and socio-economic characteristics of the respondent. Respondents were also asked about the number of visits they had made to each of the five sites during the last twelve months. In addition, information was collected regarding their scenic preferences in relation to forest management intensity. This involved choosing one of three pairs of pictures. Each pair featured two photographs of a pine stand, one at the age of 40 years and one at the age of 120 years. The three pairs of pictures differed in terms of the intensity of forest management and the scenery. Experts in forest management and biodiversity science assisted in creating these photos. The three levels of intensity were no management, medium management (i.e. close to the current forest practices at the sites) and intensive management.

In the analysis of these data, information about the most frequently visited site of a respondent was combined with one's choice of their favourite scenery. Checking to see whether a management alternative included a visitor's favourite scenery at their most frequently visited site allowed for the construction of a site-specific dummy variable, assessing their favourite scenery at the site they visited most often. Along with the attributes of the choice set in the instrument, the dummy variable was

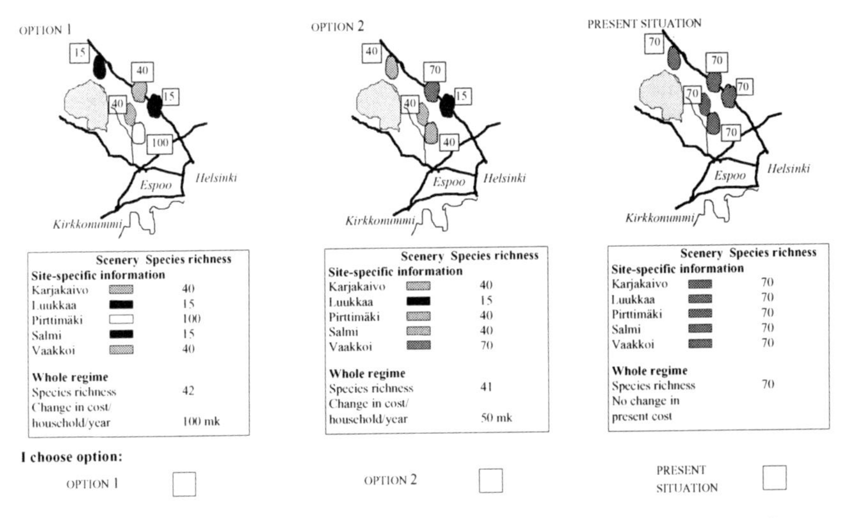


Figure 13.1. Example of a choice set used in the choice experiment instrument

used as an additional variable to determine the choice of management alternative.

Prior to answering the choice tasks, the respondents were given written information on the attributes used in the choice experiment. Information was provided on the current forest management practices at the sites and how they were financed, on the meaning of and threats to forest biodiversity, and on the forest type presented in the photos. The same information was given at all interview sites.

The choice experiment involved the presentation of choice sets, each consisting of three alternative forest management regimes for the five recreation sites (see Figure 13.1). One of the three alternatives was always the so-called status quo that refers to the present management regime where all five sites are under the same management practice. Respondents were instructed to choose the preferred management alternative in each choice set.

Each alternative forest management regime was described using an information box together with a map that featured the five recreation sites and the Nuuksio National Park. The park was included because its presence might affect preferences for forest management in the municipal recreation system. The information box and the map outlined the levels of impact that the alternative management regimes had on the chosen attributes: species richness at each site, scenery at each site and the overall management costs for the regime (see Table 13.1). Scenery was assumed

Table 13.1 *Summary of the attributes and their levels used in choice instrument*

| Attribute | Levels |
|---|---|
| Species richness at each site | 15 |
| | 40 |
| | 70 |
| | 100 |
| Average species richness | Calculated on the basis of site-specific species richness at the five sites |
| Variance of species richness | Calculated from the species richness at the five sites |
| Scenery at each site | No management |
| | Medium management |
| | Intensive management |
| Annual costs of management | +50 (8.4) |
| (change in municipal taxes: | 0 |
| FIM/household/year) (Euros) | –50 (–8.4) |
| | –100 (–16.8) |
| | –200 (–33.6) |
| | –300 (–50.5) |

Notes: The species richness and scenery attributes are designed as a single six-level attribute where species richness levels of 15 and 100 have unique scenery levels (two levels of the joint attribute), while richness levels 40 and 70 each share two levels of scenery (four levels of the joint attribute). This joint attribute is specific to each site, thus each alternative is designed from five- to six-level richness – scenery attributes and one six-level price attribute

to have non-consumptive use value to visitors, while species richness, composed mainly of species not readily observable to a typical visitor, was assumed to be a proxy for biodiversity. The types of scenery that would be found at each site were marked with a colour code. There were three colours, corresponding to the types of scenery shown to the respondents earlier in the pairs of photographs. Each site was also given an index level of species richness. The index levels had been derived using the number of tree species and the amount of decayed wood on the ground (S. Paavola, Personal Communication). The index levels of species richness were restricted to two for each scenery option in order to have plausible combinations of the two attributes (i.e. no management with indices 70 and 100; medium management with indices 40 and 70; intensive management with indices 15 and 40). This design forces the levels of species richness and forest scenery to be correlated. While both species richness and scenery were in the experimental design, the scenery variable was not significant in an initial estimation of the model and was excluded from further analysis. Thus, scenery and species richness are related and changes in richness appear to capture the preferences for scenery. There are no difficulties associated with excluding scenery as an attribute since the experiment is orthogonally designed.

The average level of species richness across the regime was calculated from the site-specific indices and presented in the information box for each management alternative. In addition, the cost of management was expressed as a change in annual municipal taxes per household in Finnish marks. Municipal taxes were chosen as the payment vehicle because the work of the Green Area Division of the Public Works Department, which manages the sites, is financed through the budget of the city of Helsinki. These attributes were used to assess preferences for the 'overall' measures of biodiversity and costs of management.

The attributes and levels shown in Table 13.1 result in $(65 \times 61) \times (65 \times 61)$ different combinations. Since the number is large, a fractional factorial, main effects design was used to minimise the number of choice combinations presented to respondents. In main effects designs, interactions among the attributes are assumed to be statistically insignificant (Ben-Akiva and Lerman 1985). Given the number of attributes and their levels in this study, the main effects design produced seventy-two different choice sets. Since the number of choice sets for each respondent should not exceed the cognitive capacity of the respondent (Swait and Adamowicz 1997), the seventy-two choice sets used in the study were blocked into twelve versions of the choice experiment, each containing six choice sets. One of the choice sets that was used to gather the choice data is shown in Figure 13.1.

## 3 Results and discussion

### *3.1 Choice experiment results*

A total of 431 respondent questionnaires were used to estimate preferences. Of these respondents, ninety (21 per cent) always selected the status-quo alternative in their choices. Inclusion of the status quo provides respondents with something familiar and also offers a means to say no change is preferable. This alternative thus provides the initial state for economic welfare analysis (Carson *et al.* 1994). This feature may be the result of status quo bias that characterises difficult choices, with the result that people are reluctant to move from the current situation. Adamowicz *et al.* (1998), who analysed a passive-use value issue, described issues surrounding the analysis of data containing large numbers of status-quo choices. However, large frequencies of status-quo choices can also arise

Table 13.2 *Estimated model parameters (and standard errors) using site-specific and average species richness*

| | Parameters (SE) | | |
|---|---|---|---|
| Variable | Site-specific model 1 | Site-specific model 2 | Average model |
| Constant | 0.4645*** | 0.4235*** | 0.3338*** |
| | (0.0979) | (0.0988) | (0.1007) |
| Tax | −0.0018*** | −0.0018*** | −0.0018*** |
| | (0.0003) | (0.0003) | (0.0003) |
| Favourite scenery at favourite site | | 0.4533*** | 0.4521*** |
| | | (0.0563) | (0.0562) |
| Karjakaivo species richness | 0.4370*** | 0.4469*** | |
| | (0.0639) | (0.0642) | |
| Luukkaa species richness | 0.1766*** | 0.1925*** | |
| | (0.0613) | (0.0618) | |
| Pirttimäki species richness | 0.2053*** | 0.2077*** | |
| | (0.0618) | (0.0624) | |
| Salmi species richness | 0.3188*** | 0.3312*** | |
| | (0.0610) | (0.0617) | |
| Vaakkoi species richness | 0.1437** | 0.1495** | |
| | (0.0590) | (0.0593) | |
| Average species richness | | | 1.5473*** |
| | | | (0.1700) |
| Variance of species richness | 0.0002** | 0.0002** | −0.00005 |
| | (0.0001) | (0.0001) | (0.00009) |
| Log-likelihood | −2233.26 | −2200.76 | −2210.05 |
| R-square adjusted | 0.1234 | 0.1360 | 0.1331 |

*** Significant at $p < 0.01$, ** Significant at $p < 0.05$

when respondents genuinely prefer the present state of management and may not want to trade this for any other management regime (Carlsson and Martinsson 2001). Since this latter condition is likely to characterise the recreation-use data presented in this study, the choices of these individuals remained in the choice data.

Three choice models were estimated, two using site-specific attributes for species richness and one using the average measure of species richness over the recreation system. In each model all species-richness variables were converted to natural logarithms since we found their effects to be non-linear in the indirect utility function. To account for status-quo effects, an alternative specific constant (ASC), representing the current situation, was included in each estimation (Table 13.2). The ASC in each model is significant at the 1 per cent level and positive, indicating that respondents approve of the present state of management and prefer no changes when the levels of the other attributes are constant. The parameters on the tax variables are negative as expected, suggesting that the respondents favour alternatives with low tax burden on themselves.

The variance in species richness over the system of sites is significant only in the two site-specific models. This means that holding the other attributes constant, respondents prefer variability in species richness across the system, suggesting that management should differ among the recreation sites. This is an interesting finding, in that it appears that the respondents prefer that preservation should not occur everywhere in this system. In other words there are preferences for heterogeneous biodiversity and scenery levels across the system that could call for changes in management strategies at the sites.

In site-specific model 2, if an alternative had the respondent's favourite scenery at one's most visited site, a dummy variable representing that site was added to the model. This parameter is positive and significant in both the site-specific model 2 and the average model, indicating that an alternative with a respondent's favourite scenery at their most often visited site would have a higher probability of being chosen. Comparison of the parameter vectors of site-specific models 1 and 2 suggests that the addition of this dummy variable to the model has little impact on the other estimated parameters.

In both site-specific models, the indices of species richness are significant and positive for all sites. However, there is a clear difference in the magnitude of the parameters. Coefficients for species richness in Karjakaivo and Salmi are larger than those of the other sites. An alternative with a high biodiversity index for these sites has a higher probability of being chosen as the best alternative. This suggests that greater gains in economic welfare would result from higher indices of species richness at Karjakaivo and Salmi than at the other three sites.

In the average model, the average species richness parameter is positive and significant. The magnitude of this coefficient reflects the importance of species richness at all sites considered together. This result conveys the importance of species richness overall, but the information on the spatial nature of preferences for this feature of forest management is missed when one uses this model. Thus, the remaining analysis reports the results only for site-specific model 2.

### *3.2 Economic welfare impacts of changing management at the different sites*

Species richness could be enhanced in the recreation system explored in this study by leaving some patches unmanaged at all the sites to create natural habitats, while managing the rest of the forest areas for recreational purposes. Alternatively, management practices could differ among sites, with some sites enhancing species richness and others focusing on recreational use. To examine this latter approach, we consider a scenario where Salmi and Karjakaivo are left unmanaged while the other three sites remain under the present management regime. These two sites were chosen because the choice model parameters suggested that higher welfare gains could be met by changing management at these sites. The parameters from site-specific model 2 were used to examine the economic welfare impacts of forest management options on visitors with heterogeneous preferences for scenery. It should be noted that when management is changed to enhance biodiversity conservation, the scenery changes simultaneously.

The dummy variable for the presence of favourite scenery at favourite site was used as a proxy for the use value of scenery to the visitors. Thus, with a change in management, there would be three different groups of respondents based on whether or not they would gain or lose their favourite scenery at the site they visited most often. One group of respondents would not be affected by changes in management at Salmi and Karjakaivo because they visit some other site more often. This group included 81 per cent of respondents. The second group, which consisted of about 13 per cent of respondents who visit either Salmi or Karjakaivo most frequently, would lose their favourite scenery at these sites with the management change. Finally, the remaining 6 per cent visited Salmi or Karjakaivo most often and would gain their favourite scenery by the management change at these sites.

The welfare impact on visitors of the new management scenario with the changes in scenery and the levels of species richness was calculated using the Hicksian compensating variation measure (Eq. (1)). For the sample as a whole, the resulting welfare change was a loss of −€10.36. This negative welfare impact is due to the changes in management. Since the ASC is so strongly positive, any change in management would need to bring large benefits to compensate for the negative impact of moving away from the current situation. However, this overall welfare measure masks the distribution of welfare across the three different groups of individuals. Each group benefited from the higher index of species richness in the new scenario, but impacts differed when tempered with the change in scenery. The group that had no change in scenery at their favourite site had a negative welfare impact of −€7.92. For the group that lost their favourite scenery from their most often visited site, there was a welfare loss of −€50.11. The group that obtained their favourite scenery at their most often visited site had a welfare gain of +€34.27. Comparison of the welfare changes across the three groups of respondents identifies the importance of examining the heterogeneity of preferences in the respondents.

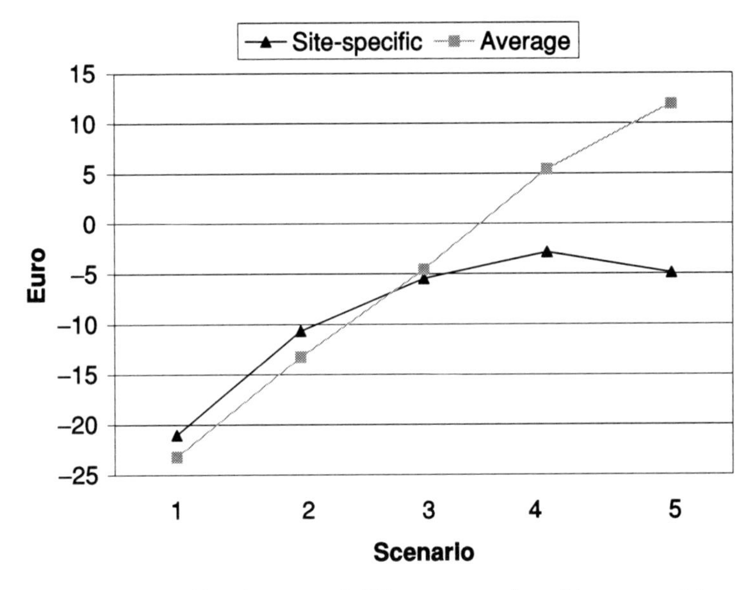


Figure 13.2. Welfare impacts of different scenarios with two models

### 3.3 *Further comparison of management scenarios*

To examine the benefits of enhancing species richness levels using a site-specific model instead of an average model, a management strategy was introduced to increase species richness across the system by incrementally changing forest management site by site. The strategy was to increase species richness by changing the forest management site by site. In order to determine the order of change by site, the average welfare impact, associated with changing management to enhance species richness and the consequential effects on the respondents' favourite scenery at favourite site, was estimated. These welfare impacts were −€21.14, −€27.10, −€30.45, −€31.06 and −€33.92 for Karjakaivo, Salmi, Pirttimäki, Vaakkoi and Luukkaa, respectively. Changes were introduced into the overall management regime starting with Karjakaivo, the site where change would be most readily accepted (i.e. with the lowest welfare impact). Next, management of the second preferred site, Salmi, was similarly changed. This continued by changing the management at each additional site, in order of increasing welfare impact.

The cumulative welfare impact of the site-specific model clearly shows how the different sites contribute different levels of impact to the recreation community who visit the sites (Figure 13.2). The cumulative welfare impact initially suggests that management changes provide a loss to the average visitor. However, as management is changed at more sites, the cumulative welfare effect increases in a positive direction. With the addition of changing management at the fourth site, Vaakkoi, the welfare effects level off. Management changes at Luukkaa apparently do not add welfare in this management change scenario. As the visitors preferred variability in management, managing all the sites similarly lowers the welfare impact. Also Luukkaa is the most popular site to visit and thus changing its management would result in more visitors losing their favourite scenery than if the change took place at other sites.

The average model of species richness did not capture this pattern of welfare enhancement (Figure 13.2). The exclusion of the site-specific nature of preferences for biodiversity and variation in species richness over the system of sites provides remarkably different economic results. The average model suggests that changing management to enhance species richness over the sites provides increasingly more economic benefits to visitors. Furthermore, this increase appears to be linear in nature.

## 4 Conclusions

This chapter presents an empirical example of using spatial patterns of attributes in applying choice experiments to forest management decisions. This Finnish application considered tradeoffs between non-use and use-valued recreational goods. The choice experiment method provided detailed information on visitor preferences for forest attributes resulting from management changes. One of the main benefits of the method for drawing practical policy conclusions is the knowledge of the explicit tradeoff among attributes. Horne and Ovaskainen (2001a) reported data from the same survey respondents who had been asked to rate the importance of different elements in their recreation environment. Nearly all respondents in their study stated that virtually all the elements were important – respondents considered scenery and biodiversity to be very important aspects of local forest management. The results of the choice experiment reported in this present study, however, demonstrate the existence of tradeoffs between forest benefits, when they cannot be supplied simultaneously. These tradeoffs result in impacts on the economic welfare of recreationists who visit the sites.

The survey was conducted on-site in a recreation system consisting of five adjacent outdoor recreation sites. Horne and Ovaskainen (2001b) reported that visitors to this system had clear preferences for which site they chose for their outdoor experience and were not particularly interested in forest management at the other sites. However, when these visitors were faced with biodiversity consequences resulting from management changes at their less preferred sites, their interests in management at the other sites became stronger. This demonstrates the importance of the non-use benefits provided by forest management.

Choice models were constructed to examine the importance of the spatial nature of preferences of Finnish recreationists. Site-specific models included statistically significant variables for species richness levels at all sites and a measure of the variance in species richness across the system. A model using the average measure of species richness across the system provided different conclusions than the site-specific model, which was found to be statistically more efficient and to provide more information on the preferences for management. Comparison of the two models illustrates the benefit of including spatial information as a variable in understanding preferences for forest management.

## REFERENCES

Adamowicz, W., Boxall, P., Williams, M. and Louviere, J. 1998. Stated preference approaches for measuring passive use values: choice experiments versus contingent valuation. *American Journal of Agricultural Economics*. **80** (1). 64–75.

Adamowicz, W. L., Swait, J., Boxall, P. C., Louviere, J. and Williams, M. 1994. Perceptions versus objective measures of environmental quality in combined revealed and stated preference models of environmental valuation. *Journal of Environmental Economics and Management*. **32**. 65–84.

Akabua, K., Adamowicz, W. L. and Boxall, P. C., 2000. Spatial non-timber valuation decision support systems. *Forestry Chronicle*. **76**. 319–327.

Ben-Akiva, M. E. and Lerman, S. R., 1985. Discrete Choice Analysis: *Theory and Application to Travel Demand*. London: The MIT Press.

Boxall, P. C. and MacNab, B., 2000. Exploring the preferences of wildlife recreationists for features of boreal forest management: a choice experiment approach. *Canadian Journal of Forest Resources*. **30**. 1931–1941.

Boxall, P. C., Adamowicz, W. L., Swait, J., Williams, M. and Louviere, J. J. 1996. A comparison of stated preference methods for environmental valuation. *Ecological Economics*. **18**. 243–253.

Carlsson, F. and Martinsson, P. 2001. Do hypothetical and actual marginal willingness to pay differ in choice experiments? *Journal of Environmental Economics and Management*. **41**. 179–192.

Carson, R. T., Louviere, J. J., Anderson, D. A., Arabie, P., Bunch, D. S., Hensher, D. A., Johnson, R. M., Kuhfeld, W. F., Steinberg, D., Swait, J., Timmermans, H. and Wiley, J. B. 1994. Experimental analysis of choice. *Marketing Letters*. **5** (4). 351–68.

Hanemann, W. M. 1982. Applied Welfare Analysis with Qualitative Response Models. Working Paper No. 241. Berkeley, CA: University of California.

Horne, P. and Ovaskainen, V. 2001a. Metsän ominaisuuksien arvottaminen virkistysalueilla. In J. Kangas and A. Kokko (eds.), Metsän Eri Käyttömuotojen Arvottaminen ja Yhteensovittaminen. Metsän Eri Käyttömuotojen Yhteensovittamisen Tutkimusohjelman Loppuraportti. Metsäntutkimuslaitoksen tiedonantoja 800. 242–249.

Horne, P. and Ovaskainen, V. 2001b. Luonnon monimuotoisuuden suojelu virkistysalueilla kävijöiden näkökulmasta. In J. Siitonen. (ed.). Monimuotoinen Metsä. Metsäluonnon Monimuotoisuuden Tutkimusohjelman Loppuraportti. Metsäntutkimuslaitoksen tiedonantoja 812. 223–226.

Hytönen, M. (ed.). 1995. *Multiple-Use Forestry in the Nordic Countries*. Finnish Forest Research Institute. Jyväskylä, Finland: Gummerus Printing.

Karjalainen, E. 2001. Metsänhoitovaihtoehtojen arvostus ulkoilualueilla. In J. Kangas and A. Kokko (eds.). Metsän Eri Käyttömuotojen Arvottaminen ja Yhteensovittaminen. Metsän Eri Käyttömuotojen Yhteensovittamisen Tutkimusohjelman Loppuraportti. Metsäntutkimuslaitoksen tiedonantoja 800. 183–185.

Louviere, J. J., Hensher, D. A. and Swait, J. D. 2000. *Stated Choice Methods. Analysis and Applications*. Cambridge: Cambridge University Press.

Miettinen, A. and Horne, P. 1999. Nuuksion Ulkoilualueiden Kävijätutkimus 1998. Helsingin kaupungin rakennusviraston julkaisuja 1999:2. Helsingin kaupungin rakennusvirasto, Helsinki, Finland.

Savolainen, R. and Kellomäki, S. 1984. The scenic value of the forest landscape as assessed in the field and the laboratory. *Communicationes Instituti Forestalis Fenniae*. **120**. 73–80.

Sievänen, T. 1992. Aulangon ja Ahveniston ulkoilualueiden käyttö ja kävijät. Metsäntutkimuslaitoksen tiedonantoja 415. Helsinki, Finland.

Swait, J. and Adamowicz, W. 1997. Choice task complexity and decision strategy selection. Staff paper 97–08. University of Alberta, Edmonton: Department of Rural Economy.

Tanguay, M., Adamowicz, W. L. and Boxall, P. 1995. An economic evaluation of woodland caribou conservation programs in Northwestern Saskatchewan. Project Report 95–01. University of Alberta, Edmonton: Department of Rural Economy.

Wellstead, A. M., Stedman, R. C. and Parkins, J. R. 2003. Understanding the concept of representation within the context of local forest management decision making. *Forest Policy and Economics*. **5**. 1–11.

*Part III*

# Policies for biodiversity conservation

*Section A*

# Contracts

# 14 Auctioning biodiversity conservation contracts: an empirical analysis

*Gary Stoneham, Vivek Chaudhri, Loris Strappazzon and Arthur Ha*

## 1 Introduction

A century ago in Australia, food and fibre were scarce relative to the supply of habitat. Today the opposite could be argued. Governments now face the problem of encouraging landholders to provide public goods, such as habitat conservation, in the face of an economic environment that facilitates the production of private goods. Governments, both in Australia and overseas, have used a wide range of policy mechanisms to influence private land management including fixed-price grants, tax incentives and voluntary schemes. Latacz-Lohmann and Van der Hamsvoort (1997) propose, however, that auctioning conservation contracts as a means of creating markets for public goods has many theoretical advantages. They argue that competitive bidding, compared with fixed-rate payments, can significantly increase the cost-effectiveness of conservation contracting because of the cost revelation advantages of bidding processes.

In this chapter we explain how the now extensive economic literature on auction and contract design, and new approaches to measuring habitat quality, can be incorporated into a practical field trial conducted under the name of BushTender©. Results from two pilots conducted in two different regions of Victoria, Australia, are presented and discussed.

The first BushTender© pilot was conducted in two areas of Northern Victoria and the second in three areas within West and East Gippsland (see Figure 14.1). Although we report the results of two BushTender© pilots, there have been several other applications of this approach in Victoria and more recently across Australia. These other applications focus on a variety of environmental goods and services including riverine habitat, native grasslands and carbon sequestration, and more recently a large pilot incorporating multiple environmental outcomes (carbon, dryland salinity, water quality, stream flow and terrestrial biodiversity) has been completed. Following these pilots, the Victorian Government

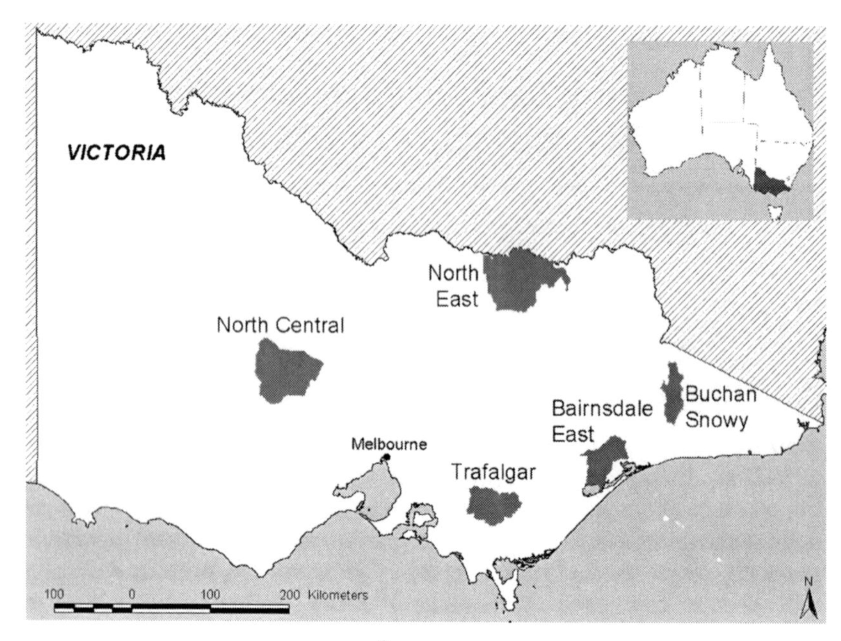


Figure 14.1. BushTender© pilot areas in Victoria, Australia

has endorsed BushTender© as a state-wide policy programme for habitat conservation.

## 2 Conservation of biodiversity on private land

### 2.1 *The problem of habitat loss on private land*

There are over a million hectares of native vegetation remaining on private land in Victoria. Crowe *et al.* (2006) report that managing native vegetation on private land is important for the conservation of native flora and fauna. Twelve per cent of Victoria's remaining native vegetation is on private land, 60 per cent of which is a threatened vegetation type (i.e. its conservation status is either endangered, vulnerable or depleted); and private land supports 30 per cent of Victoria's threatened species populations. There are also important land and catchment protection reasons for improving the management of vegetation on private land including the benefits for salinity, water quality, soil erosion and greenhouse emissions.

Conserving biodiversity on private land has been an important, but elusive, objective for government agencies. Despite government programmes, many important biodiversity assets on private land remain subject to degradation due to land-use practices such as livestock grazing, firewood collection and weed and pest invasion. Generally, it has not been feasible to include remnant vegetation on private land in the national reserve system through land purchase. Remnants are often of small scale and are spatially dispersed so that incorporating them into the public reserve system would involve high maintenance and protection costs and would not take advantage of local knowledge, expertise and resources. Public reservation will not protect all, or even most, biodiversity and 'off reserve' conservation will be required to protect biodiversity.

### 2.2 *Nature conservation programmes*

#### 2.2.1 *Australia*

In Australia, both State and Commonwealth Governments allocate large budgets to environmental and natural resource management. The Australian National Audit Office report on the Natural Heritage Trust in 2000–01 shows that this programme will have committed approximately US$2 billion to environmental works by June 2007 (Australian National Audit Office 2001). A further US$1.1 billion has been allocated to the National Action Plan for Salinity and Water Quality over a seven-year period by State and Federal Governments (Australian National Audit Office 2001). These and other environmental and natural resource management programmes employ a combination of intervention mechanisms including community and catchment-based planning, voluntary programmes, fixed-price subsidies and grants, education programmes and capital works programmes.

Although there is general acknowledgment that these programmes have altered community awareness about environmental issues, there is not a widespread belief that they have cost-effectively achieved significant on-ground outcomes. For example, the Australian National Audit Office (2001) commented on the Natural Heritage Trust by saying that the programme has been successful in raising awareness, but that it has not been so successful in achieving long-term landscape outcomes, and that cost-sharing, monitoring and administration have been poor. Thus, while achieving attitudinal shift, these programmes have been less effective at delivering and demonstrating improvements in the environment.

In Australia, State governments have substantial legislative responsibility for private land. Legislation controlling clearing of native vegetation – such as Victoria's Planning and Environment Act 1987 – is used in most states. Different states also offer financial incentives, such as assistance for fencing of remnants and weed and pest control (Denys Slee and Associates 1998). State governments sometimes make targeted purchases of land to address critical gaps in the reserve system, and revegetation programmes operate through grants to community groups. A range of voluntary programmes is supported by State governments. For example, a Victorian programme called Land for Wildlife aims to establish non-binding agreements with landholders for biodiversity conservation. The states, or state-based organisations, also offer programmes such as the Voluntary Conservation Agreement Programme in Queensland and Conservation Agreements (administered by the Trust For Nature) in Victoria (Denys Slee and Associates 1998). These schemes are legally binding and often have offsetting concessions such as rate relief, cash offsets or fencing concessions.

### 2.2.2 *International experiences*

In other countries, environmental agencies have implemented a number of policy mechanisms to deal with nature conservation on private land. The US has employed essentially two approaches: farmland protection easements and mechanisms that involve payments to landholders. The latter includes the Conservation Reserve Programme (CRP) and the Wetlands Reserve Programme (WRP) which are funded under the US Farm Bill. These programmes evolved partly from concerns over soil erosion and partly as assistance programmes for farmers. Predecessors to the CRP, such as the Soil Bank Programme, were introduced to divert land from crop production in order to reduce commodity inventories as well as to establish protective cover for land taken out of production (Wiebe, Tegene and Kuhn 1996). The CRP commenced in 1985 with broad environmental objectives and with a requirement that funds be allocated on a competitive basis. Currently, farmers bid for public funds based on an environmental benefits index (EBI), which scores landholders based on six environmental factors (wildlife; water quality; erosion; enduring benefits; air quality; and conservation priority areas) and a cost factor. The United States Department of Agriculture (USDA) selects contracts based on the EBI, but it has a reserve price based on the rental value of land adjusted for its productive capability. Other programmes, such as the USDA's Water Quality Incentive Program (WQIP), involve stewardship payments and the provision of technical information about surface and groundwater management. The WQIP employs fixed payments to landholders (Cooper 1997).

In Canada a Permanent Cover Program (PCP) has been introduced to encourage soil conservation and other environmental outcomes on farmland. The PCP employs a fixed payment approach with participating landholders required to engage in long-term contracts (including a buy-out option). Payments are determined on the basis of the length of the contract and the area involved.

Fraser and Russell (1997) provide an overview of agri-environmental schemes in the UK, three of which are relevant to nature conservation on private land: Environmentally Sensitive Areas (ESAs); the Conservation Stewardship Scheme (CSS); and the Nitrate Sensitive Areas (NSAs). The ESA and the NSA target farmers in specific geographic areas. The ESA focuses on the 'maintenance or enhancement of the environmental and landscape quality' and the NSA focus is on reducing the presence of nitrates in water. Both schemes offer fixed payments for undertaking certain actions. Wynn (2002) analysed the ESA scheme in Scotland and found that it did not target farms with high biodiversity, nor focus on low-cost producers. Wynn notes that better targeting would increase the cost-effectiveness of the scheme. The CSS targets environmental features, not geographic areas. The CSS offers a fixed payment for pre-specified actions, although, not all farmers who submit an offer are accepted. Instead, the CSS agency chooses farmers who offer the best quality management plans.

## 3 The economics of nature conservation on private land

### *3.1 Missing markets*

It is widely acknowledged that existing markets and institutions misallocate resources to environmental goods and services. While markets are generally efficient in allocating resources to commodity production, they may be ineffective or nonexistent with respect to creating 'environmental value'.

Ideas about why markets are missing or inefficient have changed over time. Coase (1960) argued that when property rights are clearly defined, market players will bargain to achieve an efficient solution (create a market), assuming that transaction costs are zero. However, when transaction costs are positive, the institutional arrangement that minimises these costs should be preferred. Thus the boundaries of the firm, and by extension, the market, are found by identifying the organisational form that minimises transaction costs.

The role of information in markets was first highlighted by Akerlof (1970). Subsequently, many economists have refined our understanding of how the distribution of information affects market players, and how these players may or may not respond to the problem (see, for example, Laffont 1990). It is now appreciated that information problems can destroy markets in extreme cases, or render markets inefficient because transaction costs diminish scope for value creation. The literature on information economics has forced economists and policy-makers to reassess policy mechanisms employed for many public policy problems. Likewise, there are new insights into policy mechanism design that arise from the application of information economics to environmental problems.

Using an information perspective, it can be seen that the problem with environmental goods is with the revelation of preferences for these goods (willingness to pay). The free rider problem associated with public goods hinders accurate revelation of preferences. One of the key solutions to this problem in the past has been via government stepping in and acting as demander on behalf of society. There has been much debate about what information should inform government allocation decisions, with many economists advocating valuation techniques such as contingent valuation. Whatever the manner in which demand-side preferences are expressed, we will argue in Section 5 that good information about the supply side (the focus of this chapter) is a complement to more efficient decisions regarding the creation and protection of environmental goods.

There are also important information effects that can be observed with respect to the supply side. Latacz-Lohmann and Van der Hamsvoort (1997) explain how information asymmetry affects the functioning of markets for environmental goods and services associated with private land. They note that there is a clear presence of information asymmetry in that: 'farmers know better than the program administrator about how participation (in conservation actions) would affect their production plans and profit' (Latacz-Lohmann and Van der Hamsvoort 1997: 407). Likewise, environmental experts, not landholders, hold information about the significance of environmental assets that exist on farm land. Further, landholders may not have all the relevant information about government priorities and are unlikely to understand how this information might influence subsequent contracts. Hence, although flat-rate Pigouvian taxes and subsidies may 'correct' market failures in circumstances where information asymmetry is not evident, other policy mechanisms will be needed when information is hidden. Latacz-Lohmann and Van der Hamsvoort (1998) conclude that: 'some institution other than a conventional market is needed to stimulate the provision of public goods from agriculture' (p. 334). They argue that auctions are: 'the main quasi-market institution used in other sectors of the economy to arrange the provision of public-type goods by private enterprises' (p. 335).

Auctioning conservation contracts is, therefore, a means of creating missing markets for nature conservation. The basic proposition is that markets for nature conservation are missing because of the asymmetric information problem and that policy mechanisms can be designed to reveal hidden information needed to develop meaningful transactions (markets) between government and landholders. It is contended that this process will facilitate price discovery and allow resources to be allocated where this has been difficult and inefficient in the past. The following sections draw on auction and contract design literature to identify the key features of this approach.

### 3.2 *Auction design*

Formal analysis of auctions in the economic literature is relatively new. While a complete literature review on the many design aspects of auctions is beyond the scope of this chapter, a broad understanding of the underpinnings of current theory is instructive. Early work on auctions stems from the seminal papers of Friedman (1956) for the case of a single strategic bidder, and Vickrey (1961) for the equilibrium game theoretic approach. The development of appropriate game theoretic tools has made auction theory an increasingly researched topic. The three broad models studied are: the independent private value model of Vickrey (1961), the symmetric common value model of Rothkopf (1969) and Wilson (1969, 1977) and the asymmetric common value model of Wilson (1969). Several survey articles summarise the auction design literature (see McAfee and McMillan 1987; Wolfstetter 1996; and Klemperer 2002).

#### 3.2.1 *Sealed bids*

The possibility of collusion between landholders bidding in an auction is always an important consideration in the choice of auction format. Repeated open, ascending and uniform-price auctions are generally more susceptible to collusion than a sealed-bid approach (see Klemperer 2002). Moreover, where bidders are risk averse, as we might well expect with private landholders, a first-price sealed-bid auction will facilitate lower bids because landholders can reduce commodity and weather-related income variability by adding a regular income stream from conservation payments (Riley and Samuelson 1981).

#### 3.2.2 *Single round*

Latacz-Lohmann and Van der Hamsvoort (1997) argue that a single round of bidding is preferred to multiple rounds because landholders are assumed to have independent private values rather than common values. In a private values model agents know their own valuations with certainty but make predictions on the values of others. While in the common values world, players have identical valuations but form their estimate on the basis of private information. In a common values world, agents will be able to learn about the 'common value' of the asset through the bidding strategies of all the other agents (as each agent has private information on the value of the asset). Thus, multiple rounds of bidding can facilitate information aggregation in the market and enable bidders to get a better sense of the true (common) value of the asset. Absent some mechanism for the efficient aggregation of such information, common value auction formats suffer from the 'winner's curse' (where the item is sold to the person with the most 'optimistic' private estimate of the true common value). However, where values are private and specific to each individual, information aggregation does not yield superior outcomes. Variation from farm to farm with respect to soil quality, rainfall, production systems etc. suggests that each landholder would base their bid on private, rather than common, information about opportunity costs and would be unlikely to alter this bid when given information about other landholders' valuations. In some contexts, we may expect affiliated values where there are both private and common value components in the bidding behaviour. This requires further attention in auction and policy design.

### *3.2.3 Discriminative price*

Where bidders draw valuations from different distribution functions, Myerson (1981) argues that optimal auction design is achieved by awarding contracts to the lowest bidders. Note that the performance of the auction format can be thought of from two perspectives. First, as in the Myerson (1981) case, which format maximises the value created, and second, how value is divided between the buyer and the sellers.

These questions lead to consideration of whether a one-price or price discriminating auction should be employed. Though the theory on optimal bidding strategies in a discriminatory price auction versus a one-price auction is inconclusive, it is worth noting that if both formats are successful in achieving truthful revelation, a discriminatory price auction is analogous to a first-degree price-discriminating monopolist. As such, there will be a change in the distribution of value, not the quantum of value created. Similarly, in the context of an auction of nature conservation contracts, the discriminatory price auction would, subject to the caveat highlighted above, achieve the same outcome as the one price approach, but at lower cost. Cason and Gangadharan (2005) examine the use of one-price versus discriminative-price auctions in an experimental setting. They find that bidding does change, and that the discriminative auction is more cost-effective. However, their assessment does not explicitly examine which auction format is superior with respect to economic efficiency.

### 3.2.4 *Hidden information*

Cason *et al.* (2003) used laboratory experiments to examine bidder behaviour in an auction when the value of their output was known, compared with when it was not. These experiments indicate that when bidders did not know the value of output, their bids tended to be based on the opportunity costs of land-use change. By contrast, when bidders were given information about the significance of their biodiversity assets, they tended to raise bids and appropriate some information rents.

### 3.2.5 *Budget constraint and no reserve price*

A reserve price strategy is a key element of auction design. While a reserve price will be less important where there is a budget constraint (see Myerson 1981; Riley and Samuelson 1981), this will not hold for repeated auctions. In repeated auctions it would be possible to transfer funds between rounds to maximise the nature conservation outcomes presented in other regions, or in subsequent auctions. An appropriately designed reserve price strategy would have implications for inter-temporal resource allocation as well as providing a means of spatially allocating conservation funds.

### 3.2.6 *Auction design in BushTender©*

The key design elements chosen for BushTender© auctions include: first-price, sealed bid, single round, price minimising and price discriminating format. A budget constraint applied to the auction and a reserve price was not formulated *a priori*. In the pilot auction, the exact value of the landholder's biodiversity asset was withheld from the landholder to improve the auction's cost-effectiveness. There are, however, other considerations that may influence this strategy. These are discussed later in the chapter (see section 5).

## 3.3 *Contract design*

There are many design issues that arise in the development of contracts between government (the principal) and landholders (agent) for the purpose of conserving biodiversity on private land. From contract theory, the main problems of contract design relate to incentives and asymmetric information. Specifically these problems are manifested as adverse selection, moral hazard and observability. Other problems of contract design include commitment, credibility and incomplete contracts (Salanie 1997).

Adverse selection refers to situations where agents have private information on their types that would be valuable to the principal in terms of contract design. In the case of nature conservation contracts, the opportunity cost of land-use is hidden from the principal but will be important in the selection of successful contracts and in the price associated with conservation services offered. The problem with adverse selection here is the payment of information rents to induce the agent to reveal private information (Salanie 1997).

Moral hazard refers to the problem of hidden actions. It arises where the principal is unable to observe the actions of an agent who in this case carries out the requirements of a conservation contract on farms that are often in remote locations. It leads to consideration of contracts that mitigate against agents 'shirking' their commitments (Laffont and Martimort 2002).

Even if contracts can be designed to prevent adverse selection and moral hazard, outcomes may still be unobservable. Observability is a problem with nature conservation contracts because it is difficult to measure and monitor the status and resilience of habitat for native plants and animals. For example, monitoring the impact of changes to land management in terms of the improvement in the stock and quality of fauna and flora would be very costly and subject to dispute. The level of observability has implications for monitoring and enforcement of contracts and their subsequent incentive effects on agents' behaviour (Laffont and Martimort 2002). An alternative strategy would be to specify a contract on the basis of inputs, such as fencing, weed control and understorey protection, that can be expected to improve habitat quality. These inputs are known to improve habitat status and resilience, but the transformation function that maps these actions (inputs) into outcomes is not known with certainty, even if the actions were carried out diligently. Further, the effect of unexpected events, such as drought and floods, could not reasonably be predicted by the agent (landholder), nor the principal (government).

These two problems (unobservability of outcomes and imperfect knowledge about the transformation function) were considered by Ouchi (1979), and explained in the context of the public sector by Wilson (1989). Williamson (1985) has characterised this as the problem of 'measurement'. The principal-agent literature has considered one or both of these problems to varying degrees (see, for example, Holmstrom and Milgrom 1991, 1994). This literature has recommended a host of ways to deal with these difficult problems, including: organising activities inside the firm; using fixed pay arrangements (again inside the firm); and contracting on the basis of inputs.

### *3.3.1 Contract design in BushTender©*

Conservation contracts for the pilot were developed based on inputs rather than outcomes, individual management agreements, menus of actions, progress payments and a monitoring and enforcement strategy. Input contracts were chosen because there were no low-cost means of measuring outcomes on which to base (enforce) these contracts. Because environmental benefits vary from site to site (non-standard benefits), individual management agreements specifying a schedule of management commitments were employed with progress payments made on the basis of agreed actions. This allowed the government scope to identify what actions were valuable, from a nature conservation perspective, and for landholders to choose a menu of actions that they preferred. For example, on some sites regenerating understorey was an imperative, whereas on others agreeing to not collect firewood (this action disturbs habitat) was relatively important.

Landholders were required to self-report on an annual basis. If a landholder did not report, or the report flagged non-performance, a counselling process was initiated before financial penalties were employed. If compliance was not forthcoming, payments were withheld, or in a worst-case scenario the contract terminated.

This type of contract has implications for risk bearing. Specifically, the government agency bears most of the risk associated with structural parameters where contracts are specified in terms of inputs. This was considered sufficient for the pilot, where the main purpose was to test the auction mechanism and the supporting information systems. However, improvements in knowledge (for example, new technology that allows lower-cost monitoring of species prevalence) may enable a government agency to base at least part of its payments on output.

## *3.4 Ecological service assessment*

Before transactions can occur between environmental agencies and landholders, certain information will be needed to avoid the lemons problem noted by Akerlof (1970). Two types of information will assist government to distinguish between different bids and different conservation actions that might be taken – information about the significance of habitat and information about service (habitat improvement).

### *3.4.1 Biodiversity significance*

Landscapes that have been modified for agricultural purposes will not necessarily retain a representative mix of habitat types. One way of expressing the conservation value of different types of habitat is with a Biodiversity Significance Score (BSS) where $BSS_i$ represents the biodiversity value of landholder i's remnant vegetation. The Biodiversity Significance Score draws on information about the scarcity of vegetation types and its Ecological Vegetation Classification[1] (NRE 1997).

### 3.4.2 *Habitat improvement*

There are a number of actions that landholders can take to improve the condition of habitat on private land. These include fencing to exclude stock from remnant vegetation, controlling environmental weeds and pests and minimising habitat disturbance by not harvesting firewood. The value of these habitat management actions, in terms of the improvement in habitat condition, can be expressed as a Habitat Services Score (HSS) where $HSS_i$ represents the change in quality of habitat from landholder i's habitat management actions. Parkes *et al.* (2003) developed a new metric called a 'habitat hectare' (referred to in section 4.2) to express the change in quantity and quality of habitat improvement.

Information about significance and habitat improvement was summarised in a Biodiversity Benefits Index (BBI) for each landholder i:

$$BBI_i = \frac{BSS_i \cdot HSS_i}{b_i} \tag{1}$$

In equation 1, ($b_i$) represents the nominal bid submitted by i to protect and enhance the remnant vegetation offered into an auction.

## 4 Results

Following advertisement of the auction, expressions of interest, assessment of proposed sites by ecologists and measurement of BBI, landholders submitted bids based on agreements that stipulated their selected actions. Bids were then ranked in ascending order according to BBI and contracts were written with the successful bidders up to a budget constraint. Tables 14.1 and 14.2 summarise the number of participants and sites assessed in the auctions.

Information about bids in the auctions is shown in Tables 14.3 and 14.4. Bids relate to either individual sites or, where landholders had multiple sites, bidders were given the option of submitting a combined bid for all their sites. A number of landholders chose to submit a combined bid. Joint bids between two or more landholders were not allowed in these trials.

[1] Ecological Vegetation Classes indicate whether vegetation is presumed extinct, endangered, vulnerable, depleted etc.

Table 14.1 *Northern Victoria BushTender© pilot – participation*

| Pilot phase | North Central | North East | Total |
|---|---|---|---|
| Expressions of interest (in pilot areas) | 63 | 63 | 126 |
| Properties assessed | 61 | 54 | 115 |
| Sites assessed | 104 | 119 | 223 |
| Hectares assessed | 1833 | 2006 | 3839 |

Table 14.2 *Gippsland BushTender© pilot – participation*

| Site assessment | Trafalgar | Bairnsdale East | Buchan Snowy | Total |
|---|---|---|---|---|
| Expressions of interest (in pilot areas) | 55 | 35 | 11 | 101 |
| Properties assessed | 37 | 22 | 9 | 68 |
| Sites assessed | 68 | 52 | 15 | 135 |
| Hectares assessed | 531 | 1134 | 702 | 2367 |

Table 14.3 *Bids for the Northern Victoria pilot*

| | North Central | North East | Total |
|---|---|---|---|
| Number of bidders | 50 | 48 | 98 |
| Number of bids | 73 | 75 | 148 |
| Number of sites | 85 | 101 | 186 |
| Number of successful bidders | 37 | 36 | 73 |
| Number of successful bids | 47 | 50 | 97 |
| Number of successful sites | 61 | 70 | 131 |
| Area under agreement (ha) | 1644 | 1516 | 3160 |

Table 14.4 *Bids for the Gippsland pilot*

| | Trafalgar | Bairnsdale East | Buchan Snowy | Total |
|---|---|---|---|---|
| Number of bidders | 27 | 19 | 5 | 51 |
| Number of bids | 43 | 25 | 5 | 73 |
| Number of sites | 48 | 42 | 9 | 99 |
| Number of successful bidders | 16 | 14 | 3 | 33 |
| Number of successful bids | 19 | 16 | 3 | 38 |
| Number of successful sites | 21 | 30 | 6 | 57 |
| Area under agreement (ha) | 262 | 906 | 516 | 1684 |

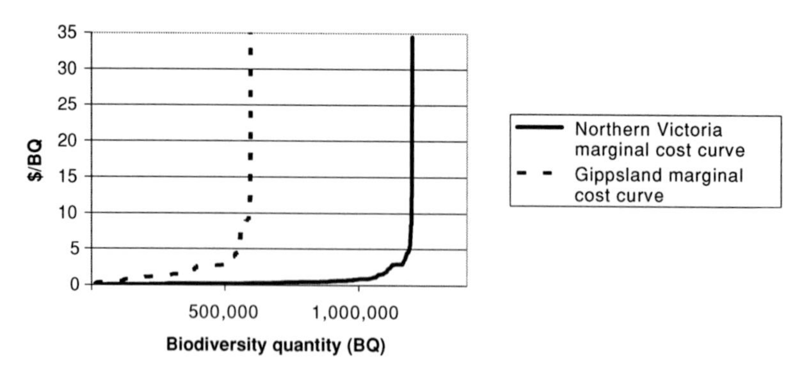


Figure 14.2. Supply curves from BushTender©

## 4.1 *Analysis of bids*

Drawing on information from the bids, Figure 14.2 illustrates the cost of generating additional units of biodiversity. We will henceforth refer to the curves in Figure 14.2 as supply curves for biodiversity.[2] The horizontal axis depicts the total quantity of biodiversity supplied (BQ). This measure is adjusted for biodiversity quality and is the numerator of the BBI as given in (1): the biodiversity significance score times the habitat services score. The vertical axis displays the (bid) price per unit BQ. As shown in Figure 14.2, the supply curves for biodiversity are relatively flat over much of the quantity range, but then transform to relatively steep as the quantity of BQs rises. All auctions of conservation contracts tend to generate supply curves that display the same general profile. This occurs because supply prices are derived by combinations of the distribution of opportunity costs and the distribution of habitat gains. Regional differences in opportunity cost tend to shift the supply curve but not to change its general characteristics. The steeply rising section of the supply curve tends to be due to declining biodiversity benefits of bids rather than rising offer prices.

Although it is difficult to compare the results from the auction with other mechanisms, it has been possible to examine how a hypothetical fixed-price scheme would perform compared with the discriminative price auction used in the pilot. To make this comparison, we must assume that the fixed-price scheme would operate as a one-price auction and

[2] The bids shown in Figure 14.2 are inclusive of any 'information rents' that bidders may have included in their bid price. We assume here that opportunity costs and information rents make up bids. This is different from the characterisation of Latacz-Lohmann and Van der Hamsvoort (1998), who differentiate the supply curve on account of it being exclusive of rents.

Table 14.5 *Comparison of fixed-price scheme to discriminating auction*

| | Northern Victoria | Gippsland |
|---|---|---|
| *Comparison holding biodiversity quantity constant* | | |
| Actual Budget (US$) | 325,817 | 629,403 |
| Budget required in fixed-price scheme (US$) | 2,113,600 | 1,632,900 |
| Proportionate increase in cost of fixed-price scheme | 6.5 | 2.6 |
| *Comparison holding budget constant* | | |
| Actual BQ | 1,165,019 | 530,099 |
| BQ of fixed-price scheme | 874,412 | 371,679 |
| Percentage fall in quantity from fixed-price scheme | 25 | 30 |

that bidder behaviour would not change in both auctions.[3] We justify this assumption by first recognising that landholders' behaviour would change with different schemes but theory or empirical evidence does not allow us to assume how the supply (bid) curve would change. In a fixed-price scheme, an agency would pay each successful landholder the same price: the price of the marginal offer. This is the price that an agency would need to offer to all landholders to generate the same supply of biodiversity made available from the price discriminating auction.

The results of this hypothetical comparison are given in Table 14.5. For the Northern Victoria pilot, a fixed-price scheme would require a budget of approximately US$2.1 million (over six times more than the budget allocated through the auction) to elicit the same quantity of BQ units as the discriminative price auction. Looked at another way: for the same budget of around US$325,000, a fixed-price scheme would give an agency approximately 25 per cent less biodiversity. A similar analysis for the Gippsland pilot shows that a fixed-price approach would require over 2.5 times the funding allocated through the auction.

Different contracts were employed in the two pilots reported. For the Northern Victoria pilot, landholders were offered a menu of actions but once actions were selected, standard contracts were used each with a three-year time span. In the Gippsland pilot a menu of contracts was offered including three- or six-year contracts, with the further option of ten-year or permanent protection following the active management period. The options chosen by landholders are summarised in Table 14.6.

[3] Assuming that the hypothetical fixed-price scheme is a one-price auction is probably going to overstate the benefits of a fixed-price scheme, but understate its administrative costs. This would be because most fixed-price schemes do not quantitatively score biodiversity outcomes, do not conduct landholder visits and do not advise on a range of possible landholder actions.

Table 14.6 *Management agreements taken up in Gippsland pilot*

| Successful contract type | per cent |
|---|---|
| Three years | 2.5 |
| Three years plus 10 years' protection | 0 |
| Three years plus permanent protection | 0 |
| Six years | 49 |
| Six years plus ten years' protection | 28 |
| Six years plus permanent protection | 20.5 |

Table 14.7 *Area of habitat secured under contracts: Northern Victoria*

| | Conservation significance | | | | |
|---|---|---|---|---|---|
| | Very high | High | Medium | Low | Total |
| Area secured under management agreements (in hectares) | 666 | 1540 | 934 | 20 | 3160 |
| Vegetation quality under management agreements (in habitat hectares) | 371.3 | 831.7 | 509.1 | 6.4 | 1718 |
| Habitat hectare rating | 0.56 | 0.54 | 0.55 | 0.32 | 0.54 |
| Change in quality (in habitat hectares) | 22.8 | 92.6 | 53.4 | 0.1 | 168.8 |

### 4.2 *Improvements in habitat quality and quantity*

Tables 14.7 and 14.8 summarise the expected improvements in habitat quality procured through the auctions reported. These data indicate that conservation contracts were allocated over 3,160 hectares of land in Northern Victoria and 1,684 hectares in Gippsland. For the pilot in Northern Victoria, the existing stock of vegetation was assessed at 1,718 habitat hectares which is expected to increase by 168.8 habitat hectares due to the interventions specified in contractual agreements with landholders. Habitat hectares is a metric developed by ecologists to measure the quality of vegetation relative to its pristine condition (see Parkes *et al.* 2003). A score of 1 represents pristine condition. As shown in Tables 14.7 and 14.8, the habitat hectare score in the Northern Victoria pilot averaged 0.54 and in Gippsland 0.69. The expected increase in habitat quality in the Gippsland pilot was 217.9 habitat hectares compared with the existing assessment of vegetation quality of 1,158 habitat hectares.

Table 14.8 *Area of habitat secured under contracts: Gippsland*

| | Conservation significance | | | | |
|---|---|---|---|---|---|
| | Very high | High | Medium | Low | Total |
| Area secured under management agreements (in hectares) | 610 | 231 | 744 | 99 | 1684 |
| Vegetation quality under management agreements (in habitat hectares) | 423.4 | 164.6 | 516 | 53.9 | 1158 |
| Habitat hectare rating | 0.69 | 0.71 | 0.69 | 0.54 | 0.69 |
| Change in quality (habitat hectares) | 81.4 | 37.7 | 113.2 | 13.3 | 217.9 |

### 4.3 *Awareness and participation in the auction*

Following the pilot auctions, surveys (380 samples) were conducted of participants and non-participants to collect information about: attitudes towards the auction; attitudes toward environmental issues; landholder demographics; and enterprise characteristics (see Ha *et al.* 2003). Logistic regressions were used to model awareness and participation (see Appendix).

In general, we conclude that membership of environmental groups or participation in environmental activities is the only consistent factor influencing awareness and participation. Demographic variables were important in both regions but there were no consistent trends across the two regions. Male and older participants displayed higher levels of awareness. Enterprise orientation was significant in determining awareness in one but not both regions. From the Gippsland results, it would seem that those who thought that BushTender© was a 'good idea', and landholders who believe native vegetation management is the responsibility of landholders, were more likely to participate. Other physical and demographic variables showed inconsistent trends across both the pilot regions. These data suggest that awareness and participation in auctions (at least in the two regions investigated) does not seem to be consistently influenced by the characteristics of landholders.

## 5 Discussion and conclusions

The pilot auction has shown that it is possible to create at least the supply side of a market for nature conservation: with a defined budget, prices can be discovered and resources allocated. Characterising nature conservation on private land as a problem of asymmetric information has improved our understanding of why this and related environmental markets are missing or ineffective and has introduced an alternative policy mechanism to those currently available. Auctioning nature conservation contracts offers many advantages over planning, command and control, voluntary approaches and fixed-price policy mechanisms.

The standard approach of dealing with missing markets by adopting coarse policy tools such as taxes, regulation and voluntarism has been questioned in the context of the information problems inherent in environmental landscapes. Policy mechanisms, such as auctions, that reveal and aggregate relevant information efficiently are likely to yield efficiency advantages and have implications for other environmental problems.

Many important design issues have been addressed in the process of implementing the auction. Besides choices about auction format, contract design and the specification of biodiversity preferences, many practical choices arise concerning communication with landholders, skills required to successfully run an auction and timing of activities. These factors all influence the performance of the auction.

Perhaps the most important finding from the pilot auction of nature conservation contracts is that where there are heterogeneous agents and non-standard environmental benefits, an auction offers significant cost savings over fixed-price schemes, such as subsidies and tax concessions. This comparison is made on cost-effectiveness, rather than economic efficiency grounds. For the budget available and the bids received, it has been shown that a price-discriminating auction would reduce by a large proportion the cost of achieving the same biodiversity improvement using a fixed-price approach. Moreover, a fixed-price approach – such as a fixed price per metre of fencing – essentially reveals the wrong information from the parties involved. It requires landholders to reveal the actions that they believe will improve the environment (when this information is perhaps held by environmental agencies); and agencies to reveal price that will be paid for these actions (when this information is often held by landholders).

The attraction of an auction of nature conservation contracts rests in the value of information revelation. The pilot auction was designed to reveal specific but previously hidden information from the agency responsible for nature conservation and from landholders. As part of the auction, the government agency had to derive a metric expressing the impact of land use change on the stock of biodiversity. The agency had to consider how to score the improvement in biodiversity associated with changes in land management (the Habitat Services Score) and the relative conservation status of different areas of vegetation (the Biodiversity Significance

Score). This information would significantly improve priority setting for nature conservation, whatever the mechanism employed.

Because the auctions were one-off pilots, it was not possible to observe their impact on other markets in the economy. However, there would be dynamic effects that could be important if such institutions became fully embedded in the economy. For example, land markets would reflect native vegetation 'value' in addition to production and location effects.

### 5.1 *Future directions*

Although the auctions described above were successful and popular with landholders, there remain many interesting design and implementation issues that deserve further consideration.

#### 5.1.1 *Repeated auctions*

The pilot auctions were constructed essentially as a 'one-shot game' between the government and private landholders. Design of a sequential auction, however, would be more complicated than the pilot because landholders could be expected to learn through rounds of the auction. Under these circumstances, landholders could change their bidding strategies and possibly raise the cost of nature conservation to the agency. For example, Riechelderfer and Boggess (1988) found that bidders in the Conservation Reserve Program – which is a sequential auction – revised bids from previous rounds by offering bids at the reserve price. The reserve price in this case was set as a per-hectare rate and when landholders learned this reserve price, they anchored their bids accordingly.

#### 5.1.2 *Multiple environmental outcomes*

Another interesting development would be to design auctions capable of dealing with multiple environmental outcomes from landscape change where these outcomes are complementary and/or competing. Revegetation of parts of the landscape may, for example, improve habitat quality and address land degradation. Auction theory is starting to make inroads into questions of how complementarities make market design difficult. Milgrom (2000) shows that complements to some bidders but not to others pose a threat to the existence of equilibria. Roth (2002) also notes that this problem arises in labour markets, such as the medical internship placement system, where couples prefer co-placement.

A second generation auction that includes multiple environmental goods and services (water quality, stream flow, land salinisation and terrestrial biodiversity), involving combinations of policy mechanisms (auctions and tradeable permits for carbon) has now been piloted in Victoria. This pilot (EcoTender) has raised a number of further considerations particularly with respect to the way scientific information is included in auctions, buyer aggregation procedures, methods of determining preferences for environmental goods and services and the way that other environmental markets – such as tradeable permit systems – interact with auction mechanisms in the future (see Strappazzon *et al.* 2003).

### 5.1.3 *Information hidden versus information revealed*

One of the most interesting design issues with the pilot auction of conservation contracts was the extent to which information was made known to landholders prior to formulation of their bids. In the pilots reported above, some of the information about the biodiversity metric was withheld from landholders: they knew the Habitat Services Score but not the Biodiversity Significance Score. While this strategy was empirically supported on cost-effectiveness criteria by laboratory experiments (see Cason *et al.* 2003), other considerations suggest that full disclosure of information about biodiversity significance may be appropriate. In the short-run, withholding some information limits the scope for landholders to extract information rents from the auction. Clearly, if landholders knew that they had the only remaining colony of some plant or animal, they would be able to raise their bid well above opportunity cost, compared with a situation where this information were not known. The alternative strategy also has merit in that (i) the information rents that accrue to landholders would influence land markets and encourage investment in nature conservation; and (ii) landholders would know exactly what scarce biodiversity assets they have and could self-select into the auction process improving the matching between government priorities and the bidders in an auction.

### 5.1.4 *Reserve prices and demand valuation*

The purpose of the auctions to date has been to develop an understanding of the cost of obtaining the next units of biodiversity (the supply side of the market). However, government will always have to make decisions about budgetary expenditure in the light of information about both costs and benefits – the demand-side of the equation. This raises the question of whether society is prepared to pay for the next unit of biodiversity and at what price. Viewed in this way, a reserve price strategy requires a government to bring together notions of both supply (opportunity cost) and demand (willingness to pay).

As stated in 3.1 it is notoriously difficult to elicit truthful revelation of non-priced values. However, allocating a budget to a biodiversity auction implies that there will be some bids that are rejected, and hence there will be an implicit valuation of a contract price that is too high.

The best way to consider the demand side of biodiversity is still very contentious. Myriad articles have been written on how to explicitly value goods such as biodiversity. Some have argued that the aim of explicitly valuing biodiversity is futile and that governments – as representatives of society – should make choices about variables such as the biodiveristy budget. Whatever prevails, it is clear that an approach such as BushTender© helps to reveal information about the supply price of the next unit of biodiversity, determined on a competitive basis. This information could be used to populate part of the information space in which resource allocation decisions are made.

Reserve prices will also become more important as subsequent auctions are run. In this case, governments would face the decision of purchasing within the current auction where marginal costs rise (as demonstrated in figure 14.2) or differing purchases to subsequent auctions to take advantage of better conservation contracts.

#### 5.1.5 *Funding models*

The BushTender© approach opens up several avenues for increased funding for biodiversity services. The first is that government may be more likely to fund conservation activities if the outcomes are more visible, reportable and cost-effective. The auction approach is rich in terms of the data that it provides to decision-makers and should therefore make it easier to convince decision-makers about its value. In other words, it provides the potential buyers with a more accurate description of the type and quantity of product that can be procured with a given budget.

Governments in Australia are beginning to see this approach as a useful tool and are starting to allocate funds to it. For example, the Victorian Government has recently endorsed BushTender© as an official government programme with an on-going financial commitment. A number of projects funded under the Commonwealth's programme of US$3.9 million on Market Based Instruments were influenced by the BushTender© approach. Another application has occurred with the national biodiversity stewardship component of the Commonwealth of Australia's Biodiversity Hotspots programme.

#### 5.1.6 *Information to facilitate cross-programme comparisons*

Other indirect benefits could arise from the application of auctions and other market approaches to environmental management. For example, information about the marginal cost of habitat conservation would assist public sector decision-makers in allocating resources between conservation investments on public (eg. national parks) and private land. Similarly, the emergence of more formalised and quantitative methods of expressing relative preferences for alternative environmental actions may facilitate development of more robust offset and trading schemes.

### *5.1.7 Site synergies*

Auctions employ contracts to facilitate transactions with individual landholders. However, the aim of habitat conservation schemes may at times involve many landholders whose actions are interdependent (site synergies). Currently, the index for biodiversity developed for BushTender© attempts to take site synergies into account by using a 'landscape context' scoring element. Further research is needed to refine ways of representing habitat interdependencies. There may also be benefits from further research into alternative auction format and contract design problems. With respect to auction format, there appears scope to consider and pilot a combinatorial auction that would assist bidders to interact and discover efficient package of contracts.

## REFERENCES

Akerlof, G. A. 1970. The market for 'Lemons': quality uncertainty and the market mechanism, *Quarterly Journal of Economics*. **84**. 488–500.

Australian National Audit Office. 2001. Performance Information Commonwealth Financial Assistance under the Natural Heritage Trust Audit Report No. 43, Commonwealth Government of Australia, Canberra.

Cason, T., Gangadharan, L. and Duke, C. 2003. A laboratory study of auctions for reducing non-point source pollution. *Journal of Environmental Economics and Management*. **46**, 446–471.

Cason, T. and Gangadharan, L. 2005. A laboratory comparison of uniform and discriminative price auctions for non-point source pollution. *Land Economics*. **81**. 51–70.

Coase, R. 1960. The problem of social cost. *The Journal of Law and Economics* **3**. 1–44.

Cooper, J. C. 1997. Combining actual and contingent behaviour data to model farmer adoption of water quality protection practices. *Journal of Agricultural and Resource Economics*. **22**. 30–43.

Crowe, M., Todd, J., Parkes, D., Macfarlane, S., Stoneham, G. and Strappazzon, L. 2006. The BushTender Trials Evaluation Report. East Melbourne: Department of Sustainability and Environment.

Denys Slee and Associates 1998. Remnant Native Vegetation – Perceptions and Policies: A review of legislation and incentive programs Environment Australia Research Report 2/98.

Fitzsimons, J. and Wescott, G. 2001. The role and contribution of private land in Victoria to biodiversity conservation and the protected area system. *Australian Journal of Environmental Management*. **8**. 142–57.

Fraser, I. and Russell, N. 1997. The economics of UK agri-environmental policy: present and future developments. *Economic Issues*. **2**. 67–84.

Friedman, L. 1956. A competitive bidding strategy. *Operations Research*. **4**. 104–112.

Ha, A., O'Neill, T., Strappazzon, L. and Stoneham, G. 2003. BushTender Participation in First Bidding Round: What are the Characteristics of Rural Landholders who Participated? *Paper presented to the 47th Annual Conference of the Australian Agricultural and Resource Economics Society, Fremantle*. February.

Holmstrom, B. and Milgrom, P. 1991. Multitask prinicipal-agent analyses: incentive contracts, asset ownership, and job design. *Journal of Law, Economics, and Organisation*. 7 (Special Issue). 24–52.

Holmstrom, B. and Milgrom, P. 1994. The firm as an incentive system. *American Economics Review*. **84** (4). 972–991.

Klemperer, P. 2002. What really matters in auction design. *Journal of Economic Perspectives*. **16** (1). 169–189.

Laffont, J. 1990. *The Economics of Uncertainty and Information*, Cambridge, MA: MIT Press.

Laffont, J. and Martimort, D. 2002. *The Theory of Incentives: The Principal-Agent Model*. Princeton, NJ: Princeton University Press.

Latacz-Lohmann, U. and Van der Hamsvoort, C. 1997. Auctioning conservation contracts: a theoretical analysis and an application. *American Journal of Agricultural Economics*. **79**. 407–418.

Latacz-Lohmann, U. and Van der Hamsvoort, C. 1998. Auctions as a means of creating a market for public goods from agriculture. *Journal of Agricultural Economics*. **49** (3). 334–45.

McAfee, R. P. and McMillan, J. 1987. Auctions and bidding. *Journal of Economic Literature*. **25**. 699–738.

Milgrom, P. 2000. Putting auction theory to work: the simultaneous ascending auction. *Journal of Political Economy*. **108**. 245–272.

Myerson, R. B. 1981. Optimal auction design. *Mathematics of Operations Research*. **6**. 58–63.

NRE 1997. *Victoria's Biodiversity: Directions in Management*. The State of Victoria, Department of Natural Resources and Environment, Melbourne.

NRE 2002. *Victoria's Biodiversity: Directions in Management*. The State of Victoria, Department of Natural Resources and Environment, Melbourne.

Ouchi, W. 1979. A conceptual framework for the design of organizational control mechanisms. *Management Science*. **25** (9). 833–48.

Parkes, D., Newell, G. and Cheal, D. 2003. Assessing the quality of native vegetation: the habitat hectares approach. *Ecological Management and Restoration*. **4**. S29–S38.

Riechelderfer, K. and Boggess, W. G. 1988. Government decision making and program performance: the case of the conservation reserve program. *American Journal of Agricultural Economics*. **70**. 1–11.

Riley, J. and Samuelson, W. 1981. Optimal auctions. *American Economics Review*. **71**. 381–392.
Roth, A. E. 2002. The economist as engineer: game theory, experimental economics and computation as tools of design economics. Fisher Schultz lecture. *Econometrica*. **70** (4). July. 1341–1378.
Rothkopf, M. 1969. A model of rational competitive bidding. *Management Science*. **15**. 774–777.
Salanie, B. 1997. *The Economics of Contracts: A Primer*, Cambridge MA: The MIT Press.
Strappazzon, L., Ha. A., Eigenraam, M., Duke, C. and Stoneham, G. 2003. Efficiency of alternative property right allocations when farmers produce multiple goods under the condition of economies of scope. *Australian Journal of Agricultural and Resource Economics*. **47**. 1–27.
Vickrey, W. 1961. Counter-speculation, auctions, and sealed tenders. *Journal of Finance*. **16**. 8–37.
Wiebe, K., Tegene, A., and Kuhn, B. 1996. Partial interests in land – policy tools for resource use and conservation. *Agricultural Economic Report*. **744**. USDA.
Williamson, O. 1985. *The Economic Institutions of Capitalism*. New York: Macmillan, Inc.
Wilson, J. Q. 1989. *Bureaucracy: What Government Agencies Do and Why They Do It*. New York: Basic Books.
Wilson, R. 1969. Competitive bidding with disparate information. *Management Science*. **15**. 446–448.
Wilson, R. 1977. A bidding model of perfect competition. *Review of Economic Studies*. **44**. 511–518.
Wolfstetter, E. 1996. Auctions: an Introduction. *Journal of Economic Surveys*. **10** (4). 367–420.
Wynn, G. 2002. The cost-effectiveness of biodiversity management: a comparison of farm types in extensively farmed areas of Scotland. *Journal of Environmental Planning and Management*. **45**. 827–40.

## APPENDIX

### *Modelling*

Using information from a survey of landholders (both participants and non-participants), data on awareness and participation was modelled using logistic regressions. Tables 14.A1–14.A4 report results from regression models of the factors that influence awareness of the BushTender© scheme. For participation, the results report on the factors that determine the decision to participate.

Variables were chosen by first testing whether the variable was a significant explanator of the dependent variable in question in a univariate regression. All variables are then included in an initial model which was refined using significance tests until a minimum Akaike information criterion value was reached. See Ha *et al.* 2003 for more details.

Table 14.A1 *Northern Victoria awareness model*

| Dependent variable: awareness of BushTender© scheme (380 observations) | |
|---|---|
| Variable | Coefficient |
| *Action variables* | |
| Actively increasing or managing remnant vegetation by: | |
| – cleaning up or maintaining area | 0.837** |
| – establishing soil erosion measures | 1.625** |
| – undertaking good farm practices | 2.852** |
| – controlling and/or monitoring rabbits | 1.591* |
| Member of either: Alpine Valley, Land for Wildlife, VSS, Grasslands Society, Meat and Livestock Corporation, Target 10, Women in Agriculture, Landcare, Sustainable Grazing Systems, North Eastern Stud Breeders, Olive Growers Association, Agricultural Society and other agricultural groups | 1.431*** |
| Respondent or spouse is a member of an organisation concerned with land protection or the environment | 1.224*** |
| In the past three years participated in either a Heartlands, Country Fire Authority, 20/20, fencing, soil erosion, salinity control, native planting, Bushcare, Project Platypus, Hindmarsh Biolink, duck boxes, organic farming, Murray River Action Group, wildlife monitoring/rescue, local groups, Target 10, roadside management, other or unknown environmental groups | 1.011*** |
| Regularly reads the Chronicle | −2.117** |
| Native vegetation, bushland or unimproved pasture is used for: | |
| – grazing of livestock | −0.698*** |
| – weed control | 0.937*** |
| *Perception variables* | |
| The amount and quality of native vegetation within 10–15 km of property is very good | −0.665** |
| Decline of wildlife due to habitat loss | 0.969* |
| *Demographic variables* | |
| The respondent is male | 0.742*** |
| Respondent's spouse is aged sixty years or more | 0.884*** |
| *Constant* | −1.799*** |
| Akaike Information Criterion (AIC) | 1.116 |
| McFadden R-squared ( per cent) | 24.4 |

Note: All estimates rounded to the nearest third decimal place unless otherwise shown
'*', '**', '***' denotes estimate significant at the 90, 95 and 99 per cent level respectively
'^' denotes estimate was insignificant at the 95 per cent level

Table 14.A2 *Gippsland awareness model*

| Dependent variable: awareness of BushTender© scheme (380 observations) | |
|---|---|
| Variable | Coefficient |
| *Action variables* | |
| Actively increasing or managing remnant vegetation by: | |
| – fencing off vegetation areas | 0.564* |
| – protecting remnant vegetation | 0.597** |
| Member of: | |
| – Landcare | 1.167*** |
| – Land for Wildlife | 1.25^ |
| Participated in an environmental programme run by: | |
| – Landcare | 0.923* |
| – Land for Wildlife | 2.211* |
| Participated in an environmental programme in the last 3 years | −0.84** |
| Native vegetation, bushland or unimproved pasture is used for sheltering stock | 2.215*** |
| *Perception variables* | |
| Strongly disagree that it is difficult to find useful information on native vegetation and biodiversity | 0.818** |
| Believe that landowners should take more responsibility for managing native vegetation on their properties | |
| – disagree | 1.626*** |
| – neither agree nor disagree | −1.738*** |
| Observed decline in local vegetation due to other reasons | 0.973^ |
| Observed decline in wildlife due to urban sprawl | −1.661^ |
| *Physical variables* | |
| Type of farm enterprise: | |
| – Sheep | 0.676* |
| – Beef | 0.491* |
| – Cropping | −3.294** |
| Trafalgar locality | −0.621** |
| Proportion of property that is improved pasture ( per cent) | −0.001^ |
| *Demographic variables* | |
| Respondent's education: | |
| – formal training in agricultural or land management | 0.715** |
| – post-graduate qualification | 1.932*** |
| Respondent is male | 0.737** |
| Years in locality | −0.032*** |
| Respondent's age is less than 30 | −1.887*** |
| Spouse has attained a trade qualification | −1.033** |
| *Constant* | −0.466^ |
| Akaike Information Criterion (AIC) | 1.078 |
| McFadden R-squared ( per cent) | 31.7 |

Note: All estimates rounded to the nearest third decimal place unless otherwise shown
'*', '**', '***' denotes estimate significant at the 90, 95 and 99 per cent level respectively
'^' denotes estimate was insignificant at the 95 per cent level

Table 14.A3 *Northern Victoria participation model*

| Dependent variable: participation in BushTender© scheme's expression of interest stage (167 observations) | |
|---|---|
| Variable | Coefficient |
| *Action variables* | |
| Heard about BushTender© from a radio programme | −2.53** |
| Respondent or spouse is a member of an organisation concerned with land protection or the environment | 1.224*** |
| Regularly reads industry journals | −2.064* |
| Native vegetation, bushland or unimproved pasture is used for planting trees or shrubs | 0.518^ |
| *Perception variables* | |
| Respondent does not think about native vegetation management and biodiversity very much | −1.176** |
| *Constant* | −0.96** |
| Akaike Information Criterion (AIC) | 1.208 |
| McFadden R-squared ( per cent) | 16.4 |

Note: All estimates rounded to the nearest third decimal place unless otherwise shown
'*', '**', '***' denotes estimate significant at the 90, 95 and 99 per cent level respectively
'^' denotes estimate was insignificant at the 95 per cent level

Table 14.A4 *Gippsland participation model*

| Dependent variable: participation in BushTender© scheme's expression of interest stage (189 observations) | |
|---|---|
| Variable | Coefficient |
| *Action variables* | |
| Heard about BushTender© from a radio programme | −4.53** |
| Involved with Land for Wildlife as a: | |
| – programme participant | 5.883*** |
| – member | 3.101** |
| Actively increasing or managing native vegetation by: | |
| – fencing off native areas | 1.987*** |
| – controlling weeds | 1.113^ |
| Native vegetation, bushland or unimproved pasture is used for livestock grazing | −1.754** |
| *Perception variables* | |
| BushTender© involves the community, which is a: | |
| – very good idea | 3.942*** |
| – good idea | 3.961*** |
| Strongly agree that is its own responsibility to manage native vegetation and biodiversity | 1.953** |
| Strongly agree that enthusiastic when it comes to native vegetation and wildlife protection | −2.073** |
| Amount and quality of native vegetation within 10–15 km of property is good | −1.515** |
| Strongly disagree with learning more about native vegetation and biodiversity management | −5.559** |
| Thinks BushTender© is a good idea | −2.818*** |
| Somewhat agree that had a positive impact on the quality and quantity of native vegetation on my property | −1.882** |
| Local vegetation is very good for other reasons | 7.116*** |
| Somewhat agree that it is very difficult to find useful information on native vegetation and biodiversity | 2.457*** |
| *Physical variables* | |
| Trafalgar locality | −3.44*** |
| Proportion of property that is unimproved pasture ( per cent) | 0.026* |
| *Demographic variables* | |
| Spouse has attained tertiary education | 1.168^ |
| Years in locality | −0.044** |
| Children are involved in land management decisions | 2.872** |
| *Constant* | −0.539^ |
| Akaike Information Criterion (AIC) | 0.69 |
| McFadden R-squared ( per cent) | 62.1 |

Note: All estimates rounded to the nearest third decimal place unless otherwise shown
'*', '**', '***' denotes estimate significant at the 90, 95 and 99 per cent level respectively
'^' denotes estimate was insignificant at the 95 per cent level

# 15 An evolutionary institutional approach to the economics of bioprospecting

*Tom Dedeurwaerdere, Vijesh Krishna and Unai Pascual*

## 1 Introduction

There is a significant strategic interest by 'Northern' industries of accessing and using genetic resources (GR) and associated traditional knowledge (TK) from the South. Such repository of bioresources in the South co-evolves through the development of TK and the continuous GR refinement adaptations in natural and managed ecosystems. The North/South debates over ownership, intellectual property rights (IPR) and access to the GR-TK stock were crystallized in the negotiations of the United Nations' Convention on Biological Diversity (CBD) which came into force in 1993, and now establishes the legal framework for the reciprocal transfer of bioresources between countries (Bhat 1999). In fact, the CBD stands as the only major international negotiated instrument that makes explicit provisions for the special link between TK, biodiversity and local and indigenous communities by granting rights to the latter in order to protect TK (Bodeker 2000).[1] The CBD also regulates bioprospecting activities carried out by industrial (usually Northern) firms and it assigns a formal protocol for sharing the benefits from bioprospecting activities based on the 'access and benefit sharing' (ABS) agreement to GR-TK between the parties.[2] In addition, it also calls for a free prior informed consent to be obtained from the holders of GR-TK prior to the bioprospecting activities taking place (Berlin and Berlin 2003). In addition,

[1] The recently ratified UN Convention to Combat Desertification (UNCCD) also makes explicit such provisions. Interestingly such provisions are included directly in the UNCCD text, while the CBD itself becomes more detailed in the later, and still non-binding, Bonn guidelines. The FAO International Treaty on Plant Genetic Resources for Food and Agriculture, adopted in 2001, also makes explicit this link. However, it covers only the plant genetic resources for food and agriculture and the treaty only attributes to the local communities the right to participate in the decision-making processes at the national level. Moreover, the 'access and benefit sharing' provisions become legally binding only if transited into national legislation.

[2] Here we use the term 'bioprospecting' following the definition by ten Kate and Laird (1999, p. 19), i.e. the research, collection and utilisation of biological and genetic resources, for purposes of applying the knowledge derived from it for scientific and/or commercial purposes.

the CBD effectively asserts the property rights of the bioresources and GR in particular to the source country (c.f. CBD Article 15: Access to Genetic Resources).

However, in many instances the rights of GR-TK holders, including the source country governments and indigenous/local communities, are being erased and replaced by those who have exploited their biogenetic and TK through prospecting endeavours. Such cases of biopiracy are being reported more frequently (Sheldon and Balick 1995; Shiva *et al.* 1997; Drahos 2000; Dutfield 2002a; Verma 2002).[3] The CBD acknowledges that when effective ABS systems are removed, it creates disincentives for *in-situ* conservation of the GR-TK stock. Against this backdrop, the debate on the conflicting approaches to IPR with regard to domesticated and wild bioresources and associated TK is re-emerging in order to devise ways of defensive protection against the misappropriation by bioprospectors (Dutfield 2002b).

In order to evaluate the potential contribution of benefit sharing systems to local communities and other relevant parties, a number of studies have focused on estimating the value of bioprospecting using a wide array of approaches (Principe 1989; Pearce and Purushothaman 1992; Simpson *et al.* 1996; Rausser and Small 2000; Craft and Simpson 2001). Broadly speaking, these studies assess the value of bioprospecting using standard cost-benefit analysis, in which the opportunity cost of land conservation, among others, is weighted to assess the expected benefits related to the discovery of a new useful property of a bioresource (net of the associated R&D costs such as biological material screenings). In the light of the debate on how to address the IPR problem, this chapter addresses the question whether such 'static' analyses are appropriate to approximate the social welfare loss from depreciating the GR-TK stock through non-adequate or absent North-South bioprospecting contracts and ABS agreements. We draw insights from contemporary economic analyses of contracts and property rights based on (evolutionary) institutional economics. The aim of this chapter is to address the challenge to build concepts that are better adapted to the specific character of the bioresources and that take into account their evolving

[3] The word 'biopiracy' was first introduced by Pat Mooney of the Rural Advancement Foundation International (now known as ETC, Action Group on Erosion, Technology and Concentration). RAFI defined biopiracy as 'the use of intellectual property laws (patents, plant breeders' rights) to gain exclusive monopoly control over genetic resources that are based on the knowledge and innovation of farmers and indigenous peoples' (RAFI 1996, p. 1).

(versus static) nature and the collective character of the associated traditional knowledge.

An important challenge in static valuation analyses of the costs and benefits from bioprospecting activities to local economies is the diffuse character of the values, both monetary and/or non-monetary, created by biodiversity within (intrinsically) complex and adaptive socio-ecological systems. The added value of biological resources does not arise at the final stage of the innovation process only. Instead, added value is created at each step of the innovation process – from the ecosystem itself creating the diversity of GR, through the contributions of the local communities' TK, the research laboratories and to final industrial applications and marketing (Swanson 2000). This implies that the existing IPR mechanisms that are associated with the property of the final stage of the innovation chain only address the tip of the iceberg. It thus remains insufficient as a mechanism for rewarding and adding value in all the other stages (Goeschl and Swanson 2002; Laird 2002). Furthermore, the current IPR mechanism remains insufficient for addressing the wider social values associated with the flow of resources and information generated by biodiversity (Brush 1996). For instance, in the case of TK, IPRs may conflict with the collective nature of indigenous knowledge and the importance of cultural and religious values towards nature.

Under such conditions, it seems appropriate to adopt a 'dynamic' approach to assess the use value of biodiversity in terms of conserving GR stocks and associated TK when benefits through bioprospecting can be realised (Dedeurwaerdere 2004). Such an approach incorporates the notion of bounded rationality and a broader vision of economic rationality (Driesden 2003), alongside the dynamics of economic and cultural change outside the view of a static (equilibrium) situation (North 1990). Accordingly, the focus shifts away from a narrow concern about the optimal allocation of existing resources (based on a static cost-benefit analysis mentality), to one about issues of dynamic efficiency. This entails focusing on knowledge acquisition throughout the entire process of value creation and incentives for the preservation of future possibilities of innovation and use of GR-TK under conditions of uncertainty.

By arguing in favour of a dynamic approach, new questions arise which have to be addressed in the implementation of any governance mechanism that is adopted, be it of a market, communal or public nature. That is, any mechanism that aims at valuing the diversity of GR and associated TK through bioprospecting needs to address the question regarding the creation of institutions for coordinating the diversity of social values associated with biodiversity and the enabling of collective learning processes in situations of intrinsic uncertainty.[4] This implies that an analysis of the full chain of innovation is necessary to assess the potential benefits from bioprospecting. Such analysis should address in a comprehensive and systematic manner the interconnected roles of the ecosystem, the local communities, the research community as well as private companies. Within this framework, this chapter attempts to show key shortfalls of static valuation approaches in the context of designing efficient benefit sharing agreements (e.g. through monetary compensation in terms of the current IPR framework) to the holders of valuable GR-TK sought by bioprospectors.

We use the case of a unique ABS biodiversity contract in India as an example of how the monetary valuation of TK/GR may be assessed directly from the perspective of the TK holders themselves. This analysis allows us to identify some of the key gaps in static analysis of similar ABS cases that tend to focus primarily on the final stages of the innovation chain. The case study presented in this chapter is based on a widely acclaimed model of ABS that involves the Kani tribe of the Western Ghats (WG) in India (Anuradha 1998; Moran 2000). The WG is a 160,000-km$^2$ eco-region shared by six southern Indian states: Gujarat, Maharashtra, Goa, Karnataka, Tamil Nadu and Kerala. It is one of the 25 biodiversity hotspots that have been identified globally with an estimated 10,000–15,000 plant and animal species, of which about 40 per cent are endemic (WGF 2003).

Besides being a hotspot for biodiversity, the Kani tribe of the WG has become well known for its model of benefit sharing. The Kani model of benefit sharing (KMBS henceforth) is recognised as the first instance in which payments have been made to the TK holders for a successfully developed pharmaceutical product with therapeutic properties.[5] This product is based on *Trichopus zeylanicus*, a small perennial herb that is distributed in Southern India, with the subspecies *Travancoricus* being found only at an altitude of approximately 1000 m (Anuradha 1998). After the incidental *in-situ* 'discovery' by a group of scientists of the therapeutic properties of the herb, the local botanical garden formulated a herbal tonic, known as Jeevani or 'the ginseng of the Kani people', that bolsters the human immune system. The production technology was then transferred to a private Indian pharmaceutical company for

[4] For an overview of the literature on institutional economics and the analysis of bioprospecting, cf. the special issue of *Ecological Economics* on Access and Benefit Sharing (Siebenhüner *et al.* 2005).

[5] The KMBS received the 'Equator Initiative award' from the UNDP for developing a novel benefit sharing model during the World Summit on Sustainable Development at Johannesburg in 2002.

its commercialisation and the company agreed to compensate the Kani community through the intermediation of a locally established welfare trust.

The rest of the chapter is structured as follows: first, we address some key questions that point towards the reasons for the inadequacy of the incentive mechanisms under current ABS regimes that lead to socially sub-optimal levels of investment in biogenetic resources as a source of innovation. The discussion is then applied to qualify the degree of 'success' of a unique bioprospecting case based on the KMBS. After briefly describing this bioprospecting case, the KMBS is analysed from a wider institutional angle. This allows us to address a specific question regarding the actual CBD-based access and benefit sharing system drawing on the acclaimed KMBS case: how does the realised KMBS agreement compare to the value of the compensation implicitly requested by the local community for sharing their traditional knowledge? We finally draw some policy conclusions from the analysis.

## 2 From a static to a dynamic IPR framework in access and benefit sharing contracts

The existing mechanisms for the regulation of bioprospecting contracts involve two main parties, the industrial sector in the 'North' (mainly the biotechnology and pharmaceutical sectors) and the providers of the biogenetic resources in the 'South' (mainly local communities, botanical gardens and government administrations). Two basic features are inherent in the contracts. Firstly, the contracts aim at providing an incentive for innovation through the IPR on the finished product at the end of the production line. Secondly, they aim at protecting the providers' rights through the insertion of clauses in the contract with regard to the *free prior informed consent* to be obtained from the holders of GR-TK and the equitable sharing of the benefits from the development of commercial applications, i.e. the 'access and benefit sharing' clause. Since the CBD came into force, numerous ABS agreements have been signed and analysed (see, for example, Mulligan 1999; Svarstad and Dhillion 2000; Peña-Neira *et al.* 2002).

CBD and ABS agreements are dependent on a static notion of efficiency that has characterized the classical economic analysis of regulation (Dedeurwaerdere 2005). This notion is linked to the idea of optimal allocation of existing resources under ideal conditions of perfect rationality. Moreover, it has characterized environmental policy during the last two decades, resulting in an intensive application of cost-benefit analysis in the determination of the objectives of environmental regulation and the recourse to economic incentives as the means to achieve these objectives, increasingly through the creation of markets for environmental goods and services (Driesden 2003; Pearce 2006). By contrast, a dynamic conception of efficiency focuses on the acquisition of new knowledge allowing the maximisation of the range of future choices of the product development processes.

In the context of regulations for the conservation of GR-TK, the actual approach by the CBD is largely based on the static approach ultimately seeking to provide the 'right' incentives to effective GR-TK conservation through market creation. The problem is that the actual IPR mechanisms rely in valorising (i.e., adding value) to GR-TK at the final stage of the innovation process. By contrast, the dynamic approach seeks to address each step of the innovation process from the ecosystem as the repository of co-evolutionary GRs to the industrial applications, and through the added value of local communities' TK and scientific research laboratories. This implies that there is a need to create incentives for innovation along the entire chain of the innovation process. In the broader field of biodiversity governance, there is already an increasing recourse to tools aiming to implement such a dynamic approach. Such mechanisms can include the creation of trust funds dedicated to the conservation of biological diversity[6] or certification schemes monitoring the flow of resources along the process of value creation (Barber *et al.* 2003; Gulbrandsen 2004), such as the International Plant Exchange Network (IPEN) for the exchange of biological resources between botanic gardens[7] or the unique identifier system for transgenic plants developed by the OECD.[8]

[6] The most recent example of such a fund on a global scale is the Global Crop Diversity Trust established in 2004 as a public-private partnership of FAO and the 15 Future Harvest Centres of the Consultative Group on International Agricultural Research (CGIAR). An early case for Trust Funds in the field of biodiversity has been made by Thomas Eisner (Eisner and Beiring 1994); other prominent examples (though with mixed success) are the Genetic Recognition Fund established at the University of California Davis (Gupta 2004) or the Healing Forest Conservancy of Shaman Pharmaceuticals. For an overview of the different types of trust funds in the ABS field and their use, cf. Guerin-McManus *et al.* (2002). For a discussion on the design principles of a biodiversity trust fund, cf. Swanson (1997).

[7] All plant material supplied by an IPEN member needs to be accompanied by an IPEN number that remains connected with the material and its derivatives through all generations to come. With the aid of this number it is possible to track where and under which conditions the plant entered the network.

[8] See OECD documents: (ENV/JM/MONO (2001) 5; 2001 and ENV/JM/MONO (2002) 7; 2004). OECD describes the unique identifier as being a key attributed to a biotech product, which could unlock information from a range of databases, as well as an harmonised unique entry point enabling information management related to that product.

This brings us to the debate on the necessity to move beyond the actual ABS provisions of the CBD. This debate also joins discussions about the new forms of governance that emerged in the 1990s as being linked to an overly simplified conception of the path of application of the norms of regulation, both in economic theory and in the theory of legal regulation. In particular, if the evolutionary economics approach of Nelson and Winter (1982) and Dosi (1988) is followed, the conception of efficiency at work in the emerging regime of ABS can be criticized (Driesden 2003). For instance, expanding on theoretical insights by evolutionary institutional economics (Dopfer 2005), a broader vision of the rationality governing the economic decisions of parties engaged in bioprospecting agreements (e.g. government agencies and businesses) can be obtained. A key factor is the analysis of how institutional objectives have to cope with behavioural routines and partial information.[9]

The actual ABS agreements based on a static idea of efficiency have a double limitation as regards providing effective incentives for biodiversity conservation in the context of the actual IPR mechanisms. The first limitation is situated at the level of the short time-scale considered in the ABS agreements, which is inappropriate for dealing with a long-term investment in biological resources. The static approach tends to lock in the innovation process by providing only institutional incentives related to the current market opportunities and not addressing the future options of development. The second limitation is that the static view of ABS is incapable of dealing with the integration of the 'distributed knowledge' generated along the entire innovation chain. Instead, it focuses on IPRs where benefits and ownership can more easily be established. These two limitations are addressed in more detail in the following paragraphs.

The first limitation in the actual IPR model that constrains CBD as regards bioprospecting activities arises due to the overwhelming attention to those products that are '*currently*' interesting to the industry, making the bilateral contract mechanisms considered in the ABS regime inadequate from a social perspective which is concerned with the long-term investment in biological resources. The main reason for this inadequacy from a broader socio-economic point of view is the lack of investment in biogenetic resources that will potentially be productive in the future. Hence, the actual IPR mechanism is inadequate regarding a resource that is itself evolutionary by definition (Swanson and Goeschl 1998). An

[9] Examples of behavioural routines and bounded rational behaviour are developed more extensively in Dedeurwaerdere (2005). For example, conservation policies for agricultural genetic resources should take into account cooperative habits and insurance mechanisms in rural communities, such as informal seed exchange amongst farmers (Brush 1998).

illustrative example is that of the agricultural sector in which a highly productive, competitive seed that is resistant to pathogens is introduced. This introduction induces an adaptation in the population of pathogens in a way that can make them more 'aggressive', therefore enhancing the relative fitness of successful mutants adapted to intensively cultivated crops (Swanson and Goeschl 1998) or by increasing resistance of the pathogens to pest control technologies (Goeschl and Swanson 2002). As a result, the resistance of these newly introduced productive seeds decreases with time and its latent competitive disadvantage needs to be taken care of permanently by adapting the seeds and/or the means of production in reaction to the adaptation of the population of pathogens in the environment. Similar mechanisms operate in the pharmacological field, where one observes, for example, a decrease in the effectiveness of antibiotics and anti-malarial products (*Ibid.*).

Moreover, coupled to the evolutionary nature of GR, the associated TK and know-how also co-evolves with the bioresources (Brush 1996), adding another layer of complexity to the process of generating and using biological diversity. Yet the IPR mechanism creates an artificial monopoly on a productive seed or an effective drug, in the present, but it does not stimulate the investment in potentially useful biological resources able to cope with new populations of pathogens in the future. In order to maintain the innovation process over the long term, an incentive for the maintenance of a population of biogenetic resources that is potentially productive in the future needs to be established, for example satisfying the constant need for new innovations which can thwart the dynamics of natural evolution of pathogens.

The second limitation arises due to the focus of the ABS agreement on the 'end of the pipeline' of the knowledge generation process, where benefits and ownership can clearly be established, and not addressing the other stages of the innovation process, where ownership in knowledge is distributed amongst different players and benefits are highly uncertain.

Solving the problem of the uncertainty about the potential value of these contributions to knowledge generation by only compensating the few fortunate cases of bioresources that make it to the marketplace is a poor strategy from an economic perspective. Figure 15.1 represents the problem of uncertainty by adapting the scheme proposed for analysing a four-step industry (Swanson 2000) to the case of knowledge generation for research/industry input through bioprospecting. The latter depends on an investment in the resource at the level of (1) ecosystems that produce GR diversity; (2) communities of local users (traditional farmers, healers, etc.) that co-evolve and manage the bioresource stock; (3) the scientific community doing research into new properties; and (4) product

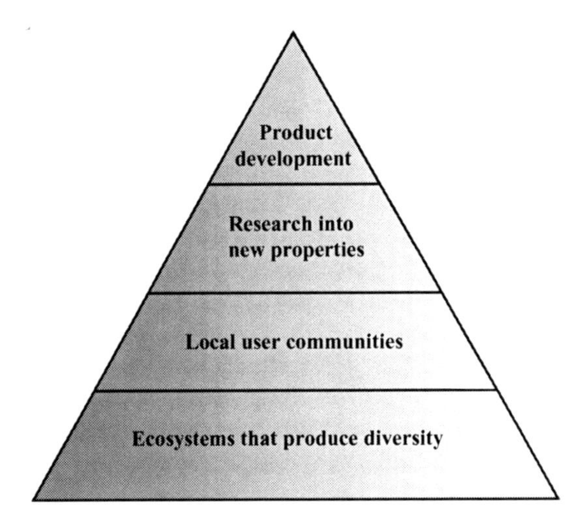


Figure 15.1. The bioprospecting chain

development. At each of these steps, the outcome of the investment is uncertain and, moreover, the investment at each stage is motivated by a broader set of social values than merely utilitarian values related to potential monetary benefits.

It is clear that the current ABS mechanism does not address the entire innovation chain. The contracts generally regulate those cases in which the development of an effective marketable product is likely (ten Kate and Laird 1999), or the case of a specific sector (e.g. cancer research) where such a development can be anticipated. But the problem is that in reality all of those involved in the initial stages of the innovation process are in a period of intense experimentation, knowledge gathering, exchange of materials and information, etc. with outcomes that are difficult to predict. Innovative biotechnological applications reach only so far because they are standing on the shoulders of giants, e.g. the scientific merits of researchers, the cultural heritage of so many years of traditional seed and other bioresource improvements and the social networks of exchange of knowledge and resources (Brush 1998). The point is that bioprospecting depends on initiatives at different stages of the innovation chain to guarantee a permanent flow of creation and regeneration of valuable biogenetic resources.

The double limitation addressed here leads effectively to sub-optimal levels in biodiversity conservation investment as a source of innovation. This idea fits with Goeschl and Swanson's (2002) view point about the three main kinds of insufficiencies that result from actual ABS regimes, all based on incentives relying on the existing IPR mechanisms. First, the IPR mechanism does not offer sufficient incentives to invest in products that have a short life span. It thus creates an underinvestment in GRs with high adaptability. Second, the IPR mechanism creates a trend of monopolisation and is therefore not compatible with the requirements of an innovation process based on diversity. Third, the IPR mechanism acts at the level of individual companies and does not create an incentive to invest in the other stages of value creation whose benefits are diffuse. In particular, it produces an underinvestment at the level of the ecosystem and its local or indigenous users.

The interest of addressing this triple insufficiency from the point of view of a dynamic approach is to show the necessity to change the static efficiency notion underlying the actual IPR framework which in turn governs bioprospecting and constrains actual ABS systems for GR/TK conservation. We argue that there is a need to progress towards a conception that better accounts for the collective character of the innovation process and the relationship between the natural evolution of GRs and efficient markets for such resources. The question that arises then is: what are the consequences of using a dynamic framework for the economic analysis of bioprospecting contracts?[10]

As shown by North (2005), dynamic efficiency ultimately depends on the cognitive belief structure of the broader community involved, such as the beliefs underlying science and democracy, which have played an important historical role in organising processes of permanent inquiry and social learning. However, these beliefs are the evolutionary product of centuries of path-dependent institutional change. The complexity of this historical process is beyond the scope of any empirical relevant model of dynamic efficiency, so that no general dynamic theory that is useful is likely to be developed (North 2005, p. 71–78, p. 125–126). Nevertheless, North also indicates some more modest and pragmatic goals that should be the object of an economic analysis of dynamic efficiency (*Ibid*: p. 163–164). These involve four complementary goals: (1) analysing why dynamic efficiency locks in suboptimal development paths; (2) understanding the cultural heritage of a society and the margins at which the belief system may be amenable to changes; (3) developing the institutional and organisational framework for capturing the productivity potential inherent in integrating the dispersed knowledge essential to efficient production in a world of specialisation; and (4) analysing the conditions for more effective monitoring of the political system. While

[10] Several general methodological consequences have been drawn from these insights on dynamic efficiency, most importantly in Aoki (2001, p. 387); Eggertsson (2005, p. 184) and North (2005, p. 155–165). Here we follow in particular the cognitive framework put forward by North (2005).

the second and fourth goals are beyond the scope of this chapter, the first and the third have been the focus of our analysis thus far, and it is to these two objectives that we turn in the next two sections.

In the next section, we address the reasons why bioprospecting contracts are not able to realise the full potential for biodiversity conservation as a condition for economic development. In order to do so a unique bioprospecting case from the Western Ghats of India is introduced. After this bioprospecting case is described, an institutional 'fitness' analysis is carried out in order to point out how even such seemingly successful cases may prove unsustainable and therefore fail their long-run conservation and development goals. We argue that in order to achieve a more sustainable contract design an alternative institutional design needs to be adopted. This is the one in which the holders of TK (i.e. the Kani community) also need to be full 'owners' of their TK. This would effectively allow them to directly enter into the ABS contract to obtain what they would perceive to be a fair amount of compensation for sharing their TK with any given bioprospecting company or alternatively decide upon another (and from their point of view, legitimate) use of their TK.

Then section 4 carries out an assessment of the way in which this alternative institutional context is perceived as being a source of economic progress by the different contractual agents (i.e. the TK holders, the private commercial company and the State). The two alternative situations that are addressed are (1) the full transfer of IPRs to the private company, as is the case in current contract described in section 3, versus (2) a situation in which the TK holders retain full ownership of their TK. We argue that the latter case would provide a more sustainable contractual design because it takes into account the perception of a key agent in the innovation chain, i.e. the local community, which is not taken into account in the actual contract design.

## 3 The Kani model of benefit sharing (KMBS): An institutional fitness analysis

This section introduces and then analyses the widely acclaimed Kani model of benefit sharing (KMBS) in the Western Ghats of India from an institutional economics perspective. The focus is on addressing the appropriateness of the 'evolutionary rules in use' in such ABS cases (even in those qualified as 'successful') drawing on the idea of 'institutional fitness' (Folke *et al.* 2002; Brown 2003). Such fitness is largely determined by flexible and open institutions that allow for multi-scale governance systems which in this case could facilitate the adaptive capacity of ABS systems within the CBD framework.

Before the institutional 'misfits' of the KMBS that limit the scope of a more complete ABS system are addressed, let us first describe in a nutshell this ABS case that prides itself on being a unique case in which actual payments have been made to the TK holders for a successfully developed commercial therapeutic product (Anuradha 1998).

The Kani community comprises around 18,000 people spread across 30 settlements and villages mostly in the forests of the Agasthiyar Hills of the Western Ghats in Kerala (some few households are also located in the border state of Tamil Nadu). This area is designated as a reserved forest, rich in biodiversity and strictly regulated by the Forest Department of the State Government. Following a visit to the reserve by a group of scientists, an *in-situ* 'incidental discovery' of the therapeutic properties of a small perennial herb, *Trichopus zeylanicus*, known as *Sathan Kalanja* or *Arogyappacha*, locally and traditionally consumed to reduce fatigue (Pushpangadan *et al*, 1988) took place.[11] On the basis of the discovery, the Tropical Botanical Garden and Research Institute (TBGRI) from Kerala standardised a herbal tonic to bolster the immune system and provide energy known as Jeevani ('provider of life') and formulated with *T. zeylanicus* in combination with three other medicinal plants. Then in 1996 the production technology was transferred to an Indian pharmaceutical company, Arya Vaidya Pharmacy Coimbatore Ltd (AVP). The TBGRI licensed *Jeevani* to AVP, and it agreed to share the licence fee of Rs 1 million (about US$23,000) and a royalty of 2 per cent on the profits with the Kani community on a one-to-one basis.

This was then followed by the creation of a local Trust Fund for the Kanis known as the 'Kerala Kani Community Welfare Trust', first registered with members from the Kani tribe. In 1997 the amount due to the Kanis was transferred to the Trust with the understanding that it was to be used for welfare-enhancing activities of the Kanis (Sahai 2000). More specifically, under the establishment of the Kani Welfare Trust, the KMBS was based on the transfer by AVP of Rs 519,000 to the account of the Trust (Rs 500,000 as the 50 per cent of the licence fee and the rest the first instalment of royalties from the sale of the drug, which up to 2003 generated Rs 100,000).[12] The mode of expenditure of the Trust was decided by majority voting.

[11] The phytochemical and pharmacological studies of *T. zeylanicus* have revealed the presence of certain rare glycolipids and non-steroidal polysaccharides with profound adaptogenic, immuno-enhancing, antifatigue properties.

[12] The inadequate supply of the leaves of the herb was the main reason for the relatively low amount of royalty accrued during this period. Subsequently, the pharmaceutical firm AVP began to use a limited quantity of raw drug collected from another Western Ghat region of the nearby state of Tamil Nadu.

Once Jeevani started to be marketed, the fast proliferation of domestic and international markets for the herbal tonic necessitated regular supply of fresh leaves of *T. zeylanicus*. Since the wild collection was both inadequate to meet the market requirements and could create ecological overexploitation due to being habitat-specific (the therapeutically active compounds are produced only when the herb is cultivated in and around its natural habitat), AVP proposed a plan for the cultivation of *T. zeylanicus* to the Kerala Forest Department, part of the State Government, and the Tribal Welfare Department. According to this plan, the AVP would enter into a buy-back arrangement with the local community to buy the leaves harvested from the cultivated plants. The firm was prepared to buy five tonnes of leaves per month and the TBGRI trained fifty Kani households for a pilot-level cultivation season in 1996 by availing a subsidy of Rs 1,000 for each cultivating household. However, due to the lucrative nature of the leaf sale of *T. zeylanicus*, the local community began to collect the whole plant from its natural forest habitat. This induced the Forest Department to proscribe its cultivation, fearing the ultimate extinction of the wild varieties through overexploitation.[13] It was not until several years of negotiation concluded in 2003 that the Forest Department re-issued consent to cultivate the herb and the Kanis were in a position to bargain for a better price for their 'cash crop'. However, the contract with AVP lasted only another six months, and the pharmaceutical firm was unwilling to negotiate a new price contract. The monetary benefit flow from the KMBS is illustrated in Figure 15.2.

Despite the acclamation of the KMBS, we argue that it has not yet achieved its full potential due to various institutional impediments. These are based on the conflict of interests and coordination problems between the local botanical garden (TBGRI), the Forest Department, the pharmaceutical firm and the Kani local community. For example, whereas the TBGRI as a part of the State Government licensed AVP to manufacture the drug, the Forest Department did not facilitate the manufacturing process (Anuradha 1998). Hence, improper coordination amidst various governmental bodies led to partial execution of the scheme. Moreover, the major source of income from the ABS would have come from the supply of *T. zeylanicus* leaves for drug manufacturing. However, the Kanis could harvest only two crops in 1996, and their effective bargaining made AVP offer a threefold increase in the price of the raw drug (from Rs 25/kg of fresh leaves to 75/kg). But due to fear of overexploitation of the herb

[13] TBGRI tried with only limited success to develop a propagation technique through tissue culture seedlings.

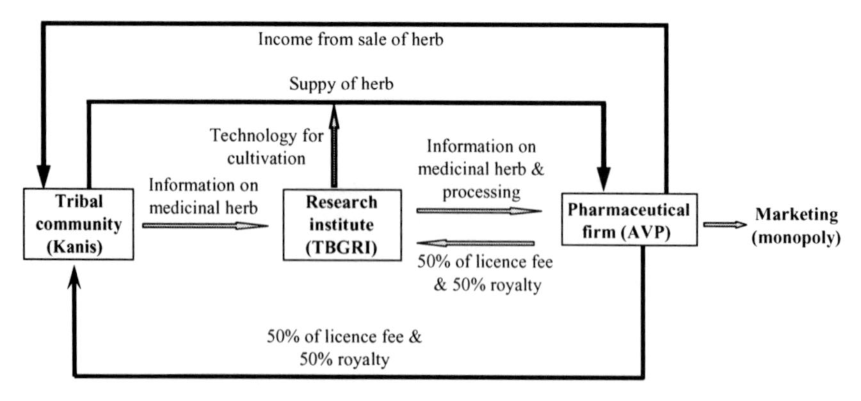


Figure 15.2. The monetary benefit flow from the KMBS
Source: Adapted from TBGRI Scientists, the TRUST trusties and Anuradha (1998), Moran (2000), Pushpangadan (2000) and Gupta (2002)

from its natural forest habitat, the State Forest Department banned its cultivation.[14]

The 50 households which first cultivated the herb witnessed a significant increase in income given the low opportunity cost of family labour. As a result, more households began to cultivate the plant in the next growing season. Despite the small size of the area for cultivation by each household (average of 0.1 ha), its cultivation allowed households to generate an average net revenue of Rs 1,123 and Rs 849 respectively during the two harvests in 1996 (the Rs 1000 subsidy given by the ITDP being primarily responsible for the higher figure for the first crop). Hence, had the scheme been implemented according to the proposal by AVP (in which a monthly demand of 5 tonnes of fresh leaves was anticipated), the community could have earned a minimum of Rs 4.5 million annually at a fresh leaf price of Rs 75/kg. Even without taking into account the associated increase in royalty (due to the increased raw drug supply and resulting higher level of production and sale), the income forgone by the Kanis is significantly greater than what they had achieved.

But this begs the question of whether the cultivation in the forest reserve would have been ecologically sustainable. Moran (2000) has expressed concern over the present system of sourcing *T. zeylanicus*, since there is no information on sustainability studies connected to methods of managing and harvesting the herb. There are countless examples of why mere

[14] It bears a resemblance to the harvest of the entire adult population of *Maytenus buchananni* (a source of anticancer compound Maytansine) by the US National Cancer Institute in Kenya for testing its drug development programme (Oldfield 1984; Reid *et al.* 1993).

market creation for bioresources need not always facilitate conservation (Barrett and Lybbert 2000). In fact, in this case unregulated bioprospecting and drug development could speed up the destruction of the resource. This experience points out that the question of the control and sanction mechanisms for dealing with overexploitation of the wild variety and illegal trade should also be addressed.

The ABS agreement with the Kani was established on a voluntary basis and not on a broader legal framework for regulation of bioprospecting, specifying the rights and duties of the TBGRI and private companies. In this situation, even with a clear incentive for the Kani members involved in the contract to adopt sustainable management practices, there could be no guarantee that other groups would not free-ride on the contract through exploitation of the wild variety or, alternatively, that the pharmaceutical company would not look for other providers of the same plant under less restrictive conditions (as, in fact, it subsequently did).

Moreover, the appropriate protection of the rights of the indigenous community over its TK also depends on the existence of such guarantees. In the case of the Kanis, the disclosure of their traditional knowledge to the Indian scientists was entirely based on trust and good faith. It was based on the belief that they would honour their promise of benefit sharing in case of the development of a new product. Hence, it is not possible to replicate the contract automatically to other situations, where these relationships of trust may not be robust. Under these conditions, the incentive to disclosure TK by other communities remains limited to situations where personal relations and informal guarantees that their property rights will be protected and that the contract will lead to appropriate benefit sharing exist.[15]

Lastly, looking at the Kani example we can explore whether the focus on the issues of IPR and the associated ABS system has not shifted the attention away from the question of the involvement of other actors in the negotiation of the contract. In the Kani case, the contract is clearly the outcome of an agreement negotiated between scientists from the TBGRI and the AVP pharmaceutical company, which in turn was initially based on a confidential agreement between the scientists and the Kanis. The property right holders of the physical asset, the forest administration and the members of the tribal community, seem to have been involved only marginally in the drafting of the terms of the contract, and consequently the legitimacy of the agreement is not recognised with the same intensity by all the actors. In particular, as Ramani (2001) shows, different

[15] In other cases, such as the Costa Rican InBio-Merck agreement, an ABS agreement is already signed at this stage.

perceptions subsist between the younger and the older tribal Kani members, the latter caring more about the loss of cultural identity.[16] This lack of legitimacy may be due to the fact that the focus of the TBGRI had been on the bilateral contract between the pharmaceutical company and those Kani guides who transmitted the local knowledge to the Indian scientists, thus by *de-facto* acknowledging them as the original providers of the TK-GR. This clearly begs the question of the possible disregard of the role of the majority of the community members and that of the Forest Department.

## 4 Valuing the bioresource from the TK holders' perspective

The classic model of bioprospecting in the case of a GR-TK system, such as in the KMBS case, involves three main actors: (1) the ecosystem as the natural repository of the GR base, (2) the indigenous community acting as stewards of the ecosystem and thus the GR-TK base, and (3) the commercial firm interested in the search for new chemicals from nature. Here we pay special attention to the second node of the chain: the local community as the custodian of TK. We seek to provide an approximate estimate of how the Kani community values its role in the innovation chain leading to the successful commercialisation of Jeevani. That is, the interest is in shedding light on the Kanis' willingness to pay (WTP) for protecting their TK with regard to the external appropriation of bioresources and on the various household socio-demographic and economic characteristics that affect their implicit valuation. The results can be interpreted more directly as the level of compensation that representative members of the Kani community demand for their involvement in the *T. zeylanicus* bioprospecting activities by the botanical garden and the pharmaceutical firm. We carry out this analysis by employing a contingent valuation study.

The monetary benefits realised from the current Kani ABS scheme reach the community in the form of cash payments to the Trust. Since the rights to the service under consideration (the use of TK) are held by the local community, compensation for participating in the biodiscovery process by disclosing its traditional ethnobotanical knowledge would be the appropriate format for value elicitation (Shyamasundar and Kramer 1996). One difficulty of using the WTA format is that the local community receives indirect payments through the provision of public goods to the community by the Trust, making direct elicitation of WTA less precise

[16] Concerns have been raised by the elder tribe members that the expected welfare benefits could be outweighed by the loss of traditional medicinal practices (Ramani 2001).

in reflecting households' preferences. Hence, the question posed to the Kani community members is based on the maximum WTP to protect their traditional knowledge from outside illegal appropriation.[17]

The survey for the statistical analysis was carried out in 2004 in the Western Ghats. The statistical sample is made up of 68 households randomly selected from ten settlements of the Kanis and stratified into cultivators (50 per cent) and non-cultivators of *T. zeylanicus* (50 per cent). Using the local language (Malayalam), household heads were invited to report on households' socio-economic characteristics, the management of *T. zeylanicus* cultivation, and various aspects concerning the knowledge and attitudes towards the implementation of the bioprospecting contract and protection of their traditional knowledge.

The average annual per capita income of the surveyed household was found to be Rs 7,727 (about US$176 at 2005 prices) with 68 per cent of income arising from homestead farming in about one hectare of land that includes crops such as coconut, tapioca, banana, betel nut, black pepper and rubber. Approximately 20 per cent of income accrues from wage labour and 12 per cent from selling various permitted non-timber forest products (NTFP) such as wild gooseberries, asparagus, honey and nutmeg.

The contingent valuation study is based on a dichotomous choice model and the results are shown in Table 15.1 together with a description of variables. The hypothetical scenario presented and the question posed to the households is the following: 'Suppose a pharmaceutical firm markets a herbal medicine using the traditional knowledge of the Kanis without asking for your prior consent. In this regard, the Trust or any other NGO (dealing with Kani welfare) has decided to bring this particular firm to court. If the Trust/NGO wins the case, the right to the use of this particular traditional knowledge will rest within the community only, or alternatively the community may get a fair amount of compensation for sharing the knowledge. The Trust/NGO decides to collect money from Kani tribes to meet the court expenses. In this regard, would you be willing to donate Rs X to the fund?'[18] This dichotomous choice question was followed up by two more questions which asked respondents whether they would be willing to pay a higher or lower amount, setting upper or

[17] The estimated Kanis' WTP value for protecting their TK through the CV study is possibly a lower bound of the true compensation required, as suggested by most studies comparing WTP and WTA values (e.g. Adamowicz *et al.* 1993; Shogren *et al.* 1994; Morrison 1997).

[18] The bids of the first WTP question ranged from Rs 50 to Rs 400 with a constant interval of Rs 50. The amount was specified as a one-time payment.

Table 15.1 *Variable definitions and estimated double bounded dichotomous choice model*

| Variables | Description and measurement (mean ± std. deviation) | Coefficient (Std. error) Model I | Model II |
|---|---|---|---|
| Constant | | −706.46 (535.16) | −414.67* (230.53) |
| Per capita income# | Per capita annual income of household in 000 rupees (7.73 ± 6.61) | 98.38** (42.16) | 71.47** (33.66) |
| Age# | Chronological age of the respondent in years (33.31 ± 12.00) | 45.54 (108.32) | – |
| Education# | Formal education attained by the respondent in years of schooling (4.00 ± 4.14) | −95.79** (49.19) | −95.83** (47.21) |
| Household size# | Number of members in the household of respondent (4.03 ± 1.47) | 23.62 (99.00) | – |
| Farm size# | Size of farm managed by the household of respondent in acres (2.97 ± 1.94) | −20.84 (52.01) | – |
| Wage labour | 1 if respondent participates in the non-farm labour market, 0 otherwise (63 %) | −159.62** (64.68) | −146.71*** (57.35) |
| Remote# | Distance between respondent's household to public transport facility in kilometres (9.27 ± 3.89) | 107.99 (89.49) | 83.81 (76.33) |
| City# | Frequency of visiting nearby city by respondent in number per month (8.34 ± 4.91) | 126.79** (64.67) | 112.50* (58.62) |
| Adults | Proportion (0–1) of adult members in the family size (0.77 ± 0.32) | −50.38 (114.61) | – |
| Community development | 1 if the respondent actively engaged in community development activities, 0 otherwise (32 %) | −5.42 (67.50) | – |
| Read | 1 if respondent read newspapers regularly, 0 otherwise (46 %) | 182.53 (116.12) | 173.48* (100.52) |
| Radio | 1 if respondent listened to radio programmes regularly, 0 otherwise (78 %) | 94.46 (67.74) | 70.85 (61.09) |
| Television | 1 if respondent watched television programmes regularly, 0 otherwise (54 %) | 166.52*** (63.14) | 174.59*** (59.15) |
| Cultivator | 1 if respondent was engaged in *Trichopus* cultivation, 0 otherwise (50 %) | 106.26* (59.03) | 97.50* (55.07) |
| NTFP | 1 if respondent engaged in non-timber forest product collection, 0 otherwise (81 %) | 70.84 (80.53) | – |
| Herb Consumption | 1 if respondent consumed *Trichopus* fruits regularly, 0 otherwise (87 %) | 113.71 (87.76) | 153.99** (78.68) |
| Log likelihood function | | −65.04 | −65.88 |
| $\chi^2$ | | 36.86 | 35.17 |

Notes: Sample size, N = 68. Coefficients can be directly interpreted as marginal effects
*, **, and ***: statistically significant at 0.1, 0.05 and 0.001 levels, respectively
#Variables are taken in their natural logarithmic form

lower bounds. A double-bounded dichotomous choice (DBDC) format is thus used (Hanemann *et al.* 1991). These questions intend to capture the Kanis' view of the prior-informed consent aspect within the ABS system. It should be noted, however, that even prior-informed consent was granted, the question does not help to resolve how this is obtained or who decides that it is obtained in a legitimate way (Berlin and Berlin 2003). Here, the DBDC model just tries to capture the effect of various socio-economic factors on Kanis' willingness to donate to the proposed fund as a proxy to their efforts to protect their TK from misappropriation.

Answers to the two sequential WTP questions of DBDC format are sorted into four intervals: $(-\infty, P^L)$, when the first and second answers are both 'NO'; $(P^L, P^*)$, when a discount offer is accepted at the second bid; $(P^*, P^H)$, when the premium is rejected; and $(P^H, +\infty)$, when both answers are 'YES', where $P^*$, $P^L$ and $P^H$ denote initial price bid, lower price bid (bid with a discount) and higher price bid (bid with premium), respectively. The probabilities for the above choice indices can be specified as:

$$\begin{aligned} Prob(yes/yes) &= Prob(WTP \geq P^H) \\ Prob\,(yes/no) &= Prob(WTP \geq P^H) - Prob(WTP \geq P^*) \\ Prob(no/yes) &= Prob(WTP \leq P^*) - Prob(WTP \leq P^L) \\ Prob(no/no) &= Prob(WTP \leq P^L) \end{aligned} \tag{1}$$

Correspondingly, the log-likelihood function for this WTP model is,

$$\begin{aligned} \ln L = \sum_{i=1}^{n} I^{YY} \; & \ln\left[1 - \Phi\left(\frac{P^H - \beta' x}{\sigma}\right)\right] + I^{YN} \\ & \ln\left[\Phi\left(\frac{P^H - \beta' x}{\sigma}\right) - \Phi\left(\frac{P^* - \beta' x}{\sigma}\right)\right] + I^{NY} \\ & \ln\left[\Phi\left(\frac{P^* - \beta' x}{\sigma}\right) - \Phi\left(\frac{P^L - \beta' x}{\sigma}\right)\right] + I^{YY} \\ & \ln\left[\Phi\left(\frac{P^L - \beta' x}{\sigma}\right)\right] \end{aligned} \tag{2}$$

where the $I$ symbols denote binary indicator variables for the four response groups. The coding of our likelihood model allows one to estimate $\beta$ directly and the coefficients can be interpreted as the marginal effects of the $x$ variables on WTP in rupee terms. The socio-economic variables assumed *a priori* to have a bearing on respondents' WTP are included in the DBDC model and are presented as Model I in Table 15.1. Some of the estimated $\beta$ parameters associated with the explanatory variables are found to be insignificant, and hence to save degrees of freedom those variables having $z$ values less than unity were omitted and the model was re-estimated (Model II).[19]

We incorporated household level variables including household income, education, age structure of the household, as well as variables related to their livelihood activities (e.g. whether households cultivate and are direct consumers of the herb, whether they engage in the collection of non-timber forest products) and how connected they are to the 'outside world' geographically and through the media.

As expected, the data suggest that the Kanis' (per capita) income controls their ability to pay. In other words, poorer households are less able to afford the proposed voluntary contribution to the community's fund to protect their TK. On average, a 1 per cent increase in per capita income increases the (latent) WTP to the hypothetical fund by Rs 71 (i.e. 0.9 per cent of per capita income). Hence, the income per capita is close to being unit elastic with respect to the (latent) WTP to protect TK by the Kanis. Interestingly, the *a priori* expectation that older tribe members would be more likely to donate for TK conservation as they may be assumed to be more attached to traditional community values is not reflected by the data (albeit its positive sign) given its low statistical significance. Further, although the level of formal education among Kani members is associated with a lower willingness to donate to the fund, other forms of information channels, such as access to newspapers and television and through direct visits to nearby cities, increase their WTP considerably. Regarding the livelihood activities carried out by the households, it is suggested that households which cultivate the herb are willing to donate a higher amount to the fund than non-cultivators, which could possibly be the result of direct experience by the former with respect to deriving a tangible use value from trading with the herb. This result may also be associated with the positive effect on WTP of having the direct experience of consuming the *Trichopus* fruit. Lastly, the results from the DBDC-CV model suggest that the households which participate in the non-farm labour market and derive daily wages in that sector are less willing to donate for the community's TK protection cause. This may also be associated with their lower attachment to the agro-ecosystem and the values that they derive from it.

[19] A log-likelihood test conducted to verify whether the coefficients of the omitted variables were jointly zero, failed to reject the null hypothesis, implying that dropping of variables is statistically justified. The test statistics are defined as $-2(L_0 - L_{max})$, where $L_0$ and $L_{max}$ are the values of the log-likelihood functions for the restricted and unrestricted models respectively. The unrestricted and restricted models are statistically significant at the 0.01 level with $\chi^2$ values of 36.86 and 35.17 respectively, i.e. $\chi^2_{16} = 1.68$. Thus, the null hypothesis that omitted variables are jointly not different from zero cannot be rejected at any meaningful significance level.

Table 15.2 *Mean WTP of Kani households (in Indian rupees#)*

| Household group | Mean WTP (std. error) |
|---|---|
| Cultivators | 409.71* (113.82) |
| Non-cultivators | 246.39 (103.85) |
| Average WTP (weighted by the share of cultivating and non-cultivating sample) | 251.29 (174.98) |

* Significantly different from mean WTP of non-cultivators at 0.01 level
# 1 US$ = Rs 44 (exchange rate of 2005)

Using the estimated coefficients in Model II, the mean WTP is Rs 410 (Rs 246) for a representative household that cultivates (does not cultivate) the herb (c.f. Table 15.2). The difference, also depicted in Figure 15.3, is significant at the 1 per cent level. The weighted mean WTP is Rs 251 (around US$ 5.7) per household which is about 3.3 per cent of their annual per capita income. Notwithstanding the possibly lower bound with regard to the implicit true WTA value, this amounts to 1 million rupees by the whole Kani community. If this is compared to what the pharmaceutical AVP offered which was shared on a one-to-one basis between the Kani community (through the Trust) and the TBGRI, it is clear that the community obtained just half of the minimum benefit that is perceived as appropriate compensation for engaging in the bioprospecting contract.

Hence the valuation study sheds some light on the degree of the inadequacy of compensation levels in ABS cases, as we displayed that even in those cases hailed on being successful such as the KMBS there is a significant disparity between actual payments and what is perceived as appropriate and necessary compensation by the local TK holders.

It is important to note, however, that these types of valuation studies need to be complemented by a broader analysis to fully address the stake of preserving future possibilities of use and innovation and the contributions of the other actors involved in the entire innovation chain. For instance, the danger of the extinction of the herb from the forest ecosystem is not addressed in the bilateral (Kani-AVP) relationship, and it is the conservation policy of the Forestry Department that takes into account this preservation value within the conservation of the habitat.

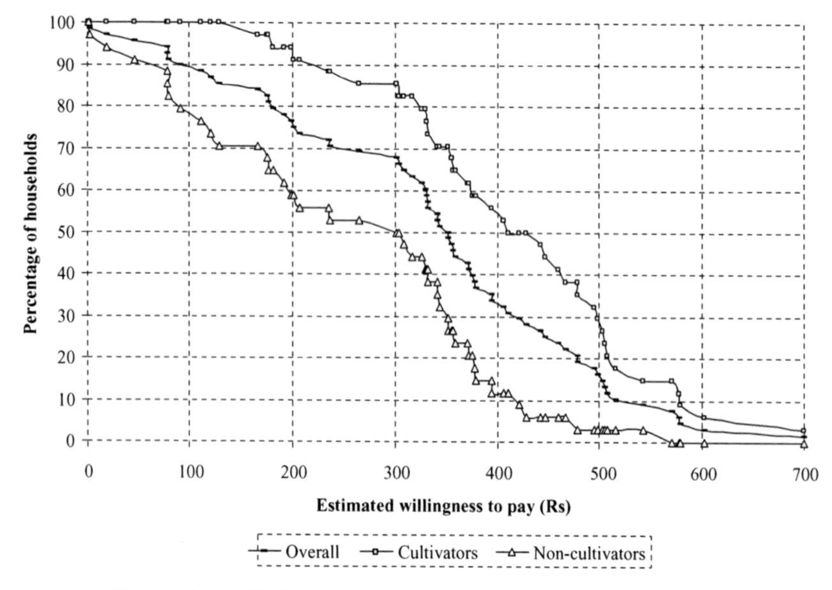


Figure 15.3. Estimated cumulative distribution of farmer households and their respective WTP, controlling for cultivating and non-cultivating households of *Trichopus*

Additionally, the value of the GR and associated TK for scientific research into taxonomy and plant-related medicine also needs to be accounted for, even though public actors are investing in it through government or international cooperation.[20] What is apparent is that in the actual institutional setting there is no real integration of these different players throughout the innovation process, and the ABS scheme has not been able to live up to its full economic potential. The estimated gap between the Kanis' perception of the value of sharing their TK and the actual compensation offered to them by the bioprospecting company is just one of the manifold manifestations of the suboptimality of these kinds of ABS contracts.

A full dynamic valuation approach should focus on a more balanced assessment of the different biodiversity related values, under conditions of strong uncertainty and evolving social preferences, hence addressing

[20] TBGRI benefits from several international projects for cooperation of research and value addition. In particular, a collaborative research project entitled 'Ethnopharmacology of Indian Medicinal Plants' is carried out between the TBGRI and the Department of Medical Chemistry at the Royal Danish School of Pharmacy, Copenhagen, Denmark, sponsored by the Danish International Development Agency. It is in the framework of this collaborative research that the components of *arogyapaacha* (the local name of *T. zeylanicus* ssp. *Travancoricus*) were isolated, some of them having been sent to Copenhagen for characterisation (Gupta 2004).

not only the market value of the final product of the innovation chain, but also other social values including the cultural values of the Kani community, the public good value of the preservation of a diverse genetic stock or the value for scientific research. This would imply combining the CV approach with qualitative information on the motives and attitudes underlying the local people's statements on the value of bioresources (O'Connor 2000; Spash 2000).

Furthermore, the WTP for keeping full property rights over TK analysed here is an aggregate measure of value, covering all the Kani tribe members who individually may have very different attitudes towards the Trust fund and the endeavour. Indeed, for some members, the estimated welfare measure covers the compensation for licensing the property rights on the TK, while for others it may also consist of the anticipated monetary return from engaging in cultivation and selling of *T. zeylanicus*. For others, it may represent the importance of preserving the traditional culture values attached to a broader notion of indigenous traditional healthcare.[21]

In summary, the CV carried out here shows a clear shortcoming of the static approach to ABS as it considers only the potential market value of the product at the end of the pipeline of its development process. In doing so, our analysis brings into the foreground another biodiversity-related value derived from the community which is the bearer of the TK.

## 5 Conclusion

This chapter has focused on the economic incentives for *in-situ* knowledge-sharing in the context of bioprospecting. We have adopted a dynamic approach to the economic institutions of contracts and property rights. In this dynamic framework, the focus is not on the *ex ante* determination of the optimal allocation of resources under conditions of perfect rationality, but on issues of dynamic efficiency, such as knowledge acquisition and incentives for the preservation of future possibilities of use under conditions of uncertainty. By applying this (institutional) evolutionary approach to the process of bioprospecting, the chapter has attempted to address the importance of analysing the full chain of bioprospection in the innovation processes.

[21] The importance of the preservation of culture value in *in-situ* conservation is also confirmed by an interesting case study of Dyer *et al.* (2000) on local seed markets in Mexico. The introduction of new crop varieties caused a diversification of farmers' activities. Nevertheless, because of local traditions and culture, they continue to grow the classical varieties, despite the fact that from a financial point of view one can show that they have no reason to do so.

In this way, the current analysis moves away from the position that considers the difficulties posed by the actual IPR system on genetic resources as being merely a technical and legal issue. At present, in the field of genetic resources, one sees a tendency to create new laws for each sector of activity. This results in the emergence of many specific legal regimes for the protection of genetic resources and related traditional knowledge. These include, for example, patents for processes relying on genetic manipulation, plant breeders' rights for plant varieties resulting from genetic selection, farmers' rights for traditional farmers' varieties and national sovereignty governing the rights to access and use the natural resources from ecosystems that maintain and create biodiversity. Nonetheless, the multiplication of different sector-based laws still falls in a static conception of efficiency and does not really meet the need for an integrated approach to the process of value creation through the whole innovation chain.

This problem calls for a more differentiated approach towards institutional mechanisms for promoting conservation and sustainable use of bioresources. For instance, in the case of bioprospecting they include the financing of plant-genetic resource conservation by research institutions such as the International Plant Genetic Resources Institute and community management of risk in agrarian societies based on a system of reciprocity allowing for the preservation of a high level of agro-biodiversity (Brush 1998). National programmes for the development of biotechnological capabilities can also play a key role in the sustainable use of bioresources and contribute to a differentiated institutional approach (Artuso 2002).

In the case of the Kani model of benefit sharing, the trust fund is an example of an institution for coordinating the different social demands coming from the community. However, as we have seen, it largely remains insufficient, because limited social learning is generated for bridging the conservation interests of the Forest Department and the interests of (a part) of the community involved in the benefit sharing agreement. Further, within the community different perceptions subsist between the younger and the older tribal Kani members regarding the appropriate protection of their traditional knowledge, the latter caring more about the loss of cultural identity.

Other means for enhanced institutional coordination that are currently being considered in international fora are the creation of an international system of certification of origin for monitoring the flow of genetic resources (Barber *et al.* 2003), the establishment of 'collection institutions' for traditional knowledge registries (Drahos 2000) or the creation of partnerships between research institutions and community-based breeding programmes (Brush 2002). In the field of IPR, Reichman (2000) proposes that policy formation evolves from a paradigm that functions by hybridisation of existing tools, based essentially on patent and copyright, to a paradigm based on a system of liability regimes, allowing the *ex post* compensation of the prior link in the innovation chain. These proposals include mechanisms that aim at diffusing incentives through the entire production chain and maximising the future choices of development. These consider the necessity of new legal tools and governance mechanisms, but also the importance of the associated institutional means for social learning and information sharing.

This chapter has argued that there is a need for major reforms or the use of alternative mechanisms to the existing bilateral market approaches to bioprospecting contracts and the voluntary mechanisms of benefit sharing. The reformed bilateral market approach should at least be based on a more 'dynamic' approach to the assessment of the use value of biodiversity (in terms of conserving GR stocks and associated TK) in the case when benefits through bioprospection can be realised. Further, in a second-best world, it is important to design alternative institutional means, including informal norms and formal legal regulations, that allow the effective coordination of the different actors involved in the innovation chain. This should allow for appropriate sanctioning of opportunistic behaviour and collective learning.

## REFERENCES

Adamowicz, W. L., Bhardwaj, V., and Macnab, B. 1993. Experiments on the difference between willingness to pay and willingness to accept. *Land Economics*. **69**. 416–427.

Anuradha, R. V. 1998. *Sharing with the Kanis: a case study from India*. Secretariat to the Convention on Biological Diversity, Fourth Meeting of the Conference of the Parties to the CBD, Bratislava, Slovakia.

Aoki, M. 2001. *Toward a Comparative Institutional Analysis*. Cambridge, MA: MIT Press.

Artuso, A. 2002. Bioprospecting, benefit sharing, and biotechnological capacity building. *World Development*. **30** (8). 1355–1368.

Barber, Ch-V., Johnston, S. and Tobin, B. 2003. *User Measures*. Tokyo: United Nations University Institute of Advanced Studies Report.

Barrett, C. B. and Lybbert, T. J. 2000. Is bioprospecting a viable strategy for conserving tropical ecosystems? *Ecological Economics*. **34**. 293–300.

Bhat, M. G. 1999. On biodiversity access, intellectual property rights and conservation. *Ecological Economics*. **29**. 391–403.

Berlin, B. and Berlin, E. A. 2003. NGOs and the process of prior informed consent in bioprospecting research: the Maya ICBG project in Chiapas, Mexico. *International Social Science*. **55** (4). 629–638.

Bodeker, G. 2000. *Indigenous medical knowledge: the law and politics of protection.* Oxford Intellectual Property Research Centre seminar. St Peter's College, Oxford. 25 January.

Brown, K. 2003. Integrating conservation and development: a case of institutional misfit. *Frontiers in Ecology and the Environment.* **1** (9). 479–487.

Brush, S. B. 1996. *Is common heritage outmoded?* In S. B. Brush and D. Stabinsky (eds.). *Valuing Local Knowledge: Indigenous People and Intellectual Property Rights.* Washington, DC: Island Press.

Brush, S. B. 1998. Bio-cooperation and the benefits of crop genetic resources: the case of Mexican maize. *World Development.* **26**. 755–766.

Brush, S. B. 2002. *The Lighthouse and the Potato: Internalizing the Value of Crop Genetic Diversity.* University of Massachussets at Amherst: Political Economy Research Institute Working Paper No. 37.

Craft, A. B. and Simpson, R. D. 2001. The social value of biodiversity in new pharmaceutical product research. *Environmental and Resource Economics.* **18** (1). 1–17.

Dedeurwaerdere, T. 2004. Bioprospecting, intellectual property law and evolutionary economics: the stake of a theory of reflexive governance. In M. M. Watanabe, K. I. Suzuki and T. Seki (eds.). *Innovative Roles of Biological Resources Centres.* Tsukuba, Japan: World Federation for Culture Collections. 389–395.

Dedeurwaerdere, T. 2005. From bioprospecting to reflexive governance. *Ecological Economics.* **53** (4). 473–491.

Dopfer, K. 2005. *The Evolutionary Foundations of Economics.* Cambridge: Cambridge University Press.

Dosi, G. 1988. Sources, procedures and microeconomic effects of innovation. *Journal of Economic Literature.* **26** (3). 1120–1171.

Drahos, P. 2000. Traditional knowledge, intellectual property and biopiracy: is a global bio-collecting society the answer? *European Intellectual Property Review.* **6**. 245–250.

Driesden, D. M. 2003. *The Economic Dynamics of Environmental Law.* Cambridge, MA: MIT Press.

Dutfield, G. 2002a. Literature survey on intellectual property rights and sustainable human development Geneva: UNCTAD www.iprsonline.org/unctadictsd/docs/bioblipr.pdf

Dutfield, G. 2002b. 'Protecting traditional knowledge and folklore: A review of progress in diplomacy and policy formulation. UNCTAD/ICTSD Capacity Building Project on Intellectual Property Rights and Sustainable Development.

Dyer, G. A., Taylor, J. E. and Yúnez, A. 2000. Who pays the cost of in situ conservation?. Paper presented at the Symposium 'Scientific basis of participatory plant breeding and conservation of genetic resources'. Oaxtepec, Mexico.

Eggertsson, T. 2005. *Imperfect Insitutions: Possibilities and Limits of Reform.* Ann Harbor, MI: University of Michigan Press.

Eisner, T. and Beiring, E. A. 1994. Biotic exploration fund- protecting biodiversity through chemical prospecting. *BioScience.* **44** (2). 95–98.

Folke, C., Carpenter, S., Elmqvist, T., Gunderson, E., Holling, C. S., Walker, B., Bengtsson, J., Berkes, F., Colding, J., Danell, K., Falkenmark, M., Gordon, L., Kasperson, R., Kautsky, N., Kinzig, A., Levin, S., Mäler, K-G., Moberg, F., Ohlsson, L., Olsson, P., Ostrom, E., Reid, W., Rockström, J., Savenije, H. and Svedin, U. 2002. Resilience and Sustainable Development: Building Adaptive Capacity in a World of Transformations. Scientific Background Paper on Resilience for the process of The World Summit on Sustainable Development on behalf of The Environmental Advisory Council to the Swedish Government.

Goeschl, T. and Swanson, T. 2002. On the economic limits of technological potential: will industry resolve the resistance problem? In T. Swanson (ed.). *The Economics of Managing Biotechnologies*. Dordrecht/London/Boston: Kluwer Academic Publishers. 99–128.

Guerin-McManus, M., Nnadozie, K. C. and Laird, S. A. 2002. 'Sharing financial benefits: trust funds for biodiversity prospecting'. In S. A. Laird (ed.). *Biodiversity and Traditional Knowledge: Equitable Partnership in Practice*. London: Earthscan Publications Ltd. 333–359.

Gulbrandsen, L. H. 2004. Overlapping public and private governance: can forest certification fill the gaps in the global forest regime?. *Global Environmental Politics*. **4** (2). 75–99.

Gupta, A. K. 2002. Value Addition to Local Kani Tribal Knowledge: Patenting, Licensing and Benefit-Sharing. IIMA Working Paper No. 2002–08–02. Ahmedabad: Indian Institute of Management.

Gupta A. K. 2004. The role of intellectual property rights in the sharing of benefits arising from the use of biological resources and associated traditional knowledge. World Intellectual Property Organization publications No. 769 (E).

Hanemann, M., Loomis, J. and Kanninen, B. 1991. Statistical efficiency of double bounded dichotomous choice contingent valuation. *American Journal of Agricultural Economics*. **73**. 1255–1263.

Laird, S. A. (ed.). 2002. *Biodiversity and Traditional Knowledge. Equitable Partnerships in Practice. People and Plants Handbook*. London: Earthscan.

Moran, K. 2000. Bioprospecting: lessons from benefit-sharing experiences. *International Journal of Biotechnology*. **2**(1/2/3). 132–144.

Morrison, G. C. 1997. Willingness to pay and willingness to accept: some evidence of an endowment effect. *Applied Economics*. **29**. 411–417.

Mulligan, S. P. 1999. For whose benefit? Limits to the sharing in the bioprospecting regime. *Environmental Politics*. **8** (4). 35–65.

Nelson, R. R. and Winter, S. G. 1982. *An Evolutionary Theory of Economic Change*. Cambridge, MA: Belknap.

North, D. C. 1990. *Institutions, Institutional Change and Economic Performance*. Cambridge: Cambridge University Press.

North, D. C. 2005. *Understanding the Process of Economic Change*. Princeton, NJ: Princeton University Press.

O'Connor, M. 2000. Pathways for environmental evaluation: a walk in the (Hanging) Gardens of Babylon. *Ecological Economics*. **34**. 175–193.

Oldfield, M. L. 1984. *The value of conserving genetic resources*. Washington, DC: US Department of Interior. National Park Service.

Pearce, D. W. 2006. Do we really care about biodiversity? In A. Kontoleon, U. Pascual and T. Swanson (eds.). *Frontiers in Biodiversity Economics*. Chapter 1. Cambridge: Cambridge University Press.

Pearce, D. W. and Purushothaman, S. 1992. Preserving biological diversity: the economic value of pharmaceutical plants. Discussion Paper: 92–27. CSERGE, London.

Peña-Neira, S., Dieperink, C. and Addink, H. 2002. *Equitability Sharing Benefits from the Utilization of Natural Genetic Resources: The Brazilian Interpretation of the Convention on Biological Diversity*. Paper presented at the 6th conference of the parties of the Convention of Biological Diversity. The Hague, The Netherlands.

Principe, P. 1989. *The Economic Value of Biodiversity among Medicinal Plants*. Paris: OECD.

Pushpangadan, P. 2000. Pushpangadan model of benefit-sharing. Lucknow, India: National Botanical Research Institute. http://www.nbri-lko.org.

Pushpangadan, P., Rajasekharan, S., Ratheesh Kumar, P. K., Jawahar, C. R., Velayudhan Nair, V., Lakshmi, N., and Saradamma, L. 1988. Arogyappacha (*Trichopus zeylanicus*): The ginseng of Kani tribes of Agasthyar Hills (Kerala) for evergreen health and vitality. *Ancient Sciences of Life*. 7. 13–16.

RAFI 1996. *Biopiracy Update: RAFI Communique*. Ottawa: Rural Advancement Foundation International.

Ramani, R. 2001. Note and comment: market realities versus indigenous equities. *Brooklyn Journal of International Law*. **26**. 1147–1175.

Rausser, G. C. and Small, A. A. 2000. Valuing research leads: bioprospecting and the conservation of genetic resources. *Journal of Political Economy*. **108** (1). 173–206.

Reichman, J. H. 2000. Of green tulips and legal kudzu: repackaging rights in subpatentable innovation. *Vanderbilt Law Review*. **53**. 1743.

Reid, W. V., Laird, S. A., Gamez, R., Sittenfeld, A., Hanzen, D. H., Gollin, M. A. and Juma, C. 1993. *Biodiversity Prospecting: Using Genetic Resources for Sustainable Development*. Washington, DC: World Resource Institute.

Sahai, S. 2000. Commercialisation of Indigenous Knowledge and Benefit Sharing. UNCTAD Expert Meeting on Systems and National Experiences for Protecting Traditional Knowledge, Innovations and Practices. Geneva.

Sheldon, J. W. and Balick, M. J. 1995. Ethnobotany and the search for balance between use and conservation. In T. Swanson (ed.). *Intellectual Property Rights and Biodiversity Conservation: An Interdisciplinary Analysis of the Values of Medicinal Plants*. Cambridge: Cambridge University Press.

Shiva, V., Jafri, A., Bedi, G., and Holla-Bhanu, R. 1997. *The Enclosure and Recovery of the Commons*. New Delhi: Research Foundation for Science, Technology and Ecology.

Shogren, J. F., Shin, S. Y., Hayes, D. J., and Kliebenstein, J. B. 1994. Resolving differences in willingness to pay and willingness to accept. *American Economic Review*. **84**. 255–270.

Shyamasundar, P. and Kramer, R. A. 1996. Tropical forest protection: an empirical analysis of the costs borne by local people. *Journal of Environmental Economics and Management*. **31**. 129–144.

Siebenhüner, B., Dedeurwaerdere, T. and Brousseau, E. 2005. Biodiversity conservation, access and benefit sharing and traditional knowledge. *Ecological Economics*. **53** (4).

Simpson, R. D., Sedjo, R. A. and Reid, J. W. 1996. Valuing biodiversity for use in pharmaceutical research. *Journal of Political Economy*. **104**. 163–185.

Spash, C. 2000. Ecosystems, contingent valuation and ethics: the case of wetland recreation. *Ecological Economics*. **34**. 195–215.

Svarstad, H. and Dhillion, S. 2000. *Responding to Bioprospecting: From Biodiversity in the South to Medicines in the North*. Oslo: Spartacus Vorlag.

Swanson, T. 1997. *Global Action for Biodiversity*. London: Earthscan.

Swanson, T. 2000. Property rights involving plant genetic resources: implications of ownership for economic efficiency. *Ecological Economics*. **23** (1). 75–92.

Swanson, T. and Goeschl, T. 1998. The management of genetic resources for agriculture: ecology and information, externalities and policies. Centre for Social and Economic Research on the Global Environment, CSERGE Working Paper, University College London.

ten Kate, K. and Laird, S. A. 1999. *The Commercial Use of Biodiversity – Access to Genetic Resources and Benefit-Sharing*. London: Earthscan.

Verma, I. M. 2002. Biopiracy: distrust widens the rich-poor divide. *Molecular Therapy*. **5** (2). 95.

WGF 2003. Western Ghats Forum website: http://www.westernghatsforum.org/ Last accessed November 2005.

# 16 An ecological-economic programming approach to modelling landscape-level biodiversity conservation

*Ernst-August Nuppenau and Marc Helmer*

## 1 Introduction

The intention of this chapter is to introduce a new unified approach to economic and ecological modelling of field sizes, farming intensities, landscape patterns and nature elements. The approach aims to enable governments to specify objective functions for biodiversity and landscape management. These functions will include ecologically retrievable criteria for payments, shall help to determine payments and must fit into farmers' concerns of capturing economies of scale to maintain competitiveness. In the chapter we first outline why this is a problem and review, to a certain extent, what has been accomplished with respect to integrated modelling so far. We then show, as an innovation, how a geometrical presentation of land use may help in improving the specification of interfaces between economic and ecological modelling compartments. Then we turn to elaborate on a nature production function linked to landscape elements and demonstrate how farmers can be paid for these elements. Using this analysis, it can be shown how farm modelling can be redirected to landscape design. Finally we outline how the previous deliberations can be used to achieve a principal-agent specification of objective functions for farmers seeking to maximise income from land and a government that wants to optimise the level of biodiversity in agricultural landscapes. A special reference is made to farm size and the question of how diversity in land use and landscapes is linked to the number of farmers and field sizes. The approach itself seeks maximum flexibility in nature provision, searches for modes to reduce complexity and can be considered a tool that helps in preserving cultural landscape and biodiversity in the face of economic pressure.

Let's turn to the specific problem addressed in this chapter. There are many deliberations suggesting that nature services can be provided in cultural landscapes if farmers care about landscapes. Primarily, it is believed that farmers should not convert their lands into a purely production-oriented cultural steppe of large fields, but rather recognise ecological concerns, mainly expressed as diverse landscapes and heterogeneity of farming (Dramstad *et al.* 2001). The provision of diverse landscapes has become a major public concern as payments for nature services are frequently discussed nowadays, even by ecologists (Johst *et al.* 2002). There remain conflicts and problems, however, and it is hoped that ecological economic modelling might contribute to their solution. The idea is, specifically, to bridge human behaviour and natural science (Tress *et al.* 2001) through an integrated approach based on a workable interface between ecology and economy. Landscape modelling, for instance, shall deliver locations (focal areas of nature production) and measures (buffer strips) so that ecological services (wildlife) occur at minimal economic costs (Wossink *et al.* 1998). Ecological services, based on a rich biota, will in all instances require high overall diversity of landscapes (Dauber *et al.* 2003).

It is necessary, however, to acknowledge the economic pressure farmers face. Farmers want to economise landscapes, expressed as prevalence of large fields and monotonous cropping patterns. To achieve jointness, governments have to recognise the increase in farm and transaction costs resulting from biodiversity preservation. At least minimising costs in wildlife provision requires a formal approach (Peerlings and Polman 2004), which can be very detailed on cost types and approaches. There have already been several attempts to integrate ecological concerns into practical landscape modelling. To mention only some more recent attempts from a broad literature, van Wenum *et al.* (2004), for instance, have worked with a fixed outlet of representative farms in a Dutch polder. This study is based on concepts by Bockstael *et al.* (1995) defining proper interfaces. A modification of Bockstael *et al.*'s (1995) concepts can also be found in the work of Weber *et al.* (2001). The polder approach works with fixed fields as larger basic units that are jointly addressed by ecological and economic modelling. In other approaches, units are broken down to pixels or grid cells. Such explicit spatial recognitions are also applied by Matthews *et al.* (1999) and Rounsevell *et al.* (2003). Another issue to be noted in achieving jointness in optimisation is that mostly, in meta-modelling approaches, separate procedures for solving sub-models are applied and only later are they brought together through iteration (Weber *et al.* 2001). Also, sometimes, merely trade-off functions are envisaged (Lichtenstein and Montgomery 2003) to leave governments discretionary choices.

Given a desired level of biodiversity, governments and farmers are considered able to select locations where cost minimisation can be achieved (at least theoretically, as represented by optimisation in computer models). Then (practically, as an extension tool of decision-support systems), farmers should integrate biodiversity in their business plans on a regional scale. But do these models really address the problems farmers face? In the farmers' opinion, economising production and cost minimisation are possible by increasing field sizes, further use of economies of scale and making nature provision in separate areas if requested. A solution is to acknowledge that nature provision is location-specific (Verburg *et al.* 1999), and that it makes sense to assign priority to special locations, as well as giving some farmers special tasks (Münier 2004). Farmers, however, will respond and there is a danger that non-assigned locations, i.e. those not given priority for conservation, will suffer an even greater extent of landscape transformation. If the system-wide implications of the behaviour of many farms are not included, integration becomes weak.

The question remaining is how to make payments both more efficient and targeted towards biodiversity conservation. As a new way out, farmers may seek to maximise revenues from real, marketable goods (e.g. wheat, yielding income) but also ecological goods and services (e.g. species, giving diversity) if they receive compensation that fits into their mode of decision making. Yet, the obvious problem is that no clear criteria exist defining what service farmers should be paid for, how payments should be organised and, as far as the product, a landscape, is concerned, how the issue of multiple users is addressed.

The general objective of this chapter is to show how landscape-oriented payments can be introduced to address ecological objectives by taking into account sizes of farm fields, buffer strips, and various nature elements. In order to do so, a new way of modelling the provision of landscapes is introduced, which is characterised by diversity, contains nature elements and controls the intensity of farming (as far as this is related to the provision of biodiversity). In so doing, we address the following questions simultaneously: 1) what to pay for (area or species); 2) whether to give priority to certain areas of ecological importance; and 3) how to promote farmers' participation. The approach is innovative as it addresses heterogeneity in landscapes and shows how non-linear programming (e.g. Howitt 1995) as a tool of farm economics can be used for landscape modelling. Furthermore, the modelling strategy aims at providing three interfaces to biodiversity within landscapes:

- The size of a field as an edge-driven delineation, which is modelled by that field's longitudinal and transversal dimensions. Farm size is transversal and price policy dependent. Then, cropping patterns determine longitudinal field sizes on existing farms. This way of treating land

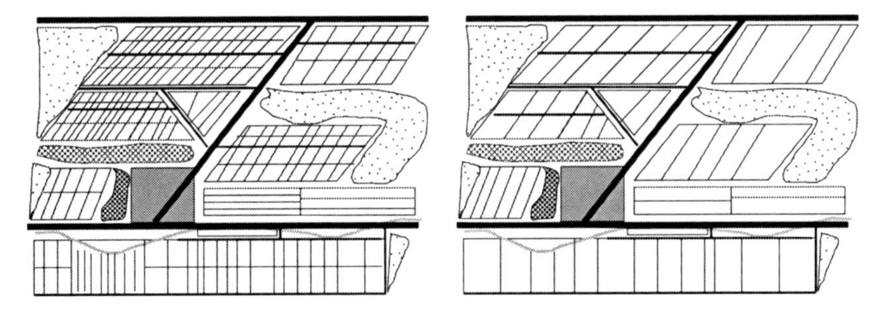

(a) Traditional land use structure (b) Modern land use structure

Figure 16.1. Landscape appearance as dependent on farming and land use structure

is similar to a vector presentation. In particular, it can be shown how price policy impacts on field sizes as a reference for payment design.

- Farming intensity, which can be geared by optional strategies targeting input uses that are labour interdependent. The intensity of farming with respect to chemical inputs depends on labour costs and chemical input prices. This aspect has to be reflected in payments.
- Nature elements such as buffer strips can be determined in relation to field sizes. The idea is to acknowledge farmers' concerns for income and behaviour with respect to in(de)creases in field sizes, as ecologically warranted, and to compensate losses through payments, for instance for unfarmed strips.

## 2 Outline of the theoretical approach

### *2.1 A general framework for landscape modelling*

Normally, farms and ecosystems, and also their interactions, are self-organising systems (Naveh 2000), but they may be related to governments that may want to maximise biodiversity: notably at reasonable costs and through instruments that influence behaviour. To model interactions and self-organisation, system compartments have to be related explicitly by defining interfaces and specifying instruments to determine options for intervention. We start with land use, suggesting a geographical presentation (Figure 16.1), and provide a realistic approach for interventions at field level.

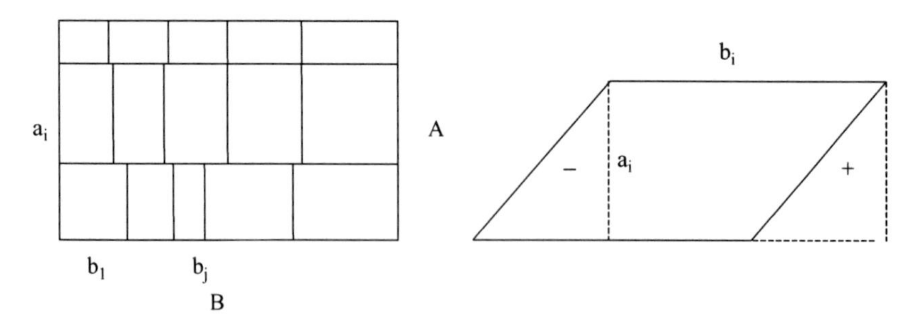


(a) Stylised for mathematical presentation (b) Calculation of area

Figure 16.2. Stylized structure of a landscape for the construction of a mathematical interface between spatial representation and geometric measures of field sizes
where:
$a_i$ = transversal stretch of a field (farm size) and number of farms 'n' (for instance A/a = n for equal size)
$b_{ij}$ = longitudinal stretch of a field (cropping pattern)

While considering payments and regulations that apply to the land use pattern, a problem is to keep things tractable, i.e. to maintain a balance between complexity and operability. The suggestion is to work with a natural clustering of fields as shown in Figure 16.1. For illustration, we deliberately begin with the stylised landscape in Figure 16.1a, which, perhaps by chance, is still 'traditional' and has not yet been transformed into a 'modern landscape' (Figure 16.1b). In Figure 16.1 the landscape is composed of sub-units in which fields are organised around (agronomically) suitable entities, so that the land use pattern serves a large number of farms. Modernisation (Figure 16.1b) would mean a radical switch with the result that few farms survive (perhaps eight farms in Figure 16.1b) while field sizes increase (consolidation). A comparison of Figures 16.1a and 16.1b provides a rough picture addressing heterogeneity, fragmentation, etc. It shows that diversity and field structures are linked.

Next, we suggest identifying farm sizes by the number of fields, the stretch on a transversal 'y'-axis, and the field sizes on a longitudinal 'x'-axis (see Figure 16.2). Practically, in the case of Figure 16.1a, we still have 24 farms. In Figure 16.1b, we assume that eight farms survived and their fields are larger than before. Possibly, only roads are maintained as fixed structures as field size is changed. Decisions on field size and organisation can be made visible and mathematically treatable. The problem is redefined in Figure 16.2. An ideal for a geometric presentation is shown in Figure 16.2a.

Variables such as farm and field size can be defined by $a_i$, $b_{ij}$ components, respectively. For instance, a farm of one hundred hectares (ha) may have ten fields of ten ha each. As a rectangular farm of a distance of 500 m on the 'y'-axis and 2000 m (200 m per field) on the 'x'-axis, this gives 1 million square metres. This defines the farm and field sizes of one farm. In a landscape, by definition, the number of farms follows vertically (further stretch in 'y') and the number of fields horizontally on corresponding axes (further stretch in 'x'). The representation of a landscape as a map in Figure 16.1 just needs a transformation procedure to enable translation and explanation of land use as land quantity: for instance, the semi-irregular map (Figure 16.1) is transformed into a geometrical representation (Figure 16.2b), and in turn Figure 16.2b is then translated into mathematical expressions. For the mathematical presentation we use the distances (distance in $b$ multiplied by distance in $a$ equals $l$, the size of a field). A transformation such as that in Figure 16.2 helps us to understand interactions between spatial analysis and programming farm economics. The approach will be discussed sequentially. Note that we aim to achieve an integrated representation of ecological quality and payments, with matching scales, scope and aggregation levels (Vermaat *et al.* 2005). Also, the number of farms $n$ is determined by $a$; for instance, with equally sized farm: $n = A/a$ ($A$ is the absolute length of a sub-unit of the landscape, see Figure 16.2).

### 2.2 *Mathematical representation of farm-land use, landscape planning and ecology*

Land, as spatial entity, is a prerequisite for farming. Harvest can be defined as yield multiplied by size (in hectares). Technically, we address the size of a field as $l_i = a_i\ b_i$ and use a Taylor expansion of second order around a reference point (Mood *et al.* 1974). Taylor expansion is a technique to make non-linear functions linear. A reference point may be a simple field structure of equal size (for instance 1 ha). It is important to note that fields can be adjusted in size. Also, fields require a double-sided optimisation: they can expand in both directions as distance, in $a$ and $b$. For the moment, we will try to linearise as much as possible. Equations (1a) and (1b) express land use in terms of farm size $a_i$ and field size $b_j$.

$$l_{i,j} = a_i \cdot b_j = a^*_{i,0} \cdot b_j + b^*_{j,0} \cdot a_i = A/n^* \cdot B/n^* + A/n^*(b_j - B/m) + B/m^*(a_i - A/n) \tag{1a}$$

A coefficient with a zero in the subscript and an asterisk is a representation of a fixed distance unit (for instance, 100 m of one ha) as a given starter.

By starting in this way we arrive at a function that is linear in grid length and width:

$$l_{i,j} = \kappa_0 + \kappa_1^*(b_j - b_{j,0}) + \kappa_2^*(a_i - a_{i,0}) \tag{1b}$$

The $\kappa$-coefficients are combinations of the preset unit distances (for instance: $\kappa_2 = B/m^*$, note: $B$ is treated similarly to $A$) and represents an artificially perceivable farm structure, where every field is of equal size. Given this spatial delineation of land use, we can formulate a farm-optimisation, an ecological and a payment module. The central idea in economics will be to work with the size of a farm first, and then decide on the fields' structure. This implies optimisation along $a$, firstly, to determine the width of a strip to be farmed and, secondly, to recognise the length of a field by determining $b$ afterwards. Then, the final structure of a landscape is determined by iteration.[1]

### 2.3 *Ecological aspects*

Ecological improvements are based on a deliberate planning for ecosystem services and are associated with changes in the field structure of landscapes and nature elements. This has to be discussed in some detail, because we have to link it to land use and a concrete payment scheme. Since nature is a self-organising system, it does not seem worthwhile to pay farmers for actual appearance of storks, for example (Johst *et al.* 2002), or rare plants, amphibians etc., but rather services. We see the contribution of fields to habitats as a key element and will model this as interface. The assertion follows two steps. First, we assume a conversion of land into habitats, and then, second, of habitats into species. This might raise a lot of questions, which we might want to, but cannot, avoid. An important question is how to set specific objectives (species level) and aggregate these to an overarching objective (landscape level). We think that, if biodiversity conservation is a goal, perhaps one can start with a Shannon Index $D_b$ (Lichtenstein and Montgomery 2003):

$$D_b = s' \ln(s) \tag{2}$$

where $s$ is the species vector. Given such a measurement, we then must relate it to the spatial organisations of habitats. Including habitats means to address landscape appearance by exploring field structures, specifically

[1] Iteration implies sequential optimisation of farm size '$a$' and field sizes '$b$' in PMP according to Howitt (1995). For further explanation see the structure of first derivatives in the appendix (A1.1). For instance, given a fixed 'a', the field structure 'b' is optimised and, given 'b', 'a' can adjust, so we finally approach optimal farms.

and interactively. Since habitats can be related to a probability matrix $\Xi_h$ of species appearance, habitats translate into species and vice versa, i.e:

$$s = \Xi_h h \tag{3}$$

where $h$ is a habitat vector. Habitats may decompose into fields $b$ (as vector). Thus, $h = \Xi_b b$. For further clarification, we introduce a new transfer matrix $\Xi_t$ (i.e. $\Xi_t = \Xi_h \Xi_b$), translating species into shares of fields (via habitats). Through approximation, in equation (4), we then can represent species by: 1) areas set aside, i.e. buffer strips, $c$; 2) a new landscape structure, i.e. deviations from economically optimal field sizes $b$; and 3) an indicator of yields, $y$, reflecting the intensity, i.e. use of inputs. It has been shown by Dauber *et al.* (2003) that these are relevant elements of an ecologically most favourable landscape for biodiversity conservation. It follows in our model out of spatial delineation that:

$$\begin{aligned} s = {} & \Xi_{t,1}[a_0^* \cdot c + a_0^* \cdot c_0 \cdot u] + \Xi_{t,2}/A/B[a_0^* \cdot (b - b_0) \\ & + (a - a_0)^* \cdot b_0] + \Xi_{t,3}[1' a_0 [1'b] + 1'a[1'b_0]] + \Xi_{t,4}(y - y_0) \end{aligned} \tag{4}$$

Note in equation (4), that 1) $c$ is introduced as a part of the distance of $b$ which is now not farmed (similar approaches to use buffer strips have been suggested: Lankoski and Ollikainen (2003); 2) $u$ is introduced as labour on a buffer strip that is necessary to maintain the quality of the strip; 3) $b$ and $a$ are departures from purely economically optimal farming (explained later: it shows the reduction in field size); and 4) $y$ is equally a reduced yield. The sub-matrices $\Xi_{t,i}$ give the partial influence in $\Xi_t$. Especially in $\Xi_{t.3}$ we portray the effect of fragmentation, i.e. short distance in adjacent field structures which create multiple edges (Dauber *et al.* 2003).

Some remarks on the possibility of determining function (4) are necessary. For a depiction of different species scenarios and a calibration of the above ecological relationship, we need observations and experiments. The structure of function (4) can be made very close to the probability structure of a cellular automata and simulation methods can be used to calibrate relationships (Steiner and Köhler 2003). Cellular automatas are used as statistical methods to retrieve probabilities of appearance of species as dependent on habitat structures. Although elements are only attainable with a certain likelihood, we can either take the initial situation or artificially simulate situations as reference. An artificial reference situation might be a situation of no interventions in favour of the cultural landscape. For a further understanding we explain each element in the above formula. The steps are as follows:

1. In segment 1, an increase of buffer strip size $c$ (a vector as for fields $j$ on farm $i$) augments the availability of suitable habitats for $s$.
2. In $\Xi_{t.1}$ we consider strips $c$ (a separate distance added to $b$) as important for species prevalence and ask how nature improves if labour $u$ is added (planting hedgerows, etc.). Labour is proportional on strips. 'Strips' may currently exist in landscapes and a most simple strategy is to take this as a reference. For convenience, we work with linear formulations. Labour both creates costs and must be paid for, but farmers can decide what better fits their preference: to give up labour or land.
3. On the size of fields, we take into account the important notification from landscape ecology that with an increase in the number of fields (a decline in size) an increase in the numbers of edges occurs. A maximum of edges can be an indicator for high landscape diversity (Dauber *et al.* 2003). We model this on a micro and a macro scale. $\Xi_{t.2}$ gives the micro (field) impact and $\Xi_{t.3}$ (scalar) reflects the macro (length of edges) impact.
4. We include a measure, $\Xi_{t.4}$, for yields as an indicator for intensity. By this we assume a negative correlation such that increased competition for space between crops and herbs, within a field, leads to a reduction in herbs.

Inserting equation (4), i.e. the species/nature 'production' function, into (2), i.e. the objective function, and looking for constraints on the availability of ecological 'input' variables, as in equation (4), we basically deliver interfaces. Land constraints, labour and finance constraints can be reasons for limitations to nature development. Farmers who purely see nature development as competing with farming and will only go for nature development if they receive compensation.

## 2.4 *Scaling aspects and organisational matters*

Correctly addressing the scales of investigation is another issue (Dabbert *et al.* 1999). What should be noted is that:

- A landscape is comprised of fields and farmers. Modelling of landscapes with respect to field margins should reveal the number of farms as $a_i$'s, and fields as $b_j$'s (see above); but the farm sizes, $a_i$'s, are dependent on the prevalent price levels for products and inputs.
- If we want to increase or keep constant the numbers of farmers who should pursue ecologically friendly farming by adopting practices according to payment schemes, we have to specify the number of farms and the structure simultaneously.

- In programming, this requires an algorithm linking the number of farms remaining vacant $a_i = 0$ and the level of the payment. Alternatively, we might look at how payments contribute to farm survival. The same applies to the ecology. We have to find algorithms for modelling, where habitats and fields exist and what the contribution to biodiversity is. Perhaps even extinction and reintroduction matter. We pursue this idea, later, by introducing optional payment schemes, assigning payments to fields and farms. As regards ecology, however, the problem is that species prevalence and richness only count at the landscape level, i.e. all fields. We specify vectors $b$ at the level of farm $i$ and fields $j$ in equation (5):

$$b = [b_{1,1}, \ldots, b_{1,j}, \ldots, b_{1,m}, b_{2,1}, \ldots, b_{2,j}, \ldots, b_{2,m}, \ldots, b_{i,j}, \ldots, b_{n,1}, \ldots, b_{n,j}, \ldots, b_{n,m}] \quad (5)$$

Importantly, the logic is to look at necessary field sizes, $bi_j$'s, from an ecological point of view, and this forms the early stages of modelling. So ecology must tell us potential fields. Equation (5) represents a spectrum of activities in a landscape that are necessary if field design is to contribute to a desired species vector. Similar vectors are introduced for labour, strips and farms as per equation (4). It should be noted that slots may be empty depending on ecological substitution possibilities for habitats. Lastly, the sum of fields determines farm location; farms are composed of fields close by, since a farm faces transport costs for accessing fields. Therefore, fields will be linked to farms as if they were in their backyards. To simplify matters and contrary to ownership fragmentation possibly existing in reality, we assume that farmers have their strips, for which they get compensation, in a sub-landscape division. Ownership is another issue.

## 3 Land use modelling at farm level

Farmers have a certain degree of freedom to decide on the sizes of their fields. Currently, they expand field sizes due to price and cost pressures and this endangers biodiversity (see above). To address this problem, the agricultural economics literature offers two distinct modes of analysis: one is a purely spatial modelling of agronomic practices. The methodological background is a raster point; and only in a second step are rasters recombined to fields (Weber *et al.* 2001). The alternative is a farm enterprise-oriented model (Röhm and Dabbert 1999). Both approaches are limited in actions with respect to sizes and change of land use. Here, we explain what the problems of modelling and programming are in the given context. First, we address the existing tools for modelling and taking into account farmers' objectives. Then we turn to what may be deemed as 'new tools' for economic land use modelling.

### 3.1 *Given tools for modelling and farmers' objectives*

To solve some of the analytical problems arising in economic land use modelling, we suggest the following approach: we start with the usual objective of profit maximisation for farmers. A first decision on how to approach profits has to be made; specifically addressing land is important in the formulation of revenue minus costs. There are several methods for including land aspects into revenues and costs. A very simple way is to take a cross section for gross margins on land of equal quality and then keep the underlying complexity as simple as possible. This approach has been traditionally used in the linear programming literature where gross margins are maximised given certain constraints on homogenous factors such as labour, land, capital, etc. It is similar to the first step of a linear programming compartment in a positive mathematical programming (PMP) (Howitt 1995; Paris and Howitt 2001). In addition, since land quality varies, the dual solution has to be calculated complementing the linear approach to achieve continuous supply functions. Those familiar with programming tools have trained their minds according to the logic of linear programming (LP) in such a way that reality becomes a construct that fits into LP logic. However, a recent debate has come up with strong criticisms given the significant simplifications implied in the traditional LP approach.

Since PMP emerged, the advantages of quadratic programming have been demonstrated (Paris and Howitt 2001). Critical elements are substitution elasticities and limitations to factor availability. For instance, an LP assumes perfect substitution of land on a farm within an absolute limit of farm size. Indeed, although land purchases can change it, it still remains a constrained problem. The limitations of LPs equally apply to their use for landscape modelling. There is also the danger that if landscapes are modelled within a given set-up or number of farms, field structure and labour availability will replicate the past. The topic of heterogeneity in field design is essential. For that we need the PMP approach (Howitt 1995).

However, taking into consideration the current developments of increases in farm and field sizes in many farming regions of the world, especially in peripheral regions with high biodiversity, the very immediate question is how such developments can be modelled in order to identify driving forces beyond changes in the relative prices of crops. As said, one direction of research would be not to ignore institutional and socio-economic backgrounds, assuming that large tracks of land are farmed according to highest land rents (Weber *et al.* 2001). Then, obviously, any preservation of small farms would mean compensation payments for ecological reasons because, assuming certain types of economies of scale, an immediate result is a preferred landscape that offers large fields to farmers and recognises no ecological concerns. Any restrictions on field size impact negatively on land rents and, ultimately, there is a request to compensate farmers for their losses. But if there is a societal and political will arising from other reasons than maximising land rents, one has to specify types of landscapes and farm operations that work with small fields and/or with buffer strips based on ecological insights.

### *3.2 New tools for economic land use modelling*

We suggest a middle course between a very strict programming (farm-oriented) approach, given all temporary constraints of farm size, labour, machinery, etc. as is typical in LP models, and a spatial landscape (raster-pixel-oriented) approach including agronomic and socio-economic conditions. The basic idea is to return to the spatial concept that includes limited substitution possibilities as introduced above. The purpose is twofold: 1) to have a tool that allows prediction of changes in farm and landscape composition along current trends of full utilisation of economies of scale, driven by cost pressures due to changes in agricultural prices and decoupled income payments (Weinmann *et al.* 2005), and 2) to have a tool that enables landscape planners to determine compensation payments for ecological services if money is short. It is important to find tools that help us optimise instruments that create landscapes based on 'suitable' field sizes, field and crop diversity, necessary agronomic differentiation of behaviour on plots (fields), etc. In doing so, we will head for a mixture of a quadratic cost approach and yield determination.[2]

We start with one farmer $i$, but it should be noted that we want to extend the approach to $m$ farmers and interactions. Also, we have to decide on variables to be included and on appropriate relationships between variables. Most variables of our farm modelling are in accordance with standard theory of cost function approaches. However, a special treatment is needed for production, land and yields. As mentioned earlier, in a natural

[2] This approach can be set up using GAMS (Brooke *et al.* 1995).

science-oriented approach, plot yields are considered as partly fixed. We can reduce the problem to land use.

$$q_{ij} = y_{ij}l_{ij} = \left[y_{ij,o} + \Theta^1_{ij}i_{i,j}\right]' l_{ij} \tag{6}$$

where $q_{ij}$ stands for the production on field $j$ of farm $I$; $y_{ij}$ is the yield on field $j$ of farm $i$; $i_i$ refers to input use vector as intensity and $l_{ij}$ is the size of field $j$ on farm $i$. $\Theta$ is a matrix of impacts of chemical inputs raising yields above the natural yield $y_{ij,o}$.

Using a combination of yields, determined partly by site-specific conditions and partly by input use (intensity), we can make use of detailed field-specific agronomic information such as water availability, evaporation, soil fertility, etc. (Weber *et al.* 2001). Land in field *ij* is considered an interior constraint to a cost function and costs are defined at production (crop) level.[3] Furthermore, from the beginning, a vector approach will be used, qualifying summation of items as vector products. Then, total revenue on farm $i$ using vectors is:

$$R_i = \sum_j p_{ij}q_{ij} = p_i' q_i \tag{7}$$

where, $q_i = [q_{i,1}, q_{i,2,\cdots}, q_{i,m1}]$' represents a quantity vector of outputs per hectare and $p_i = [p_{i,1}, p_{i,2,\cdots}, p_{i,m1}]$' the gross margin vector. In addition, $p = p^* - k^*$ with $p^*$ and $k^*$ being the crop price and the variable unit cost, respectively. It follows that profits can be defined as revenue $R$ minus costs $C$: $P_i(.) = R_i(.) - C_i(.)$. For further deliberations see appendix 1. Land and additional labour demand/supply conditions prevail explicitly. For labour to stay on a farm, a criterion of reference is earning more (or at least) the same as a salary $m_i$, obtainable if the farmer fully resigns agriculture:

$$\begin{aligned} m_i &\leq w_i h_i = w_i^i 1' h_{i,j} + w_i^0 o_i \\ m_i &= \left[w_i^i - w_i^o\right]\left[1'\left[H_i^f + i_i'\Theta_{i,1}^2\right]l_i\right] + w_i^o h_i^t \\ &\qquad \text{whereas } o_i = h_i^t - 1' H_i^f l_i \end{aligned} \tag{8}$$

The criterion is dependent on hours of off-farm work and wage as part-time farmer where, $w_i^o$ is the exogenous off-farm wage rate, $o_i$ is off-farm labour, $h_i^t$ is total hours of labour available to a farm family, and $w_i^i$ represents the endogenous returns to labour on the farm.

[3] The approach can be extended to animal production, including the interactions between crops and animals (including manure, etc.). But, for the moment, let us assume fodder crops have a certain internal price.

Payments are income components and only a proper balance of requests and effective payments enables the cost effectiveness of programmes. The expectation regarding the level of $m_i$ is individual. A farmer has to work an amount of hours $H_i^f$ per hectare which is internally priced at $w_i^i$. By changing intensity $i$, labour can be augmented. Equation (8) requires that we calculate the shadow price of labour per hectare, $w_i^i$. Labour requirements in hours are assigned to a single farm proportionally. A potential full-time salary, as opportunity cost if a farmer quits farming $m_i$, is given to a farmer from outside. It must be, at least, obtained by farming and additional wage payments for part-time off-farm activities, otherwise the farmer changes his profession. Normally, we have off-farm labour $o_i$ as balancing income expectations (positive $o_i$). The concept given spells out the case for family labour, though sales and purchases can occur. To assure income, farmers think about changing intensity on farms (augment labour by investment), which is why $i$ (chemical input stands for intensity) appears and which is also important for the specification of yields.

The above specification enables a necessary and powerful inclusion of labour aspects in landscape planning. Labour aspects become involved because, most likely, ecologically preferred small fields, small farms, and non-consolidated, heterogeneous field structures require more labour. Labour aspects are like a two-sided coin: on the one side, a certain labour-land combination provides a specific income, but it also limits farming to land availability and farm sizes. On the other hand, labour is a cost factor. Labour costs, as composed of the volume of labour (hours to work on a hectare and on a farm) and the price of labour (opportunity costs of labour, i.e. salary comparison), determine the competitiveness of farms on world markets: the less mechanised the farming, the higher the costs of production. It is crucial, in the model presented, to appreciate that income aspirations exist. With the existing formulation, we can specify different income levels which include compensation payments and shadow (land) price, as aspirations. In taking such an approach, we are not only maximising land rents (Weber *et al.* 2001), but also considering farm income as relevant. By varying income constraints, policy-makers get information on the acceptance of certain conservation concepts.

Technically in programming, at the landscape scale, a given level of needed income provision is the result of summing up individual income aspirations of farmers. Farms can continue to exist (if calibrated at the current farm structure), will emerge or close down. In programming, we can cater for the existence of a certain number of farms, as being associated with a certain landscape structure and labour intensity. An immediate question is, how can we determine 'correct' levels of farms, labour, and farmers' income aspirations. One way of dealing with this question is to construct a reference scenario where the current income distribution prevails. Note that one cannot avoid dealing with income aspirations in biodiversity management. For instance, given a certain ecological orientation of a landscape as a minimal level of biodiversity which, technically, is a constraint in optimisation, we have to look at the amount of compensation required per farm and the number of farms. Summing up compensation payments per farm over the number of farms gives the total amount. With an increase of the number of farms, payments increase but also biodiversity. The model outline can only approach the problem of solving the conflict between ecological and economic aspirations by reaching a compromise on an investigated trade-off between: number of farms, field structure and ecology; it cannot give an ultimate result. To get started, we suggest a mathematical solution that offers the potential to build scenarios since it goes for an analytical solution showing the dependency of landscape structures on willingness to pay and price policies.

### *3.3 Background for a mathematical solution*

In this sub-chapter we will briefly outline some major aspects for retrieving a mathematical solution. The solution itself is given in appendix 1. When solving the problem minimising payments for a given biodiversity, i.e. mathematically, it is important to remember that, in the tool described so far, land $l_{ij}$ is defined as a presentation (1') of farm sizes (transversal stretch $a_i$) and field sizes (longitudinal stretch $b_{ij}$). So we have to optimise both simultaneously. In the model, we try to capture the observation that farm sizes $a_i$ increase with opportunity costs of labour and that we have a general tendency to increase the size of fields due to income aspirations. Also, due to heavy investment in machinery, farm sizes can and certainly do increase. These observations can be translated into a mathematical framework for endogenously determining product mixes of farms, farm and field sizes, labour allocation, etc. through linear response functions (c.f. appendix 1, equations A1.1). Let us express this function as a vector-driven profit function (equation 9, explained in appendix 1, cf. equation A1.3).

$$
\begin{aligned}
P_i(.) = {} & p_a^{/} y_i^o f_{i,1}(a_i, b_i) - C\left(y_{ij}^o, b_i, a_i, r_i^{/}\right) \\
& + \lambda_{i,2}\left[m_i - \left[w_i^i - w_i^o\right] f_{i,2}(a_i, b_i, u_i)\right]
\end{aligned} \tag{9}
$$

where $\lambda_{i,2}$ is a shadow price for the salary-income balance constraint as well as $f_{i,1}$ and $f_{i,2}$ are implicit functions for the above-introduced explicit relationships between landscape organisation ($a$ and $b$) and labour requirements $u$, respectively (appendix 1).

For understanding the background: we need response functions. Specified linear response functions are the first derivatives of a quadratic objective function of a farm. Exogenous factors are output and input prices as well as income aspirations and we use a constrained optimisation applying a Lagrange approach (PMP: Howitt and Paris 2001). Determination of field size in the context of quadratic programming (PMP) means a spatial representation, as addressed by $a_i$'s and $b'_{ij}s$, is possible. Since binding to the orthogonal character of landscapes saves mathematical efforts, an objective function for landscape planning that caters for individual farms, number of farms, and structures should become a special case where we seek to depict observations such as:

- farm size increases as the opportunity cost for farming, given by the off-farm wage rate, increases, and
- field sizes increase, though single fields may be comprised of land of variable quality.

## 4 Compensation payments, financial costs and scenarios

Our concurrent problem is now to specify payments and link them to things that can be changed by farmers. Apparently a major reason for the above specification of a landscape as a spatial unit of fields and farms, and bringing them into programming model, was to find payment modes. Also we considered the link to ecology. Then a payment scheme requires three aspects: 1) payment criteria, 2) determination of the size of contribution per criteria, and 3) the volume of an overall payment. Equation (10) gives a set of criteria that are derived from previous modelling: field size $b$, size of buffer strips $c$, yields $y$, labour $u$, and farm size $a$.

$$g_{i,k} = g_{0,i,k} + g_{1,i,k}a_{j,i,0}(b_{j,i,k} - b_{j,0,i,k}) + g_{2,i,k}a_{j,i,0}c_{j,i} + g_{3,i,k}(y_{j,i,k} - y_{j,0,i,k}) + g_{4,i,k}u_{j,i,k} + g_{5,i,k}b_{j,0,i,k}a_i \quad (10)$$

As usual, a total payment $g_{i,k}$ is the sum of products of quantities (criteria; $b$, $c$, $y$, $u$, and $a$) multiplied by 'monetary incentives' (prices) $g_{i,j,k}$. We have five criteria and 'monetary incentive' $g$ that need to be determined. Note that field sizes come as changes. Basically, this means the above farm and landscape modelling provides a reference scenario of no-intervention, first, and payments are in accordance with changes in behaviour desired by a government secondarily introduced. Furthermore, we can test whether it makes sense to pay beyond field level, i.e. on farm level, where the size of a farm matters. The problem is that, from an ecological point of view, perhaps it would make sense to address only fields, but fields are with farms and farms need money for existence. The equation is an attempt to combine the prevalence of special habitats (fields, buffer strips and structures) and their importance in species appearance with farm income, i.e. a mixed scheme should be used. The farm approach, measured by size $a$ and labour $u$, should have a strong impact on ecology if ecologists prefer small farms.

The advantage of the above specification is that payments can be individually assigned. To maximise participation in schemes and to find out individual farmers' bargaining positions, differentiated payments can be tested. This is important, since we have to show that the product 'biodiversity' can only be created by addressing strong requests for income by farmers, notably under the pressure to simultaneously pursue ecological goals. An immediate step towards the reconciliation of goals of ecologists, farmers and governments is to include payments in the income constraint (i.e. for $m_i$). From the perspective of modelling and programming of farm behaviour, this inclusion means that, in order to provide nature services for the society, farmers have to work for nature and devote land to nature, thereby losing money from farm products. In exchange, they will request money for doing the work for nature. Two issues are associated with this paradigm: 1) who decides, controls and coordinates the exchange and 2) what is his objective function? For the first aspect, the spatial and ecological landscape planning unit (a principal) must balance costs and benefits and try to solve conflicts. For the second aspect, a transparent objective function is needed. But according to what rules, what costs and what benefits? Furthermore, what responsibilities are to be set?

Finding an objective function involves several aspects. We have to reconsider that any interference in a competitive, world market price-oriented determination of landscapes implies economic losses. These losses have to be either minimised, if certain ecological goals shall be obtained, or the ecological goals, having monetary values, must become competitive. Additionally, we have to be cautious of price policies, i.e. how far variables such as prices really are exogenous and payments endogenous. A reference may be a situation of competitive pricing according to world market prices and then payments are the sole instrument. However, as can be shown from behavioural analysis, pricing also matters with respect to farmers' design of fields. Only the joint recognition of price and payment policies may enable us to reconstruct an overall policy that favours diverse landscapes! Then expenditure may not be the right measure for social costs. Since they contain a transfer element, which is not welfare-reducing, payments are financial costs. Real costs can be retrieved if we have an economically 'ideal' situation of no interferences and we know the value-added in alternative scenarios. Then social costs are a decline in welfare. Payments are tools, not directly welfare measures!

To investigate the payment system correctly, we must develop a reference scenario without interventions. Only then can we state that we want to pay for active changes. Pure economic optimisation will provide us with such a reference scenario. By that reference, we can show how farming is optimal under economic conditions, i.e. we receive $b_{j,0,i,k}$ and $y_{j,0,i,k}$. By 'economic farming' we primarily mean a spatial organisation of fields that comes close to a consolidated landscape of large fields and few farms. Deviations from the economically optimal field structure through payments $g_{j,j}$ will create losses in value-added and we define a new profit function (loss function: see A1.4 and A1.5).

However, we now come to the question: what is really the goal in landscape design? In the above setting we were looking for a public good $D_b$. In this respect, the goal is exogenous: i.e. bio-diversity. Alternatively, we could reckon that ecological functions of landscapes benefit farmers. Ecological-economic functions of landscapes, as eco-system services, can either be appreciated purely by the public or the public and the farmers. At the moment, it seems as if farmers do not recognise ecological functions other than the immediate productivity of their soils; i.e. fields are seen as a resource. Maybe things such as ecological activities are irrelevant for them for good reasons. A low recognition of the profit-enhancing functions of 'good' landscapes is an issue that has to be clarified with respect to a reference system for payments. Otherwise, payments may be lower. We could model a second reference scenario, where we include the effects of a positive impact of the ecosystem, i.e. $s$, on cost functions and substitute eco-system services for chemicals. Technically, such reference scenarios are possible and should include the species vector $s$ in the cost functions of individual farmers as dependent on landscape provision; for simplicity we start to sketch conditions for a first likely scenario.

## 5 Objective functions

### *5.1 Farm objectives and scaling*

Introducing the above deliberations on landscape structure and payments as key elements, we firstly have to modify the farm objective functions to include the payment scheme (9). For farmers a new objective function means that there is a trade-off between: 1) loss in production, because they comply with regulations set by planners. It means, essentially, new field sizes $b$, the introduction of buffer strips $c$, additional labour $u$, having a small farm $a$, and having lower yields $y$. This can be jointly written as a new vector $e = [\Delta b, c, \Delta u, a, y]$. The vector $e$ can be interpreted as an effort vector in the frame of principal-agent-approach (Richter and Furubotn 1997) which is explained later. 2) gain on the ecological side because of compensation payments $g\, e$. 3) being flexible enough regarding ecological concerns in decision-making because a variable incentive structure exists. We can apply a farm objective function by defining the objective dependent on length and width of fields, i.e. we rewrite all terms in (11) using a two-dimensional expression. Here for general notice, only an implicit function is given. (In appendix 1 the objective is written as an explicit quadratic function: A1.4.)

$$P_i(.) = p_a^{\prime} y_i^o [f_{i,1}(a_i, b_i, c_i) - e] + g^{\prime} e - C(y_{ij}^o, b_i, a_i, c_i, r_i^{\prime}, e) \\ + \lambda_{i,2}[m_i - [w_i^i - w_i^o]^{\prime} f_{i,2}(a_i, b_i, c_i, u_i) - g^{\prime} e] \qquad (11)$$

For the first time, as a strategic element, we have included a vector $c$ for buffer strips. Next, we have to upscale and look at a whole landscape. Primarily, this means a summing up (or higher order) of vectors adding farms. It also implies that we have to include the size of the landscape (limit of $a_i$'s as $A$ and $b_i$'s as $B$) as further constraints. This summing up ensures that individual farm sizes fit into landscapes. Taking a simple approach and optimising for all farmers jointly, it is nevertheless possible to obtain an aggregated solution. For a mathematically and technically oriented approach, we suggest working with a complete presentation in vector and matrix form (appendices). For instance, as has been suggested for landscape interactions, vectors such as the angles of fields in $b$ can be used throughout the landscape analysis. As an example, if adding up shall apply, a long vector $b$ must be written:

$$b = [b_{1,1}, \ldots, b_{1,j}, \ldots, b_{1,m}, b_{2,1}, \ldots, b_{2,j}, \ldots, b_{2,m}, \ldots, \\ b_{i,j}, \ldots, b_{n,1}, \ldots, b_{n,j}, \ldots, b_{n,m}]^{\prime} \qquad (12)$$

Note, by this notation we receive, as a summary of $B$, $B = \Sigma_j^n B_i = \Sigma_j^n \Sigma_i^m b_{ij} = 1' \, b$ (in vector presentation (11), $b$ is composed of $m$- and $n$-vector elements; for instance, five farms of ten fields each, meaning fifty vector elements. Things are treatable only if vectors are used.

### 5.2 *Landscape level planning*

Following the above notation, three further and major aspects re-emerge. 1) A reformulation of landscapes as a profit maximisation problem means that the outcome is a 'desired' or planned landscape. This landscape is influenced by ecological and economic goals. Though planning can impede farmers, they may cooperate, since they are compensated. If they lost income as a result of planning, they would be reluctant to get involved. This is excluded now, since the landscape depends on $m$, the income aspirations. 2) From the logic of behaviour, prices $p$ for crops $q$ and the amount $g$ paid for services are planned. 3) Hence $g$ has also to be accepted: i.e. $g_1$ as service $\Delta b$ (change in length), $g_2$ as service $c$ (buffer strip), $g_3$ as service $\Delta y$ (reduction in intensity), $g_4$ as service $\Delta u$ (additional labour) and $g_5$ as service $a$ (small farms) are clear to farmers. Payments are calculated and depend on $p$.

We have now reached the point where it is necessary to decide what is exogenous and what is endogenous to the system. Payments are exogenous to farmers. But since they are to be calculated at landscape level, they are endogenous to a government. But what happens to the gross margin (output price) $p$? High $p$s may reduce $g$ and vice versa. Finally, two approaches are possible for a complete optimisation of instruments. The first one could start with a full integration of ecological and economic objectives in one unified objective function for a landscape. In this case, both the quantitative interactions as well as the price and payment are endogenous. The second approach is a principal-agent approach (Richter and Furubotn 1997) where an ecological target is reached at minimal expenditures. We will follow the second one. Agent-based approaches especially are popular in landscape management and payment schemes (e.g. see Happe *et al.* 2004 although approaches differ greatly in intent and scope).

## 6 A principal-agent approach to the provision of nature elements

In this chapter we pursue a special approach, which rigorously follows the idea of deriving payment schemes, as related to maximising biodiversity, endogenously. However, to solve the problem in a principal-agent (PA) framework, 1) an objective function of an agent is needed. For this purpose we can use the above specification of a profit-maximising farm community that wants to satisfy income aspirations. 2) We need an objective of the principal. As said in the beginning, we think that, if biodiversity conservation is a goal, perhaps one can start with the Shannon Index. 3) We must include as a restriction the availability of a budget $g$ for the government. 4) The payment scheme must be communicated since it provides the incentive constraint. We can go for the above specification of services, especially for the problem of modelling the government as principal and the farmers as agent. However, this creates problems with respect to non-linearity. $g_1$, $g_2$, $g_3$, $g_4$, and $g_5$ are part of the description of the budget as well as multipliers of the instruments in the farmers' income constraint. In appendix 2 we show how to treat that problem and to get a final solution for an objective function of the government as principal, based on parameters derived from the PMP. Finally the efforts $e$ must be clearly specified. We have already done that by stating how changes in farm behaviour towards a more ecologically sound landscape can be the basis for both, a payment and incentive scheme. That scheme improves or maintains farm incomes on one side, and an ecologically oriented landscape, that maximises biodiversity, on the other hand.

Let us briefly touch on the procedures to solve biodiversity provision within the framework of a principal-agent-approach. A more detailed outline is given in appendix 2 since it involves some mathematical arguments like matrix optimisation that go beyond the scope of an introductory explanation. However, all elements of the approach have been developed. One starts with the objective function of the agent (farm community), which is maximising profits from farming and payments for ecological services simultaneously. A response function to service payments emerges that is the first derivative of maximising farm profits given payments, revenues and costs (appendix A2). Broadly described, the agent takes the incentives $g$ as given and responds with $e$. (For a more detailed description including farm production see A2.3).

$$e = \varepsilon_0 + \varepsilon_1 g + \varepsilon_2 p \tag{13}$$

The task is to specify $\varepsilon_0$, $\varepsilon_1$, and $\varepsilon_2$ as reaction coefficients in an optimal payments scheme through a rigorous mathematical procedure. For the mathematical procedure we again refer to the principal-agent literature (Richter and Furubotn 1997 as well as Holmstrom and Milgrom 1991 for a more general exposition). In a second step this function can be inserted into the objective function of the government and the government finally optimises $g$ whereas a budget constraint forces it not to opt for a maximum of ecological services. Rather the optimisation aims at a compromise between income support and needed money on the one side and biodiversity and needed changes in land use on the other side. As is shown in appendix 2, the biodiversity index has scope to be expressed as a newly formulated quadratic Lagrange function (13). That function includes the government budget as constraint.

$$E = [\Xi^1 x + \Xi^2 e]' \, [\mathrm{I} + [\Xi^1 x + \Xi^2 e]] - \rho' [g_g - g^* e] \tag{14}$$

Here $x$ stands for a vector of all exogenous factors. The government should maximise the function $E$ which is a combination of the biodiversity index $D$ and a financial constraint (see appendix 2: A2.4 and

A2.5). Inserting equation (13) into (14) gives the objective of the government in biodiversity management $E$ finally as a function of g (A2.6 in appendix 2).

$$E = [\Xi^1 x + \Xi^2[\varepsilon_0 + \varepsilon_1 g + \varepsilon_2 p]]' \, [\mathrm{I} + [\Xi^1 x + \Xi^2[\varepsilon_0 + \varepsilon_1 g + \varepsilon_2 p]]] - \rho'[g_g - g \cdot [\varepsilon_0 + \varepsilon_1 g + \varepsilon_2 p]] \quad (15)$$

Thus, to conclude, the government (principal) can finally see the problem of biodiversity management as a matter of optimising the budget. Optimising the budget means to decide where exactly to spend money aiming at a maximum of biodiversity. Having an optimal budget structure $g$, we can recursively determine the intermediaries of landscape design and services.

## 7 Summary

The objective of this chapter has been to show how payments for environmental services can be integrated into landscape planning to facilitate a cost-effective provision of biodiversity. Various aspects are worth mentioning: 1) We have touched on the modelling of a landscape design that goes along the geometric presentation of landscape elements like fields, farms and nature. 2) We have shown how this translates into a description of nature prevalence as dependent on landscape elements. The chapter included a measurement of biodiversity as an index to be maximised. 3) We have also discussed the necessary elements of a farm programming model for a landscape, in order to derive an objective function. 4) Then we switched to the objective of a government. In particular, an objective function for a government has been derived that captures economic aspects of land user concerns and ecological aspects of public concerns for biodiversity conservation. For deriving a government objective function and integration of farm objectives and behaviour into landscape planning we used a principal-agent approach. As an instrument we introduced a payments scheme that compensates and encourages farmers to offer changes in land use and nature elements that are conducive to biodiversity maximisation given a certain budget. Thereby, a jointly planned and used cultural landscape is arrived at from a given budget and maximal biodiversity or, vice versa, minimal payments and given biodiversity. A necessary condition is that farmers are compensated for interventions in the design of landscapes. To generalise: 5) We assumed that property rights were with farmers and farmers need incentives. Further, 6) to enable planning, we developed a unified modelling approach that is based on a quadratic programming approach and that is extended to a principal-agent approach. Lastly, 7) it has been shown that such modelling is possible, although it requires a certain reduction of complexity.

### REFERENCES

Bockstael, N., Constanza, R., Strand, I., Boynton, W., Bell, K. and Waiger, L. 1995. Ecological economic modelling and valuation of ecosystems. *Ecological Economics*. **14**. 145–159.

Brooke, A., Kendrick D. and Meerhaus, A. 1995. *GAMS: A User's Guide*. New York: Van Nostrand Reinhold.

Dabbert, S., Herrmann, S., Kaule, G. and Sommer, M. 1999. Landschaftsmodellierung und Planung: Methodik, Anwendung und Übertragung am Beispiel von Agrarlandschaften. Berlin: Springer.

Dauber, J., Hirsch, M., Simmering, D., Waldhardt, R., Wolters, V. and Otte, A. 2003. Landscape structure as an indicator of biodiversity: matrix effects on species richness. *Agriculture, Ecosystem and Environment*. **98**. 321–329.

Dramstad, W. E., Fry, G., Fjellstad, W. J., Skar, B., Helliksen, W., Sollund, M.-L. B., Tveit, M. S., Gelelmuyden, A. K. and Framstad, E. 2001. Integrating landscape-based values – Norwegian monitoring of agricultural landscapes. *Landscape and Urban Planning*. **57**. 257–268.

Happe, K., Balmann, A. and Kellermann, K. 2004. An agent-based analysis of different direct payment schemes for the German region Hohenlohe. In G. van Huylenbroeck, W. Verbeke and L. Lauwers. *Role of Institutions in Rural Policies and Agricultural Markets*. Amsterdam: Elsevier. 171–182.

Holmstrom, B. R. and Milgrom, P. 1991. Multitask principal agent analysis: incentive contracts, asset ownership, and job design. *Journal of Law, Economics and Organisation*. 7 (sp.). 24–52.

Howitt, R. E. 1995. Positive mathematical programming. *American Journal of Agricultural Economics*. 77 (1). 329–342.

Johst, K., Drechsler, M. and Wärtzold, F. 2002. An ecological-economic modelling procedure to design compensation payments for efficient spatio-temporal allocation of species protection measures. *Ecological Economics*. **41**. 37–49.

Lankoski, J. and Ollikainen, M. 2003. Agri-environmental externalities: a framework for designing targeted policies. *European Review of Agricultural Economics*. **30**. 51–75.

Lichtenstein, M. E. and Montgomery, C. A. 2003. Biodiversity and timber in the coast range of Oregon: inside the production possibility frontier. *Land Economics*. **79**. 56–73.

Matthews, K. B., Sibbald, A. R. and Craw, S. 1999. Implementation of a spatial decision support system for rural land use planning: integrating geographic information system and environmental models with search and optimisation algorithms. *Computers and Electronics in Agriculture*. **23**. 9–23.

Mood, A., Graybill, F. and Boes, D. 1974. *Introduction to the Theory of Statistics*. Third Edition. Auckland: McGraw-Hill.

Münier, N., Birr-Pederson, K. and Schou, J. S. 2004. Combined ecological and economic modelling in agricultural land use scenarios. *Ecological Modelling*. **174**. 5–18.

Naveh, Z. 2000. The contribution of landscape ecology to the sustainable future of postindustrial landscapes. In Ü. Mander and R. H. G. Jongman (eds.). *Landscape Perspectives of Land Use Changes. Advances in Ecological Sciences 6*. Southampton: Computational Mechanics Publications 191–209.

Paris, Q. and Howitt, R. E. 2001. The multi-output and multi-input symmetric positive equilibrium problem. In T. Heckelei, H. P. Witzke and W. Henrichsmeyer. *Agricultural Sector Modelling and Policy Information Systems*. Proceedings of the 65th European Seminar of the European Association of Agricultural Economists, 29–31 March 2000, Bonn. Kiel, Germany: Vauk. 88–100.

Peerlings, J. and Polman, N. 2004. Wildlife and landscape service production in Dutch dairy farming; jointness and transaction cost. *European Review of Agricultural Economics*. **31** (5). 427–449.

Richter, G. and Furubotn, E. G. 1997. *Institutions and Economic Theory: An Introduction to and assessment of New Institutional Economics*. Ann Arbor, MI: University of Michigan Press.

Röhm, O. and Dabbert, S. 1999. Modelling regional production and income. In G. V. Huylenbroeck and M. Whitby (eds.). *Countryside Stewardship: Farmers, Policies and Markets*. Oxford: Pergamon and New York, Amsterdam: Elsevier Science. 113–133.

Rounsevell, M. D. A., Annetts, J. A., Audseley, E., Mayr, T. and Reginster, I. 2003. Modelling the spatial distribution of agricultural land use at the regional scale. *Agriculture, Ecosystems and the Environment*. **95**. 465–479.

Steiner, N. C. and Köhler, W. 2003. Effects of landscape patterns on species richness – a modelling approach. *Agriculture, Ecosystems and Environment*. **98** (1–3). 353–361.

Tress, B., Tress, G., Decamp, H. and d'Hautescerre, A.-M. 2001. Bridging human and natural science in landscape research. *Landscape and Urban Planning*. **57**. 137–141.

van Wenum, J. H., Wossink, G. A. A. and Renkum, J. A. 2004. Location-specific modelling for optimizing wildlife management on crop farms. *Ecological Economics*. **48**. 395–407.

Verburg, P. H., de Koning, G. H. J., Kok, K., Veldkamp, A. and Bouma, J. 1999. A spatial allocation procedure for modelling the patterns of land use change based upon actual land use. *Ecological Modelling*. **116**. 45–61.

Vermaat, J. E., Eppink, F., Bergh, J. C. J. M., v. d. Barendregt, A. and Bellen, J. V. 2005. Aggregation and the matching of scales in spatial economics and landscape ecology: empirical evidence and prospects for integration. *Ecological Economics*. **52**. 229–237.

Weber, A., Fohrer, N. and Möller, D. 2001, Long-term land use change in a meso-scale water- shed due to socio-economic factors – effects on landscape structures and functions. *Ecological Modelling*. **140**. 125–140.

Weinmann, B., Sheridan, P., Schroers, J. O. and Kuhlmann, F. 2005. Modelling the CAP reform at the regional level with ProLand. Contributed Paper, EAAE Conference. 24–27 August. Copenhagen. CD-ROM.

Wossink, A., Jurgens, C. and Wenum, J. van. 1998. Optimal allocation of wildlife conservation areas within agricultural land. In S. Dabbert, A. Dubgaard, L. Slangen and M. Whitby. *The Economics of Landscape and Wildlife Conservation*. Wallingford: CAB International. 205–216.

## Appendix 1: Agent behavioural functions and profits

A system of linear farm behavioural equations, with respect to the spatial design $(a, b)$ as factor demand for land and various other factor demand and output supply functions, is the basis for programming. Coefficients can be retrieved by various statistical methods. Paris and Howitt (2001), for instance, suggest maximum entropy. The system is farm price dependent:

$$\frac{\partial P_i(.)}{\partial q_i} = p_a - \psi_i - \Psi_{i,1}q_i + \Psi_{i,2}(a_{i,0}b_{i,j} + b_{i,j,0}a_i) + \Psi_{i,4}r_i + \Psi_{i,6}x_i + \psi_i a_i = 0 \quad \text{(A1.1a)}$$

$$\frac{\partial P_i(.)}{\partial b_i} = \psi_i a_i^* + a_i^*\Psi_{i,2}q_i + \Psi_{i,3}(a_{i,0}^* b_{i,j} + b_{i,j,0}^* a_i) + \Psi_{i,6}a_i^* x_i - 1 a_i^* \lambda_{i,1} - w_i \Theta_{i,1} a_i^* \lambda_{i,2} = 0 \quad \text{(A1.1b)}$$

$$\frac{\partial P_i(.)}{\partial r_i} = \psi_i^{/} + \Psi_{i,4}q_i + \Psi_{i,5}r_i = i_i \quad \text{(A1.1c)}$$

$$\frac{\partial P_i(.)}{\partial a_i} = \psi_i^{/} q_i + B\lambda_{i,1} = 0 \quad \text{(A1.1d)}$$

$$\frac{\partial P_i(.)}{\partial h_i^f} = \Psi_{i,6}q_i^{/} + \Psi_{i,6}(a_{i,0}b_{i,j} + b_{i,j,0}a_i) = w_i^i \quad \text{(A1.1e)}$$

$$\frac{\partial P_i(.)}{\partial \lambda_{i,1}} = a_i B - 1^{/}(a_{i,0}b_{i,j} + b_{i,j,0}a_i) = 0 \quad \text{(A1.1f)}$$

$$\frac{\partial P_i(.)}{\partial \lambda_{i,2}} = m_i - [w_i^i - w_i^o][1'[H_i^f + i_i'\Theta_{i,1}^2](a_{i,0}b_{i,j} + b_{i,j,0}a_i) + w_i^o h_i^t = 0 \quad \text{(A1.1g)}$$

Also, this system is in vector and matrix expression. Hereby we can address ecological, field and farm structure aspects simultaneously, and call it a bio-economic system. System A1.1 is linked to a corresponding quadratic profit function of a farm (A1.1) is a derivative of function (A1.2); note the profit function is the dual of cost minimisation:

$$
\begin{aligned}
P_i(.) = {} & p_a^{/} q_i - \psi_i^{/} q_i - 0.5 q_i^{/} \Psi_{i,1} q_i + q_i^{/} \Psi_{i,2} l_i + \psi_i^{/} l_i + 0.5 l_i^{/} \Psi_{i,3} l_i \\
& + q_i^{/} \Psi_{i,4} r_i + \psi_i^{/} r_i + 0.5 r_i^{/} \Psi_{i,5} r_i + q_i^{/} \Psi_{i,6} x_i + l_i^{/} \Psi_{i,6} x_i \\
& + q_i^{/} \psi_i z_i + \lambda_{1i} [z_i - 1^{/} l_i] + \lambda_{2i} \left[ m_i - \left[ w_i^i - w_i^o \right] \right. \\
& \left. \times \left[ 1' \left[ H_i^f + i_i' \Theta_{i,1}^2 \right] l_i + w_i^o h_i^t \right] \right]
\end{aligned} \qquad \text{(A1.2)}
$$

This expression of profits, by land use, was firstly given with $l_i$ as a land vector for crops (land in field $ij$ is: $a_{i,0} b_j$) since this is the usual expression in programming, but amended to $a$ and $b$; $z_i$ is the size of farm ($Ba_i$), $r_i$ stands for the input price vector (including labour costs), $o_i$ stands for off-farm employment, and $x_i$ account for exogenous factors including fixed labour availability per hectare and agronomic conditions.

Secondly and essentially, the price dependency of behaviour shows how farm price policy impacts on the landscape appearance; this is important for reference scenarios and policy mix. Further note in detail: the behavioural expression (A1.1 and A1.2) of profits enables a calibration. For this, we can explore the programming aspects linked to PMP (Howitt 1995). Technically, we foresee a behavioural system of multiple conditions (including $a$ and $b$), which can be used as response system, though we still have to include payments. For the time being, system (A1.1) is a reference scenario. Endogenous variables are $q$, $a$, $b$, $i$, $w$ and $\lambda_1$ and $\lambda_2$. The variable input $i$ is individual to farms (a vector of industrial inputs, such as fertilizers, pesticides, etc. being applied to different crops). System A1.1 has some non-linear components, primarily because of internal valuation of labour. Exogenous variables are agronomic conditions and prices. Additionally, in cases of landscapes comprised of sub-units of farms, we have to add land constraints or sub-divisions of land parcels (Figure 16.1).

Furthermore, as we have inserted orthogonal grid lengths in a specific cost function giving starting points, then non-linearities can be solved. Optimisation takes place along $a_i$ and $b_i$, which, however, might result in numerical problems that can be solved through start values. In numerical research, one should work with a programming software that is only slightly above linear programming (quadratic in GAMS) and that avoids too complicated constraint structures. There is a great danger of spending excessive effort on numerical algorithms used for optimisation. To prevent this, we suggest either using a current structure of fields for calibration or normalising the spatial organisation by starting with a certain qualitatively justifiable number of farms or fields, for instance given by a specific description of the landscape. If we have a certain number of farms, maybe thirty as in the example, then we can divide the distance A by this number. This results in equally sized farms as a reference point and an acceptable way of calibration is found. Then we re-iterate, i.e. change $a$ first, $b$ second, etc.

Again, since we consider a field vector $l_i = a_i\, b_{ij}$ as a land measure the numerical profits are spatially expressed:

$$\begin{aligned} P_i(.) = {} & p_a^/ q_i - \psi_i^/ q_i - 0.5 q_i^/ \Psi_{i,1} q_i + q_i^/ \Psi_{i,2} [a_i^* b_i] + \psi_i^/ [a_i^* b_i] \\ & + 0.5 [a_i^* b_i]^/ \Psi_{i,3} [a_i^* b_i] + q_i^/ \Psi_{i,4} r_i + \psi_i^/ r_i + 0.5 r_i^/ \Psi_{i,5} r_i \\ & + q_i^/ \Psi_{i,6} x_i + [a_i^* b_i]^/ \Psi_{i,6} x_i + q_i^/ \psi_i a_i + \lambda_{i,1} [a_i B - 1^/ [a_i^* b_i]] \\ & + \lambda_{i,2} \left[ m_i - w_i^i \left[ 1' \left[ h_i^f + \Theta_{i,1}^2 i_i \right] [a_i^* b_i] + w_i^o o_i \right] \right] \end{aligned} \quad \text{(A1.3)}$$

Next, we have to convert profits from a spatial and geometric expression into a linearised system to which we can add payments for change in landscape design. Given the representation (1b) and inserting them in the profit functions payment (A1.4) criteria are linearised.

$$\begin{aligned} P_i(.) = {} & \left[ p_a^/ y_{ij}^o - \psi_i^/ y_{ij}^o + \psi_i^/ [\kappa_{0,i} + \kappa_{1,i} a_i + \kappa_{2,i} b_i] - y_{ij}^/ [0.5 \Psi_{i,1} y_{ij} \right. \\ & + \Psi_{i,2} + .5 \Psi_{i,3}] + r_i^/ \Psi_{i,4} y_{ij}^o + r_i^/ \psi_i^/ + x_i^/ \Psi_{i,6} y_{ij}^o + x_i^/ \Psi_{i,6} \\ & + z_i^/ \psi_i] [\kappa_{0,i} + \kappa_{1,i} a_i + \kappa_{2,i} b_i] + 0.5 r_i^/ \Psi_{i,5} r \\ & + \lambda_{i,1} [a_i B - 1^/ [\kappa_{0,i} + \kappa_{1,i} a_i + \kappa_{2,i} b_i]] \\ & + \lambda_{i,2} \left[ m_i - [w_i^i - w_i^o] \left[ 1' \left[ H_i^f + i_i' \Theta_{i,1}^2 \right] [\kappa_{0,i} + \kappa_{1,i} a_i + \kappa_{2,i} b_i] \right. \right. \\ & \left. \left. + w_i^o \left[ h_i^t - u_i \right] \right] \right] \end{aligned} \quad \text{(A1.4)}$$

Now changes in behaviour, for instance $\Delta b$, and payment functions become included, so we get A1.5. To programme and get behavioural equations is, in principle, similar to A1.1. As before, we start with fixed (i.e. zero) payment. Then, knowing landscape structures, we introduce payments for departure from the initial optimality. Departures are indicated by $\Delta$ in (12):

$$\begin{aligned} P_i(.) = {} & \left[ p_a^/ y_{ij}^o - \psi_i^/ y_{ij}^o + \psi_i^/ - [\kappa_{0,i} + \kappa_{1,i} a_i \right. \\ & + \kappa_{2,i} [b_i - \Delta b_i] y_{ij}^/ [0.5 \Psi_{i,1} y_{ij} + \Psi_{i,2} + .5 \Psi_{i,3}] + r_i^/ \Psi_{i,4} y_{ij}^o \\ & + r_i^/ \psi_i^/ + x_i^/ \Psi_{i,6} y_{ij}^o + x_i^/ \Psi_{i,6} + z_i^/ \psi_i] [\kappa_{0,i} + \kappa_{1,i} a_i \\ & + \kappa_{2,i} [b_i - \Delta b_i]] + 0.5 r_i^/ \Psi_{i,5} r_i - w_i \Delta u_i + g_{i,0} + g_{i,1} a_{i,0} \Delta b_i \end{aligned}$$

$$+ g_{i,2}a_{i,0}c_i + g_{i,3}\Delta i_i + g_{i,4}\Delta u_i + g_{i,5}b_{i,0}a_i$$
$$+ \lambda_{i,1}[a_i B - 1'[\kappa_{0,i} + \kappa_{1,i}a_i + \kappa_{2,i}[[b_i - \Delta b_i] + c_i]]]$$
$$+ \lambda_{i,2}[m_i - [w_i^i - w_i^o][1'[H_i^f + i_i'\Theta_{i,1}^2][\kappa_{0,i} + \kappa_{1,i}a_i$$
$$+ \kappa_{2,i}b_i] + w_i^o[h_i^t - u_i]]]] \qquad \text{(A1.5)}$$

This numerical revelation of the profit function (A1.5) is the basis for the agent behaviour.

## Appendix 2: Principal-Agent Modelling

### *A2.1 Farm objectives, landscapes and agent description*

In this appendix, we suggest how to exemplify a functional approach built along the PMP concept for a principal-agent optimising of landscapes for biodiversity management. Taking the above explicit functions from appendix 1, A1.5, practically, a vector presentation can be chosen A2.1, which meets the need of capturing structural and numerical components of change in farm behaviour conducive for biodiversity management $e$ and payments $g$; note, with a basic argument of complete vector presentation of field $b$ and farm size $a$, buffer strips $c$, yields (intensity) $y$, labour $u$, we receive

$$A(p, g, a, b, i, \Delta b, \Delta u, \Delta i)$$
$$= p'q + g'e - \pi'q - 0.5[q + e]\Pi[q - e] + z'\Omega[q - e]$$
$$+ \lambda'[\Phi[q + e] + g'\mathrm{I}e + Zx] \qquad \text{(A2.1)}$$

where additionally: $q$ = production-oriented reference landscape design
$e$ = ecologically motivated response function to payment as deviation from economic optimal

Equation (A2.1) is a comprehensive representation (for the full matrix expression see the notation in A2.2) of objectives of farmers in a landscape. As has been outlined, this quadratic vector presentation is sufficient to provide response functions that are landscape design-oriented.

$$A(p, g, a, b, i, \Delta b, \Delta u, \Delta i)$$
$$= pq + ge - \pi'q - 0.5[q' + e]\Pi[q - e] + z'\Omega[q - e]$$
$$+ \lambda'[\Phi[q + e] + g'\mathrm{I}e + Zx]$$

$$= \begin{bmatrix} p \\ \psi \\ y_o \end{bmatrix}' \begin{bmatrix} a \\ b \\ i \end{bmatrix} + \begin{bmatrix} g_5 \\ g_1 \\ g_3 \\ g_4 \\ g_2 \end{bmatrix}' \begin{bmatrix} a \\ \Delta b \\ \Delta i \\ u \\ c \end{bmatrix} + \begin{bmatrix} \pi_{1,0} \\ \pi_{2,0} \\ \pi_{3,0} \end{bmatrix}' \begin{bmatrix} a \\ b \\ i \end{bmatrix} - .5 \begin{bmatrix} a \\ b + \Delta b \\ i + \Delta i \end{bmatrix}'$$

$$\times \left[ \begin{bmatrix} \pi_{1,1}^{1,1} & \pi_{1,2}^{1,1} & \pi_{1,3}^{1,1} \\ \pi_{2,1}^{1,1} & \pi_{2,2}^{1,1} & \pi_{2,3}^{1,1} \\ \pi_{3,1}^{1,1} & \pi_{3,2}^{1,1} & \pi_{3,3}^{1,1} \end{bmatrix} \begin{bmatrix} a \\ b - \Delta b \\ i - \Delta i \end{bmatrix} + Z^1 x_1 \right]$$

$$+ \begin{bmatrix} \lambda_1 \\ \lambda_2 \\ \lambda_3 \end{bmatrix}' \left[ \begin{bmatrix} aB \\ m \\ g \end{bmatrix} - \begin{bmatrix} \pi_{1,1}^{2,1} & \pi_{1,2}^{2,1} & \pi_{1,3}^{2,1} \\ \pi_{2,1}^{2,1} & \pi_{2,2}^{2,1} & \pi_{2,3}^{2,1} \\ \pi_{3,1}^{2,1} & \pi_{3,2}^{2,1} & \pi_{3,3}^{2,1} \end{bmatrix} \begin{bmatrix} a \\ b + \Delta b \\ i + \Delta i \end{bmatrix} \right] + [\lambda_4]'$$

$$\times \left[ \begin{bmatrix} g_5 \\ g_1 \\ g_3 \\ g_4 \\ g_2 \end{bmatrix}' \begin{bmatrix} 0 & 0 & 0 & 0 & 0 \\ 0 & 0 & 0 & 0 & 0 \\ 1 & 1 & 1 & 1 & 1 \end{bmatrix} + \begin{bmatrix} \pi_{1,1}^{3,1} & \pi_{1,2}^{3,1} & \pi_{1,3}^{3,1} & \pi_{1,4}^{3,1} & \pi_{1,5}^{3,1} \\ \pi_{2,1}^{3,1} & \pi_{2,2}^{3,1} & \pi_{2,3}^{3,1} & \pi_{2,4}^{3,1} & \pi_{2,5}^{3,1} \\ \pi_{3,1}^{3,1} & \pi_{3,2}^{3,1} & \pi_{3,3}^{3,1} & \pi_{3,4}^{3,1} & \pi_{3,5}^{3,1} \end{bmatrix} \right.$$

$$\left. \times \begin{bmatrix} a \\ \Delta b \\ \Delta i \\ u \\ c \end{bmatrix} + Z^2 x_2 \right] \tag{A2.2}$$

## *A2.2 Agent behaviour*

As function (A2.1) is a quadratic representation of objective functions of farmers in a landscape, its first derivative is a linear function and we obtain equation (A2.3) as response function to payments. In equation (A2.3), the landscape design criteria, i.e. payment and ecosystem goals, are given by $e$ and $g$; i.e. again $e$ is the 'ecosystem service notification' and $g$ is the payment. From equation (A2.2), two cases can be distinguished 1. We consider prices to be exogenous, i.e. we accept market prices and use payments as a single instrument. This yields an approximation of the response function on eco-services where $q^*$ is given:

$$e = [\Pi + \Phi]^{-1}[[\lambda_0 \mathrm{I} + \mathrm{I}]g + \Phi q^* + \Omega z + Zx] \tag{A2.3}$$

where $q^*$ is a function of $p: q^* = \Pi^1[p + \pi + \Omega z'] + \cdots$ determining structures, off-set by payments. 2. The other alternative is to optimise $p$ and $e$ whereas $p$ influences $q$. Again, this requires a more detailed discussion of objective functions since welfare has to be specified. Further deliberations may reveal that price is a secondary instrument since higher prices can reduce payments.

### *A2.3 Government as principal and its objective function*

To be clear, we simplify and drop 'p' as an instrument in equation (A2.3). Hereby, the willingness to change landscape structures according to payments $g$ (price levels are exogenous) is given as needed payments per service unit expected. Price matters, though: the lower $p$ is, the higher are payments $g$. Furthermore, our government shall be a purely ecologically oriented government focusing on biodiversity and payments, not farm welfare. To integrate, we can specify the objective of the principal as maximising $D_b$ given a budget $g$. This is the least complicated and most easily perceivable, ad hoc-delineation of a principal's goal. Function (A2.4) aims to model the provision of a certain level of biodiversity (as Shannon Viener Index) by a landscape, designed according to payments (the efficacy problem).

Steps are as follows: 1. We set $D_b$ as $D_b = s' \ln[s] \approx [s]' [1 + [s]]$ (since for small numbers the natural logarithm is 1 plus a per cent). 2. We split structural and response components in equation (4) and get: $s = \Xi^1 x + \Xi^2 e$. 3. By this expression, we receive biodiversity as a combination of initial landscape structure $q$ and responses $e$. 4. Inserting into the objective function gives:

$$D_b = \lfloor \Xi^1 x + \Xi^2 e \rfloor' \lfloor \mathrm{I} + \lfloor \Xi^1 x + \Xi^2 e \rfloor \rfloor \tag{A2.4}$$

Now, we add the budget constraint and obtain (A2.5), where $D_b$ is maximised under a given budget, i.e. we specify the problem round a fixed budget (also possible vice versa):

$$E = \lfloor \Xi^1 x + \Xi^2 e \rfloor' \lfloor \mathrm{I} + \lfloor \Xi^1 x + \Xi^2 e \rfloor \rfloor - \rho' \lfloor g_g - g \cdot e \rfloor \tag{A2.5}$$

Then vector $\rho$ contains shadow prices and the vector $e$ represents the principal's flexibility. Inserting (13) in (14), the problem can be solved for $g$, providing the payment structure.

$$\begin{aligned} E = {} & [\Xi^1 x + \Xi^2 [\Pi + \Phi]^{-1} [[\lambda_0 \mathrm{I} + \mathrm{I}] g + \Phi q^* + \Omega z + Z x]]' \\ & \times [\mathrm{I} + [\Xi^1 x + \Xi^2 [\Pi + \Phi]^{-1} [[\lambda_0 \mathrm{I} + \mathrm{I}] g + \Phi q^* + \Omega z + Z x]]] \\ & - \rho' [g_g - [\Pi + \Phi]^{-1} [[\lambda_0 \mathrm{I} + \mathrm{I}] g + \Phi y^* + \Omega z + Z x]] \end{aligned} \tag{A2.6}$$

Equation (A2.6) is the derived objective function of a government as principal. The principal decides on incentives *g*, given market prices *p* in landscape planning and *x* as structural variables of the system. Ways to solve the objective, by taking first derivatives including participation constraints, are given in Richter and Furubotn (1997). A starting point is *e* and then an iterative optimisation of *q* and *e* occurs. Thereby, we can provide a final solution that includes income concerns and ecology.

*List of variables*

| | |
|---|---|
| a | = distance of vertical length of a farm (with fixed b's also a measure of the farm size) |
| b | = distance of horizontal length of a field (with fixed a also a measure of the field size) |
| c | = buffer strips |
| e | = ecologically motivated response function to payment as deviation from economic optimal |
| g | = monetary incentive |
| $h_i^t$ | = total hours of labour available to a farm family |
| i | = intensity |
| $k^*$ | = unit costs |
| l | = field size |
| $m_i$ | = salary (for a farmer fully resigned from agriculture) |
| $o_i$ | = off-farm labour |
| p | = gross margin |
| $p^*$ | = price |
| q | = production/production-oriented reference landscape design in programming |
| r | = input price vector |
| u | = labour on buffer strips |
| $w_i^o$ | = exogenous off-farm wage rate |
| $w_i^l$ | = endogenous returns to labour on the farm |
| x | = vector of all exogenous factors |
| y | = yield |
| z | = size of farm in programming |
| A | = the absolute length of a sub-unit of the landscape |
| B | = the absolute width of a sub-unit of the landscape |
| C | = costs |
| E | = Lagrange objective function (combination of the biodiversity index *D* and a financial constraint) |
| $H_i^f$ | = amount of hours per hectare |
| P | = profit |

R = revenue
$D_b$ = Shannon Index
s = species vector
h = habitat vector
$\lambda$, $\rho$ = shadow prices

*Coefficients*

$\varepsilon$ = reaction coefficients in an optimal payments scheme
$\kappa$ = preset unit distance
$\Xi$ = probability matrix
$\psi$ = matrix of unified coefficients in objective function
$\Theta$ = matrix of impacts of chemical inputs
$\Phi$, $\Pi$, and $\Omega$ = matrices in a stylised objective function; coefficients retrievable from sub-models
Z = matrix representing coefficients from appendix

*Other symbols used*

f(.) = implicit function
$\Delta$ = change
i, j, k = indices
m, n = upper limits of indices

*Section B*

# Implementation

# 17 The effectiveness of centralised and decentralised institutions in managing biodiversity: lessons from economic experiments*

*Jana Vyrastekova and Daan van Soest*

## 1 Introduction

The world's stock of biodiversity is declining at an alarming rate: species are thought to be going extinct at rates unprecedented in the history of mankind. To the extent that species have an intrinsic value, biodiversity decline constitutes a welfare loss, but also because of the resulting decrease in the world's stock of genetic material. Opinions differ with respect to the actual rate of extinction as well as with respect to how costly it is for society when species disappear. But the fact remains that many governments nowadays implement policies specifically aimed at conserving individual species as well as species diversity in general, which suggests that there is widespread consensus about the urgency of protecting biodiversity.

The main direct threats to biodiversity conservation are excessive harvesting of specific species and habitat destruction (see Reid and Miller 1989). A common feature of these two types of threats is that there is little incentive for individual agents to take into consideration the impact of their (economic) activities on species conservation. In many instances, property rights with respect to the use of natural resources are either simply lacking or not well enforced. Hence, agents who invest in larger future stocks of a particular species or resource – by reducing current extraction – cannot claim an exclusive right to the future benefits of their investment. Although this problem can be at least partly side-stepped by increasing the efforts to capture the genetic information of threatened species as well as by working towards their *ex-situ* conservation, uncontrolled access to natural resources is expected to result in such high rates of species loss that *in-situ* conservation is indispensable.

The problem of biodiversity loss thus consists of essentially two components. First, benefits of conservation accrue to the world as a whole, as the stock of genetic information is essentially a public good. Second, biodiversity conservation cannot be viewed separately from the conservation of natural resources: species are threatened by extinction as either the species themselves are overharvested, or because their habitat decreases because of expanding economic activity. These two considerations are relevant when trying to determine how much biodiversity should be conserved. The remaining stock should at least be equal to the size that maximises the long-run welfare of the group of resource users as a whole. That means that if governments implement policies that force resource users to internalise the negative impact of their harvesting behaviour on the welfare of other resource users, they take an important step towards the global socially optimal level of biodiversity conservation. But depending on whether or not governments also consider the information value of species, the optimal stock size is likely to be even higher.

The obvious strategy for policy-makers aiming to regulate natural resource use is that of centralised intervention. By announcing extraction norms combined with a system of fines and monitoring activities, regulators can implement optimal levels of resource conservation, at least in theory. But in the actual practice of resource conservation in developing countries (which are, after all, richest in biodiversity), centralised enforcement may not be very effective. Although many factors can be identified as possible causes of intervention failure, the lack of sufficient financial means to adequately monitor the use of (sizable) resources is especially prominent in developing countries. Moral hazard problems also play an important role in the sense that the government institutions implementing actual resource management do not always face appropriate incentives to actively prevent excessive harvesting.

So, the crucial question that arises is the following. If straightforward centralised enforcement policies may not be very effective in inducing biodiversity conservation by means of protecting natural resources in situ, what alternative strategies are available? One such alternative is self-regulation by the resource users themselves. Overharvesting of natural resources not only harms the global community but is also welfare decreasing for the individuals involved in the harvesting activity and hence users themselves also have a stake in preventing resource depletion. Resource users can, in principle, induce their peers to choose cooperative harvesting levels; both sanctions and rewards may be used to obtain the desired actions by one's peers. The question is then to what extent resource users are willing to incur costs (when punishing or rewarding) to induce cooperative behaviour by one or more of their fellow resource users. This question is pertinent because whereas the costs of rewarding or sanctioning are private, the benefits (if the resource user sanctioned or rewarded adjusts his/her behaviour) accrue to the group as a whole.

Whereas complete decentralisation may be able to induce resource users to internalise the negative externalities their activities have on their fellow resource users, such a system is not able to induce resource users to also internalise the transboundary externalities associated with resource and biodiversity conservation. Therefore, some sort of centralised enforcement may be desirable after all, but one which involves the community members in the actual implementation. By involving the regulated agents in the management of the resource, policy-makers may still be able to exploit self-regulatory mechanisms within the group.

The policy relevance of analysing self-regulation is obvious for the resource-rich developing countries, given the very mixed results obtained by centralised enforcement in the field. However, empirically analysing the effectiveness of various types of enforcement regulation is very difficult, because of the many confounding factors that affect success and failure in the real world. It is seldom that one finds natural experiments in the sense of two identical resources confined to identical regions, but where two different types of regulations are in place. If such natural experiments are lacking, one may turn to analysing the effectiveness of various types of regulations in more controlled environments, such as in computer labs where subjects can be brought in to make decisions very similar to the ones made by resource users in the real world. And indeed, this is the route that the bulk of the literature on regulatory design has chosen.

In this chapter, we review the implications for biodiversity preservation regulation that follow from the experimental economics literature. We focus on two types of games, which mimic the decision confronting problem resource users in practice. One is the common pool resource game, where increased extraction effort by one resource user decreases the other group members' return to their extraction effort. The other is the (linear) public goods game, where investments made by one individual give rise to benefits that accrue to the group as a whole.

This chapter is organised as follows. In section 2, we discuss the insights from the literature on the centralised enforcement of laws by means of pecuniary punishment (fines) and we discuss whether and to what extent efficiency of resource conservation can be enhanced by including local users in the management of the resource. In section 3 we turn our attention to the role self-regulation might play in preserving natural resources and, thus, biodiversity. So, in the former section we address formal laws and their effectiveness in inducing socially optimal levels of extraction or investment and in the latter we do the same for informal institutions. The differences between formal and informal institutions are essentially of origin and stability. Regarding origin, formal ones are designed by the policy-maker while the informal ones are the outcome of group dynamics and evolution of social norms in a community. With respect to stability, the formal institutions have to be backed by incentives provided to their employees (e.g. the agents policing resource users' behaviour), but the informal institutions have to be self-enforcing, relying on the norms and restraints imposed by the stakeholders themselves.

The lessons to be taken from the experimental research come from two sources: from experimental laboratories, using as subjects predominantly university students, and from experiments performed in the field, where subjects are real natural resource users. These two data sources complement each other and each of them has its advantages and disadvantages. On the one hand, while the experiments performed in the field provide data collected from the relevant subject pool (the actual resource users themselves), they also suffer from looser experimental control than the rigorously designed laboratory experiments. On the other hand, where population composition is crucial for the response to the investigated institution, data on student pools have to be compared against the field conditions. In the following, we report experiments on public good provision and common pool resource management both from the field and from the laboratory.

## 2 Centralised enforcement and the efficient management of common pool resources

Centralised enforcement requires the regulator to establish utilisation rules with respect to the resource, as well as sanctions in case these rules are violated. Obviously, the optimal level of extraction specified differs depending on whether the regulator takes into account only the local benefits of resource conservation, or the global benefits as well. In section 2.1, we address the extent to which resource user actual behaviour is sensitive to the design of the enforcement institutions, and in section 2.2 we discuss to what extent this design affects the regulated individuals' support for the institutions in place.

### *2.1 On the efficient design of enforcement of conservation rules*

Whereas specifying the rules and regulations with respect to natural resource use may be fairly straightforward, actually enforcing them is not. Indeed, centralised enforcement requires the active involvement of agents in the field monitoring the harvesting activities of all people in the region, which is by definition costly. When thinking about the deterrence effect of norms and regulations, norm violation is less attractive – all else equal – the more severe the sanctions and the higher the probability of detection. That means that assuming rational behaviour, agents can be prevented from overharvesting the resource if the marginal benefits of doing so fall short of the expected marginal costs, which depends on both the probability of detection and the severity of the sanction. So, the question is whether indeed the regulator can save on enforcement costs (resulting in lower probabilities of detection) – while achieving the same level of resource conservation – by increasing the severity of the punishment on violations of the extraction norms.

The economic theory of law enforcement (as pioneered by Becker 1968) states that if the regulated agents are rational and risk-neutral, the same 'crime rate' can be achieved by having either high probabilities of being punished and less severe punishments, or vice versa, as long as the expected level of punishment remains constant.[1] If indeed it is only the level of expected punishment that matters, the costs of law enforcement can be reduced by choosing the appropriate combination of the punishment levels and probabilities of conviction. Increasing the probability of conviction is typically expensive, as it requires more effort to catch suspects, more effort to gather sufficient evidence to get suspects convicted, etc. But especially in the case of financial sanctions (as opposed to imprisonment, cf. Polinsky and Shavell 1984), increasing the severity of the punishment is to a large extent an administrative affair where the costs of executing it is largely independent of the fine level. Focusing on financial sanctions, this implies that the costs of achieving a specific crime rate can be reduced by choosing higher fines with lower conviction probabilities (cf. Garoupa 2000; Polinsky and Shavell 1979 and 2000).

If these considerations of minimising the costs of law enforcement hold in general, they are *a fortiori* relevant in case of biodiversity conservation, as already hinted at in the introduction to this chapter. As the most valuable natural resources are located in remote areas of developing countries, enforcing utilisation rules is difficult for two reasons. First, monitoring is often highly imperfect because of the lack of adequate financial means. Second, enforcement itself is hampered as government institutions to which enforcement is entrusted do not always face appropriate incentives to prevent excessive harvesting. In these circumstances, it is tempting to design laws that enable the imposition of large fines rather than invest in increasing the conviction probability.

[1] Given that more severe (financial) punishments as well as higher probabilities of detection result in lower levels of overextraction, one may conjecture that the optimal punishment should be infinitely high. Whereas this may be true in theory, in practice severity of sanctions is kept in check by considerations that the punishment needs to fit the crime (Becker 1968; Polinsky and Shavell 1979, 1968; Polinsky and Shavell 2000).

The question is to what extent higher fines and higher probabilities of detection are indeed perfect substitutes in practice. Theory predicts that if individual resource users are risk averse rather than risk neutral, a 1 per cent increase in the fine level is more effective in deterring norm violation than a 1 per cent increase in the probability of conviction, whereas for risk-loving resource users the opposite holds (Becker 1968).

The empirical validity of this theoretical prediction has been tested by Anderson and Stafford (2003), who address the question of optimal fine/probability mix in an experiment using students as subjects. They find that indeed the size of the sanctions has a bigger deterrence effect than the probability with which they were assigned. However encouraging, this observation has to be taken as one possible outcome, given the impact risk preferences are likely to have. If the population we are interested in has a different distribution of risk attitudes than the student population used in the experiment, then our conclusion might be reversed (see, for example, Block and Gerety 1995). Indeed, Cardenas (2004) observes that at a fixed (low) probability of imposing a fine, a considerable increase in the monetary fine has only a minor impact on free-riding in common pool resource use. The presence of fines is incorporated into the decision-making process, but the size of the fine does not have a considerable effect and the outcomes still fall short of the social optimum.

Another study that focuses on explicitly testing the extent of higher fines and higher detection probabilities is one by van Soest and Vyrastekova (2005). They use a common pool resource game to capture the negative externalities of one individual user's harvesting activities on the payoffs to all other members of his/her group having access to a single resource. The game is a finitely repeated one and static in the sense that current harvesting affects the state of the resource in the current period, but not in the next. Groups consist of five subjects harvesting one resource and a sixth one who plays the role of the enforcer of an official norm with respect to the maximum extraction effort. This norm is set equal to the individual extraction effort level that maximises the payoff to the group as a whole; if an individual resource user puts in more than this particular level of effort, he/she faces a probability of being fined. The fine is a constant amount per unit of excess extraction effort and hence the total amount paid increases linearly with excess extraction effort.

To analyse whether higher fines and higher probabilities of detection are perfect substitutes, van Soest and Vyrastekova study two cases. For a fixed expected fine per unit of excess extraction effort, they compare the impact on harvesting behaviour in the case where the probability of free-riders' actually being fined is 50 per cent (with the fine equal to a certain amount per unit of overextraction) and 90 per cent (with a concomitant decrease in the per-unit fine). Here it is important to note that the combinations of per-unit fines and probabilities of free-riders being fined are set such that risk-neutral resource users would reduce their extraction effort level towards the socially optimal level (the norm), but not all the way down to that level. Thus, the researchers can identify whether there is indeed a difference between the two combinations of fines and probabilities in terms of their effectiveness in changing behaviour.

The results from this study are straightforward. When comparing the aggregate extraction effort in the absence of an enforcer to the aggregate effort level that results when fines can be imposed, van Soest and Vyrastekova find that the threat of enforcement does improve resource conservation. The aggregate extraction is closer to – but still higher than – the socially optimal level when an enforcement institution is present. And when comparing the subjects' extraction and enforcement behaviour if the probability of being fined is 50 per cent or 90 per cent, they find that these differences are too small to result in significantly different levels of harvesting or enforcement. This research therefore suggests that indeed higher probabilities and higher fines are (fairly) good substitutes in terms of their direct impact on individual extraction effort. Hence, van Soest and Vyrastekova conclude that when individual extraction effort is all that matters, governments can indeed economise on the implementation costs of biodiversity conservation programmes by increasing fines while saving expenditures on monitoring effort (thus lowering the probability of free-riders being fined).

### 2.2 *Institutional crowding out*

We have, thus, seen that even though actual results are likely to depend on the composition of the group of resource users under consideration (especially with respect to their risk preferences), the direct effect of lower probabilities of being fined can be compensated by roughly proportional increases in fines. But this conclusion is based on the assumption that in the absence of formal intervention, there are no alternative mechanisms giving rise to at least some level of resource conservation above the non-cooperative level. However, the real-world validity of this assumption cannot simply be taken for granted; many examples can be cited of communities that are fairly successful in managing common pool resources even in the absence of formal government intervention. In such circumstances, introducing a centralised enforcement agency might reduce the community members' involvement in natural resource management, thus crowding out the informal rules, norms and regulations that were in place (see, for example, Wit 1999).

An example of this phenomenon of formal rules crowding out informal norms is provided by Cardenas *et al.* (2000). These authors conducted experiments with people who are confronted with a common pool problem in everyday life. When comparing these subjects' behaviour in a common pool resource game without any restrictions with those obtained after imposing an extraction norm enforced by a mild probabilistic fine, Cardenas *et al.* find that subjects extract more aggressively in the second case than in the first. As a result, their payoffs are significantly lower when they are confronted with a formal norm than in its absence; the weak official norm interacted with the internal norms of the subjects and crowded out the incentives to cooperate. While the authors do not investigate in detail the complex question of crowding out, their paper presents a warning towards indiscriminately introducing regulatory intervention without a proper understanding of how it might undermine norms already operating in the field.

Having established that indeed formal intervention may be counterproductive in terms of resource conservation, the question arises as to whether government policies can be designed such that formal and informal institutions are mutually reinforcing. Some recent papers suggest that including regulated individuals in the process of designing formal intervention is key. In this respect, several papers have looked at the extent to which allowing regulated subjects to vote on the details of the enforcement institution's design can substantially improve the effectiveness of the enforcement institution under consideration. Voting serves a dual purpose. First, the voting outcome (based on, for example, a majority voting rule) affects the design of the institution and hence its direct effectiveness. But voting outcomes also provide information about the intentions and preferences of the community's majority to effectively protect the resource and to maximise group payoff (as opposed to trying to non-cooperatively maximise one's individual payoffs). Obviously, the consequences for actual resource conservation depend crucially on the voting outcome.

An example where a failure to gain majority agreement for the socially optimal action is detrimental to social welfare is provided by Sutter and Weck-Hannemann (2004). In their study, subjects have the possibility to vote on a minimal contribution level to a public good, upon which they make their decisions over how much to contribute to the public good. The results make intuitive sense. Subjects contribute more when a majority vote was achieved in favour of a certain minimum contribution level, but when the group failed to achieve such a consensus, contributions are significantly lower. This suggests that the simple signalling effect of majority voting outcomes affects individual decision making, even though the voting outcome itself does not have any formal consequence.

Obviously, the consequences of not achieving a majority vote are even more detrimental if the voting outcome results in the abolition of formal institutions, as is uncovered in a study by Tyran and Feld (2001). In this study, subjects can vote on the level of a (deterministic) sanction in a public goods environment. As is the case in Sutter and Weck-Hannemann (2004), subjects tend to contribute significantly less (more) when the majority vote was against (in favour of) the presence of an enforcement institution empowered to impose fines on those who contribute less than a certain level.

Having established that introducing voting with respect to the details of the design of the enforcement institution can either improve or reduce welfare (and conservation) depending on the voting outcome, the question arises as to what factors determine voting behaviour. An attempt to answer this question is made by Vyrastekova and van Soest in two related papers (Vyrastekova and van Soest 2003; van Soest and Vyrastekova 2005). In these two papers, subjects were allowed to vote on whether the enforcement institution should be provided with sufficient incentives to actively sanction excessive extraction, or not. More specifically, subjects could vote on whether or not the agent representing the enforcement institution is allowed to pocket the fine revenues. If a majority votes against pocketing fines, the enforcer is not expected to actively impose fines when observing violation of the formal norm because there are fixed costs associated with punishing. In this setting the weakly dominant strategy is to vote in favour of the enforcer receiving the fine revenues. Voting itself is costless and in equilibrium aggregate extraction effort is closer to the socially optimal level, even though the enforcer does not actively enforce the norm because of the fixed costs of enforcement. In other words, the threat of the enforcer engaging in norm enforcement reduces aggregate extraction effort towards the norm, but in equilibrium the enforcer never actually imposes fines because it is not profitable to do so (but only just so).

In Vyrastekova and van Soest (2003), the hypothesis was tested as to whether allowing resource users to vote on the incentive structure of a natural resource management institutional regime enhances the efficiency of resource use. This was examined experimentally by comparing two treatment groups. In the first, the policy enforcer always receives the revenues of her sanctioning activity (i.e. the fines imposed on those resource users who extract more than is prescribed by a norm); in the other treatment, the incentives faced by the enforcer are decided upon by the resource users themselves by means of a majority voting rule as described above.

Vyrastekova and van Soest find support for the hypothesis that voting actually *improves* efficiency of resource use as compared with the treatment in which incentives are assigned exogenously. Casting their vote serves as a means for resource users to communicate their stance with respect to the need for reduced aggregate extraction. Conditional on a majority having voted in favour of implementing an appropriate incentive structure, the extraction behaviour was significantly more cooperative in the voting treatment than in the treatment where the enforcer is always allowed to pocket the fine revenues.

In a companion paper, van Soest and Vyrastekova (2005) analyse to what extent actual voting outcomes depend on the characteristics of the enforcement institution. The specific characteristic they focus on is the probability that when engaging in enforcement, the institution is indeed able to successfully impose fines. Keeping the expected fine constant, they compare the impact on voting behaviour of a 50 per cent chance of conviction (and a specific fine level) with the case of a 90 per cent chance of conviction (and a lower fine level). In both cases, the weakly dominant strategy is to always vote in favour of the enforcer receiving the fine revenues, because of the arguments given above. Van Soest and Vyrastekova actually find marked differences between the 50 per cent and 90 per cent probability treatments. Whereas in the latter treatment resource users almost always vote in favour of the enforcer receiving the fines, a favourable majority voting outcome is achieved in less than 40 per cent of the cases in the former treatment. These results are striking and clearly indicate that the higher fines and higher conviction probabilities may be fairly good substitutes in terms of their direct impact on extraction behaviour (see the conclusion of the previous section), but that trying to save on enforcement costs by reducing the probability of conviction (with a concomitant increase in the fine level such that the expected fine is kept constant) is hazardous if the enforcement institution's effectiveness is at least to some extent dependent on the support of the regulated individuals.

## 3 Self-regulation and common pool management

In this section, we address the effectiveness of informal institutions in governing the behaviour of natural resource users. A substantial amount of research has been dedicated to exploring the extent to which self-governance can prevent overharvesting of a common property resource. The main instruments individual community members have to influence the behaviour of their peers are (pecuniary) sanctions, rewards and outright eviction from the community. These instruments are discussed in turn in the next three subsections.

### 3.1 *Peer sanctioning*

The form of self-governance that has received most attention in the experimental economics literature is decentralised sanctioning. Here, individual resource users can affect the extraction behaviour of their peers by inflicting social or pecuniary punishment on those who overexploit the resource (e.g. Ostrom *et al.* 1994; Baland and Platteau 1996). Anecdotal evidence about real-world relevance of this form of self-regulation includes, for example, Brazilian fishermen in the Bahia region who destroy the nets of fellow fishermen who do not respect the catch quotas (Cordell and McKean 1992).

Very often, sanctioning of this type not only involves costs inflicted upon the person punished but also the person imposing the sanction incurs costs. The costs of punishing may consist of direct opportunity costs in terms of the time spent on monitoring the activity of others, or indirect costs of forgone cooperation between the sanctioned and the sanctioning individual in some other economic activity. A third reason why peer sanctioning may be costly to the punisher is because some forms of sanctioning (like destroying the personal property of the free-rider) are actually illegal. Hence, imposing punishments can be expensive because of possible confrontations with the formal legal system.

To economists, the observation that individuals are willing to engage in sanctioning their peers when they themselves also incur costs is surprising. The reason for this is that whereas the costs of sanctioning are incurred by the individual, the benefits (reduced extraction by the punished individual, if sanctioning is effective) accrue to all individuals having access to the resource under consideration. So why should one provide a public good rather than free-ride on the provision of that good by other resource users? If all individuals having access to a resource followed the '*homo economicus*' way of reasoning, the possibility of being sanctioned would not discipline resource users' behaviour because punishments would never be imposed.

Still, there is ample experimental evidence that indeed self-regulation by means of (pecuniary) punishment is effective in reducing overextraction in common pool resource games, or in inducing higher levels of contributions in public goods games (Ostrom *et al.* 1992; Fehr and Gächter 2000a, 2000b, 2002; Falk *et al.* 2001; Anderson and Putterman 2003; Carpenter 2004a, 2004b). One general explanation for this 'anomaly' is that not all humans are *homo economicus*; indeed, about half of us are 'reciprocal' individuals (Fischbacher *et al.* 2001). Reciprocal individuals act cooperatively if they observe or expect others to act cooperatively, but they are willing to act non-cooperatively (or even punish) if they observe or expect others to act non-cooperatively.[2] That does not mean that economic considerations do not play a role with these individuals; the more costly it is to sanction another individual, the fewer sanctions are assigned (Anderson and Putterman 2003). It suggests only that reciprocal individuals are willing to trade off some costs incurred in sanctioning non-cooperative behaviour by their peers for experiencing the warm glow of 'disciplining those who deserve it'.

The above finding suggests that governments might rely on peer enforcement to ensure resources are conserved at a level above the stock size that would materialise if *homo economicus* was the uniformly correct description of human nature. Obviously, when relying on self-regulation, governments can save substantial amounts of money associated with monitoring and policing activities. To the extent that individuals can better monitor their peers' behaviour than outsiders (such as an enforcement institution), relying on self-regulation may even be more efficient.

There is a considerable body of research on the circumstances under which self-regulation is likely to be most effective. One of the more evident prerequisites for enhancing effectiveness that has been studied is that individual resource users themselves have a direct stake in conservation. But self-regulation can also be more effective if communication among resource users is improved. For example, face-to-face communication among common pool resource users is found to be very effective in enhancing net efficiency of resource use, as it substantially facilitates cooperation (Ostrom *et al.* 1992, 1994; Isaac and Walker 1998; Cardenas *et al.* 2000; Cardenas 2004; Bochet *et al.* 2005). The effect is similar to that of voting in centralised enforcement. By being informed about the views of one's peers over the necessity to reduce aggregate extraction, it becomes easier for individuals to coordinate their actions on a certain type of behaviour. Similarly, there is some evidence that just the expression of disagreement (rather than the imposition of a monetary punishment) is enough to discipline the behaviour of individuals. Recently, Masclet *et al.* (2003) showed that the message that somebody does not agree with one's behaviour is able to induce individuals to increase their contributions in a public goods game.

Governments may also try to exploit the fact that individuals care about their reputation. Denant-Boemont *et al.* (2005) revisit a classical paper by Wilson and Sell (1997) to investigate the channels along which cheap talk announcements of own intended actions work in a social dilemma

[2] For this reason, reciprocal individuals are also sometimes referred to as conditional cooperators.

situation. Giving feedback to all participants about the actual behaviour of individuals as compared with the actions they announced beforehand is found to stimulate cooperation in the public goods game. The fact that others may note discrepancies between promises and actual behaviour is sufficient to induce individuals to contribute more.

But unfortunately self-regulation by means of pecuniary punishment is not a panacea. First, whereas this mechanism may be able to reduce aggregate extraction towards the socially optimal level (or increase contributions in a public goods setting), it is not necessarily welfare improving. Indeed, many studies find that when taking into account the deadweight loss of sanctioning, the possibility of imposing sanctions actually decreases the total payoff to the group as a whole, as compared with the situation in which a resource is exploited without any means to discipline one's peers' behaviour (Ostrom *et al.* 1994).

Second, pecuniary punishment may turn out to be ineffective if the persons imposing the sanctions may be exposed to retaliation by the persons punished. Interactions related to the harvesting of a particular natural resource (such as lake or forest) usually take place locally and among a stable group of individuals. These individuals therefore interact repeatedly with each other and have ample opportunities to engage in positive but also in negative reciprocal behaviour. Nikiforakis (2004) shows that incorporating the option of retaliation in sanctioning in an experiment actually mitigates the effectiveness of having the option to sanction in increasing the provision of the public good. The threat of a negative response to an imposed sanction decreases the motivation of individuals to actually impose punishments in the first place.

### 3.2 *Peer rewards and selective exclusion*

As we have seen that self-governance by means of sanctioning may give rise to negative side effects, the question as to whether it can ever be an effective instrument for biodiversity conservation essentially remains open. There is, however, another type of peer pressure that is based on rewards rather than sanctions. Refusing to give rewards is also a sanction, but one which is more realistic than the direct imposition of (financial) sanctions. Indeed, whereas instances of self-regulation by means of punishment are known, everyday experience suggests that it is not very common. Ordinary citizens do not usually have the right to destroy another person's property, nor do they have the authority to impose fines; it is the government that has the exclusive right of coercion in most societies. Also denying individuals the right of access to natural resources is often not legal or feasible in practice (McCarthy *et al.* 2001).

What citizens can do, however, is decide either not to reward their peers who do not deserve it, or to cease interacting with free-riders and to refuse to cooperate with them in other social or economic circumstances in which they meet. We discuss these two possibilities in turn.

In the experimental economics literature, much less attention has been paid to the effectiveness of rewards (compared with sanctions) in inducing desired behaviour. Sefton *et al.* (2003) introduce the possibility of transfering part of one's earnings to other subjects in a public goods game. Unlike sanctions, such rewards represent a pure transfer in that one monetary unit deducted from the subject giving the reward results in the receiver gaining one monetary unit. Sefton *et al.* observe that there is little agreement with respect to who should receive the rewards and although rewards are used more often in the early rounds of the game than sanctions, their incidence decreases as the game progresses. Sefton *et al.* therefore come to the conclusion that rewards do not provide efficient incentives for individuals to act cooperatively with respect to their contributions to the public good. Gürerk *et al.* (2004) come to a similar conclusion.

The relative ineffectiveness of rewards (as compared with sanctions) may, however, be attributable to the fact that Sefton *et al.* and Güreck *et al.* analyse a mechanism where the person receiving (giving) a rewards sees his/her payoff increase (fall) by the same amount. These so-called transfer rewards are unable to equalise payoffs between those who cooperate in the public goods game with those who free-ride (unlike with the sanctions discussed above). Why would an individual who acts cooperatively him/herself give money to a fellow cooperator, increasing that person's payoff while decreasing one's own payoff by exactly the same amount? If, for example, rewards can be given that increase the receiver's welfare more than it costs the person giving the reward (the so-called efficiency-improving rewards), there is reason to reward each other's behaviour, whereas there is none in the case of pure transfers. Indeed, Andreoni *et al.* (2003) do find some support for this view. In the context of a bargaining-like game, they observe that the ability to assign efficiency-increasing rewards actually improves levels of cooperation (even relative to their treatment with sanctions). Similar evidence is available for common pool resource games (see Vyrastekova and van Soest 2005).

Rewards are one alternative mechanism to sanctions, but interactions between alternative economic activities provide another. Indeed, individuals do not usually interact in just one type of activity, such as resource harvesting. Behaviour is embedded in a system of interpersonal relations (Granovetter 1985) and social dilemmas often occur in communities which are, by definition, characterised by the presence of multiple forms of interaction that require cooperation by two or more individuals (cf. Bowles and Gintis 2002). Ceasing cooperation in these other activities is a natural sanctioning device to discipline behaviour of one's peers in the social dilemma situation.

This idea seems very much applicable to the communities harvesting common pool resources in developing countries. In this context the welfare of individual users does not depend on the behaviour of their fellow users with respect to resource harvesting alone, as full specialisation of labour is rare. Resource users are generally also active in alternative types of economic activity, especially agriculture. Maximisation of agricultural returns often implies dependence on the cooperation of fellow community members, in the form of either provision of labour (for activities that need to be completed within a certain time period such as harvesting) or shared use of equipment. Therefore, common pool resource users have the option to sanction excessive harvesting by fellow resource users by excluding them from profitable means of cooperation with respect to agriculture. For example, it has been observed that what Japanese villagers, Irish fishermen and inhabitants of the Solomon Islands have in common is that they cease contact with fellow villagers who free-ride with respect to fishing, thus denying them the benefits of cooperation in other economic activities (Taylor 1987; McKean 1992; Hviding and Baines 1994).

This type of peer pressure exerted by selectively excluding free-riders from the common pool resource problem from cooperation in another game has been studied by van Soest and Vyrastekova (2004). In their experiments, subjects participate in a finitely repeated game. Its stage game consists of two games that are played sequentially, first the common pool resource game and then a game which requires bilateral cooperation (the alternative economic activity). The repeated game has only one subgame perfect Nash equilibrium: rational money-maximising individuals always overharvest the common pool resource and never engage in the alternative economic activity with any other individual. This prediction is independent of whether individuals interact with the same group of individuals in both constituent games or not.

However, the presence of reciprocal individuals invalidates this prediction as these subjects may be willing to engage in the alternative economic activity that requires bilateral cooperation. Then, aggregate efficiency of resource use may be higher if subjects interact with the same group of individuals in both activities (the community treatment, or linked treatment) than if they do not (the disjoint groups treatment, or unlinked treatment). Individuals in the linked treatment have the option to refuse to cooperate in the alternative activity with those who overharvest the natural resource ('selective exclusion'), whereas there is no such possibility in the unlinked treatment. By comparing the linked and unlinked treatments, one can assess the viability of the selective exclusion mechanism and its effect on the efficiency of resource use. Indeed, the authors find that embedding the common property game into a larger social framework (by adding the efficiency-improving alternative economic activity), aggregate extraction effort is much closer to the socially optimal extraction effort level. The authors find that the more a subject deviates from the average extraction effort level, the less help he/she receives in the alternative economic activity.

This paper therefore demonstrates that embedding a social dilemma situation in a wider economic context gives rise to more cooperation in the social dilemma situation than predicted by economic theory. Moreover, whereas the pecuniary punishment mechanism results in a decrease in net efficiency (because of the 'deadweight loss' associated with the costs of imposing sanctions – see Ostrom *et al.* 1992, p. 176; Ostrom *et al.* 1994), the selective exclusion mechanism uncovered in van Soest and Vyrastekova results in a pure efficiency gain. Whereas aggregate extraction effort in the common pool resource game is closer to the social optimum level in the linked treatment than in the unlinked treatment, the aggregate payoff from the alternative economic activity is identical across these two treatments. Thus, these experiments suggest that strengthening community ties can give rise to powerful pro-social incentives with respect to cooperation in social dilemma situations and hence improves community welfare.

### *3.3 Ostracism*

In the case of selective exclusion as analysed by van Soest and Vyrastekova (2004), individuals unilaterally decide to no longer interact with excess harvesters in *other* economic activities that require bilateral cooperation. However, another well-known mechanism is ostracising free-riders. Ostracism occurs when individuals are excluded from the benefits of the common pool resource *itself*. This is the subject of a paper by Masclet (2003), who analyses whether the possibility of ostracising fellow participants increases contributions to the public good. Subjects can participate in two public goods games that are played consecutively. Having played the first with the entire group, individual participants may decide to unilaterally prevent one or more of their peers from participating in the second public goods game, which is played next. Masclet finds that individuals who do not contribute sufficient amounts in the first game are excluded from the second game and that the threat of this sanction raises contributions in the first game (but not in the second). Net, ostracism is also not found to actually increase efficiency of the game (as was the case with sanctions). Contributions in the first game increase but not in the second, and the positive effect of the former impact is mitigated by the welfare loss associated with some users being excluded from the benefits of the public good in the second game. In another study, Cinyabuguma *et al.* (2004) show that ostracism provides an equally effective threat as sanctions in the public goods game: when subjects can 'vote out' one or more members of the group on the basis of a simple majority voting rule, contributions to the public good are close to the efficient level. Those who free-ride on the group's effort are expelled.

## 4 Conclusions

In this chapter we reviewed experimental evidence on the formal and informal institutions in the provision of public goods and management of common pool biodiversity resources. Both of these game-theoretic constructs fall into the category of social dilemmas. Privately optimal behaviour differs from the socially optimal one due to externalities that individual actions impose on others (positive externalities in the case of public goods and negative externalities in the case of common pool resources). We discussed (i) the role of centralised formal institutions in biodiversity conservation, (ii) the consequences of decentralisation of power by giving the right to the local users to decide on these institutions by voting, and (iii) the way informal self-regulatory institutions impact on the behaviour of the local users.

We found that the standard theory predicts well the agents' response to changes in the two main ingredients of formal government intervention (the fine level and the probability of actually being forced to pay the fine). But we also found that formal intervention may crowd out informal norms. A poorly enforced norm with a low probability of conviction is found to be a worse arrangement than no enforcement at all. Moreover, when agents themselves are given a say in the design of the enforcement institution, such participation is welfare enhancing in case the enforcement institution is likely to be fairly effective in enforcing the norm, but welfare decreasing otherwise. The lesson to be drawn is, thus, that policy-makers should be aware of hidden costs of low conviction probabilities in their enforcement institutions.

An alternative approach to biodiversity conservation may be to rely on self-regulation, exploiting the fact that resource users also have a stake in conserving the resource at a level above the one that materialises when all users pursue their own private objectives. Most attention has been paid to self-regulation by means of pecuniary punishments, but relying on self-regulation is likely to be more effective in closely knit societies, where the welfare of community members depends on cooperation not only with respect to resource harvesting but with respect to other economic activities as well.

* The authors are grateful to the Netherlands Organisation of Scientific Research for financial support as part of the Program on Evolution and Behaviour.

REFERENCES

Anderson, Ch.M. and Putterman, L. 2003. Do non-strategic sanctions obey the law of demand? The demand for punishment in the voluntary contribution mechanism. Brown University Department of Economics Working Paper 2003–15.

Anderson, L. R. and Stafford, S. L. 2003. Punishment in a regulatory setting: experimental evidence from the VCM. *Journal of Regulatory Economics*. **24**. 91–110.

Andreoni, J., Harbaugh, W. and Vesterlund, L. 2003. The carrot and the stick: rewards, punishments, and cooperation. *American Economic Review*. **93** (3). 893–902.

Baland, J. and Platteau, J. P. 1996. *Halting Degradation of Natural Resources: Is there a Role for Rural Communities?* Oxford: Clarendon Press.

Becker, G. 1968. Crime and punishment: an economic approach. *Journal of Political Economy*. **76**. 169–217.

Block, M. K. and Gerety, V. E. 1995. Some experimental evidence on differences between student and prisoner reactions to monetary penalties and risk, *Journal of Legal Studies*. **24**. 123–138.

Bochet, O., Page, T. and Putterman, L. 2005. Communication and punishment in voluntary contribution experiments. *Journal of Economic Behaviour and Organization*. **60** (1). 11–26.

Bowles, S. and Gintis, H. 2002. Social capital and community governance. *Economic Journal*. **112**. 416–436.

Cardenas, J. C. 2000. How do groups solve local commons dilemmas? lessons from experimental economics in the field. *Environment, Development and Sustainability*. **2**. 305–322.

Cardenas, J. C. 2004. Norms from Outside and from Inside: An Experimental Analysis on the Governance of Local Ecosystems. Manuscript.

Cardenas, J. C., Stranlund, J. and Willis, C. 2000. Local environmental control and institutional crowding-out. *World Development*. **28**. 1719–1733.

Carpenter, J. P. 2004a. Punishing Free-Riders: How Group Size affects Mutual Monitoring and the Provision of Public Goods. Manuscript.

Carpenter, J. P. 2004b. The Demand for Punishment. Manuscript.

Cinyabuguma, M., Page, T. and Putterman, L. 2004. Cooperation under the threat of expulsion in a public goods experiment. Working Paper 2004–2005. Brown University, Department of Economics.

Cordell, J. C. and McKean, M. A. 1992. Sea Tenure in Bahia, Brazil. In D. W. Bromley and D. Feeny (eds.). *Making the Commons Work*. San Francisco: ICS Press. 183–205.

Denant-Boemont, L., Masclet, D. and Noussair, Ch. 2005. Announcement, Observation and Honesty in the Voluntary Contributions Game. Manuscript.

Falk, A., Fehr, E. and Fischbacher, U. 2001. Driving Forces of Informal Sanctions. IEW Working Paper 59. University of Zurich.

Fehr, E. and Gächter, S. 2000a. Fairness and retaliation: the economics of reciprocity. *Journal of Economic Perspectives*. **14**. 159–181.

Fehr, E. and Gächter, S. 2000b. Cooperation and punishment in public goods experiments. *American Economic Review*. **90**. 980–994.

Fehr, E. and Gächter, S. 2002. Altruistic punishment in humans. *Nature*. **415**. 137–140.

Fischbacher, U., Gächter, S. and Fehr, E. 2001. Are people conditionally cooperative? Evidence from a public goods experiment. *Economics Letters*. **71** (3). 397–404.

Garoupa, N. 2000. Optimal Magnitude and Probability of Fines. Universitat Pompeu Fabra Working Paper 454. Barcelona.

Granovetter, M. 1985. Economic action and social structure: the problem of embeddedness. *American Journal of Sociology*. **91**. 481–510.

Gürerk, O., Irlenbusch, B. and Rockenbach, B. 2004. On the Evolvement of Institution Choice in Social Dilemmas. Manuscript.

Hviding, E. and Baines, G. B. K. 1994. Community-based fisheries management, tradition and challenges of development in Marovo, Solomon Islands. *Development and Change*. **25**. 13–39.

Isaac, R. M. and Walker, J. M. 1988. Communication and free-riding behaviour: the voluntary contribution mechanism. *Economic Inquiry*. **26**. 585–608.

Masclet, D. 2003. Ostracism in work teams: a public good experiment. *International Journal of Manpower*. **24** (7). 867–887.

Masclet, D., Noussair, Ch., Tucker, S. and Villeval, M. C. 2003. Monetary and nonmonetary punishment in the voluntary contributions mechanism. *American Economic Review*. **93** (1). 366–380.

McCarthy, N., Sadoulet, E. and De Janvry, A. 2001. Common pool resource appropriation under costly cooperation. *Journal of Environmental Economics and Management*. **42**. 297–309.

McKean, M. A. 1992. Management of traditional common lands (Iriaichi) in Japan. In D. W. Bromley and D. Feeny (eds.). *Making the Commons Work*, San Francisco: ICS Press. 66–98.

Nikiforakis, N. S. 2004. Punishment and Counter-Punishment in Public Goods Games: Can we still govern ourselves?. Discussion Paper Series 2004–2005. Royal Holloway, University of London.

Ostrom, E. 1990. *Governing the Commons: The Evolution of Institutions for Collective Action*. Cambridge, New York: Cambridge University Press.

Ostrom, E., Gardner, R. and Walker, J. 1992. Covenants with and without sword: self-governance is possible. *American Political Science Review*. **86**. 404–417.

Ostrom, E., Gardner, R. and Walker, J. 1994. *Rules, Games and Common-Pool Resources*. Ann Arbor: The University of Michigan Press.

Polinsky, A. M. and Shavell, S. 1979. The optimal tradeoff between the probability and magnitude of fines. *American Economic Review*. **69** (5). 880–891.

Polinsky, A. M. and Shavell, S. 1984. The Optimal Use of Fines and Imprisonment. *Journal of Public Economics*. **24**. 89–99.

Polinsky, A. M. and Shavell, S. 2000. The economic theory of public enforcement of law. *Journal of Economic Literature*. **38**. 45–76.

Reid, W. V. and Miller, K. R. 1989. *Keeping Options Alive: The Scientific Basis for Conserving Biodiversity*. Washington, DC: World Resources Institute.

Sefton, M., Shupp, R. and Walker, J. 2003. *The Effect of Rewards and Sanctions in Provision of Public Goods*. Indiana University Working Paper, Bloomington, IN.

Sutter, M., and Weck-Hannemann, H. 2004. An experimental test of the public–goods crowding-out hypothesis when taxation is endogenous. *Finanzarchiv*. **60**. 94–110.

Taylor, L. 1987. The river would run red with blood: community and common property in Irish fishing settlement. In B. J. McCay and J. M. Acheson (eds.). *The Question of the Commons: The Culture and Ecology of Communal Resources*. Tucson: University of Arizona Press. 290–307.

Tyran, J. R. and Feld, L. P. 2001. Why People obey the Law: Experimental Evidence from the Provision of Public Goods. Working Paper, Department of Economics, University of St.Gallen.

van Soest, D. P. and Vyrastekova, J. 2004. Economic Ties and Social Dilemmas. CentER Discussion Paper 2004–55, Tilburg University, Tilburg.

van Soest, D. P. and Vyrastekova, J. 2005. *On the Economics of Law Enforcement: Higher Fines, or Higher Probabilities of Conviction?* Tilburg: Tilburg University, mimeo.

Vyrastekova, J. and van Soest, D. P. 2003. Centralized common pool management and local community participation. *Land Economics*. **79** (4). 500–514.

Vyrastekova, J. and van Soest, D. P. 2007. On the (in)effectiveness of rewards in sustaining cooperation. *Experimental Economics*, forthcoming.

Wilson, R. and Sell, J. 1997. Liar, liar . . .: reputation and cheap talk in repeated settings. *Journal of Conflict Resolution*. **41**. 695–717.

Wit, J. 1999. Social learning in a common interest voting game. *Games and Economic Behaviour*. **26** (1). 131–156.

# 18 Conserving species in a working landscape: land use with biological and economic objectives

*Steve Polasky, Erik Nelson, Eric Lonsdorf, Paul Fackler and Anthony Starfield*

## 1 Introduction

Loss of habitat is perhaps the single largest factor causing the decline of biodiversity (e.g. Wilson 1988; Wilcove *et al.* 2000). The widespread conversion of natural habitat to human-dominated land uses has left smaller and more isolated islands of natural habitat in a growing sea of agriculture, pasture, managed forests and urbanised areas. About half of the earth's useable land is devoted to pastoral or intensive agriculture (Tilman *et al.* 2001). Other lands are managed forests or are developed for housing or industrial use. In response, conservation biologists have called for the establishment of a system of formal protected areas to preserve key remnants of remaining natural habitat.

While formal protected areas play a vital role, many conservation biologists and ecologists recognise the need for conservation beyond the boundaries of protected areas (e.g. Franklin 1993; Hansen *et al.* 1993; Miller 1996; Reid 1996; Wear *et al.* 1996; Chapin III *et al.* 1998; Daily *et al.* 2001; Rosenzweig 2003). Nearly 90 per cent of land across the globe lies outside formal protected areas (IUCN categories I–VI, see WRI 2003), and protected status may arise on lands for reasons other than biodiversity conservation, such as aesthetics or low economic values (Pressey 1994; UNDP *et al.* 2000; Scott *et al.* 2001). For these reasons, the consequences of land use and land management decisions in working landscapes outside protected areas are vital. As Miller (1996, p. 425) stated: '. . . biodiversity will be retained to the extent that whole regions are managed cooperatively among protected areas, farmers, foresters and other neighboring land users.'

Acknowledgements: The authors would like to acknowledge Blair Csuti, Jimmy Kagan, Mark Lindberg, Claire Montgomery, Xuejin Ruan, Nathan Schumaker and Denis White for help in GIS and assembling biological and economic data for the Willamette Basin.

While some land uses are clearly incompatible with some conservation goals, many elements of biodiversity can tolerate at least some level of human disturbance and alteration of the landscape (e.g. Redford and Richter 1999, Currie 2003). A key question for conservation is whether the entire landscape, including both protected areas and areas devoted to economic uses outside protected areas, provides a sufficient likelihood that elements of biodiversity will persist on the landscape. The flip side of this question is whether conservation plans that provide a sufficient likelihood of biodiversity persistence will be acceptable to landowners and other decision-makers. Conservation planning is never done in a vacuum isolated from economic and political factors. Conservation plans that prove costly to the bottom-line of landowners or other decision-makers (at least in the short-term) will engender more political opposition and are less likely to be implemented.

In this paper we develop a spatially explicit model for analysing the consequences of alternative land use patterns on the persistence of a suite of species and market-oriented economic returns. The biological model uses habitat preferences, habitat area requirements and dispersal ability for each species to predict the probability of species persistence given a land use pattern. The economic model uses characteristics of the land unit and location to predict the value of commodity production given a land use pattern. We use the combined biological and economic model to search for efficient land use patterns in which the conservation outcome cannot be improved without lowering the value of commodity production.

We illustrate our methods with an example that includes three alternative land uses, managed forestry, agriculture and biological reserve (protected area), and a set of 97 terrestrial vertebrates on a modelled landscape whose physical, biological and economic characteristics are based on conditions found in the Willamette Basin in Oregon. Prior work evaluating both species persistence and economic returns focuses on a single or small set of species and a single economic activity such as forestry (Montgomery *et al.* 1994; Haight 1995; Hof and Bevers 1998; Marshall *et al.* 2000; Calkin *et al.* 2002; Moilanen and Cabeza 2002; Nalle *et al.* 2004). Using this example, we find land use patterns that achieve a large fraction of potentially achievable species conservation with little reduction in the value of commodity production.

We contrast our approach with a more traditional analysis of reserve site selection in which a set of reserves is chosen to represent a target set of species in as few sites as possible (e.g. Margules *et al.* 1988; Saetserdal *et al.* 1993), or to represent as many species from the target set as possible given a constraint on the number of sites selected or the conservation budget (e.g. Church *et al.* 1996; Faith and Walker 1996; Ando *et al.* 1998). Reserve site selection implicitly assumes that only reserve sites contribute to conservation objectives (and only non-reserve sites contribute to economic objectives). In our model, land in managed forestry and agriculture provides some habitat value while also generating valuable commodities.

## 2 The biological and economic models

The land use pattern is input for both the biological and economic models. The land use pattern is determined by land use decisions made on each land parcel in the study area and the characteristics of those parcels. In general, parcels may be defined as irregularly shaped polygons determined by land ownership, land cover or other criteria, or as cells in a grid. In the illustrative example developed in section 4, land parcels are 400-hectare squares. The land use pattern and characteristics of land parcels determine habitat patterns that are used by the biological model to determine species persistence probabilities in the study area. The land use pattern and characteristics of land parcels are used in the economic model to determine economic returns in the study area.

### *2.1 The biological model*

The biological model predicts the probability of persistence for a large suite of species given a land use pattern. Each species' appraisal of a land use pattern depends on three species-specific traits: the amount of land area required for a breeding pair, compatibility with habitat in the land use pattern and the species' ability to disperse between suitable patches of habitat.

We begin by calculating a suitability score for each land parcel $j$ for each species $s$. The suitability score, $Z_{sj}$, defines the number of breeding pairs of species $s$ that the land parcel could support given its land use, $X_j$:

$$Z_{sj} = \frac{A_j C_{sj}(X_j)}{AR_s} \tag{1}$$

where $A_j$ is the area of parcel $j$, $C_{sj}(X_j)$, is the habitat compatibility score of parcel $j$ for species $s$, and $AR_s$ is the amount of area needed by a breeding pair of species $s$. The habitat compatibility score, $C_{sj}(X_j)$, ranges from 0 to 1, where 0 represents unsuitable habitat and 1 represents prime habitat. This score scales actual area of parcel $j$ to 'effective parcel area'. The habitat compatibility score for a parcel depends upon its land use, $X_j$; the score may be quite different for different land uses (e.g. agricultural use versus natural habitat). Dividing by $AR_s$ yields the number of breeding pairs of species $s$ that can use the effective area of parcel $j$.

For a given land use pattern, we aggregate adjoining land parcels – parcels that share a common border rather than just touching at a point – that contain suitable habitat for a species into habitat patches for that species. We define 'suitable habitat' for species $s$ as those parcels that have a habitat compatibility score above a threshold value: $C_{sj}(X_j) \geq \underline{C}_s$, for $0 < \underline{C}_s \leq 1$. Because $C_{sj}(X_j)$ values differ by species, each species potentially has a uniquely defined set of habitat patches. The suitability score for habitat patch $n_s$ for species $s$, is defined as the sum of the parcel suitability scores for all adjacent parcels that constitute the habitat patch:

$$Z_{sn_s} = \sum_{j \in n_s} Z_{sj}. \tag{2}$$

We use the habitat patch suitability scores and the location of habitat patches for each species to determine the landscape suitability score for that species. There are several steps in determining the landscape suitability score. We first calculate a range of possible landscape suitability scores assuming unlimited dispersal among patches (habitat as one single patch) and then assuming no dispersal among patches (complete isolation of all habitat patches). The landscape suitability score with no dispersal limitations for species $s$ is defined as the sum of all of the habitat suitability scores for species $s$:

$$Lmax_s = \sum_{n_s=1}^{N_s} Z_{sn_s} \tag{3}$$

where $N_s$ represents the total number of suitable habitat patches for species $s$. The landscape suitability score for species $s$ where habitat patches are completely isolated and only contribute to the landscape score if they exceed some minimum threshold is defined as:

$$Lmin_s = \sum_{n_s=1}^{N_s} Z_{sn_s}, \text{ for the set of parcels where } Z_{sn_s} \geq \gamma_s \tag{4}$$

where $\gamma_s$ represents the minimum number of breeding pairs for species $s$ that a patch must support on its own before the habitat patch contributes to the landscape score. For high values of $\gamma_s$ the value of $Lmin_s$ can be 0. On the other hand, as $\gamma_s$ approaches zero, $Lmin_s$ approaches $Lmax_s$. In the latter case, the landscape suitability score depends only on the total amount of effective habitat and not its spatial pattern (for an example of species persistence analysis where habitat pattern does not matter for a large group of species, see Schumaker *et al.* 2004).

Whether the landscape suitability score for species $s$, $LS_s$, is closer to $Lmin_s$ or $Lmax_s$ depends on the connectivity of habitat patches and the species' dispersal ability. The connectivity score for each suitable habitat patch is defined as:

$$P_{sn_s} = \sum_{m_s=1}^{N_s} e^{-\alpha_s d_{m_s n_s}} Z_{sm_s} \tag{5}$$

where $d_{m_s n_s}$ is the distance between suitable habitat patch $m_s$ and suitable habitat patch $n_s$, and $\alpha_s > 0$ represents the reciprocal of the mean dispersal ability of species $s$. The patch connectivity score is dependent on the patch's own habitat suitability score and the habitat suitability scores for all other suitable habitat patches to which species $s$ can disperse, weighted by the distance between the patches and the species dispersal ability. The effect of distance is represented by a negative exponential distribution (Vos *et al.* 2001). Other factors besides distance may influence dispersal ability (King and With 2002; Gardner and Gustafson 2004) but are not considered here.

We then aggregate the habitat patch connectivity scores to compute a landscape connectivity score. In a completely connected landscape, all habitat patch connectivity scores for species $s$ would equal $Lmax_s$, and the aggregate patch score summing over all suitable habitat patches would be $N_s Lmax_s$. On the other hand, if all suitable habitat patches are completely isolated with no contribution from any other patch, habitat patch connectivity for patch $n_s$ would be $Z_{sn_s}$ and the aggregate patch score summing over all suitable habitat patches would be $Lmax_s$. We define the landscape connectivity score for species $s$, $LC_s$, as the observed score relative to the possible minimum and maximum values, scaled so that its value can range between 0 and 1:

$$LC_s = \frac{\sum_{N_s}^{n_s=1} P_{sn_s} - \mathrm{Lmax_s}}{(N_s - 1)\mathrm{Lmax_s}}. \tag{6}$$

The landscape connectivity score for species $s$, $LC_s$, is near zero when species $s$ has low dispersal ability in an extremely fragmented landscape. The landscape connectivity score equals one for a completely connected landscape.

We use the landscape connectivity score along with $Lmax_s$ and $Lmin_s$ to determine the overall landscape suitability score, $LS_s$, for species $s$:

$$LS_s = (1 - LC_s)Lmin_s + LC_s\, Lmax_s. \tag{7}$$

For an unconnected landscape ($LC_s = 0$), the landscape suitability score for species $s$ is $Lmin_s$. For a completely connected landscape ($LC_s = 1$), the landscape suitability score for species $s$ is $Lmax_s$.

The landscape suitability score for each species, $LS_s$, is a measure of the expected number of breeding pairs the landscape will support. To determine the expected biodiversity score for the landscape we convert $LS_s$ into a probability that the species will persist on this landscape, $LP_s$, using a saturating function:

$$LP_s = \frac{LS_s^g}{\left(LS_s^g + k^g\right)} \tag{8}$$

where $k$ is the half-saturating constant (the landscape score yielding a persistence probability of 0.5), and $g$ is a constant that determines the shape of the saturating function. Increasing $g$ leads to a more step-like function or threshold value for a viable population size.

The expected number of species that persist on the landscape, i.e. the landscape biological score, $LB$, is the sum of species probability scores over all the species:

$$LB = \sum_{s=1}^{S} LP_s. \tag{9}$$

### 2.2 *The economic model*

The economic model is used to predict the present value of commodity production for a given land use pattern. We first determine the present value of commodity production for an individual parcel based on the land use and characteristics of the parcel. We then sum these values across all parcels to generate the economic score for the landscape.

We note at the outset that we focus on the value of commodity production. In principle, the economic model should include the value of all goods and services generated by the land use pattern, including 'ecosystem services', the majority of which are not bought or sold in markets (e.g. Daily 1997; Daily *et al.* 2000). At least in theory, the general approach of the economic model discussed below can include ecosystem goods and services. We do not do so here because of the difficulty, at present, of generating reliable estimates of ecosystem service value. Our analysis, then, illustrates the degree to which there are tradeoffs between the value of commodity production and species conservation, rather than attempting to illustrate a complete set of tradeoffs among all potentially valuable goods and services generated by a landscape.

Production of commodities on a parcel is determined by the characteristics of the parcel, such as soil type and topography, and its land use. Let $y_{jc}(X_j)$ represent the annual production of commodity $c$ on parcel $j$ given land use $X_j$, $p_c$ is the market price of commodity $c$, and $Cost_{jc}(X_j)$ is the annual production costs of producing commodity $c$ on parcel $j$ associated with land use $X_j$. The present value of commodity production on parcel $j$ is:

$$V_j(X_j) = \sum_{t=0}^{\infty} \left[ \sum_{c} (p_c y_{jc}(X_j)) - Cost_{jc}(X_j) \right] \delta^t \tag{10}$$

where $\delta$ is the annual discount factor ($0 < \delta < 1$).

A parcel whose land use is a biological reserve does not produce a marketed commodity and thus is given an economic score of zero. Such parcels in fact generate valuable ecosystem services (apart from species conservation, which is captured in the biological model). Reserves may also have associated management costs. For both these reasons, the economic return to a biological reserve properly calculated is not zero. In principle it is easy to incorporate an economic score different from zero for a biological reserve; however, accurately estimating the score is difficult in practice.

The total landscape economic score, *LE*, sums the present value of commodity production of each parcel given its land use:

$$LE = \sum_{j} V_j(X_j). \tag{11}$$

In an important respect, the economic model is simpler than the biological model. The value of commodity production on a parcel is solely a function of the parcel's characteristics; nearby or adjoining parcels do not influence the economic score for a parcel. Two conditions must be true for this assumption to hold. First, prices must not be significantly influenced by local supply (in other words, local production is sold into a national or global market for which it makes up a small fraction of the total supply). Second, there must not be any 'externalities' from adjacent land uses. Examples of positive externalities include a premium for housing values for adjacency to biological reserves or open space (e.g. Tyrvainen and Miettinen 2000; Irwin 2002, Thorsnes 2002; Volser *et al.* 2003) and the effect of pollinators on crop yields (e.g. Nabhan and Buchmann 1997; Allen-Wardell *et al.* 1998). Examples of negative externalities include pollution runoff from a parcel that lowers productivity of downstream parcels, and noise or odor from nearby industrial or agricultural operations.

## 3 Optimisation problem and heuristic solution methods

We combine the biological and economic models with optimisation methods to find efficient land use patterns for which it is not possible to increase the landscape biological (*LB*) score without decreasing the landscape economic (*LE*) score, and vice versa. In general, there will be many efficient land use patterns. Finding the complete set of efficient land use patterns traces out an efficiency frontier that illustrates what is feasible and the tradeoffs between increasing biological returns and economic returns.

The combined biological and economic optimization problem can be written quite simply as follows:

$$\begin{aligned} &Max\ LB \\ &s.t.\ LE \geq \bar{L} \end{aligned} \tag{12}$$

where the maximisation is taken over the choice of land use in each parcel (i.e. the maximisation is taken over a land use pattern). In other words, the problem is to find a land use pattern with the highest possible biological score that guarantees an economic return at least as large as $\bar{L}$. By varying the required economic threshold, $\bar{L}$, a whole family of solutions can be found that trace out the efficiency frontier. The frontier can also be found by maximising the *LE* score subject to a constraint that the *LB* score meet a certain threshold.

This formulation of the problem is deceptively simple. Because the optimization problem is an integer programme involving a potentially large number of parcels each with several potential land uses, and because the biological model involves non-linear spatial considerations, finding an optimal solution to this problem can be exceedingly difficult. There are a number of heuristic algorithms that can be used to find good, though not necessarily optimal, solutions. We use six algorithms and then combine the best solutions from these algorithms to trace out the efficiency frontier. The six heuristic algorithms are summarized in Table A of the appendix in *Ecological Archives*. Each heuristic either starts at the land use pattern with the maximum value of commodity production or the land use pattern with zero commodity production (all biological reserves). Each heuristic then sequentially makes a change in land use on one parcel per step where each step maximises the increase in the biological or economic score (or minimises its loss), or maximises the ratio in the gain in one score relative to the loss in the other score. Generally, heuristics that jointly consider both biological and economic scores by looking at the ratio do best, though not always. From these six heuristic solutions, we take all solutions that are not dominated, i.e. for which there is no other solution from any of the six heuristics that yields i) a higher economic and a higher or equal biological score, or ii) a higher biological score and a higher or equal economic score. While this set of solutions is probably a good approximation of the efficiency frontier, it is not guaranteed to be identical to the true efficiency frontier because the heuristic solutions evaluate only one change at a time rather than doing a global search over all possible land use changes.

## 4 Data and methods used in the illustrative example based on the Willamette Basin

To illustrate our approach we applied our model to a simple landscape composed of 196 400-hectare square parcels arranged in a $14 \times 14$ grid with parameter values and spatial patterns similar to those found in the Willamette Basin in Oregon. This illustrative example includes 97 terrestrial vertebrate species, three land uses (managed forestry, agriculture and biological reserve) and six habitat categories (managed forestry, agriculture, shrub, hardwood, conifer and prairie/meadow). Land use uniquely determines the habitat category except for biological reserve land where habitat category is determined by the parcel's pre-settlement vegetation type.

### *4.1 Biological model*

The 97 terrestrial vertebrate species in our study are species that live in the Willamette River Basin in Oregon and do not depend on aquatic habitat (Adamus *et al.* 2000; Schumaker *et al.* 2004). Habitat compatibility scores for each of the 97 species for each of the six habitat categories (managed forestry, agriculture, shrub, hardwood, conifer and prairie/meadow) are based on Adamus *et al.* (2000). Habitat compatibility scores can take on values of 0 (unsuitable habitat), 0.5 (marginally suitable habitat) or 1.0 (prime habitat). We assume that both marginal and prime habitat count for purposes of assembling habitat patches ($\underline{C}_s = 0.5$). Parcels containing marginal or prime habitat that share a common side are combined into a habitat patch (but diagonal connections are not considered). Table B in the appendix in *Ecological Archives* contains a complete list of the 97 species and their habitat compatibility scores for each of the habitat types.

There is little systematic published information on which to base values for habitat area requirement and dispersal ability for most of the 97 species used in our model, though Brown (1985), Baguette *et al.* (2003), Joly *et al.* (2003) and Lichtenstein and Montgomery (2003) contain some useful information. Habitat area requirements and dispersal ability values are based primarily on the following assumptions: i) area requirements scale to the size of the animal (larger animals require more habitat), ii) larger animals disperse further than smaller animals (Bowman *et al.* 2002), iii) birds disperse further than mammals, and iv) mammals disperse further than amphibians/reptiles. Habitat area requirements and dispersal ability values are listed in Table B in the appendix in *Ecological Archives*. The relatively small area of our 14×14 landscape limits distances between habitat patches, which may make our results somewhat insensitive to species' ability to disperse. Dispersal ability may be important on a large fragmented landscape where distances between patches are great.

The default values of the half saturation constant ($k$) and shape coefficient ($g$) in equation (8) were chosen to create sufficiently large differences among species' evaluations of land use patterns. For small $k$, complete species persistence on the landscape results for most land use patterns. For large $k$, species cannot persist on the landscape for any land use pattern. However, because $k$ and $g$ are global rather than species-specific parameters they do not affect the relative ranking of species persistence scores across species.

On the 14×14 landscape the distance between parcels $i$ and $j$, $d_{ij}$, is given by

$$d_{ij} = \max\{\lambda_{ij} - 2000, 0\} \tag{13}$$

where

$$\lambda_{ij} = (|xc_i - xc_j|) + (|yc_i - yc_j|) \tag{14}$$

and $xc_i$ and $yc_i$ refer to the $x$-coordinate and $y$-coordinate, respectively, of a parcel $i$'s centroid (the 14×14 landscape grid is measured in metres). Finally, distance between patch $m_s$ and $n_s$, $d_{m_s n_s}$, is equal to the shortest distance between a parcel that is a member of patch $n_s$ and a parcel that is a member of patch $m_s$.

## 4.2 Economic model

Both managed forestry and agricultural land uses produce marketed commodities for which the model estimates a present value of returns. The present value of a parcel whose land use is managed forestry ($X_j = x_f$) depends on the productivity of the parcel for growing timber ($y_{jf}$, ), the price of timber ($p_f$) and the costs of harvesting timber ($Cost_{jf}$). Timber yield, measured in terms of board feet per hectare, depends upon the age of the timber stand when harvested and the parcel's forestry site index (King 1966, Curtis *et al.* 1981 and Curtis 1992). We assumed a 45-year rotation age Douglas fir forest (with commercial thinning at age 35), which is typical of commercial timber operations in the Willamette Basin. Douglas fir site index information, which is based on soil, climate conditions and other physical conditions, comes from the US Department of Agriculture (USDA-NRCS 2001a, USDA-NRCS 2001b, USDA-NRCS 2003). Timber yield is multiplied by timber price per board foot (Lichtenstein and Montgomery 2003, personal communication with Claire Montgomery) to determine timber revenue per hectare. Timber production costs equal the sum of logging and hauling costs per board foot plus an area maintenance cost. Logging costs per board foot are a function of a parcel's average slope and tree size (Fight *et al.* 1984; PNW-ERC 1999b). Hauling costs per board foot are a function of a parcel's average slope and distance to the nearest processing mill (Latta and Montgomery 2004; personal communication with Claire Montgomery). Per unit area maintenance costs of forestry production are constant across parcels (Lichtenstein and Montgomery 2003, personal communication with Claire Montgomery). We assume even-aged forestry management with 45-year rotations such that 1/45th of the parcel is harvested (and thinned) each year. Given these assumptions, the present value of economic return from a parcel whose land use is managed forestry is

$$V_j(x_f) = \sum_{t=0}^{\infty} \frac{A_j(p_f y_{jf} - Cost_{jf})\delta^t}{45}. \quad (15)$$

The present value of a parcel whose land use is agriculture ($X_j = x_a$) depends upon the parcel's crop-growing productivity ($y_{ja}$), the price of agricultural produce ($p_a$) and production costs ($Cost_{ja}$). We modelled an agricultural operation with a typical mix of crops grown in the Willamette Basin. Agricultural crop yield per hectare depends upon the parcel's soil class and whether the parcel is irrigated (PNW-ERC 1999a, OWRD 2001, USDA-NRCS 2001a, USDA-NRCS 2001b, USDA-NRCS 2003). The yield is multiplied by the market price for the agricultural produce, $p_a$, (OSUES 2002) to generate estimated revenue per hectare. Cost information ($Cost_{ja}$) comes from Oregon State University's Extension Service (OSUES 2003). Assuming that agricultural activity occurs every year, the present value of economic return of a parcel whose land use is agriculture is:

$$V_j(x_a) = \sum_{t=0}^{\infty} A_j(p_a y_{ja} - Cost_a)\delta^t. \quad (16)$$

Because a parcel in a biological reserve ($X_j = x_b$) does not produce a marketed commodity, the present value of commodity returns is zero: $V_j(x_b) = 0$.

*4.3 The landscape*

Each of the 196 parcels on the 14×14 landscape was assigned a pre-settlement vegetation cover type (shrub, hardwood, conifer or prairie/meadow) and an economic return for managed forestry and agriculture. To generate reasonable values and spatial patterns for the 14×14 landscape, we partitioned a map of the Willamette Basin (ONHP 2000) into a parcel map based on land cover (circa 1990) and a constraint that no parcel be larger than 750 hectares. This parcel map was overlaid with maps of pre-settlement vegetation cover in the Willamette Basin as described by surveyors for the General Land Office between 1851 and 1909 (PNW-ERC 1999c), soil class index, Douglas fir site index and point-of-use irrigation permits. Of the 10,372 parcels on the partitioned Willamette Basin map, 6,197 had a complete set of data.

Using the subset of 6,197 parcels with complete data, we created a probability distribution for a parcel's pre-settlement vegetation type as a function of its neighbours' pre-settlement vegetation types. We assumed that a parcel's pre-settlement vegetation (shrub, hardwood, conifer or prairie/meadow) indicates the vegetation coverage that would emerge if the parcel were a biological reserve. A pre-settlement vegetation pattern that mimics the Willamette Basin's pre-settlement vegetation pattern was generated for the 14×14 landscape using a random-number generator and the spatially explicit pre-settlement vegetation probability distribution noted above.

We used more complicated techniques to generate present values for managed forestry and agriculture on the 14×14 landscape. A forestry value for each of the 6,197 parcels was found by using equation (15), the data sources noted in section 4b and Basin parcel data. We used a spatial autoregressive (SAR) model (LeSage 1999) to explain a Basin parcel's present value in managed forestry as a function of its pre-settlement vegetation coverage and its adjacent neighbours' present value in forestry. The managed forestry present value for each parcel on the 14×14 landscape was generated using a random number generator, the estimated SAR model coefficients and the 14×14 landscape's already established pre-settlement vegetation pattern.

An agriculture value for each of the 6,197 parcels was found by using equation (16), the data sources noted in section 4b and Basin parcel data. We used the SAR model to estimate a Basin parcel's agricultural present value as a function of its managed forestry present value, pre-settlement vegetation coverage, irrigation capability and its adjacent neighbours' agricultural present value. The agricultural present value for each parcel on the 14×14 landscape was generated using a random number generator, the estimated SAR model coefficients for agriculture, modelled irrigation capability and the 14×14 landscape's already established managed forestry present value and pre-settlement vegetation pattern. Irrigation capability was placed in a subset of the 14×14 landscape's parcels such that the proportion of the modelled landscape's parcels with irrigation capability approximated the proportion of parcels in the Willamette Basin with irrigation capability.

### 4.4 *Simulation experiments*

We performed a number of simulations using the biological and economic model applied to the 14×14 landscape. We began with a 'base case' that takes a default landscape (generated as described in section 4c) and default parameter values for the biological model (as listed in section 4a and Table B in the appendix in *Ecological Archives*) and the economic model (as listed in Table C in the appendix in *Ecological Archives*). In the base case, the mean value of managed forestry is \$9,720 $ha^{-1}$ (s.d. \$1,195) and \$9,250 $ha^{-1}$ (s.d. \$9,271) in year 2000 dollars. Agricultural values vary widely with soil quality and irrigation status. The base case land use pattern that generates the highest value for commodity production is shown in Figure 18.1.[1]

We compare the results obtained in the base case with a traditional reserve site selection model, which considers only the contribution of biological reserve parcels to the biological objective. To do this, we set all species' habitat compatibility scores for managed forestry and agriculture lands equal to zero. We consider two variants of the reserve site selection model, one with dispersal and one without dispersal. In the variation that drops the dispersal ability parameter from the biology model, only reserve parcels that are contiguous contribute biological value to each other.

We also conducted a set of sensitivity analyses by changing the base case assumptions one at a time to see how such changes affect the efficiency frontier. In the first set of sensitivity analyses we generated four alternative landscapes using the same methodology used to generate the base case landscape. We also conducted sensitivity analyses by varying default parameters in the biological and economic models. We varied assumptions about: (1) the minimum amount of area needed for a breeding pair, (2) the half-saturating constant $k$ (from equation 8), (3) the power constant $g$ (from equation 8), and (4) changing the number of breeding pairs that a habitat patch must support on its own before the patch

[1] All figures appear in the appendix.

contributes to the landscape score. We also analysed the model with different net present values of economic returns in managed forestry and agriculture.

## 5 Results

### *5.1 Base case*

Using the heuristic algorithms described in section 3 to solve the optimization problem in equation (12) with the base case for our 14×14 landscape, we find the estimated efficiency frontier (Figure 18.2). The most striking feature of the efficiency frontier is its L-shape, demonstrating the existence of land use patterns that generate high scores for both biological conservation and commodity value. Relatively minor modifications of the land use pattern that maximised the *LE* score produced a 14.6 per cent increase in the *LB* score (from 74.5 to 85.3) with only a 7.1 per cent decline in the *LE* score (from $1,046 million to $972 million). Furthermore, the resulting *LB* score from this modification was near the maximum possible score, 88.5 (because the 14 × 14 landscape is relatively small, not all 97 species could persist on the landscape irrespective of the land use pattern). Further modifications of the land use pattern to increase the *LB* score from 85.3 to the maximum score of 88.5 produced dramatic declines in the *LE* score ($972 million to $270 million). Note that the *LB* score is reasonably high even when the landscape is managed to maximise economic gain. Interestingly, the maximum *LB* score did not occur when all parcels were put in biological reserves.

In moving along the efficiency frontier from lower right to upper left, land use patterns shift from maximising commodity value to maximising species persistence (Figures 18.3A to 18.3E). The parcels most likely to be converted to biological reserves initially (Figures 18.3A to 18.3C) are managed forestry parcels with a pre-settlement vegetation coverage type of shrub and hardwood. Very few agricultural parcels are converted until movement is far along the efficiency frontier (Figures 18.3D and 18.3E). Parcels with prairie/meadow pre-settlement vegetation coverage are never put into biological reserves at any of the five points along the efficiency frontier. As shown in Figures 18.3A through 18.3E, land uses tend to clump together to form larger blocks of like habitat. A measure of biological reserve 'connectivity', defined as the number of perimeter segments that form the conservation reserve network divided by the number of parcels in conservation, for various land use patterns on the base case efficiency frontier is given in Table 18.A1 in the appendix to this chapter. Smaller measures indicate more highly connected or clumped biological reserves.

### 5.2 *Comparison with reserve site selection*

The efficiency frontiers for a traditional reserve site selection model both with and without dispersal lie well within the base case efficiency frontier (Figure 18.4). For example, at an *LB* score of 77, the *LE* score in the base case ($1,044 million) is far higher than in reserve site selection with dispersal ($670 million) or without dispersal ($519 million). When all land is put into managed forestry or agriculture and none into biological reserves, the *LB* scores for the reserve site selection scenarios are zero; therefore, the efficiency frontiers extend to the horizontal axis. Because managed lands contribute nothing to the biological score under the reserve site selection scenarios, increasing the amount of land in biological reserves generally increases (can never decrease) the *LB* scores. Therefore, the reserve site selection efficiency frontiers extend to the vertical axis as well. The efficiency frontiers are more rounded (less L-shaped) under the reserve site selection scenarios than under the base case. In other words, there is more apparent tradeoff between biological and economic objectives under the reserve site selection approach.

There are major differences in land use patterns generated by reserve site selection and the base case. To illustrate the differences, consider the efficient land use pattern for each scenario at an *LB* score of 77 (Figures 18.5A–18.5C). Under the base case, 77 species can be sustained mainly through having large blocks of managed forest. There is only one parcel put into biological reserve. In the reserve site selection model with dispersal, 95 parcels need to be put into biological reserves to sustain 77 species, while 101 parcels are needed in the reserve site selection case without dispersal.

### 5.3 *Sensitivity analyses*

Comparing the efficiency frontiers for alternative landscapes shows that the efficiency frontier is largely unchanged in terms of shape or location (Figure 18.6). The only noticeable difference among landscapes comes when more randomly drawn high-value managed forestry and agricultural lands are included, which shifts the efficiency frontier a bit to the right. These results indicate that it was not the particular random draw of landscape that generates our results.

Making changes in default biological or economic parameters shifts the efficiency frontier but does not change its basic shape (Figures 18.7A–18.7C). Favourable changes in biological parameters (decreasing the minimum area required for a breeding pair, reducing the half-saturation coefficient, increasing the power coefficient in the saturation function or lowering the threshold value on the number of breeding pairs that a habitat patch must support on its own before the patch contributes to the landscape score) shift the efficiency frontier upwards. Unfavourable changes in biological parameters shift the efficiency frontier downwards. Favourable changes in economic parameters (increasing the value of agricultural or timber production) shift the efficiency frontier to the right. In general, changes in biological or economic parameters shift the efficiency frontier but do not change its L-shape. (Other changes in default assumptions, including the size and shapes of the parcels, did not change the efficiency frontier's L-shape).

## 6 Discussion

Rather than facing a stark tradeoff between conserving biodiversity and production of high-valued commodities, we find that a large fraction of conservation objectives can be achieved at little cost to the economic bottom line with thoughtful land use planning. In our example landscape, based on conditions in the Willamette Basin, there is a land use pattern that simultaneously generates 96 per cent of the maximum landscape biological score (85.3 out of 88.5) and 93 per cent of the maximum landscape economic score ($972 million out of $1,046 million). Many species are able to persist in a landscape largely devoted to economic use because they view managed forests or agricultural land as suitable habitat. The number of species that persist on the landscape can be increased by adjusting the spatial pattern of economic activity to create large blocks of forest (or agriculture), often at little economic cost. Further increases in species persistence can be achieved at relatively low cost by strategically placing biological reserves in areas with natural habitats required by some species but low economic value.

More evidence of the limited tradeoffs between conservation and economic returns in the $14 \times 14$ landscape example is shown by the fact that the efficiency frontier under the base case does not extend to either the horizontal or vertical axis (Figure 18.2). The efficiency frontier starts above the horizontal axis because many species are able to persist even when the landscape is managed to maximise economic gain because they view managed forest and/or agricultural land as suitable habitat and there are relatively large contiguous blocks of both. The efficiency frontier does not extend to the vertical axis because the maximum landscape biological score does not occur where all parcels are reserves. This initially counter-intuitive outcome occurs because a few of the 97 species depend solely on managed forestry lands or agricultural lands for habitat. A landscape comprised entirely of conservation with no economic activity, while clearly best for some species, is not the best land use pattern for maximising the sum total of persistence for all species. Other studies have found that biodiversity can be higher in slightly disturbed areas than in natural undisturbed areas (e.g. Yazvenko and Rapport 1996; Johnson *et al.* 1998).

Even the limited tradeoffs between conservation and economic objectives shown in the example may be something of an overstatement. In this example, we did not consider the economic value of ecosystem services, such as the provision of clean water, nutrient filtration, climate regulation and ecotourism (Daily 1997). Including the value of ecosystem services in economic returns would tend to increase the value of conserving land in biological reserves relative to other land uses, thereby reducing apparent tradeoffs between conservation objectives and economic returns.

Even so, there remains at least some degree of conflict between conservation and the value of commodity production. In the 14×14 landscape example, obtaining the final 4 per cent of the conservation objective (moving the landscape biological score from 85.3 to 88.5) requires a drop in commodity value of over 70 per cent ($972 million to $270 million). Similarly, pushing for maximum economic returns (moving from $972 million to $1,046 million) generates significant biological losses (from 85.3 to 74.5). Other studies have found a similar pattern in tradeoffs between conservation and economic objectives, namely that many conservation objectives can be achieved at very low cost but that full protection is often very expensive (Montgomery *et al.* 1994, Ando *et al.* 1998, Montgomery *et al.* 1999, Polasky *et al.* 2001). In general, difficult tradeoffs occur when the habitat of a species with limited range overlaps and is inconsistent with economically valuable land uses. Still, even in this case costs can be reduced by careful consideration of the type and location of economic activities that can coexist with survival of the species.

The degree of conflict between conservation and economic returns appears much greater using the reserve site selection approach than using our joint biological and economic modelling approach. Assuming no biological value in lands used for economic purposes and no economic value in lands used for biological purposes makes some degree of conflict inevitable between economic and biological objectives. A major theme of this paper is to incorporate the biological value of lands outside formal protected areas. Further work to incorporate the economic value of ecosystem services generated by formal protected areas is also needed.

The reserve site selection approach has also been criticised on the grounds that it targets current representation of species in a reserve network rather than the long-term persistence of those species (e.g. Cowling *et al.* 1999, Williams and Araujo 2000, Calkin *et al.* 2002, Moilanen and Cabeza 2002, Cabeza and Moilanen 2003). Modelling persistence requires incorporating spatial population modelling into conservation planning (Cabeza and Moilanen 2003). In this paper, we model species persistence as a function of the landscape's capacity to support species, which depends upon the extent and spatial pattern of habitat, the area requirements of the species and its dispersal ability. Hansen *et al.* (1993) and Schumaker *et al.* (2004) represent similar efforts to model persistence of a large number of species on a landscape.

The biological model developed in this paper is relatively simple so that it could be applied to a large set of species. Additional features could be added to increase biological realism of the model and to make the predictions richer and more robust. In our model, dispersal is a function of distance between habitat patches. Prior work emphasises the importance of distance between habitat patches but other factors such as the availability of prey, predation risk and dispersal barriers (e.g. highways) are also important factors in explaining dispersal (e.g. Hanski and Ovaskainen 2000, Goodwin and Fahrig 2002, Baguette *et al.* 2003, Gardner and Gustafson 2004). Inclusion of the probability of patch colonization is another promising direction for model improvement (e.g. Gustafson and Gardner 1996). We included only a subset of terrestrial vertebrates, in particular, terrestrial vertebrates not dependent on aquatic habitats. The model used in the example also does not consider different habitat needs for breeding and feeding or edge effects. We also ignored boundary effects from the landscape outside the study area. Incorporating different breeding and feeding needs, edge effects or boundary effects could be included with relatively minor extensions to the existing model. Incorporating interactions among species (e.g. competition, predator-prey interactions) would require more fundamental changes to the model.

We use the expected number of species persisting on the landscape as the biological score for the model. Other metrics could be used instead. For example, instead of giving equal weight to all species, greater weight could be given to endemic or endangered species. One could also base the landscape biological score on phylogenetic diversity, ecological diversity, ecosystem productivity, stability, resilience, ecosystem services or other measures; all one needs is a way of modelling how the desired metric changes under alternative land use patterns.

On the economic side, a broader set of economic activities such as recreation, residential and commercial land use could be included. Modelling recreation and residential development would necessitate incorporating price effects and spatial externalities in which neighbouring land use may affect economic values on a parcel. Additionally, the economic model could be expanded to include positive returns from species persistence (e.g. birdwatching) or negative returns (e.g. crop damage).

The model we developed in sections 2 and 3 is general in the sense that it can be applied to different sets of species, different economic activities and different definitions of land parcels (e.g. polygons or grid cells). How best to define land parcels presents some challenges. Ideally, parcel boundaries would match land use decision-making units (e.g. private property boundaries) and parcels would be relatively homogeneous within their boundaries. In practice, there is no perfect way to define parcels when including species with different range sizes and dispersal ability, different economic activities and land ownership patterns. There are tradeoffs between including increasingly finer-scale resolution and computational limits. However, the choice of scale is not innocuous. In the reserve site selection literature, the size of parcels can influence the choice of which parcels to include in a reserve network (Stoms 1994; Pressey and Logan 1998; Warman *et al.* 2004). In our case, the choice of which parcels to put in which land use can be affected by scale of analysis but the general conclusion about the shape of the tradeoff curve between biological and economic objectives is not dependent on scale.

Another important extension would be to explicitly include dynamics. Changing existing land use patterns entails transition costs that would change the economic returns. Species populations also respond over time to habitat changes. If dispersal is limited, species may have difficulty in colonising new habitat patches that become available. With a dynamic approach, effects of climate change and stochastic events, such as fire, drought or disease outbreaks, could also be considered.

## REFERENCES


Adamus, P. R., Baker, J. P., White, D., Santelmann, M. and Haggerty, P. 2000. Terrestrial Vertebrate Species of the Willamette River Basin: Species-Habitat Relationships Matrix. Internal Report. US Environmental Protection Agency, Corvallis, OR. (This report and its accompanying appendices can be found in the zip file wrb.species.zip on the Pacific Northwest Ecosystem Research Consortium's website, http://www.fsl.orst.edu/pnwerc/wrb/access.html)

Allen-Wardell, G., Bernhardt, P., Bitner, R., Burquez, A., Buchmann, S., Cane, J., Cox, P. A., Dalton, V., Feinsinger, P., Ingram, M., Inouye, D., Jones, C. E., Kennedy, K., Kevan, P., Koopowitz, H., Medellin, R., Medellin-Morales, S., Nabhan, G. P., Pavlik, B., Tepedino, V., Torchio, P. and Walker, S. 1998. The potential consequences of pollinator declines on the conservation of biodiversity and stability of food crop yields. *Conservation Biology*. **12**. 8–17.

Ando, A., Camm, J. D., Polasky, S. and Solow, A. R. 1998. Species distributions, land values and efficient conservation. *Science*. **279**. 2126–2128.

Baguette, M., Mennechez, G., Petit, S. and Schtickzelle, N. 2003. Effect of habitat fragmentation on dispersal in the butterfly. *Proclossiana eunomia*. *Comptes Rendus Biologies*. **326** (1). S200–S209.

Bowman, J., Jaeger, J. A. G. and Fahrig, L. 2002. Dispersal distance of mammals is proportional to home range size. *Ecology*. **83**. 2049–2055.

Brown, E. R. (ed.). 1985. Management of wildlife and fish habitats in forest of Western Oregon and Washington, Part 2. Gen. Tech. Rep.R6-F&WL-192–1985. Portland, OR: US Department of Agriculture, Forest Service, Pacific Northwest Research Station.

Cabeza, M. and Moilanen, A. 2003. Site-selection algorithms and habitat loss. *Conservation Biology*. **17**. 1402–1413.

Calkin, D., Montgomery, C. A., Schumaker, N. H., Polasky, S., Arthur, J. L. and Nalle, D. J. 2002. Developing a production possibility set of wildlife species persistence and timber harvest value using simulated annealing. *Canadian Journal of Forest Research*. **32**. 1329–1342.

Chapin III, F. S., Sala, O. E., Burke, I. C., Grime, J. P., Hooper, D. U., Lauenroth, W. K., Lombard, A., Mooney, H. A., Mosier, A. R., Naeem, S., Pacala, S. W., Roy, J., Steffen, W. L. and Tilman, D. 1998. Ecosystem consequences of changing biodiversity. *BioScience*. **48**. 45–52.

Church, R. L., Stoms, D. M. and Davis, F. W. 1996. Reserve selection as a maximal coverage problem. *Biological Conservation*. **76**. 105–112.

Cowling, R., Pressey, R., Lombard, A., Desmet, P. and Ellis, A. 1999. From representation to persistence: requirements for a sustainable system of conservation areas in the species-rich Mediterranean-climate desert of southern Africa. *Diversity and Distributions*. **5**. 51–71.

Currie, D. J. 2003. Conservation of endangered species and the patterns and propensities of biodiversity. *Comptes Rendus Biologies*. **326** (1). S98–S103.

Curtis, R. O. 1992. A new look at an old question – Douglas fir culmination age. *Western Journal of Applied Forestry*. 7. 97–99.

Curtis, R. O., Clendenen, G. W. and Demars, D. J. 1981. A new stand simulator for coast Douglas-fir: DFSIM user's guide. USDA Forest Service General Technical Report PNW-128. Pacific Northwest Forest and Range Experiment Station, Portland, OR.

Daily, G. C. (ed.). 1997. *Nature's Services: Societal Dependence on Natural Ecosystems*. Washington, DC: Island Press.

Daily, G. C., Soderqvist, T., Aniyar, S., Arrow, K., Dasgupta, P., Ehrlich, P. R., Folke, C., Jansson, A., Jansson, B. O., Kautsky, N., Levin, S., Lubchenco, J., Maler, K. G., Simpson, D., Starrett, D., Tilman, D. and Walker, B. 2000. The value of nature and the nature of value. *Science*. **289**. 395–396.

Daily, G. C., Ehrlich, P. R. and Sanchez-Azofeifa, G. A. 2001. Countryside biogeography: use of human dominated habitats by the avifauna of southern Costa Rica. *Ecological Applications*. **11**. 1–13.

Faith, D. P. and Walker, P. A. 1996. Integrating conservation and development: effective trade-offs between biodiversity and cost in the selection of protected areas. *Biodiversity and Conservation*. **5**. 431–446.

Fight, R. D., LeDoux, C. B. and Ortman, T. L. 1984. Logging costs for management planning for young-growth coast Douglas-fir. USDA Forest Service General Technical Report PNW-176. Pacific Northwest Forest and Range Experiment Station, Portland, OR.

Franklin, J. 1993. Preserving biodiversity: species, ecosystems, or landscapes? *Ecological Applications*. **3**. 202–205.

Gardner, R. H. and Gustafson, E. J. 2004. Simulating dispersal of reintroduced species within heterogeneous landscapes. *Ecological Modelling*. **171**. 339–358.

Goodwin, B. J. and Fahrig, L. 2002. How does landscape structure influence landscape connectivity? *Oikos*. **99**. 552–570.

Gustafson, E. J. and Gardner, R. H. 1996. The effect of landscape heterogeneity on the probability of patch colonization. *Ecology*. 77. 94–102.

Haight, R. G. 1995. Comparing extinction risk and economic cost in wildlife conservation planning. *Ecological Applications*. **5**. 767–775.

Hansen, A. J., Garman, S. L., Marks, B. and Urban, D. L. 1993. An approach for managing vertebrate diversity across multiple-use landscapes. *Ecological Applications*. **3**. 481–496.

Hanski, I. and Ovaskainen, O. 2000. The metapopulation capacity of a fragmented landscape. *Nature*. **404**. 755–758.

Hof, J. and Bevers, M. 1998. *Spatial Optimization for Managed Ecosystems*. New York: Columbia University Press.

Irwin, E. G. 2002. The effects of open space on residential property values. *Land Economics*. **78**. 465–481.

Johnson, G., Myers, W., Patil, G. and Walrath, D. 1998. Multiscale analysis of the spatial distribution of breeding bird species. In P. Bachmann, M. Kohl and R. Paivinen (eds.). *Assessment of Biodiversity for Improved Forest Planning*. Dordrecht: Kluwer Academic Publishers, 135–150.

Joly, P., Morand, C. and Cohas, A. 2003. Habitat fragmentation and amphibian conservation: building a tool for assessing landscape matrix connectivity. *Comptes Rendus Biologies*. **326** (1). S132–S139.

King, J. E. 1966. Site index curves for Douglas-fir in the Pacific Northwest. *Weyerhaeuser Forestry Paper Number 8*. Weyerhaeuser Forestry Research Center. Canada.

King, A. W. and With, K. A. 2002. Dispersal success on spatially structured landscapes: when do spatial pattern and dispersal behavior really matter? *Ecological Modelling*. **147**. 23–39.

Latta, G. and Montgomery, C. A. 2004. Minimizing the cost of stand level management for older forest structure in western Oregon. *Western Journal of Applied Forestry*. **19** (4). 221–231.

LeSage, J. 1999. *Spatial Econometrics*. http://www.spatial-econometrics.com/

Lichtenstein, M. E. and Montgomery, C. A. 2003. Biodiversity and timber in the Coast Range of Oregon: inside the production possibility frontier. *Land Economics*. **79**. 56–73.

Margules, C. R., Nicholls, A. O. and Pressey, R. L. 1988. Selecting networks of reserves to maximize biodiversity. *Biological Conservation*. **43**. 63–76.

Marshall, E., Homans, F. and Haight, R. 2000. Exploring strategies for improving the cost effectiveness of endangered species management. *Land Economics*. **76**. 462–473.

Miller, K. R. 1996. Conserving biodiversity in managed landscapes. In R. C. Szaro and D. W. Johnston (eds.). *Biodiversity in Managed Landscapes: Theory and Practice*. Oxford, UK: Oxford University Press, 425–441.

Moilanen, A. and Cabeza, M. 2002. Single-species dynamic site selection. *Ecological Applications*. **12**. 913–926.

Montgomery, C. A., Brown Jr., G. M. and D. M. Adams. 1994. The marginal cost of species preservation: the northern spotted owl. *Journal of Environmental Economics and Management*. **26**. 111–128.

Montgomery, C. A., Pollak, R. A., Freemark, K. and White, D. 1999. Pricing biodiversity. *Journal of Environmental Economics and Management*. **38**. 1–19.

Nabhan, G. P. and Buchmann, S. L. 1997. Services provided by pollinators. In G. Daily (ed.). *Nature's Services: Societal Dependence on Natural Ecosystems*. Washington, DC: Island Press.

Nalle, D. J., Montgomery, C. A., Arthur, J. L., Polasky, S. and Schumaker, N. H. 2004. Modeling joint production of wildlife and timber in forests. *Journal of Environmental Economics and Management*. **48**. 997–1017.

Oregon Natural Heritage Program (ONHP). 2000. Integrated Willamette Basin Landcover ARCINFO coverage and related metadata (a very similar land cover ARCINFO coverage, LAND USE/LAND COVER ca. 1990 (ec90.e00), can be found at http://www.fsl.orst.edu/pnwerc/wrb/access.html).

Oregon State University Extension Service (OSUES). 2002. Oregon Agricultural Information Network. http://ludwig.arec.orst.edu/oain/SignIn.asp.

Oregon State University Extension Service (OSUES). 2003. Oregon Agricultural Information Network Enterprise Budget web site. http://oregonstate.edu/Dept/EconInfo/ent_budget/index.cfm

Oregon Water Resources Department (OWRD). 2001. Water Rights Data for the Willamette Valley: Point-of-Use ARCINFO coverage (willpou.e00) and related metadata. http://www.wrd.state.or.us/files/water_right_data/will/

Pacific Northwest Ecosystem Research Consortium (PNW-ERC). 1999a. Soils – (SSURGO and STATSGO) ARCINFO Coverage (WRBSOILS.e00) and related metadata. http://www.fsl.orst.edu/pnwerc/wrb/access.html

Pacific Northwest Ecosystem Research Consortium (PNW-ERC). 1999b. Topographic Slope ARCINFO Coverage (SLOPEPI.e00) and related metadata. http://www.fsl.orst.edu/pnwerc/wrb/access.html

Pacific Northwest Ecosystem Research Consortium (PNW-ERC). 1999c. Vegetation – 1851 ARCINFO Coverage (VEG1851_v4.e00) and related metadata. http://www.fsl.orst.edu/pnwerc/wrb/access.html.

Polasky, S., Camm, J. D. and Garber-Yonts, B.. 2001. Selecting biological reserves cost-effectively: an application to terrestrial vertebrate conservation in Oregon. *Land Economics*. 77. 68–78.

Pressey, R. L. 1994. Ad-hoc reservations: forward or backward steps in developing representative reserve systems? *Conservation Biology*. **8**. 662–668.

Pressey, R. L. and Logan, V. S. 1998. Size of selection units for future reserves and its influence on actual versus targeted representation of features: a case study in western New South Wales. *Biological Conservation*. **85**. 305–319.

Redford, K. H. and Richter, B. D. 1999. Conservation of biodiversity in a world of use. *Conservation Biology*. **13**. 1246–1256.

Reid, W. V. 1996. Beyond protected areas: changing perceptions of ecological management objectives. In R. C. Szaro and D. W. Johnston (eds.). *Biodiversity in Managed Landscapes: Theory and Practice*. Oxford: Oxford University Press, 442–453.

Rosenzweig, M. L. 2003. *Win-Win Ecology: How the Earth's Species Can Survive in the Midst of Human Enterprise*. Oxford: Oxford University Press.

Saetserdal, M., Line, J. M. and Birks, H. B. 1993. How to maximize biological diversity in nature reserve selection: vascular plants and breeding birds in deciduous woodlands, Western Norway. *Biological Conservation*. **66**. 131–138.

Schumaker, N. H., Ernst, T., White, D., Baker, J. and Haggerty, P. 2004. Protecting wildlife responses to alternative future landscape in Oregon's Willamette Basin. *Ecological Applications*. **14**. 381–400.

Scott, J. M., Davis, F. W., McGhie, R. G., Wright, R. G., Groves, C. and Estes, J. 2001. Nature reserves: do they capture the full range of America's biological diversity? *Ecological Applications*. **11**. 999–1007.

Stoms, D. M. 1994. Scale dependence of species richness maps. *Professional Geographer*. **46**. 346–358.

Thorsnes, P. 2002. The value of a suburban forest preserve: estimates from sales of vacant residential building lots. *Land Economics*. **78**. 426–441.

Tilman, D., Farigione, J., Wolff, B., D'Antonio, C., Dobson, A., Howarth, R., Schindler, D., Schlesinger, W. H., Simberloff, D. and Swackhamer, D. 2001. Forecasting agriculturally driven global environmental change. *Science*. **292**. 281–284.

Tyrvainen, L. and Miettinen, A. 2000. Property prices and urban forest amenities. *Journal of Environmental Economics and Management*. **39**. 205–223.

United Nations Development Programme (UNDP), United Nations Environment Programme, World Bank and World Resources Institute. 2000. *World Resources 2000–2001*. Amsterdam: Elsevier Science.

United States Department of Agriculture-Natural Resources Conservation Service (USDA-NRCS). 2001a. National STATSGO (State Soil Geographic Database) Database. http://www.ftw.nrcs.usda.gov/stat_data.html.

United States Department of Agriculture-Natural Resources Conservation Service (USDA-NRCS). 2001b. National SSURGO (Soil Survey Geographic) Database. http://www.ftw.nrcs.usda.gov/ssur_data.html.

United States Department of Agriculture-National Resources Conservation Service (USDA-NRCS). 2003. Oregon Soil Survey Reports and Data. http://www.or.nrcs.usda.gov/pnw_soil/or_data.html

Vos, C. C., Verboom, J., Opdam, P. F. M., Ter, Braak, C. J. F. and Possingham, H. 2001. Toward ecologically scaled landscape indices. *The American Naturalist*. **157**. 24–41.

Volser, C. A., Kerkvliet, J., Polasky, S. and Gaintutdinova, O. 2003. Externally validating contingent valuation: an open-space survey and referendum in Corvallis, OR. *Journal of Economic Behavior and Organization*. **51**. 261–277.

Warman, L. D., Sinclair, A. E. R., Scudder, G. G. E., Klinkenberg, B. and Pressey, R. L. 2004. Sensitivity of systematic reserve selection to decisions about scale, biological data, and targets: case study from southern British Columbia. *Conservation Biology*. **18**. 655–666.

Wear, D. N., Turner, M. G. and Flamm, R. O. 1996. Ecosystem management with multiple owners: landscape dynamics in a Southern Appalachian watershed. *Ecological Applications*. **6**. 1173–1188.

Wilcove, D., Rothstein, D., Dubow, J., Phillips, A. and Losos, E. 2000. Leading threats to biodiversity: what's imperiling U.S. species. In B. A. Stein, L. S. Kutner and J. S. Adams (eds.). *Precious Heritage: The Status of Biodiversity in the United States*. Oxford: Oxford University Press.

Williams, P., and Araujo, M. 2000. Using probability of persistence to identify important areas for biodiversity conservation. *Proceedings of the Royal Society of London: Series B*. **267**. 1959–1966.

Wilson, E. O. (ed.). 1988. *BioDiversity*. Washington, DC: National Academy Press.

World Resources Institute (WRI). 2003. EarthTrends: The Environmental Information Portal. http://earthtrends.wri.org/

Yazvenko, S. B. and Rapport, D. J. 1996. A framework for assessing forest ecosystem health. *Ecosystem Health*. **2**. 40–51.

## Appendix: Figures and Table

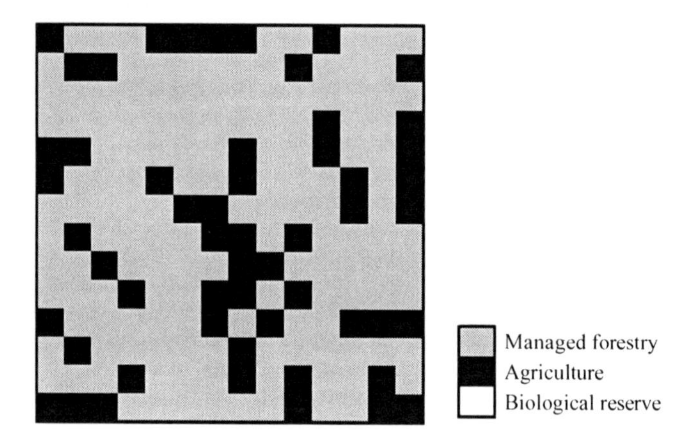


Figure 18.1. A base case land-use pattern on the 14×14 landscape where every parcel is put into its highest economic use*

*Grey parcels signify managed forestry, black parcels signify agriculture and white parcels signify biological reserve

Table 18.A1 *Summary of various base case land-use patterns (as presented in Figures 18.3a–e) that have landscape economic and biological score combinations that lie on the efficiency frontier*

| | Landscape score | | Number of parcels in land use | | | No. parcels in initial base case converted to biological reserve (see Figure 18.1) | | | Number of biological reserve parcels ($X_j = x_b$) with habitat type | | | |
|---|---|---|---|---|---|---|---|---|---|---|---|---|
| | Landscape economic (LE) score (millions of dollars) | Landscape biological (LB) score | Agriculture ($X_j = x_a$) | Managed forestry ($X_j = x_f$) | Biological reserve ($X_j = x_b$) | Agriculture ($X_j = x_b$) | Managed Forestry ($X_j = x_b$) | Biological reserve 'Connectivity' | Shrub | Hardwood | Prairie/ Meadow | Conifer |
| Land use pattern in Figure 18.3a | 1,034 | 81.09 | 54 | 138 | 4 | 0 | 4 | 3.50 | 1 | 1 | 0 | 2 |
| Land use pattern in Figure 18.3b | 954 | 85.55 | 53 | 121 | 22 | 4 | 18 | 2.82 | 9 | 10 | 0 | 3 |
| Land use pattern in Figure 18.3c | 740 | 86.91 | 45 | 81 | 70 | 10 | 60 | 1.80 | 12 | 27 | 0 | 31 |
| Land use pattern in Figure 18.3d | 498 | 87.93 | 28 | 65 | 103 | 26 | 77 | 1.67 | 20 | 29 | 0 | 54 |
| Land use pattern in Figure 18.3e | 270 | 88.48 | 13 | 52 | 131 | 35 | 96 | 1.25 | 23 | 46 | 0 | 62 |

Note: The base case initial land use pattern is formed by placing all parcels in their highest economic use (see Figure 18.1). LE is given by equation 11 and LB is given by equation 9

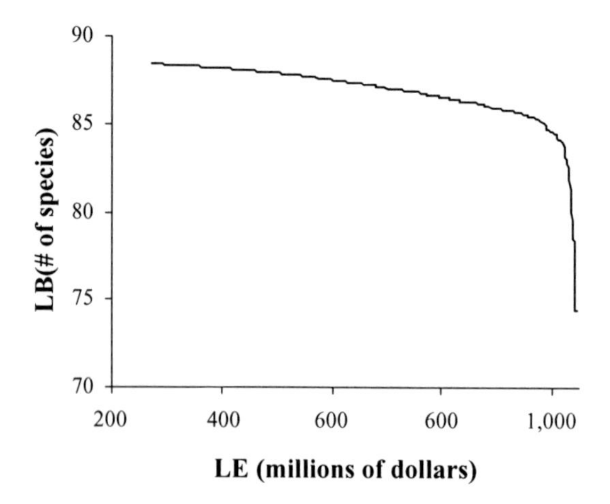


Figure 18.2. The base case landscape economic-biological score efficiency frontier*

*The *x-axis* measures the landscape economic (*LE*) score in millions of dollars for a given land-use pattern on the 14 × 14 landscape. The *y-axis* measures the landscape biological (*LB*) score for a given land-use pattern on the 14 × 14 landscape. The land-use pattern is associated with the point that forms the right-hand terminus of the efficiency frontier. The efficiency frontier has an L-shape indicating that it is possible to arrange the land pattern in a way that attains a high biological score and a high economic score. Trying to maximise either the biological score or the economic score results in large losses in the other score.

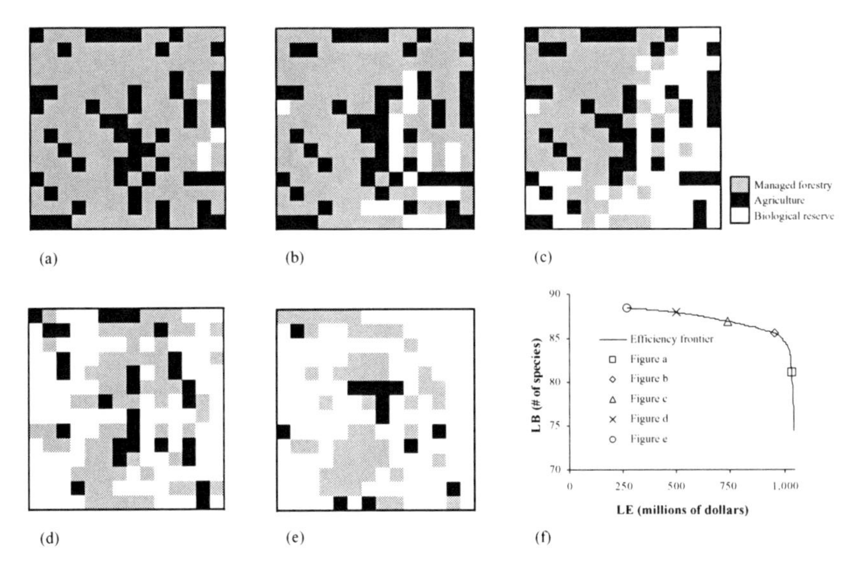


Figure 18.3. Various base case land-use patterns for points that lie on the base case efficiency frontier*

*Beginning from the right-hand terminus of the efficiency frontier, land-use patterns are shown that are 10% along the efficiency frontier (Figure a), 25% along the efficiency frontier (Figure b), 50% along the efficiency frontier (Figure c), 75% along the efficiency frontier (Figure d) and at the left-hand terminus of the efficiency frontier (Figure e). Grey parcels signify managed forestry, black parcels signify agriculture and white parcels signify biological reserve. Figure f shows the coordinate positions of the land-use patterns shown in Figures a–e

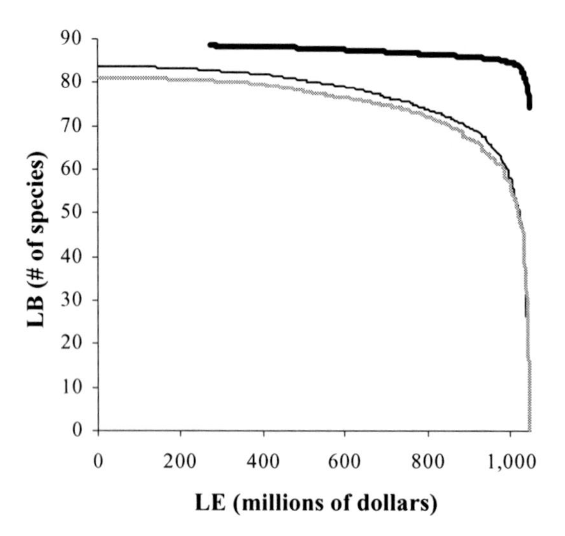


Figure 18.4. The efficiency frontiers from two reserve-site selection scenarios simulated on the default 14×14 landscape*

*For comparison purposes, the base case efficiency frontier is also presented. The thin grey line denotes the efficiency frontier of the reserve site selection scenario without dispersal. The thin black line denotes the efficiency frontier of the reserve site selection scenario with dispersal. The efficiency frontiers from the reserve site selection scenarios lie well within the efficiency frontier of the base case, which is represented by the thick black line. The efficiency frontiers associated with the reserve site selection scenarios are more rounded (less L-shaped), indicating more continuous tradeoffs between biological and economic objectives.

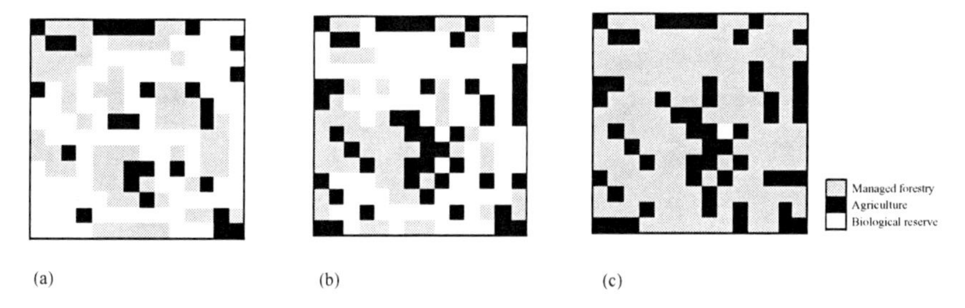


Figure 18.5. Land-use patterns on the default 14×14 landscape that have landscape biological (*LB*) scores of approximately 77 for the reserve-site selection scenarios and the base case*

*The land-use pattern for reserve site selection without dispersal scenario is shown in Figure a. The land-use pattern for reserve site selection with dispersal scenario is shown in Figure b. The land-use pattern for the base case is shown in Figure c. Grey parcels signify managed forestry, black parcels signify agriculture and white parcels signify biological reserve. A single contiguous biological reserve is chosen in the reserve site selection without dispersal scenario. With dispersal, not all biological reserves are connected. In the base case, only one parcel is put into a biological reserve.

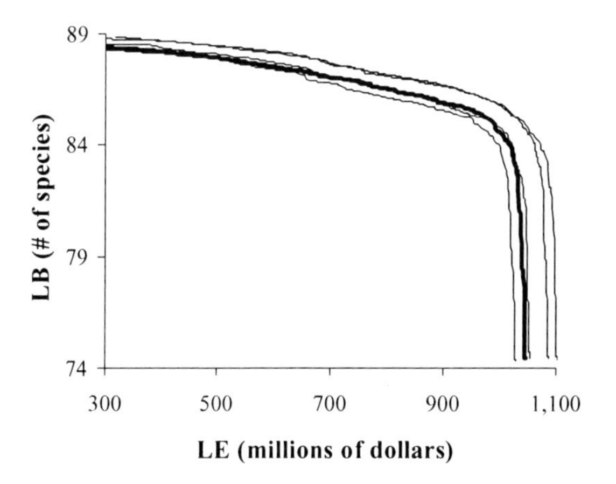


Figure 18.6. Efficiency frontiers associated with alternative 14×14 landscapes*

*The alternative landscapes were created using the same methodology that was used to create the default 14 × 14 landscape. The thick black line represents the base case efficiency frontier. All efficiency frontiers have the same basic shape. The right-hand terminus of the efficiency frontier is shifted right when the landscape has higher present values for commodity production.

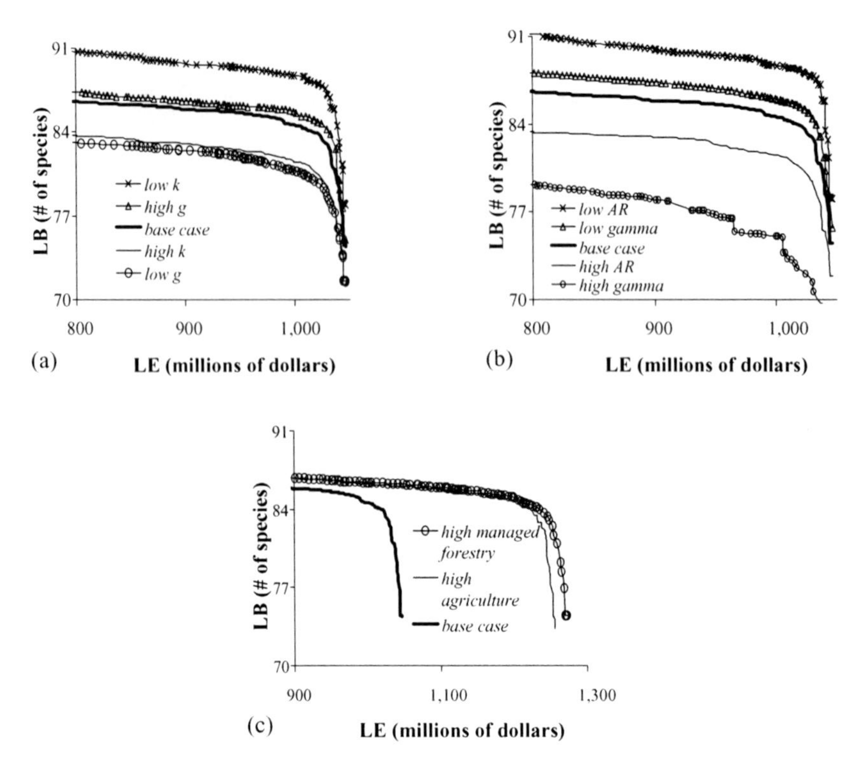


Figure 18.7. Efficiency frontiers for various sensitivity analyses on the default 14×14 landscape*

*Efficiency frontiers for simulations with a 50 per cent increase and a 50 per cent decrease from default parameter values for $g$, the constant that determines the shape of the species persistence probability function, and $k$, the half-saturating constant in the species persistence probability function (see equation 8) are shown in Figure a. Efficiency frontiers for simulations with a 50 per cent increase and a 50 per cent decrease in default parameter values for $AR_s$, the number of hectares needed by a breeding pair of species $s$ (see equation 1) and changes in the default value of $\gamma_s$ (*low* $\gamma_s = 0$, *high* $\gamma_s = 50$), the minimum number of breeding pairs of species $s$ that a patch must support on its own before the patch contributes to the minimum landscape biological score (see equation 4), are shown in Figure b. Efficiency frontiers for simulations with a 40 per cent increase in the default economic value of agriculture (*high agriculture*) or in the default economic value of managed forestry (*high managed forestry*) are shown in Figure c. A thick black line in each figure represents the base case efficiency frontier. Changes in biological parameters shift the efficiency frontier vertically (Figures a and b). Changes in economic parameters shift the efficiency frontier horizontally (Figure c). The L-shape of the efficiency frontier remains unchanged.

# 19 Balancing recreation and wildlife conservation of charismatic species

*Birgit Friedl and Doris A. Behrens*

## 1 Introduction

Out of the four components of biodiversity (UNEP 1992, Art. 2, Par. 2), two of them, namely species diversity and ecosystem diversity, focus on the fauna and flora in ecosystems and implicitly stress the importance of non-domesticated organisms living in their natural habitats. The term *wildlife* is used to scientifically describe plant and, more commonly, animal species without any serious human influence on habitats or population numbers. According to the Red List of Threatened Species, 1 per cent of all described species and 41 per cent of the species evaluated worldwide are seriously threatened by extinction (IUCN 2004). The dramatic increase in the rates of extinction can be expected to lead to a loss of half of all existent birds and mammals within the next 2–5 centuries (Wilson 1992; Stork 1997).

Species extinction is closely linked to (economically motivated) human activities such as land conversion, hunting and tourism. Irrespective of its immediate causes, species extinction is most often due to habitat modification and fragmentation (Groombridge 1992; MEA 2005).[1] Thus, measures to protect habitats or ecosystems constitute the core of wildlife (or '*in situ*') conservation. The World Conservation Union (IUCN)

Acknowledgements: Financial support from the Austrian National Bank (OeNB) (project no. 11216) is gratefully acknowledged. Thanks for comments on earlier versions of this chapter are due to Stefan Baumgärtner, Jonathan P. Caulkins, Charles Figuières, Brigitte Gebetsroither, Michael Getzner and participants of the 6th ZEW Summer Workshop for Young Economists on The Management of Global Commons, Mannheim, the 13th Annual Conference of the European Association of Environmental and Resource Economists, Budapest and the 6th Annual BioEcon Conference on Economics and the Analysis of Biology, and Biodiversity, Cambridge, where earlier versions of this chapter were presented.

[1] A reduction in size and quality of a habitat often splits an interconnected species population into smaller isolated sub-populations. If one of these (sub-) populations falls below its minimum viable population size it eventually becomes extinct. For obvious reasons this process is more dangerous if the required habitat is large. Therefore, Prins and Grootenhuis (2000) raise the question as to whether large wild animals, like grizzly bears, can survive outside protected areas.

developed a classification and certification scheme for protected area management, reaching from human exclusion (strict nature reserve; category Ia) to coexistence of conservation and recreation (national park; category II) and to sustainable use (managed resource protected area; category VI). Therefore, an *ecosystem* (or *habitat*) *approach* targets at protecting representative samples of ecosystems or habitat types through control of land use. As soon as a certain spot is classified as 'protected', it is thereafter regarded as sufficient for the conservation of species native to this spot to guarantee the maintenance of the habitat (Groombridge 1992). The main drawback of this ecosystem approach is that a particular endangered species might live outside this habitat classified as being 'representative' and therefore still be under threat or that a threatened species might need specific conservation and management measures in order to survive.

An alternative approach to conservation would be to target management directly towards species considered to be of high priority for conservation. This strategy directs conservation funds towards the most urgent cases, with the disadvantage of requiring a definition of what is considered as most urgent – species seriously threatened or highly appreciated (mostly higher vertebrates, i.e. mammals).[2] Because of this discrepancy the *species approach* is usually applied by jointly targeting areas with high diversity and particular endangered charismatic species, labelled as flagship species (Leader-Williams and Dublin 2000).

So, is the flagship approach to be preferred to the habitat approach or vice versa? As ecologists seem to be indecisive (see Groombridge 1992 or Simberloff 1998 for a discussion), economists tend to argue in favour of the flagship approach based on people's valuation of species preservation and associated larger conservation budgets. Several studies on willingness to pay (WTP) for species conservation suggest that (i) valuation is higher for mammals than for non-vertebrates or plants (even increasing with the size of the mammal; Loomis and White 1996), (ii) the aggregate WTP for many species together is not considerably higher than for a single species (Nunes *et al.* 2003) and (iii) the valuation for a single species implicitly considers the type of protection, i.e. the WTP is significantly higher when the species under concern is protected within its natural habitat (in situ) than otherwise (Kontoleon and Swanson 2003). Therefore, it is economically efficient that conservation policy exhibits a tendency towards focusing on charismatic species and on other non-scientific factors (e.g. Metrick and Weitzman 1998).

Whether the above empirical results can be supported theoretically, i.e. whether the flagship approach can effectively replace the ecosystem

[2] For a discussion of the factors influencing listing and funding decisions see, for example, Metrick and Weitzman (1996) and Dawson and Shogren (2001).

approach and if so, under which conditions, is the central topic of this chapter. It is, however, not limited to a comparison of the effectiveness of the flagship and ecosystem approaches with respect to conserving wildlife, but integrates the interactions of wildlife and humans by explicitly including non-consumptive utilisation of protected areas like nature-based tourism. Accordingly, the focus is on determining the appropriate economic limit on visitor numbers to protected areas (i.e. an *optimal visitor control strategy*) that simultaneously maximises intertemporal social welfare and guarantees the conservation of charismatic species in their natural habitat.

In order to address these issues, the chapter is structured as follows. Section 2 is devoted to a concise description of bio-economic models taking account of multi-species interactions. Thereafter, a bio-economic model is developed to address the social planner's tradeoff between a protected area's recreational value and its conservation value subject to a predator-prey-type species-habitat model. A detailed description of the optimal control model and its analytic results is given in section 4, leading to the conditions for the equivalence of the optimal visitor control strategies when decision making is based either on the flagship or on the ecosystem approach. In section 5 the model is calibrated with parameters tailored to the case of the golden eagle in the Hohe Tauern National Park (Austria) to illustrate the theoretical results derived in this chapter. Finally, section 6 concludes by summarising the results and pointing out their policy implications.

## 2 The tradeoff between wildlife conservation and economic utilisation

Multi-species modelling has an astonishingly short tradition in environmental and resource economics. Within this field, the primary focus was put on determining optimal (sustainable) harvesting rates of particular species within multi-species ecosystems, mainly in the context of fisheries (sometimes also forests), because mutually dependent species are not well managed by quotas determined independently for every species. Therefore, ecological interdependencies through competition for food or habitat, food chain (predator-prey) interactions or mutualisms have been incorporated into economic model formulations.[3] The common

[3] See, for example, Hannesson (1983), Ragozin and Brown (1985), Conrad and Adu-Asamoah (1986), Flaaten (1991), Conrad and Salas (1993), Semmler and Sievekig (1994), Ströbele and Wacker (1995), Regev *et al.* (1998), Tschirhart (2000), Bulte and van Kooten (2001), Crépin (2003), Finnof and Tschirhart (2003), Hoekstra and van den Bergh (2005) and Skonhoft (2007).

approach in these models is that human intervention through harvesting activities can be interpreted as the behaviour of a supra-predator or that people compete with other species for habitat or food.

Some authors also addressed the opposite direction of causation in multi-species models by taking account of human benefits from species conservation and biodiversity (c.f. Conrad and Salas 1993; Hoekstra and van den Bergh 2005; Skonhoft 2006). For instance, Hoekstra and van den Bergh (2005) pay tribute to the fact that an ecosystem has more to offer than a palette of species to exploit in a model where humans and a predator compete for the same prey and the predator has a protection value.

The model presented in section 3 follows a somewhat different approach by adjusting ecosystem dynamics to reflect the negative, but unintended, impact of nature-based tourism, or more precisely, the side-effects of non-consumptive wildlife-oriented recreation (for an account of these activities, see the survey article by Duffus and Dearden 1990). This damage on the ecosystem associated with nature-based tourism is particularly important for protected areas such as a national park since this management category is, according to IUCN guidelines, explicitly devoted to both ecosystem protection *and* visitor recreation (including education).

## 3 A model of balancing recreation and wildlife conservation

One way of assessing the equivalence of the flagship and the ecosystem approaches is to identify to what extent an optimal visitor control strategy for a given protected area does (or does not) vary with the specification of conservation values within the social welfare function. The way chosen to shed light on this issue is the development of a bio-economic model based on optimal control theory (see, for example, Léonard and Long 1992).

In order to do so, we assume that from the point of view of a social planner society derives welfare from recreation (due to nature-based tourism to the protected area) *and* from conservation within a national park. Moreover, social welfare is assumed to be additively separable between conservation and recreation, paying tribute to the fact that these benefits accrue to different groups of society and cannot be substituted easily (for a similar argument, see Rondeau 2001). As far as conservation is concerned, we define the intertemporal social welfare function in such a way that both conservation concepts are included as special cases regarding the relative capability of charismatic species or ecosystem to generate social welfare.[4] This capability is captured by parameter $\mu$ which takes $\mu = 1$ if welfare is merely generated by the charismatic species. Alternatively, when the entire ecosystem contributes to welfare then $\mu = 0$.

The ecosystem itself is modelled in a fairly general way, even though its design is motivated by a specific exploratory focus, namely the interaction of visitor streams and golden eagles in Austria's Hohe Tauern National Park, discussed later in the chapter. Since the eagle is a very territorial bird, its survival is supported by prey availability in its habitat area (Pedrini and Sergio 2002). Thus, the existence of the habitat – sufficient in size and, especially, quality – can be used as a proxy for the diversified diet of the golden eagle. Therefore, we can simplify the entire set of eagle-prey-habitat interactions to a two-stage eagle-habitat food chain model. Accordingly, from now on the model is explained and discussed in terms of the golden eagle as charismatic species (denoted by $S = S(t)$), its habitat (denoted by $H = H(t)$) and the current admissible number of visitors to the protected area (denoted by $u = u(t) \geq 0$) only. The notion of time is suppressed from now on to ease reading.

As mentioned above, a benevolent social planner is assumed to seek to balance the (non-use) benefits from conservation, i.e. single species values (represented by $V(S)$) and ecosystem values (represented by $W(H, S)$) and the recreational benefits (use values) from tourism (represented by $U(u)$) over time (see Figure 19.1). Accordingly, the social planner maximises intertemporal social welfare, as formalised by equation (1), following the *ecosystem approach* ($\mu = 0$) or the *flagship approach* ($\mu = 1$), subject to the visitor-inflicted eagle-habitat dynamics (equations 2a–2b).

$$\mathcal{J}^*_\mu = \max_{\substack{u \geq 0 \\ S \geq \underline{S}}} \int_0^\infty e^{-rt}(U(u) + \mu V(S) + (1-\mu) W(H,S))dt, \qquad (1)$$

with $U' > 0, V' > 0, W_H > 0, W_S > 0 \quad u, H \geq 0, S \geq \underline{S} > 0,$ and subject to

$$\dot{H} = g(H)H - f(H)S, \qquad H(0) > 0, \qquad (2a)$$

$$\dot{S} = (T(u)cf(H) - d)S, \qquad S(0) \geq \underline{S}, \qquad (2b)$$

with $g(H) \geq 0$ denoting the natural growth function of the habitat $H$, $f(H) \geq 0$ standing for the 'response' of the eagle population to habitat

[4] This approach is supported by Metrick and Weitzmann (1998) who distinguish welfare from a certain charismatic species and welfare from some overall concept of diversity (alternatively also quality of the entire ecosystem or habitat).

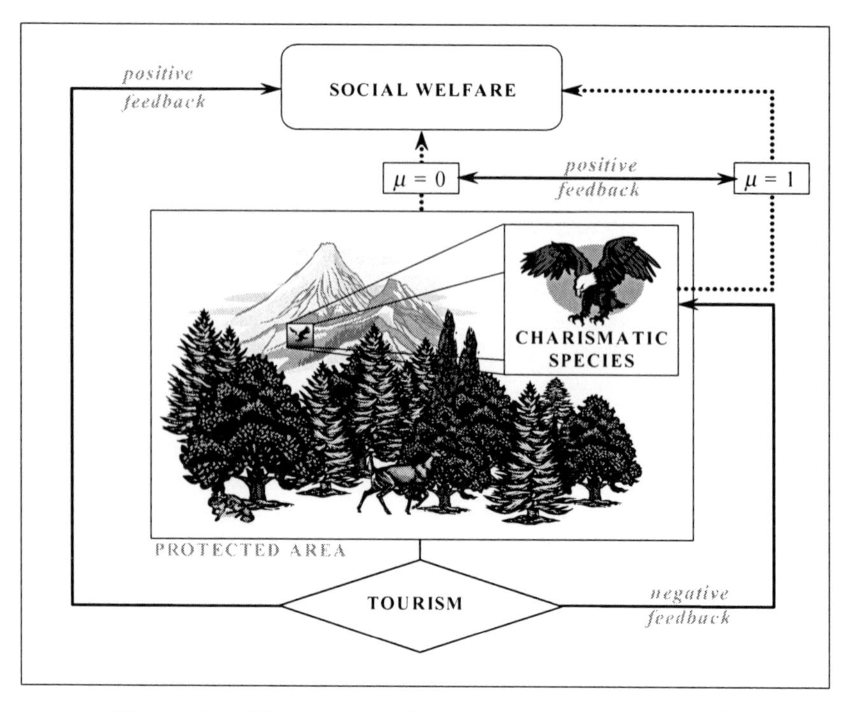


Figure 19.1. The structure of the protected area model for the ecosystem approach ($\mu = 0$) and the flagship approach ($\mu = 1$)

supply and $T(u) \epsilon [0, 1]$ describing the negative feedback of visitors $u$ on the eagle population. Moreover, parameter $c \geq 0$ measures the efficiency of 'converting' habitat into birds and $d \geq 0$ is the eagle population's (natural) mortality rate. The parameter $r \geq 0$ represents the utility discount rate which is constant over time. In order to avoid that the charismatic species falls below its minimum viable population size, we require that $S \geq \underline{S}$.

The ecosystem of concern augmented by the influence of nature-based tourism is described by a two-dimensional predator-prey-type system (Rosenzweig and MacArthur 1963) and constitutes the constraints of the social planner's optimisation. The predator-prey dynamics as described by equations (2a) and (2b) can be best understood by imagining that favourable environmental conditions lead to an increase in the supply of food for an eagle population. This short-term excess supply of food for the birds is indirectly taken account of as an increase in suitable habitat. Subsequently, within a larger bird population the likelihood of breeding success increases. This in turn has a diminishing effect on the eagle's source of food represented by a reduction of its habitat. This yields a reduction in the size of the bird population which is responsible for the recovery of the available habitat – and the 'cycle' of interaction starts all over again (for a more elaborate description, see Clark 1990; Ricklefs 1990; Begon *et al.* 1996). Thus, $H$ endogenously determines a carrying capacity for $S$.

For obvious reasons, a given habitat area cannot 'grow'. However, the protection status and the actual area size protected can increase, as was the case for the Hohe Tauern National Park over the last twenty years. However, this process cannot proceed for ever since protected areas compete with many other types of land use. These observations together motivated the modelling of habitat dynamics by means of the logistic function

$$g(H) = a\left(1 - \frac{H}{\omega}\right), \tag{3}$$

where $\omega$ denotes the carrying capacity of the habitat and $a$ stands for the natural growth rate of the habitat.

The eagle's 'demand' for habitat (or per capita prey consumed by every individual eagle) can be captured by a Holling-type II predator response function (Holling 1959)

$$f(H) = \frac{bH}{q + H}, \tag{4}$$

where parameter $b$ denotes an upper limit on daily consumption of the eagle and $q$ is the so-called half-saturation constant. While the logistic growth function $g(H)$ is well known among economists, the predator response function $f(H)$ might require some further explanation. The function $f(H)$ describes the habitat demand per individual eagle. The half-saturation constant $q$ corresponds to the habitat $H$ necessary to supply an amount of food sufficient to provide half of the maximum possible daily consumption for the entire eagle population. The size of $q$ depends on the eagle's searching and handling time for each unit of food (Myerscough *et al.* 1996). Note that $f(H)$ is reducing the size of the habitat on the one hand and increasing the population stock of the eagle on the other. Parameter $c$ accounts for the metabolic efficiency loss that translates the food of the eagle into offspring. Thus, without human influence, the net change in the eagle population size is determined by subtracting the (exogenous) death rate, $d$, from the fertility rate of the eagle, defined by $cf(H)$.

According to Grubb and King (1991), visitors disturb the nesting of bald eagles, a fact also reconfirmed by Pedrini and Sergio (2002) for the golden eagle. Assuming that this influence is representative for avifauna in general, we model the visitors' negative impact by a certain percentage reduction in the eagle population's fertility rate as given by

$1 - T(u)$, where $T(u)$ is approximated by a convex function. If there were no visitors at all, i.e. $T(0) = 1$, the fertility rate of the charismatic species would be equal to its natural rate (without any human influence). At the other extreme, if there were an infinite number of visitors, the charismatic species' fertility rate would be eradicated, i.e. $T(\infty) = 0$.

For these assumptions to hold, the visitor impact function $T(u)$ can be specified as, for example, $T(u) = e^{-mu}, m > 0$. Accurately breaking down $T(u)$ to the various individual damage factors of heterogeneous visitors goes far beyond the scope of this chapter. So, a decrease in the negative impact of visitors, $T(u)$, can be interpreted either as a decline in the number of visitors itself or as an increase in the visitors' awareness. Thus, when we call for 'a reduction of visitors', we are actually calling for 'having fewer disturbing effects of visitors on protected areas'.

## 4 An optimal short- and long-run visitor control strategy

Section 3 outlined a decision-making problem where a social planner seeks an *optimal visitor control strategy*. In this context an optimal strategy (for visitor steering) refers to measures to be continuously adapted (to the needs of the ecosystem) to limit visitor access to the protected area in a way that guarantees that welfare derived from both current and future tourism *and* the protection/survival of the species and its habitat is maximised for all current and future periods.

While we can determine the optimal limit on visitor streams over time (as the maximum level for visitors to a protected area in accordance with conservation), the visitor demand is not modelled in this approach. Therefore, the optimal visitor strategy determined in this section can be regarded as the upper limit on 'tourism' under the assumption that demand is at least as high as this capacity.

### *4.1 The solutions of the optimal control model*

Using optimal control theory, the welfare-maximising intertemporal level of visitors can be determined by solving the associated non-linear, dynamic, autonomous, infinite planning-horizon optimisation problem given by equations (1)–(4).[5] Thus, the current number of visitors is the *control variable*, $u$, which shapes the intertemporal development of the *state variables*, i.e. $S$ and $H$. The predator-prey system given by equation (2)

[5] For an excellent survey on the application of optimal control theory in environmental economics, see Feenstra *et al.* (1999). For the theory itself see Kamien and Schwartz (1991) or Léonard and Long (1992).

specifies *how* the control variable $u$ affects the interaction between the eagle population ($S$) and its habitat ($H$). To facilitate reading, mathematical derivations are cut back on a grand scale in what follows. Details can be found in Behrens and Friedl (2004).

Using Pontryagin's maximum principle, first-order conditions for an optimal intertemporal visitor control strategy can be set up. For these purposes we construct the current value Hamiltonian

$$\mathcal{H} = U(u) + \mu V(S) + (1-\mu) W(H,S) + \pi_1 \dot{H} + \pi_2 \dot{S}, \tag{5}$$

where $U$, $V$ and $W$ represent the dynamic welfare functions as implicitly defined by equation (1) following the *ecosystem approach* ($\mu = 0$) or the *flagship approach* ($\mu = 1$), $\dot{H}$ and $\dot{S}$ are the habitat-eagle dynamics as determined by equations (2a) and (2b), and $\pi_1$ and $\pi_2$ denote the shadow prices or costate variables, all in current values. The shadow price $\pi_1$ measures marginal social welfare caused by an incremental increase in $H$. The shadow price $\pi_2$ quantifies the increase in social welfare caused by a marginal increase in $S$. Obviously, having more eagles and more of their habitat contributes positively to social welfare along the optimal path ($\pi_1, \pi_2 > 0$).

At any instant of the decision-making process the social planner has the option to choose a visitor control strategy that generates immediate contributions to social welfare, mainly through visitors, $U(u)$, but also through the ecosystem, $V(S)$ or $W(H, S)$. Moreover, the planner can pick a visitor control strategy that improves the ecological quality of the ecosystem, generating contributions to social welfare in the more distant future as reflected by $\pi_1 \dot{H} + \pi_2 \dot{S}$ in (5). Thus, equation (5) can be seen as a dynamic value function taking into account the effect of the current state of the ecosystem (including the current number of visitors) on its future ecological quality. Optimal dynamic visitor control is designed in such a way that it maximises the Hamiltonian $H$ (at any instant of time) and assures the existence of the eagle population above its minimum viable population size, i.e. $S \geq \underline{S} > 0$.

As a result, for both conservation strategies ($\mu = 0$ and $\mu = 1$) an optimal interior control $u^* > 0$ corresponds to a level of visitors where the marginal social welfare of tourism given by the left-hand side of equation (6) just offsets its opportunity costs (right-hand side of the equation) along the entire optimal path:[6]

$$\underbrace{U'(u^*)}_{(+)} = \underbrace{-T'(u^*)}_{(+)} cf(H) \underbrace{\pi_2}_{(+)} S > 0. \tag{6}$$

[6] If $cf(H)\pi_2 ST''(u) < |U''(u)|$ , $u^*$ is uniquely determined along the optimal path (cf. Appendix A.1).

If $T(u) = -\eta T'(u)/U'(u)$ equation (6) can be solved explicitly for $T(u^*)$ yielding

$$T(u^*) = \eta(cf(H)\pi_2 S)^{-1}.^{7} \tag{7}$$

The opportunity cost of visitors (RHS of equation 6) is measured by the damage to the fertility rate of the charismatic species caused by an extra unit of tourism weighted by the eagle population's shadow price ($\pi_2$). Therefore, the visitor control strategy depends on the current state of the ecosystem ($H$, $S$) and takes into account the immediate recreational and ecological effects of tourism as well as the effects of (current) tourism at all future times. Moreover, $u^*$ increases with the eagles' fertility rate ($cf(H)$) and/or society's current valuation of the entire eagle population ($\pi_2 S$). The links between the evolution of optimal visitor control, the associated optimal development of the ecosystem's state and valuation over time are characterised by the so-called canonical system of the optimal control problem (see Appendix A.2).

To assure that along the optimal path the initial values for the shadow prices, ($\pi_1(0)$, $\pi_2(0)$)), are uniquely determined for any given initial state of the ecosystem, ($H(0)$, $S(0)$)), we have to specify conditions for the existence of a two-dimensional stable manifold in the four-dimensional phase-space. This task is briefly addressed next.

## 4.2 *Properties of the bio-economic equilibrium*

For both conservation strategies ($\mu = 0$ and $\mu = 1$) the bio-economic equilibrium $\hat{E} = (\hat{H}, \hat{S}, \hat{\pi}_1, \hat{\pi}_2)$ ($\hat{S} > \underline{S}$) is defined by the simultaneous solution of the following set of equations where the equilibrium level of tourism $\hat{u}^*$ is defined by equation (7):

$$\hat{H} = \{H \mid \hat{f}\hat{\pi}_1 + r\hat{\pi}_2 - \mu\hat{V}' - (1-\mu)\hat{W}_S = 0 \wedge 0 < H < \omega\} \tag{8a}$$

$$\hat{S} = \hat{S}(\hat{H}) = \frac{\hat{g}\hat{H}}{\hat{f}} \geq \underline{S}, \tag{8b}$$

$$\hat{\pi}_1 = \hat{\pi}_1(\hat{H}) = \frac{\eta\hat{f}' + (1-\mu)\hat{f}\hat{W}_H}{(r-\hat{\Psi})\hat{f}}, \tag{8c}$$

$$\hat{\pi}_2 = \hat{\pi}_2(\hat{H}) = \frac{\eta\hat{f}}{d\hat{H}\hat{g}} > 0, \tag{8d}$$

[7] If $T(u) > \eta(cf(H)\pi_2 S)^{-1}$, banning tourism ($u = 0$) would be optimal. For example, if the level of visitor control exceeded its optimum value, no more damage to the reproduction process of the charismatic species $S$ could be tolerated, since not even fully exploiting the nature reserve today could offset tomorrow's loss for society. However, if $T(u) < \eta(cf(H)\pi_2)^{-1}$, the number of visitors would be below its welfare-maximising level.

where $\hat{g}: = g(\hat{H}), \hat{f}: = f(\hat{H}), \hat{f}': = f'(\hat{H}), \hat{V}': = V'(\hat{S})$ and $\hat{W}_H: = W_H(\hat{H}, \hat{S})$.

The term $\hat{\Psi}: = \Psi(\hat{H}, \hat{S})$ is determined by equation (A.6), in Appendix A.2, and relationship $T(u) = -\eta T'(u)/U'(u)$ is required to hold.

Thus, the equilibrium size of the eagle population ($\hat{S}$) can be determined by computing the equilibrium growth of the equilibrium habitat in relation to (per capita) eagle population's response to habitat supply (as determined by equation 8b). Furthermore, $\hat{S}$ is indirectly proportional to the increase in the equilibrium value of social welfare caused by a marginal increase in its population size (see equation 6). Both marginal welfare gains, $\hat{\pi}_1$ and $\hat{\pi}_2$, decrease with the discount rate $r$. This implies that in the long run *only* a farsighted society's marginal valuation for an additional unit of either of the ecosystem's components, $H$ or $S$, is rather large.

In uncontrolled real-world ecosystems, sometimes oscillating behaviour is observed. These oscillations are not threatening *per se*, but it is possible that large amplitudes can push the species population very close (or even below) critical levels, eventually causing extinction. However, for the optimally controlled model presented in this chapter, this behaviour can be excluded (see Appendix A.3).

Thus, both the short-term and the long-term survival of the flagship population are compatible with tourism, in as much as tourism is based on joint optimisation of benefits from tourism *and* conservation (given the appropriate state constraints). To put this result in a policy context, it is crucial to have a clear understanding of the current condition of the ecosystem and to determine the degree of endangerment (or, put more technically, the 'distance' between the current state of the bio-economic system and its long-run equilibrium). Only after monitoring species and habitat can the level of visitors allowed to enter the protected area be determined and this number should be adjusted in later periods, again conditional on the state of the protected area in these periods.

### 4.3 Equivalence of the ecosystem and the flagship approach

Based on Behrens and Friedl (2004) we can identify the following conditions for equivalence of the flagship and the ecosystem approach. Given (i) a separable social welfare function in recreation and conservation, (ii) a positive marginal social welfare from tourism, $U'(u) > 0$, and (iii) a strictly negative impact of an additional unit of tourism, $T'(0) < 0$, the respective functional relationships for determining the optimal visitor control strategy (as implicitly determined by equation 6) are identical for both approaches.[8]

The optimal visitor control strategy of both approaches in one way or another includes the state of the entire ecosystem. Therefore, the *steady state values* for $S$ and $H$ are of equal magnitude if the values of the marginal social welfare of an additional eagle coincide for the two approaches, $W_S(H, S) = V'(S)$, and the marginal social welfare derived from an incremental increase in the habitat, $W_H(H, S)$, is negligible.[9] This result cannot, however, be generalised for the transition paths to the long-run equilibrium. If the values of marginal social welfare for both an incremental increase in the population size of the charismatic species and its habitat are large, the flagship and the ecosystem approaches can differ substantially in their respective ways of approaching the bio-economic equilibrium. However, if the additional social welfare coming from a marginal change in the state of the ecosystem is rather small (i.e. both $W_H(H, S)$ and $V'(S) = W_S(H, S)$ are very small), the transition to the bio-economic equilibrium does not differ for the flagship and the ecosystem approaches. Accordingly, the equivalence of the flagship and the ecosystem approaches with respect to the transition phase is dependent on the deviation of the current state of the ecosystem from the bio-economic equilibrium. The closer the state of an ecosystem comes to its long-run equilibrium level, the more similarity can be observed between the two approaches to conservation.

The equivalence result hinges, however, on a further critical assumption: the charismatic species has to be on top of the food chain, as the following thought experiment might clarify. Let us, therefore, extend the two-stage (predator-prey-type) food chain (c.f. equations 2a and 2b) to a three-stage model and assume that the charismatic species $S$ (e.g. a snow hare) is not on top of the ecosystem's food chain. Assuming furthermore that the supra- (or apex) predator $\tilde{S}$ is a non-charismatic species but would be impaired by visitors, then $\tilde{S}$ would be a nuisance in terms of social welfare. That is, attracting visitors $u$ to irritate or even destroy the $\tilde{S}$ population would increase welfare because reducing $\tilde{S}$ additionally helps to increase $S$. Thus, the flagship and the ecosystem approach can be equivalent only (if at all) if the flagship species is as high as possible in

[8] Were the welfare function not separable in conservation and recreation the equivalence result holds only if the contribution of habitat to (marginal) welfare is negligible, i.e. $\bar{W}_u(H, S, u) = \bar{V}_u(S, u)$. In other words, the marginal increase in welfare caused by more visitors does not depend on the current state of $H$ (which in turn is identical to separability). For non-separable welfare functions, it cannot be argued that the flagship and the ecosystem approach are structurally equivalent.

[9] Note that this is a condition for marginal welfare effects (with respect to the species concerned, its habitat and the number of visitors) and does not require that the absolute values of welfare are identical for both approaches.

the food chain (or, at least, does not serve as a source of living for species very sensitive with respect to human intrusion).

## 5 A case study of the golden eagle in the eastern Alps

As outlined in the previous section the optimal visitor control strategy crucially and explicitly depends on the current state of the ecosystem, given in turn by the state of $H$ and $S$, and on the current valuation (expressed in terms of shadow prices) of the eagle population, $\pi_2 S$. To further illustrate the results derived in section 4, we discuss the balance of visitors and the protection of the golden eagle in the eastern Alps, which is, due to its charismatic appearance, a 'perfect' flagship animal for its avian habitat.

The Alps, considered one of the last remote areas left in Central Europe with an astonishingly large species variety, were identified as an 'eco-region' within the WWF Global 2000 initiative. Due to the long period of being under protection, the high altitude and also the high number of bird areas located there, Austria's Hohe Tauern mountain range was classified as a priority conservation area within the Alps. Among the birds breeding above 2000 metres altitude above sea level, the golden eagle (*Aquila chrysaetos*) and the bearded vulture (*Gypaetus harbatus*) are of particular importance (WWF Germany 2004).

Severely diminished by hunting at the end of the nineteenth century, nowadays the golden eagle exhibits a rather stable population size in the Eastern Alps (AQUILALP 2003). This success has been the result of coordinated international protection efforts. At present the only threat is habitat destruction (decline in potential breeding areas). Despite stabilised population numbers, the golden eagle is listed as a rare species according to the EU Birds Directive (Pedrini and Sergio 2002). This status is important considering the fact that the eagle preys on other important endangered species – reducing the number of eagles by 'wildlife management' is therefore excluded by law.

Apart from habitat loss, considerable human threat to the eagle population comes from visitors to the national park. All other types of human utilisation, such as agriculture and forestry, are restricted to the outer zone of the national park. In the inner zone, hiking visitors are the only ones allowed. So far, the most detrimental forms of tourism are prohibited within the inner zone, such as mountain biking, paragliding or ice climbing (allowed on particular sites only). In the future, however, these activities could be allowed, at least when user fees can be collected. Obviously, this visitor behaviour could also harm the golden eagle population, for instance during nesting periods (for a discussion of potential threats, see Pedrini and Sergio 2002).

The present case study for the protection of the golden eagle population in the Hohe Tauern National Park therefore serves two aims – first, to investigate the current situation of the golden eagle population relative to the long-run equilibrium level and second, to compare the ecosystem to the flagship approach.

### 5.1 *Assigning parameters to the model for the golden eagle in Austria*

The Hohe Tauern National Park (IUCN categories II and V) is a substantial spatially connected protected area located in the eastern Alps (covering more than 2 per cent of Austrian territory) and hosts a golden eagle population of 33–35 breeding pairs (AQUILALP 2003). Since the crucial life-limiting factor for the golden eagle is its territorial demand, $H$ represents the size of the area under protection. In 2001, the Hohe Tauern National Park was $H_0 = 1{,}784.38$ km$^2$ in size (Aubrecht and Petz 2002). Since the park lies in the three Austrian counties of Carinthia, Salzburg and Tyrol, the carrying capacity of the habitat was approximated by the area classified as important bird area by the EU Birds Directive (2,554 km$^2$, according to Dvorak and Karner 1995) in the respective counties. Considering the eagles' occurrence also outside this area, we multiply this area by 1.5, yielding a carrying capacity of $\omega = 3{,}831$ km$^2$. The golden eagle population in the Eastern Alps has been monitored on a yearly basis since 2003, resulting in an estimate of 33–35 breeding pairs within the park area (AQUILALP 2003), thus we set $S_0 = 68$. According to a visitor count in 2003 (Lehar 2004), 1.74 million visitors come to the national park annually, yielding $u_0 = 1{,}740{,}100$.

The dynamics of the habitat are approximated by the change in size of the park area, which has increased steadily over the years. For an annual area increase of 32.12 km$^2$ ($\equiv g(H_0)H_0 - f(H_0)S_0$, Aubrecht and Petz 2002) and a territorial requirement of a minimum 15 km$^2$ per breeding adult ($\equiv f(H_0)$, AQUILALP 2003), the natural growth rate of habitat can be estimated as $g(H_0) = 0.58963$. This number, together with $H_0$ and $\omega$, can then be used to calibrate the natural growth factor in the logistic growth function as $a = 1.1037$. Assuming that present visitors reduce the breeding success by 10 per cent (i.e. $e^{-mu_0} = 0.9$) gives a parameter value for the tourism impact factor of $m = 6.05485 \cdot 10^{-8}$. The 'predator-response function', $f(H)$, can be calibrated against the habitat breeding requirement of 15 km$^2$, then $b_0 H_0/(1 + H_0) \equiv 15$ yields $b = 15.0084$. Finally, the eagle population is currently approximately stable, such that it grows at a very low rate (Pedrini and Sergio 2002), i.e. $ce^{-mu_0} f(H_0) \cong 0.001$ and $c = 7.44767 \cdot 10^{-5}$. Table 19.1 contains the parameter values

Table 19.1 *Base case parameter values for the flagship approach*

| Parameter | Value | Description |
|---|---|---|
| $a$ | 1.1037 | natural growth rate of the habitat |
| $b$ | 15.0084 | decay rate of the habitat (implicitly an upper limit on the eagle population's daily consumption) |
| $c$ | $7.4 \cdot 10^{-5}$ | golden eagle's prey conversion rate in the national park |
| $d$ | 0.001 | golden eagle's mortality rate in the national park |
| $q$ | 1 | half-saturation constant |
| $m$ | $6.05 \cdot 10^{-8}$ | visitor impact factor |
| $\omega$ | 3,831 | carrying capacity of the habitat, size in square kilometres |
| $\mu$ | 0 | ecosystem approach |
| | 1 | flagship approach |
| $\eta$ | 1.65 | weighting factor for tourism in the social welfare function |
| $\nu_1$ | 100 | weighting factor for the golden eagle in the social welfare function (species diversity) |
| $\nu_2$ | 1 | weighting factor for the habitat in the social welfare function (ecosystem diversity) |
| $r$ | 0.04 | discount rate |

of the bio-economic system for the golden eagle population in the Hohe Tauern National Park.

### 5.2 *Specifying the welfare functions for the flagship and the ecosystem approaches*

Merely for illustrative purposes we assume that the welfare function is also additively separable in the contributions of the eagle population and its habitat (equation 1), i.e.

$$U(u) + \mu V(S) + (1-\mu) W(H,S) \tag{9}$$

where $U(u) = \eta m u$, $V(S) = \nu_1 \ln(1+S)$, $W(H,S) = V(S) + \nu_2 \ln(1+H)$ and parameters $\nu_1$ and $\nu_2$ reflect relative scarcity of species and of habitat (degree of endangerment), respectively.[10] All relative weights are estimated such that the sub-functions initially generate benefits of 'equal magnitude', i.e. for $t = 0$ we control for all differences in size (by placing a large weight for the eagle, a small weight for the habitat and an even smaller weight for the number of visitors, see Table 19.1).

[10] Note that the separability of the welfare function with respect to recreation and conservation is crucial for the results presented here. The separability of welfare with respect to the ecosystem's components ($H$ and $S$) is, however, exclusively assumed for illustrative purposes and has no influence on the results presented in section 4.

Moreover, the social discount rate $r$ is assumed to be constant over time and set arbitrarily at 4 per cent.

Thus, the welfare function for the ecosystem approach ($\mu = 0$) corresponds to

$$U(u) + W(H,S) := U(u) + \nu_1 \ln(1+S) + \nu_2 \ln(1+H) \qquad (10)$$

Given this specification of welfare the flagship approach can be regarded as a special case of the ecosystem approach, if $\nu_1 > 0$ and $\nu_2 = 0$. Thus, starting from the flagship approach ($\nu_2 = 0$) we can gain an intuitive understanding of switching from the flagship to the ecosystem approach by doing nothing other than gradually increasing the parameter $\nu_2$ (see section 5.3).

### 5.3 *Optimal visitor control strategies for the flagship and the ecosystem approaches*

We know from section 4 that it is essential to determine the current state of the golden eagle population (in our case study of Austria's Hohe Tauern National Park) relative to its potential long-run development. Thus, we start by computing the bio-economic equilibrium according to equations (8a)–(8d) for the base case parameter values (Table 19.1). The intertemporal paths for optimal states, costates and control are calculated by backward integration from the bio-economic equilibrium in Mathematica 5.1 (Wolfram Research Ltd.).

The equilibrium for the flagship approach ($\nu_2 = 0$) and the ecosystem approach ($\nu_2 > 0$; arbitrarily chosen as $\nu_2 = 1$) is determined by

$$\hat{E}_{\nu_2=0} := \begin{pmatrix} \hat{H} \\ \hat{S} \\ \hat{\pi}_1 \\ \hat{\pi}_2 \\ \hat{u} \end{pmatrix} \cong \begin{pmatrix} 1{,}846 \\ 70 \\ 0.03 \\ 23.5 \\ 1.83 \cdot 10^6 \end{pmatrix},$$

$$\hat{E}_{\nu_2=1} := \begin{pmatrix} \hat{H} \\ \hat{S} \\ \hat{\pi}_1 \\ \hat{\pi}_2 \\ \hat{u} \end{pmatrix} \cong \begin{pmatrix} 1{,}876 \\ 70 \\ 0.03 \\ 23.5 \\ 1.83 \cdot 10^6 \end{pmatrix} \qquad (11)$$

Comparing, first, the steady-state values of $H$, $S$ and $u$ with the ones observed in 2003 (= initial conditions), only very small differences in the state of the ecosystem can be noticed. For the *flagship approach* the differences between the 2003 data and the steady-state values are literally negligible. The corresponding optimal time paths for the size of the

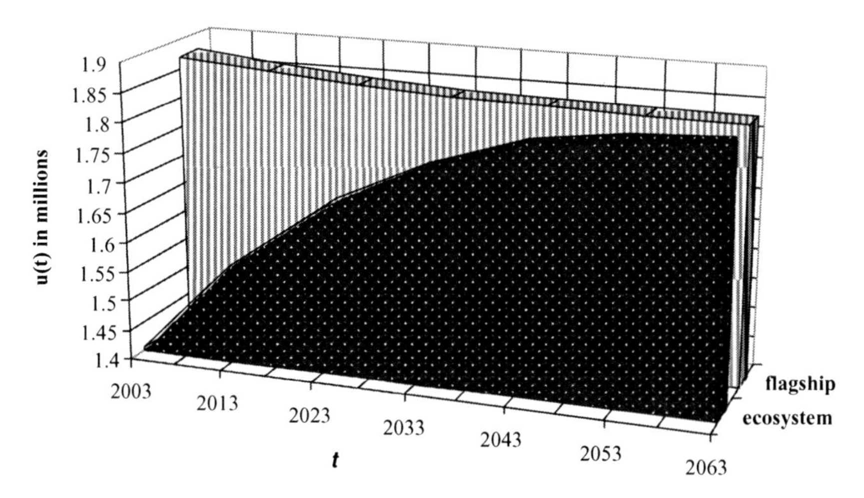


Figure 19.2. The optimal dynamic visitor control strategy for the flagship approach and the ecosystem approach with low weight on the habitat ($\nu_2 = 1$)

habitat ($H$) and of the eagle population ($S$) are both marginally decreasing towards the steady state (paths not shown here), as does the path of the optimal visitor control strategy (Figure 19.2).

For the *ecosystem approach*, the habitat increases by 1.6 per cent from $H_0$=1,846 km$^2$ to $\hat{H} = 1{,}876\,\text{km}^2$, while the golden eagle population stays (more or less) constant. This confirms that under current conditions the golden eagle population in the Hohe Tauern National Park is not threatened by extinction and neither is its habitat.

For obvious reasons it is possible to have total equivalence in the bio-economic equilibrium between the flagship and the ecosystem approaches to conservation as long as the marginal valuation for the eagle is virtually the same for both approaches and as long as the valuation of the habitat is quite low ($\nu_2 \approx 0$). By increasing the weighting factor $\nu_2$, the influence of the habitat on welfare increases. The same applies to the equilibrium size of the habitat (see top of Figure 19.3) and, for obvious reasons, the flagship and the ecosystem approaches start to diverge if $\nu_2$ starts to increase.

With the importance of habitat in contributing to social welfare increasing (=larger $\nu_2$), the equilibrium habitat size increases and so does the prey availability for the eagle. Thus, the steady-state level of eagles is increasing with $\nu_2$, as depicted by Figure 19.3. After surpassing a critical value (i.e. at $\nu_2 = 4$), the social welfare associated with the habitat gains, however, more importance than the social welfare from the eagle population and therefore the equilibrium eagle population subsequently

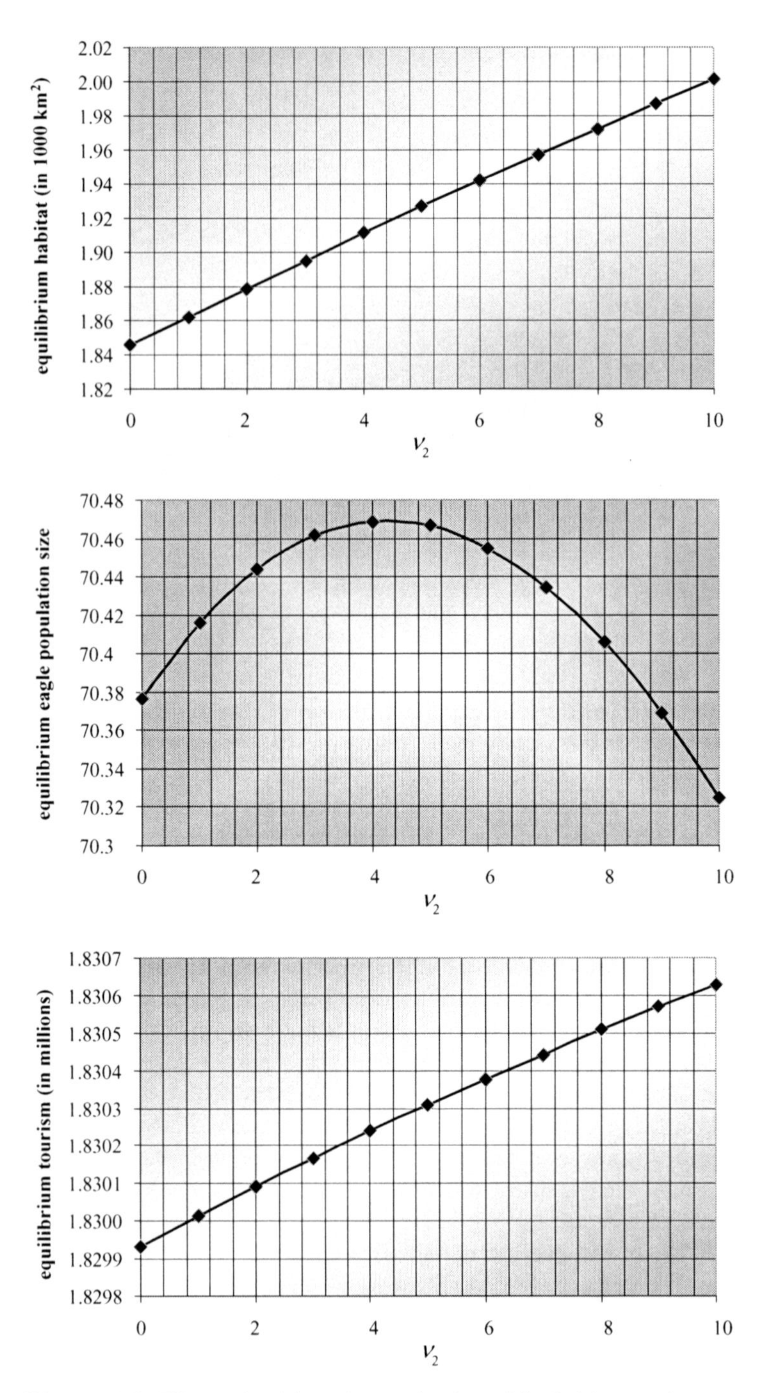


Figure 19.3. The optimal (steady-state) value of the habitat, eagle population and number of visitors for different values of $\nu_2$

decreases with a growing value of $\nu_2$. Furthermore, it is the growing number of visitors which is responsible for the decline in the eagle population. Note that the number of visitors is strictly increasing in $\nu_2$ because the optimal visitor control strategy is mainly driven by the eagle population's response to habitat supply and is, therefore, inherently and positively dependent on $H$ (c.f. equation 6). As the scaling factor $\nu_2$ further increases (in Figure 19.3) the eagle does not die out, since the optimal visitor control strategy is immediately forced to be zero as the minimum viable population size is approached.

Making the implicit valuation of the eagle population (as measured by the shadow price $\pi_2$) responsible for the surprising behaviour of $S$ would be grossly misleading. If the shadow price for one more unit of $S$ rose, $S$ would shrink and vice versa. Thus, $S$ and $\pi_2$ behave antagonistically and the joint value of the eagle population, $\pi_2 S$, has the tendency to remain fairly constant.

As given by equation (11) and Figure 19.2, the values of the long-run bio-economic equilibrium are very similar for the flagship and the ecosystem approaches. For reasons discussed in section 4.3, the transients deviate significantly, even though the respective optimal visitor control strategies do not differ structurally at all. For $t = 0$, i.e. in the year 2003, the optimal upper limit on the number of visitors is much smaller for the more comprehensive ecosystem approach than for the flagship approach, despite identical initial values for $H$ and $S$.

This might lead, however, to additional arguments in favour of the flagship approach. If the park management wants to increase the number of visitors in order to benefit from spillover effects from tourism (revenues) on conservation programmes, the flagship approach might be preferred. This compares to the argument of 'marketing' nature conservation policies which is a task much more easily achieved by promoting the survival of a charismatic species than the ecological improvement of a habitat. Moreover, for our case study the flagship approach leads to a fairly stable upper limit on visitors, implying that the flagship approach can be operated and enforced more easily than the ecosystem approach which requires gradual adjustments over time (see Figure 19.2).

## 6 Conclusions

This chapter has tried to shed additional light on the question of how a social planner can optimally manage visitor access to a protected area over time when society derives welfare from both recreation and conservation. To accomplish this goal, we introduced a dynamic quantity restriction on the number of visitors allowed to enter a national park. The economic rationale for such a quantity restriction, especially in comparison with a pricing strategy, is the following. First of all, the number of visitors can be monitored and enforced at park access points whenever the ecosystem is in danger. Second, visitor restrictions treat all visitors equally, irrespective of their income (as opposed to a pricing strategy). Third, the response of visitors to entrance fees is controversial: while fees reduce the number of visitors, the average stay per visitor increases, leading to quite low overall demand reductions (Alden 1997; Boxall *et al.* 2003).

Moreover, we were able to show that the flagship approach to conservation is sufficient to replace an ecosystem approach under certain conditions. While we agree that these conditions might not always be met in reality, we can think of particular cases fulfilling them, e.g. for highly charismatic species selected as campaign animals. Those animals are most often at the top of the food chain and, accordingly, the average citizen attaches high welfare to the existence of the species in comparison with the habitat. For this class of species, separability of recreation and conservation values is a reasonable assumption, since the majority of people value the mere existence of the species without ever being able to see them in their natural habitat.

The equivalence result between the two approaches to conservation is important because in recent years NGOs such as the World Wildlife Fund for Nature have increasingly targeted their protection campaigns at specific charismatic species (like the giant panda, the monk seal, or, for Austria, the bearded vulture) instead of at entire terrains (constituting the habitats of the respective species). More and more, this approach to conservation partially replaces the ecosystem approach because these species have the advantage of being easily recognised in picture and name as opposed to a vast number of lesser known ones (also) threatened by extinction – and these characteristics have a positive impact on people's willingness to financially contribute to conservation in general.

Another central result for both the ecosystem and the flagship approach is that an optimal dynamic visitor control strategy guarantees that neither the charismatic species nor its habitat can be driven to extinction (by operating a zero-visitor policy if necessary). The habitat is not threatened by being overpopulated and severely damaged, because it is either directly included in the society's objective (for the ecosystem approach) or serves as a carrying capacity for the charismatic species and is therefore indirectly included in the objective (for the flagship approach).

A caveat should be appended, however. Park administrations might want to pursue a flagship approach but must not, at the same time, forget about habitat protection by, for example, increasing visitor numbers beyond the limit suggested by the 'optimal path' given in section 4. It could be tempting in times of tight public budgets and park administrations' need to raise conservation budgets to increase the number of visitors und thus generate short-term revenues at the expense of long-term ecological damages. Under real-world conditions, therefore, the compatibility of tourism and conservation cannot be assured in general.

## REFERENCES


Alden, D. 1997. Recreational user management of parks: an ecological economic framework. *Ecological Economics*. **23**. 225–236.

AQUILALP 2003. AUQILALP.NET newsletter 2003. Der Steinadler in den Ostalpen (The Golden Eagle in the Eastern Alps). http://www.aquilalp.net

Aubrecht, P. and Petz, K. 2002. Naturschutzfachlich bedeutende Gebiete in Österreich, Eine Übersicht. *Monographien*. **134**. Vienna: Umweltbundesamt.

Begon, M., Harper, J. L., Townsend, C. R. 1996. *Ecology. Individuals, Populations and Communities*. 3rd ed. Malden, Oxford: Blackwell.

Behrens, D. A., Friedl, B. 2004. Is protecting the charismatic species really enough? Tourism strategies to optimally balance recreation and conservation in a predator-prey system. *ORDYS Discussion Paper No. 289*. Vienna: Vienna University of Technology.

Boxall, P., Rollins, K. and Englin, J. 2003. Heterogeneous preferences for congestion during a wilderness experience. *Resource and Energy Economics*. **25**. 177–195.

Bulte, E. H. and van Kooten, G. C. 2001. Harvesting and conserving a species when numbers are low: population viability and gambler's ruin in bioeconomic models. *Ecological Economics*. **37**. 87–100.

Clark, C. W. 1990. *Mathematical Bioeconomics. The Optimal Management of Renewable Resources*. 2nd ed. New York: Wiley.

Conrad, J. M. and Adu-Asamoah, R. 1986. Single and multi-species systems: the program of tuna in the Eastern Tropical Atlantic. *Journal of Environmental Economics and Management*. **13**. 234–244.

Conrad, J. M. and Salas, G. 1993. Economic strategies for coevolution: timber and butterflies in Mexico. *Land Economics*. **69** (4). 404–415.

Crépin, A. S. 2003. Multiple species forests – what Faustmann missed. *Environmental and Resource Economics*. **26**. 625–646.

Dawson, D. and Shogren, J. F. 2001. An update on priorities and expenditures under the endangered species act. *Land Economics*. 77 (4). 527–532.

Dockner, E. and Feichtinger, G. 1991. On the optimality of limit cycles in dynamic economic systems. *Journal of Economics*. **53** (1). 31–50.

Duffus, D. A. and Dearden, P. 1990. Non-consumptive wildlife-oriented recreation: a conceptual framework. *Biological Conservation*. **53**. 213–231.

Dvorak, M. and Karner, E. 1995. Important bird areas in Austria. *Monographien*. **71**. Vienna: Umweltbundesamt.

Feenstra, T., Cesar, H. and Kort, P. 1999. Optimal control theory in environmental economics, in J. C. J. M. van den Bergh (ed.). *Handbook of Environmental and Resource Economics*. Cheltenham: Edward Elgar.

Feichtinger, G. and Hartl, R. 1986. *Optimale Kontrolle ökonomischer Prozesse – Anwendung des Maximumprinzips in den Wirtschaftswissenschaften*. Berlin: de Gruyter.

Finnoff, D. and Tschirhart, J. 2003. Harvesting in an eight-species ecosystem. *Journal of Environmental Economics and Management*. **45**. 589–611.

Flaaten, O. 1991. Bioeconomics of Sustainable Harvests of Competing Species. *Journal of Environmental Economics and Management*. **20**. 163–180.

Groombridge, B. (ed.). 1992. *Global Biodiversity. Status of the Earth's Living Resources*. London, Glasgow, New York: Chapman & Hall.

Grubb, T. G. and King, R. M. 1991. Assessing human disturbance of breeding bald eagles with classification tree models. *Journal of Wildlife Management*. **55**. 500–511.

Hannesson, R. 1983. Optimal harvesting of ecological interdependent fish species. *Journal of Environmental Economics and Management*. **10**. 329–345.

Hoekstra, J. and van den Bergh, J. C. J. M. 2005. Harvesting and conservation in a predator-prey system. *Journal of Economic Dynamics and Control*. **29** (6). 1097–1120.

Holling, C. S. 1959. The components of predation as revealed by a study of small-mammal predation of the European Pine Sawfly. *The Canadian Entomologist*. **151**. 293–320.

IUCN (World Conservation Union) 2004. The red list of threatened species. http://www.redlist.org

Kamien, M. I. and Schwartz, N. L. 1991. *Dynamic Optimization, the Calculus of Variations and Optimal Control in Economics and Management*. Amsterdam: North-Holland.

Kontoleon, A. and Swanson, T. 2003. The willingness to pay for property rights for the Giant Panda: can a charismatic species be an instrument for nature conservation? *Land Economics*. **79** (4). 483–499.

Leader-Williams, N. and Dublin, H. T. 2000. Charismatic megafauna as flagship species. In A. Entwistle and N. Dunstone (eds.). *Priorities for the Conservation of Mammalian Diversity: Has the Panda Had its Day*? Cambridge: Cambridge University Press.

Lehar, G. 2004. *Besucherzählung, Motiv- und Wertschöpfungserhebung im Nationalpark Hohe Tauern*. University of Innsbruck: Institut für Verkehr und Tourismus.

Léonard, D. and Long, N. V. 1992. *Optimal Control Theory and Static Optimization in Economics*. Cambridge: Cambridge University Press.

Loomis, J. B. and White, D. S. 1996. Economic benefits of rare and endangered species: summary and meta-analysis. *Ecological Economics*. **18**. 197–206.

Metrick, A. and Weitzman, M. L. 1996. Patterns of behavior in endangered species preservation. *Land Economics*. **72** (1). 1–16.

Metrick, A. and Weitzman, M. L. 1998. Conflicts and choices in biodiversity conservation. *Journal of Economic Perspectives*. **12**. 21–34.

MEA (Millenium Ecosystem Assessment) 2005. *Ecosystems and Human Well-being: Biodiversity Synthesis*. Washington, DC: World Resources Institute.

Myerscough, M. R., Darwen, M. J. and Hogarth, W. L. 1996. Stability, persistence and structural stability in a classical predator-prey model. *Ecological Modelling*. **89**. 31–42.

Nunes, P. A. L. D., van den Bergh, J. C. J. M. and Nijkamp, P. 2003. *The Ecological Economics of Biodiversity – Methods and Policy Applications*. Cheltenham: Edward Elgar.

Pedrini, P. and Sergio, F. 2002. Regional conservation priorities for a large predator: golden eagles (Aquila chrysaetos) in the Alpine range. *Biological Conservation*. **103**. 163–172.

Prins, H. H. T. and Grootenhuis, J. G. 2000. Introduction: The value of priceless wildlife. In H. H. T. Prins, J. G. Grootenhuis and T. T. Dolan (eds.). *Wildlife Conservation by Sustainable Use*. Dordrecht: Kluwer.

Ragozin, D. L. and Brown, G. J. 1985. Harvest policies and nonmarket valuation in a predator-prey system. *Journal of Environmental Economics and Management*. **12**. 155–168.

Regev, U., Gutierrez, A. P., Schreiber, S. J. and Zilberman, D. 1998. Biological and economic foundations of renewable resource exploitation. *Ecological Economics*. **26**. 227–242.

Ricklefs, R. E. 1990. *Ecology*. 3rd ed. New York: Freeman.

Rondeau, D. 2001. Along the way back from the brink. *Journal of Environmental Economics and Management*. **42**. 156–182.

Rosenzweig, M. L. and MacArthur, R. H. 1963. Graphical representation and stability conditions of predator-prey interactions. *The American Naturalist*. **162**. 209–223.

Semmler, W. and Sievekig, M. 1994. On the optimal exploitation of interacting resources. *Journal of Economics*. **59** (1). 23–49.

Simberloff, D. 1998. Flagships, umbrellas, and keystones: is single-species management passé in the landscape era? *Biological Conservation*. **83** (3). 247–257.

Skonhoft, A. 2007. Modeling the re-colonisation of native species. In A. Kontoleon, U. Pascual and T. Swanson (eds.). *Biodiversity Economics*. Cambridge: Cambridge University Press.

Stork, N. E. 1997. Measuring global biodiversity and its decline. In M. L. Reaka-Kudla, D. E. Wilson and E. O. Wilson (eds.). *Biodiversity II: Understanding and Protecting Our Biological Resources*. Washington, DC: Joseph Henry Press.

Ströbele J. and Wacker, H. 1995. The economics of harvesting predator-prey systems. *Journal of Economics*. **61** (1). 65–81.

Tschirhart, J. 2000. General equilibrium of an ecosystem. *Journal of Theoretical Biology*. **203**. 13–32.

UNEP (United Nations Environment Program) 1992. *Convention on Biological Diversity*. Nairobi, Kenya: UNEP.

Wilson, E. O. 1992. *The Diversity of Life*. Harvard, MA: Harvard University Press.

WWF Germany. 2004. *The Alps: A Unique Cultural Heritage. A Common Vision for the Conservation of their Biodiversity*. Frankfurt am Main: WWF.

## Appendix

### *A.1 Concavity of control with respect to the Hamiltonian function*

According to Feichtinger and Hartl (1986), the optimal control policy is uniquely determined at any instant of time if the second-order derivative of the Hamiltonian function $\mathcal{H}$ with respect to control is negative, i.e.

$$\mathcal{H}_{uu} = c \underbrace{\pi_2}_{>0} f(H) S \underbrace{T''(u)}_{>0} + \underbrace{U''(u)}_{\leq 0} < 0 \tag{A.1}$$
$$\Leftrightarrow \quad c\pi_2 f(H) S T''(u) < |U''(u)|.$$

According to Feichtinger and Hartl (1986), a sufficient condition for the Hamiltonian $\mathcal{H}$ to be strictly concave with respect to both states ($H$ and $S$) and control ($u$) is that the Hessian of $\mathcal{H}$ is negative definite. Regrettably, for arbitrary values of $H$, $S$ and $u$ we cannot *prove* that $\mathcal{H}$ is jointly concave with respect to both states and control. Thus, we cannot generally state that the necessary conditions are sufficient but *give conditions* for sufficiency. One such condition is that the reproduction rate of the eagles diminished by the damage caused by tourism has to exceed the ratio of the shadow prices of habitat versus eagles, i.e. $cT(u)\ \pi_2 > \pi_1$. The entire set of these conditions is not discussed here, however.

### *A.2 The canonical system*

The links between the evolution of optimal tourism $u^*$ defined according to equation (7), the optimal development of the ecosystem's state and valuation (over time) can be characterised by the canonical system of the optimal control problem (given by equations 1–4) in the state/costate-space:

$$\dot{H} = g(H)H - f(H)S, \tag{A.2}$$
$$\dot{S} = \eta/\pi_2 - dS, \tag{A.3}$$
$$\dot{\pi}_1 = \pi_1[r - \Psi(H,S)] - \eta f'(H)/f(H) - (1-\mu) W_H(H,S), \tag{A.4}$$
$$\dot{\pi}_2 = \pi_1 f(H) + \pi_2[r+d] - \eta/S - \mu V'(S) - (1-\mu) W_S(H,S), \tag{A.5}$$
$$\text{where } \Psi(H,S) := g(H) + g'(H)H - f'(H)S. \tag{A.6}$$

For the derivation of the transversality conditions for the shadow prices, see Behrens and Friedl (2004).

Table 19.A1 *Classification of equilibria according to K and D > 0*

| Conditions | Equilibrium of the linearised system |
|---|---|
| $K < 0$<br>$0 > D \leq (K/2)^2$ | saddlepoint stability (due to real eigenvalues, where two are negative and two are positive); monotonous approach towards the equilibrium located on a two-dimensional manifold |
| $D > (K/2)^2$<br>$D > (K/2)^2 + r^2(K/2)$ | saddlepoint stability (due to two pairs of conjugate complex eigenvalues, one pair having negative real parts); dampened oscillatory approach towards the equilibrium located on a two-dimensional stable manifold |
| $K > 0, D > 0$<br>$D = (K/2)^2 + r^2(K/2)$ | Hopf bifurcation (due to the occurrence of two purely imaginary eigenvalues) |
| $D > (K/2)^2$<br>$D > (K/2)^2 + r^2(K/2)$ | locally unstable spiral (due to two pairs of conjugate complex eigenvalues with positive real parts) |

### *A.3 Equilibrium properties*

Determining the eigenvalues of the Jacobian matrix evaluated at the equilibrium given by equations (8a)–(8d),

$$\hat{\Im} \equiv \begin{pmatrix} \hat{\Psi} & -\hat{f} & 0 & 0 \\ 0 & -d & 0 & -\eta/\hat{\pi}_2^2 \\ \hat{\Phi} - (1-\mu)\hat{W}_{HH} & \hat{\pi}_1 \hat{f}' + (1-\mu)\hat{W}_{HS} & r - \hat{\Psi} & 0 \\ \hat{\pi}_1 \hat{f}' + (1-\mu)\hat{W}_{HS} & \eta/\hat{S}^2 - \mu\hat{V}'' - \hat{\pi}_1 \hat{f}' + (1-\mu)\hat{W}_{SS} & \hat{f} & r+d \end{pmatrix}, \tag{A.7}$$

delivers all necessary information to fully characterise the local stability properties of the long-run bio-economic equilibrium state, where

$$\hat{\Phi} = \Phi(\hat{H}) := -(2\hat{g}' + \hat{g}''\hat{H} - \hat{S}\hat{f}'')\hat{\pi}_1 + \frac{\eta}{\hat{f}^2}(\hat{f}' - \hat{f}\hat{f}''), \tag{A.8}$$

$\hat{f} := f(\hat{S})$, $\hat{f}' := f'(\hat{S})$, $\hat{f}'' := f''(\hat{S})$, $\hat{V}'' := V''(\hat{S})$, $\hat{W}_{HH} := W_{HH}(\hat{S}, \hat{S})$, $\hat{W}_{HS} := W_{HS}(\hat{S}, \hat{S})$, $\hat{W}_{SS} := W_{SS}(\hat{S}, \hat{S})$ and $\hat{\Psi} := \Psi(\hat{S}, \hat{S})$ is given by equation (A.6) and the equilibrium states $\hat{S}, \hat{\pi}_1$ and $\hat{\pi}_2$ by equations (8b)–(8d), respectively. We are able to categorise the local stability properties of the long-run equilibrium state as given by Table 19A.1, according to the indicators $K$ and $D$ as defined by Dockner and Feichtinger (1991):

$$K := (r - \hat{\Psi})\hat{\Psi} - rd - \frac{(d\hat{S})^2}{\eta}(\mu\hat{V}'' + (1-\mu)\hat{W}_{SS}) \tag{A.9}$$

$$D := |\hat{\Im}| > 0. \tag{A.10}$$

A parameter set satisfying $K > 0$ and $\hat{\Im} = (K/2)^2 + r^2(K/2)$ leads to persistent oscillation (which can also occur for standard predator-prey systems in absence of tourism as outlined by, for example, Myerscough *et al.* 1996). Thus, an appropriate choice of the discount rate $r$ could imply that persistent fluctuations of the state of the ecosystem (and of the visitor control policy) would be regarded as optimal. This behaviour is well known in the literature. However, for the optimally controlled model presented in section 4, this behaviour can be excluded for any parameter set satisfying $K < 0$.

# 20 Modelling the recolonisation of native species

*Anders Skonhoft*

## 1 Introduction

Recolonisation of native species typically represents an institutional change and reflects society's changing attitude to the species cost and benefit streams. When successful, recolonisation often influences the ecology and may come into conflict with existing economic activity. Such conflict may be particularly controversial and severe when the recolonised species are large carnivores, like wolves and grizzlies, which kill livestock and prey species with hunting and meat values. Recolonised animals may also induce conflicts with existing economic activities, like agriculture, including eating up crops and pastures and causing browsing damage. However, recolonised native species may also create hunting and trapping value or other types of consumptive values, in addition to non-consumptive values like existence value, tourist value and so forth (see Freeman 2003 for a general overview and Nunes and van den Bergh 2001 for a critical discussion of species valuation). In addition to ecology, these cost and benefit components and wildlife conflicts depend on the economic and institutional setting and there are obvious differences between, say, an East African region where people are located near wildlife with living conditions closely related to agricultural activities and, say, a region in Europe or North America where most people experience wildlife only through non-consumptive uses (Swanson 1994). The management goal will also generally differ. For these and other reasons it may seem difficult to formulate a general analytical model for studying economic impacts of species recolonisation. Nevertheless, this is actually what this chapter will attempt to do. Within such a general framework, however, several cases associated with specific economic and ecological circumstances will be considered. In the last part of the chapter, a more detailed example is studied.

Acknowledgements: Thanks to Anne B. Johannesen, Eric Naevdal, Jon Olaf Olaussen and one of the editors, Unai Pascual, for constructive criticism on an earlier draft of this chapter.

There is a difference between native recolonisation and reintroduction. While reintroduction is a man-made 'attempt to establish a species in an area which was once part of its historical range, but from which it has been extirpated or become extinct' (IUCN 1995, p. 2), recolonisation represents a species establishment in a historical habitat without *direct* man-made interventions. However, it seems difficult to make a clear-cut distinction as humans in many, if not all, instances, at least indirectly, influence recolonisations. This may typically occur when previous harvesting practices are banned, or when, say, previous production practices in agriculture and forestry influencing habitat conversion and species growth are changed. The recolonisation of the grey wolf (*Canis lupus*) in Scandinavia in the 1970s, to be considered below, is an example of recolonisation due to banning of previous harvesting practices. However, the difference between recolonisation (and reintroduction) and the existence of invasive species is clearer as invasive species represent an introduction of non-native species that generally alters an ecological system in a negative fashion and hence is an economic bad (Perrings 2007).

A recent well-known example of species reintroduction is the grey wolf in the Yellowstone National Park, North America. The first reintroduction took place in 1995 with a few wolves, followed by an additional reintroduction in 1996. The introduction was opposed by local ranchers who feared that wolves would prey on their livestock and by hunters who feared that wolves would compete with them for game. So far the wolf recovery seems to be a success story and the number of visitations to parts of the park where wolves are frequently seen has increased. However, some attacks on domestic sheep have been reported and the wolves have reduced the moose and bison population (e.g. Boyce 1997). The grey wolf has also recolonised Scandinavia during the last few decades (Wabakken *et al.* 2001). While small in number, this recolonisation has also caused several conflicts. Another example is the European bison (*Bison bonasus*) which now is found within its previous territory in Ukraine after its numbers dwindled as a result of overexploitation and agricultural expansion a long time ago. The reintroduction started in 1965 (Perzanowski *et al.* 2004) and so far the programme has resulted in eleven scattered herds, numbering about fifty animals. There seem to be few conflicts related to this reintroduction. Yet other examples are the translocations of the black-faced impala in Namibia which started in the early 1970s (Matson *et al.* 2004) and the lynx introduction in Steury and Murray (2004). In the journal *Biological Conservation*, several other recolonisations and reintroductions are reported, while Graham *et al.* (2005) provide a general overview of the various arising conflicts, making a distinction between predator–livestock conflicts and predator–game conflicts.

In what follows, we basically examine species recolonisation. However, due to the somewhat unclear terminology and lack of precise definitions, the terms recolonisation and reintroduction will often be used synonymously. Previous economic analysis of recolonisation is scarce and only a few references are reported when using the term reintroduction in *Econlit* (no citations are reported when using the term recolonisation). A key paper is that of Rondeau (2001). He formulates an optimal control model aiming to analyse the reintroduction of a white-tailed deer population with numerical examples from the USA. In this work the shadow price of the reintroduced species may be either positive or negative, depending on the cost and benefit structure as well as the biological growth conditions. Rondeau's (2001) study hence has similarities with the recent bioeconomic literature where species may be valuable but also a pest (e.g. Huffaker *et al.* 1992; Zivin *et al.* 2000; Horan and Bulte 2004; Skonhoft and Schulz 2005), but it offers a more in-depth dynamic analysis than these other papers. One reason for this, which is a special feature of the Rondeau model, is that introduction of species is explicitly considered in the population growth model; that is, the stock may grow according to new species introduced from outside areas in addition to natural growth.

In contrast to Rondeau (2001), we consider an area with no further introduction. We are also looking away from possible dispersal, or migration, from outside areas. This approach is very similar to previous studies of the cost of invasive species (Perrings 2007) where the so-called *ex ante* net benefit (the scenario without invasive species) is compared with the *ex post* benefit (the scenario with invasive species). The various cost and benefit streams related to recolonised species, as experienced by different agents, or groups of people, will be considered within a unified management scheme. It is assumed that a benevolent social planner maximises the present-value social surplus. This can hence also include values experienced outside the given area that typically may include existence values. We continue this chapter with a formulation of a general bio-economic model of species recolonisation. In section 3, various special ecological and economic cases of this general model are considered. Then in section 4 one of these versions is analysed in more depth, focusing on the recolonisation of the grey wolf in Scandinavia. This section also contains a numerical illustration. The last section concludes with a summary of the main findings and gives some policy implications. The general conclusions are that some control of the recolonised species often pays off. However, recolonised species should be kept uncontrolled when they (i) do small damage, (ii) are expensive to control and (iii) prey upon existing species that cause various types of damage, like browsing or grazing damage. Based on the grey wolf example, a general conclusion is also that the effects of economic forces often are difficult to predict when operating in an interspecies relationship.

## 2 Ecological interaction and the various cost and benefit streams

Typically, recolonisation of a native species results from banning of previous hunting practices. In some instances, this reflects that an earlier considered pest species is recognised to carry positive non-consumptive values (viewing value, existence value and so forth). The above examples of recolonisation of large carnivores are of this type. At a later stage when the recolonised species has reached a 'sustainable' stock level, it may also be considered as valuable for harvesting, or trapping. Alternatively, if the actual species does not carry any harvesting value and it is socially desirable to control the species abundance, species control will incur costs. Depending on ecological, economic and institutional circumstances, recolonisation may also cause serious conflicts and damage to existing economic activities, like preying upon livestock (cf. the above examples). Similarly, in the case of the recolonised species being a grazer, the agricultural damage may include eating up crops and pastures. But damage can also be channelled through the ecological system with existing wildlife and may be of the predator–prey type, or of the competitive type. When the existing species has hunting, or trapping, value, this value is then potentially reduced through recolonisation. The wolf–moose example considered below is of this type. The already existing wildlife species may also have positive non-consumptive values, which potentially are reduced through the recolonisation as well. However, if the existing species cause grazing or browsing damage, reintroduced species may potentially reduce such damage, as will be shown with an example below.

We start by formulating the ecological sub-model. The population growth of the recolonised species, $W$, measured in biomass, or 'normalised' number of animals, is generally given as

$$dW/dt = G(W, X) - y \tag{1}$$

with $G(..)$ as the natural growth function and where one stock $X$ (also measured in biomass) represents the existing wildlife affected by the recolonised species. Harvesting, $y$, is the control variable. As indicated, no dispersal term is included in equation (1) as we study situations where the recolonised species is established in the area and there is no further inflow from outside areas. In addition, and in contrast to Rondeau (2001), we assume $y \geq 0$, indicating that any direct man-made effort to reintroduce species is neglected as well. Accordingly, only natural growth in the given area together with possible control measures governs the population growth of the recolonised species. $\partial G/\partial X = G_X$ may be either positive (predator-prey relationship and where the reintroduced species is the predator) or negative (competitive relationship, or the reintroduced species is the prey). $G_X$ can also be zero, or close to zero, which typically happens if the recolonised species is of the opportunistic type; that is, the food intake may be grass as well as different sources of meat. The brown bear (*Ursus arctos*) may fit this category, but also the grey wolf (see below). Finally, own density dependent growth $G_W$ is generally assumed to be positive for a 'small' stock and negative for a 'large' stock. We further assume $G(0, X) = 0$ together with strict concavity, $G_{WW} < 0$.

The population growth of the existing species follows next as

$$dX/dt = F(X, W) - h \tag{2}$$

where $h \geq 0$ is the harvest, or trapping, and $F(..)$ the natural growth function. Also, $F_W < 0$ if the recolonised species competes with the existing wildlife or it is a predator–prey relationship and the recolonised species is the predator. If it is a prey, the effect will be the reverse and thus positive. However, this effect also may be weak, or even negligible. As above, $F_X$ is typically positive for a 'small' stock size and negative for a 'large' stock size and also $F(0, W) = 0$ and strict concavity in own density, $F_{XX} < 0$, are assumed.

The current net benefit, or social surplus, is given as

$$\pi = H(y, W) + R(W) - S(W) + V(h, X) + Q(X) - D(X) \tag{3}$$

where $H(y, W)$ is the benefit of controlling the recolonised species while $V(h, X)$ is the net hunting, or trapping, value of the existing species, both terms generally depending on the number of animals removed together with the species abundance, $H_W \geq 0$ and $V_X \geq 0$. These values may be positive, negative or zero. For instance, $H(y, W) > 0$ when the harvesting value is substantial and the harvesting cost is small, while $H(y, W) < 0$ when the cost of removing the recolonised stock is substantial, accompanied by a small, or perhaps negligible, harvesting value.

Furthermore, as already indicated, the production activities practised within the given area interacting with the ecology typically depend on the species density and are of various categories (Table 20.1). First of all, $S(W) > 0$, with $S(0) = 0$, is the cost of, say, predation on livestock, or grazing damage of the recolonised species, with more species implying higher costs, $S' > 0$. $S(W)$ may therefore typically reflect the cost of

Table 20.1 *Value categories*

| Recolonised species | | | Existing species | | |
|---|---|---|---|---|---|
| $H(y, W)$ | $R(W)$ | $S(W)$ | $V(h, X)$ | $Q(X)$ | $D(X)$ |
| Hunting/ trapping value | Positive stock value (existence value, viewing value, tourist value, etc.) | Negative stock value (grazing damage, livestock predation, etc.) | Hunting/ trapping value | Positive stock value (existence value, viewing value, tourist value, etc.) | Negative stock value (grazing and browsing damage, etc.) |

livestock predation if the recolonised species is a large carnivore, while being grazing damage if it is a herbivore (Zivin *et al.* 2000). $R(W)$, with $R(0) = 0$, yields the existence value, viewing value, tourist value, etc. of the recolonised species and is also generally increasing in the number of animals, $R' > 0$, but the marginal benefit may be decreasing, $R'' < 0$ (Krutilla 1967). We next have the already existing species stock values, and where $D(X)$ is the potential damage cost, also supposed to increase in the species density, $D' > 0$ with $D(0) = 0$. This cost may represent browsing, or grazing, damage, such as moose causing forestry damage (see example below). The existing wildlife also generally carries a positive stock value $Q(X)$, like existence value, with $Q(0) = 0$. Also here we typically have $Q' > 0$ together with $Q'' \leq 0$.

When the social planner aims to maximise present-value net benefit, $PV$, the problem is to find harvest and control rates of the species that maximise

$$PV = \int_0^\infty [H(y, W) + R(W) - S(W) + V(h, X) + Q(X) - D(X)]e^{-\delta t}dt \tag{4}$$

subject to the ecological growth equations (1) and (2), together with the initial stock sizes and where $\delta \geq 0$ is the (social) discount rate assumed to be constant through time. The current value Hamiltonian of this problem reads

$$\Psi = H(y, W) + R(W) - S(W) + V(h, X) + Q(X) - D(X) + \mu[G(W, X) - y] + \lambda[F(X, W) - h] \tag{4a}$$

with $y$ and $h$ as control variables, $W$ and $X$ as state variables and $\lambda$ and $\mu$ as the shadow values of the existing species and the recolonised species, respectively. It follows that the conditions (5–8) yield the necessary conditions for a maximum when it is socially desirable to keep both species and when any upper binding constraints on the control variables are neglected.

$$\partial\Psi/\partial y = H_y(y, W) - \mu \leq 0; \quad y \geq 0 \tag{5}$$

$$\partial\Psi/\partial h = V_h(h, X) - \lambda \leq 0; \quad h \geq 0 \tag{6}$$

$$d\mu/dt = \delta\mu - \partial\Psi/\partial W = \delta\mu - H_W(y, W) - R'(W) + S'(W) - \mu G_W(W, X) - \lambda F_W(X, W) \tag{7}$$

$$d\lambda/dt = \delta\lambda - \partial\Psi/\partial X = \delta\lambda - V_X(h, X) - Q'(X) + D'(X) - \mu G_X(W, X) - \lambda F_X(X, W) \tag{8}$$

The control condition (5) holds as an equality if it is optimal with control, $y > 0$, of the recolonised species along the optimal trajectory. The marginal net harvesting benefit should then be equal to the species' shadow value. Otherwise, with $y = 0$, it will be inequality. If it is optimal with no harvest of the existing species, condition (6) also holds as an inequality. In both instances of zero harvesting, the marginal benefit of control, positive or negative, should be below that of the shadow price, which may be positive or negative as well (more details below). The portfolio conditions (7) and (8) reflect the evolution of the shadow price of the recolonised species and the existing species, respectively. Dividing with $\mu$, condition (7) is the recolonised species Hotelling efficiency rule, indicating that the growth rate of the shadow price should be equal to the external rate of return as given by the discount rent $\delta$, minus the internal rate of return. Condition (8) has a similar interpretation for the existing species.

The shadow prices may be eliminated from the above system (5)–(8) and the reduced form solution together with the ecological growth equations (1) and (2) yield, in principle, a set of four interconnected differential equations between the two control variables, $y$ and $h$, and the two state variables, $W$ and $X$. However, it is not possible to say very much about the dynamics, or the steady state, of this system without further specification of the functional forms and without stating whether the ecological interaction is of the competitive or predator-prey type. Even then, the system will typically be too complex – see, for example, the much simpler two-species model in Ragozin and Brown (1985) where the predator alone is subject to harvest and there are no stock values. We therefore proceed to look at some simplified cases.

## 3 Simplified cases

Not surprisingly, loosening up the interaction between reintroduced and existing species results in more tractable situations to analyse. The same occurs if the net benefit functions of species control are given a more specific content. Altogether, four special cases are considered. First, two cases assuming negligible ecological interaction are studied. Then, two other cases are analysed with simplified harvesting functions.

### 3.1 *The case of negligible ecological interaction*

In many instances, the interaction between recolonised species and existing species is weak, or even negligible. The above example of the European bison is of this type and this may also be so when the recolonised species is of the opportunistic type (like the brown bear). The natural growth functions of the recolonised species and the existing species (1) and (2) reduce then to $G(W)$ and $F(X)$, respectively. As a consequence, there will be no economic interdependency between the species as well and the recolonised species can be managed separately from the existing one. Therefore, conditions (5) and (7a) yield the optimality conditions for the recolonised species together with $dW/dt = G(W) - y$.

$$d\mu/dt = \delta\mu - H_W(y, W) - R'(W) + S'(W) - \mu G'(W) \qquad (7a)$$

As the harvesting value may be either positive or negative and various stock values are included, this is very similar to the models considered by Horan and Bulte (2004), Skonhoft and Schulz (2005) and others. As demonstrated in these models, the shadow price, $\mu$, may be positive or negative. It will be positive if harvesting is profitable, while it is negative when controlling is a costly activity mainly for damage control. The ambiguous sign of the shadow price can result in a non-convex Hamiltonian together with possible multiple equilibria (see also Rondeau 2001 and Dasgupta and Mäler 2003). Obviously, we find the shadow price to be negative if the recolonised species (when controllable, see below) carries no trapping or hunting value, but demand effort to be controlled. It may, however, even be negative with a positive harvesting value if it, on the margin, is more costly to control the species so that $H_y(y, W)$ is negative at the optimum. Horan and Bulte (2004) analyse the dynamics of this model. When a non-linear control benefit function $H(y, W)$ is applied, they find, not surprisingly, the steady state(s) to be of the saddle point type.

When it is optimal to steer the system towards the steady state(s), condition (5) as an equity combined with (7a) gives the golden-rule condition:

$$G'(W) + H_W(y, W)/H_y(y, W) + [R'(W) - S'(W)]/H_y(y, W) = \delta \qquad (9)$$

The left-hand side of (9) yields the internal rate of species return at the optimum, which should be equal to the external rate as given by the discount rent $\delta$. This condition together with the species growth condition (1) in equilibrium determine the steady states for $W^*$ and $y^*$. If the shadow price is positive and there is a positive net harvesting benefit, $H_y > 0$, and the negative stock value dominates the positive one, $(S' - R') > 0$, it is seen that it is optimal to keep a small density of the recolonised species. If the shadow value is negative and the species may be classified as a pest, we reach the opposite conclusion. The comparative static results may also be ambiguous and typically we find that a higher rate of discount yields a higher steady-state stock $W^*$ when the shadow price is negative, which is the opposite of the standard harvesting model (Clark 1990). See also Skonhoft and Schulz (2005).

### 3.2 *The case of a fixed shadow price of the recolonised species together with negligible interaction with the existing species*

Often it may be reasonable to assume that the control cost of terrestrial animal species is density independent. This typically occurs under a hunting licence scheme (see below). If additionally the net harvesting benefit, positive or negative, is linear in the amount of animals controlled, or harvested, condition (5) indicates a constant shadow price when it is beneficial to control the species along the optimal trajectory. When still assuming a negligible ecological interaction, the recolonised species portfolio equation (7a) reduces to

$$0 = \mu\delta - R'(W) + S'(W) - \mu G'(W) \tag{7b}$$

Equation (7b) is a static one because the Hamiltonian now is linear in the control $y$ and the dynamics leading to the steady state will be of the Most Rapid Approach Path (*MRAP* dynamics, see, for example, Clark 1990). The golden rule condition (7b) also indicates that the internal rate of return, now as $G' + [R' - S']/\mu$, should be equal to the external rate, $\delta$. Another interpretation is that the net marginal value of the species 'in the forest', $(\mu G' + R' - S')$, should be equal to the marginal harvesting value 'in the bank', $\mu\delta$.

If $\mu^* > 0$, condition (7b) represents the solution of the standard one-species harvesting model missing the usual stock-dependent cost term, but extended with positive as well as negative stock values. Depending on their marginal values, the optimal number of species $W^*$ can be below or above that of the maximum sustainable yield level, $W_{msy}$. If it is costly to control and $\mu^* < 0$, we find that the optimal managed stock will be smaller when a nuisance effect is linked to it than without this effect. Therefore, reintroduced species will be left uncontrolled if they have no negative effect and are costly to control.

An even more simplified situation emerges if the harvesting benefit is small or negligible, i.e. $H(y, W) = 0$. As the marginal harvesting income is also zero, $H_y = 0$, $\mu^* = 0$ when it is still beneficial to control the species. Condition (7b) reduces then further to $-R'(W) + S'(W) = 0$ and the socially desirable number of species $W^*$ is simply determined by the equalisation of the marginal values.[1] While the optimal species number is invariant of natural growth, the steady-state harvest follows from the population growth equilibrium $G(W^*) = y^*$. The same conclusion may be reached when the existing species influences the growth of recolonised species, thus with $G(W, X)$. Of course, in this case the level of the control $y^*$ will differ.

In case it is socially desirable not to control, or harvest, the species along the optimal trajectory, condition (5) yields $\mu > 0$ when $H_y = 0$. The number of recolonised species would then approach its carrying capacity in the long term. From condition (7b), $\mu^* = (R' - S')/(\delta - G')$ is the shadow price of the unexploited stock. Because $G$ is a humped function with $G' < 0$ in the unexploited situation (section 2 above), $(R' - S') > 0$ must hold to ensure a positive shadow value. Thus, not surprisingly, it is optimal to leave the species uncontrolled because the marginal positive stock value exceeds the negative one. Under these conditions it is also seen that reintroduced species unambiguously will be left uncontrolled if they have no negative effect.

### 3.3 *The case of ecological interaction without harvesting benefit of the recolonised species*

For various reasons, the harvesting profit of the recolonised species may be zero, or close to zero. This may happen if, say, the harvesting benefit is small and negligible and the control cost is small and negligible as well (see the wolf example below). When $H(y, W) = 0$, $\mu = 0$ still holds if it pays to control along the optimal trajectory. When there is ecological interaction, the portfolio conditions yield

$$0 = -R'(W) + S'(W) - \lambda F_W(X, W) \tag{7c}$$

and

$$d\lambda/dt = \delta\lambda - V_X(h, X) - Q'(X) + D'(X) - \lambda F_X(X, W) \tag{8a}$$

[1] Concavity of the Hamiltonian requires in this case that $R'' - S'' < 0$ (c.f. section 4).

The portfolio condition (7c) of the recolonised species is also now a static equation and it can be noted that the opportunity cost of the recolonised species' biological capital is zero as the discount rent $\delta$ is not included. The optimal number of animals is found where the marginal stock value $R'(W)$ is equal to its marginal cost, comprising the damage cost, $S'(W)$, and the cost of predation evaluated at the existing species shadow value, $\lambda F_W(X, W)$.

To solve this system, the shadow price of the existing species, $\lambda$, may in a first stage be eliminated from equation (7c) and (8a) by using the control condition (6) which holds as $\lambda = V_h(h, X)$ when harvest of the existing species takes place along the optimal trajectory. $W$ can be expressed as a function of $X$ and $h$ through equation (7c). In a next step, $W$ may be substituted away from (8a). The reduced-form dynamic system is consequently steered by equation (8a) together with the population growth equation (2), comprising the variables $X$ and $h$. The dynamics of this system may be quite similar to the first case with no ecological interaction between the species, thus yielding the possibility of multiple equilibria for the recolonised species. At the steady state, it can be shown that $R'(W^*) - S'(W^*) > 0$ if $\lambda^* > 0$ and the recolonised species prey upon, or compete, with the existing species (i.e. $F_W < 0$). The opposite holds if the existing species turns out to be a pest and $\lambda^* < 0$.

### 3.4 *The case of ecological interaction without harvesting benefit of the recolonised species and with a constant harvesting value of the existing species*

In some instances the harvesting value of the existing species may simply be given by the meat value, or net hunting price $p$. Therefore, the harvesting value becomes $V(X, h) = ph$ and condition (6) reduces to $p = \lambda$ if it is profitable to harvest. If condition $H(y, W) = 0$ with $\mu = 0$ holds and it is still beneficial to control the recolonised species along the optimal trajectory as well, it turns out that

$$0 = -R'(W) + S'(W) - pF_W(X, W) \tag{7d}$$

and

$$0 = \delta p - Q'(X) + D'(X) - pF_X(X, W) \tag{8b}$$

This is a double singular system with dynamics of the *MRAP* type, or close to *MRAP* (see Clark 1990). While the natural growth of the existing species still influences the outcome, the recolonised species' natural growth does not because of the assumption that harvesting has zero profit and $\mu = 0$. If the recolonised species prey upon, or compete, with the existing species, $R'(W^*) - S'(W^*) > 0$ holds unambiguously at the steady-state equilibrium. Not surprisingly, a higher positive marginal species value yields a higher optimal stock while more damage works in the opposite direction. Even in this simplified model, however, other comparative static results are far from clear and a higher harvesting price, $p$, may either increase or reduce the optimal number of recolonised species. The stock effects of a higher discount rate $\delta$ are generally unclear as well. These results are examined in more detail in the example of the recolonisation of the grey wolf in Scandinavia which includes a wolf–moose (*Alces alces*) ecological interaction.

## 4 The recolonisation of the Scandinavian wolf

In the mid-1960s, the grey wolf was regarded as functionally extinct in Norway and Sweden (the Scandinavian Peninsula). However, due to banning of earlier hunting practices it recolonised and in the latter part of the 1970s the first confirmed reproduction in fourteen years was recorded. Since this first reproduction in Northern Sweden, all new reproductions have been located in South-central parts of the Scandinavian Peninsula. The recolonised wolf population in Scandinavia now numbers some 100–120 individuals which live in small family groups, or packs, in the Western-central part of Sweden and along the border area between Norway and Sweden (Wabakken *et al.* 2001).

Although the wolf population is still numerically small, its recolonisation is already associated with several conflicts. One is due to predation on livestock, including sheep and reindeer. Although the total loss is modest, some farmers in a few areas have been seriously affected, as in the abovementioned example from Yellowstone. In addition, predation on wild ungulates is another conflict, especially where the wolf shows a particularly strong preference for moose. As a consequence, a smaller moose population is available for hunting. In fact, while the problem of moose predation also takes place in only a few areas, it has caused great concern in rural Scandinavia because moose is by far the most important hunting game species, with about 40,000 and 100,000 animals (with a mean body weight of about 190 kg for adult females and 240 kg for adult males) shot every year in Norway and Sweden, respectively. In addition, moose hunting in September/October is an important, if not the most important, social and cultural event in many rural communities (Skonhoft 2006).

Moose–wolf ecology has been subject to several intensive studies, mostly in North America. From these studies it appears clear that wolves, when present, influence the abundance of moose (Peterson 1999). The

Scandinavian ecosystem, however, differs from the North American system as the moose density is generally higher. Additionally, the age and sex structures differ because of selective hunting schemes with a higher-proportion harvesting of calves and young males. Another important difference is that in Scandinavia harvesting accounts for a greater share of total mortality. Last but not least, wolf density in Scandinavia is also significantly lower and more patchily distributed (Wabakken *et al.* 2001). It thus follows that the moose–wolf ratio is higher in Scandinavia and the impact of wolf predation is likely to be of a more local nature. Wolf predation is focused on calves, yearlings and older females, with calves as the main food source. The predation rates reported from Scandinavia also appear to be higher than those in North America, which may indicate that predation, for a given size of wolf pack, increases with moose density (Nilsen *et al.* 2005).

Based on the studies cited above, it can be assumed that wolf predation represents an additional source of mortality for calves, yearlings and older females. In our biomass framework, the wolf population then negatively affects the natural growth of the moose population. It is assumed that the predation increases with the size and number of the wolf packs as well as the size of the moose stock. There may also be a feedback effect as the size of the moose population influences wolf population growth. However, in areas with colonising carnivore populations, this relationship will appear less interactive, meaning that the wolves are not able to respond numerically to variations in the moose population (Nilsen *et al.* 2005). Any numerical response of the wolf population is hence neglected. The ecological model of the wolf–moose interaction is therefore described by equation (2) $dX/dt = F(X, W) - h$, while equation (1) again reduces to $dW/dt = G(W) - y$.

We then have the cost and benefit streams of the considered system and we start with the wolf stock values. The livestock predation cost on sheep and reindeer of the wolf $S(W)$ is suspected to be quite small, but, as indicated, it can be of significance in a few areas (Milner *et al.* 2005). Yet the non-consumptive wolf stock value (including the intrinsic value and viewing value), $R(W)$, is suspected to be high (Boman and Bostedt 1999). However, as the stock value is highly uncertain, the effects of different assumptions need to be studied. It may be costly to control the wolf population, or it may be controlled by selling hunting licences. Another possibility is that the controlling costs more or less cover the benefits so that the net harvesting value may be small or negligible. All these possibilities are explored next when assuming that the harvesting income, or cost, increases linearly in the number of controlled animals while neglecting any stock effect.

Landowners obtain the hunting profit of the moose. The yearly hunting income is given as $V(h, X) = ph$, with $p$ as the net hunting licence price, assumed to be fixed and independent of the harvest and stock size. This is justified by the fact that there is competition among a large number of suppliers of hunting licences in Scandinavia. Following the practice in Norway (and Sweden), one licence allows the buyer to kill one animal, which is paid only if the animal is killed. The moose population also causes browsing damage to landowners, the damage on young pine being of particular importance (Wam *et al.* 2005). The damage on young pine occurs basically during the winter and varies with the quality of the timber stand and the productivity of the forest. The damage may take place immediately and damaged young pine trees may be replaced directly, but quite frequently there is a time lag between the occurrence of browsing and the economic loss of the damage. In such instances, however, discounting is not taken into account explicitly. There are also other costs connected to the moose population, the single most important being related to moose–vehicle collisions. This cost is considerable and recent estimates indicate that it may be even higher than the meat value of the moose (Skonhoft 2006). Thus, the damage cost function of the moose, $D(X)$, covers grazing damage as well as the cost of traffic collisions. There will also be a positive stock value of the moose population (viewing value, etc.). However, because of the large number of moose in Scandinavia, $Q(X)$ is suspected to be quite small, if not negligible, at the margin.

The wolf–moose example is a mix between the second and fourth cases introduced in the introduction. That is, (i) there is only a one-way ecological interaction, (ii) there is a fixed control value of the recolonised species (positive, negative or zero), and iii) the harvesting value of the existing species is not stock dependent and is linear in the amount harvested. The shadow value of the recolonised species will be constant when assuming that the wolf population is controlled along the optimal trajectory all the time. It follows that $H_y = \mu$ which is positive, negative or zero, while the control condition (6) of the existing species, the moose, is $p = \lambda$ when harvesting pays off. The dynamics of this system will therefore obey an *MRAP* path and the reduced-form steady state is given by

$$0 = \mu\delta - R'(W) + S'(W) - \mu G'(W) - pF_W(X, W) \tag{10}$$

and

$$0 = p\delta - Q'(X) + D'(X) - pF_X(X, W) \tag{11}$$

These two equations determine $X^*$ and $W^*$ simultaneously. In a next step, the number of animals removed can be derived from the equilibrium population growth conditions. The wolf population may be above or below that of the maximum sustainable yield level, $W_{msy}$, and this may also occur for the moose population. The comparative statics are also generally unclear and a higher harvesting price of the moose may either increase or decrease the socially desirable number of moose. For this reason, the wolf stock effect will be unclear as well. The effect of a higher discount rate is suspected to influence the wolf stock negatively, but this effect is also unclear because it affects the population directly as well as indirectly through the moose population equilibrium condition (11).

To shed further light on the economic and ecological forces at work, the functional forms of the various functions need to be specified. The wolf stock growth is assumed to be logistic, $G(W) = \gamma W(1 - W/L)$, with $\gamma$ as the maximum specific growth rate and $L$ as its carrying capacity. Similarly, the natural growth of the moose population in the absence of wolf predation is assumed to be of the standard logistic type, while the predation effect (the functional response) is specified in a Cobb-Douglas manner, $F(X, W) = \beta X(1 - X/K) - \alpha WX$, where $\alpha > 0$ is the predation coefficient. Therefore, the functional response of the moose population implies a fixed predation rate (as a growth rate), $\alpha W$, and indicates that the amount of predation increases linearly with the size of the moose stock.

For simplicity, it is assumed that the moose stock values are linearly increasing in stock size. Therefore, for the moose population, we have $D(X) = dX$, with $d > 0$ as the constant damage cost per moose, including browsing damage as well as traffic damage, and $Q(X) = qX$ with $q > 0$ as the fixed positive moose stock value. For the wolf population, we also assume a linear damage function with constant damage cost per wolf, $S(W) = sW$ with $s > 0$. However, quite realistically, a strictly concave function is imposed for the wolf intrinsic value. This may secure a meaningful solution of the optimisation problem even if the recolonised species shadow value is negative (see below) and the function is specified as $R(W) = r_1 W - (r_2/2)W^2$. The value of the parameters $r_1 > 0$ and $r_2 > 0$ are scaled such that the marginal value all the time is positive.[2] Inserted into the above conditions (10) and (11) it follows that

$$p\alpha X + (2\mu\gamma/L + r_2)W = \mu(\gamma - \delta) + (r_1 - s) \tag{10a}$$

and

$$(2p\beta/K)X + p\alpha W = p(\beta - \delta) + (q - d). \tag{11a}$$

[2] Typically, the moose positive stock value $Q(X)$ is suspected to be strictly concave as well. However, for simplicity, it is given as a linear function as this has no influence on the qualitative structure of the solution.

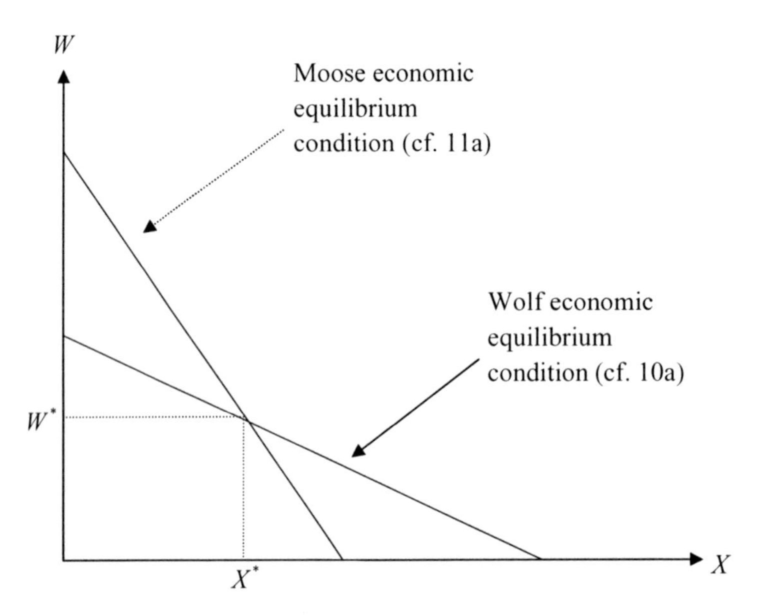


Figure 20.1. Wolf–moose economic equilibrium

These equations are straight lines in the $X$, $W$ plane. The moose equilibrium condition (11a) slopes unambiguously downwards while the wolf equilibrium condition (10a) may slope downwards as well as upwards depending on the sign and size of the shadow value. However, due to the second-order conditions for a maximum, it must slope downwards but be less negatively sloped than the moose equilibrium condition (Figure 20.1). Two parameters are of particular importance here: the ecological interaction coefficient, $\alpha$, in addition to the shadow value $\mu$. Hence, to obtain a meaningful solution of the maximum problem, the predation pressure cannot be too strong while the shadow price, if negative, cannot be too largely negative.[3]

Table 20.2 reports the comparative static results. The effects of shifts in the stock values are straightforward. If, say, the positive wolf stock

[3] The Hamiltonian must be jointly concave in the control and state variables to fulfil the second-order conditions for maximum. It can be demonstrated that this requires $\Omega = (2p\beta/K)(2\mu\gamma/L + r_2) - (p\alpha)^2 > 0$ together with $-(2\mu\gamma/L + r_2) < 0$. $\Omega$ is the determinant of the left-hand side of equations (10a) and (11a), and $\Omega > 0$ indicates that equation (11a) should be more negatively sloped than equation (10a). There must also be various restrictions on the parameter values to obtain an interior solution with positive stock sizes and stock sizes below its carrying capacities. The moose equilibrium condition (11a) must hence intersect at the $W$ axis above that of the wolf equilibrium condition (10a) while (10a) must intersect with the $X$ axis outside that of equation (11a). For a related discussion, see Skonhoft (1995).

Table 20.2 *Wolf recolonisation example – comparative static results*

| | $r_1$ | $d$ | $p$ | $\mu$ | $\delta$ | $\alpha$ |
|---|---|---|---|---|---|---|
| $W^*$ | + | + | +/(−) | −/+ | −/(+) | −/(+) |
| $X^*$ | − | − | −/(+) | +/− | −/(+) | −/(+) |

value, $r_1$, increases permanently, the social planner will keep a larger wolf population. As a result, the predation pressure will increase and the number of moose will be reduced accordingly. If the moose damage cost, $d$, increases due to, for instance, a higher frequency of moose–vehicle collisions, it will also be beneficial given a higher wolf population to increase the predation pressure and reduce the number of moose and hence the damage. Interestingly, the effects of a permanently higher rate of discount are generally unclear. However, if the wolf shadow price is positive, it can be shown that at least one of the stocks will decrease if $\delta$ increases. The effects of a more valuable moose harvest are ambiguous as well. On the one hand, a higher $p$ will increase the moose number for a given size of the recolonised wolf population. This is because the marginal damage dominates the marginal positive stock value. Therefore, the relative damage cost will be reduced. This effect will be reinforced as the cost of predation increases and the wolf equilibrium line (10a) shifts inwards. On the other hand, if the marginal moose damage is small and negligible and is dominated by the positive stock value, no clear conclusion can be drawn. In either case, the effect on the wolf number will be the opposite.

The effects of a shift in the wolf shadow value are also ambiguous. If the shadow value is positive and increases, the result will be a smaller wolf population, suggesting that the net marginal stock value $(r_1 - s)$ is positive. The predation pressure hence reduces and the moose population increases accordingly. But if $\mu < 0$ and the control cost increases further, it will be beneficial with a higher wolf population, again under the reasonable assumption that the positive marginal stock value dominates the negative one. If the shadow price is zero, (10a) simply reads $W = (r_1 - s)/r_2 - (p\alpha/r_2)X$. The effects of a higher predation pressure through $\alpha$ are also generally unclear. However, it can be shown that at least one of the stocks will decrease.

It is difficult (if not meaningless; 'what is the money value of a songbird?') to try to calculate the stock value of the recolonised species monetarily as it comprises, among others, its existence value. It is, however, possible to reveal this value indirectly by imposing a quantitative restriction on the number of reintroduced wolves. To make things simple, while capturing the main points, any net harvesting benefit is neglected so the shadow value is zero, $\mu = 0$. In addition, the wolf damage cost together with the moose existence value are assumed to be small and negligible as well, i.e. $s = 0$ and $q = 0$. It thus follows that conditions (10a) and (11a) reduce to $p\alpha X + r_2 W = r_1$ and $(2p\beta/K)X + p\alpha W = p(\beta - \delta) - d$, respectively. Therefore, for a wolf target level $\bar{W}$, the marginal stock value reads as follows:

$$r_1 = \frac{K\alpha}{2\beta}[p(\beta - \delta) - d] + \frac{K}{2p\beta}\left[\frac{2p\beta r_2}{K} - (p\alpha)^2\right]\bar{W} \tag{12}$$

The calculation is illustrated by using data from the Koppang area, some 300 km north of Oslo. A wolf pack settled in this region in 1997 in an area of 600 km$^2$, with a moose population of about 1000 individuals. Since then the number of wolves has been between five and twelve (more details are provided in Skonhoft 2006). A target level of ten wolves illustrates the calculations, $\bar{W} = 10$. The following parameter values are used. The moose carrying capacity is $K = 3,500$ (number of moose) which implies about 5.8 moose per square km. The moose maximum specific growth rate is $\beta = 0.47$, while the predation coefficient is assumed to be $\alpha = 0.005$ (1/wolf). The hunting licence price is $p = 8$ (1000 NOK/moose, 2003 prices), the marginal damage cost is $d = 1$ (1000 NOK/moose, 2003 prices) and the discount rent is $\delta = 0.05$. Finally, the baseline changing marginal wolf stock value is assumed to be $r_2 = 10$ (1000 NOK/wolf$^2$).

For these parameter values, we find $r_1 = 137$ (1000 NOK/wolf), indicating the value $R = r_1 W - (r_2/2)\bar{W}^2 = 865$ (1000 NOK) and the marginal value $R' = r_1 - r_2\bar{W} = 37$. Consequently, on these premises, the stock value of the target level wolf pack of $\bar{W} = 10$ must be at least 865 if recolonisation should be beneficial from a social point of view. Not surprisingly, $r_1$ and hence $R$ decrease if the damage cost of the moose population increases, while $r_1$ increases when the moose hunting becomes more valuable. If, say, the hunting value $p$ is doubled, we find $r_1 = 192$, while doubling the marginal damage $d$ yields $r_1 = 118$.[4]

Altogether, these calculations indicate that, depending on cost and price assumptions, the break-even wolf stock value may vary widely. Nevertheless, the calculations demonstrate a quite modest wolf value to justify recolonisation. If the moose browsing and traffic damage increase, the critical marginal recolonisation value decreases as the predation then

[4] If instead supposing that the marginal stock value reduces more slowly with $r_2 = 5$, while the other parameters are left unchanged, we find $r_1 = 86$ and $R = 615$. With $r_2 = 20$, meanwhile, the result is $r_1 = 236$ and $R = 1365$.

pays off more in the sense that it contributes to less moose damage cost. The wolf is then 'doing the job' as a damage controller. In the opposite case of a more valuable prey harvest, the predation cost increases proportionally and a higher break-even recolonisation value occurs. These values may be compared to the Scandinavian contingent value study of Boman and Bostedt (1999), which indicates (but notice the abovementioned problems with such assessments) a much higher willingness to pay for the wolf existence value.

## 5 Conclusion

Species recolonisation typically takes place in an environment where earlier harvesting practices are banned, or when previous production practices in agriculture and forestry influencing habitat conversion and species growth are changed. Therefore, recolonisation often represents an institutional change and reflects society's changing attitude to the species cost and benefit streams. When successful, recolonisation often influences existing ecology and may come into conflict with existing economic activities. However, it may also create hunting and trapping value in addition to non-consumptive values like existence and viewing value. Ecology and institutions shape these costs and benefit streams experienced by different agents and groups of people.

Correctly modelling the key interspecies relationship is the critical part of studying the economic effects of recolonised species and various situations have been considered in this chapter. Using a general model in which recolonised species interact with already existing species, like a traditional predator–prey interaction, it becomes apparent that it is difficult to explain the dynamics and also the economic and ecological forces forming the equilibrium that eventually settles. Therefore, in order to shed additional light, at the cost of generality, some simplified cases have been proposed. Not surprisingly, loosening up the interaction between reintroduced and existing species yields more traceable situations to analyse. However, even in such cases, the economic and ecological forces at work are often difficult to assess.

The general insight from these models is that some control of the recolonised species often pays off. However, recolonised species should be kept uncontrolled when they (i) do small damage, (ii) are expensive to control and (iii) prey upon existing species that cause various types of damage, like browsing or grazing damage. A calibrated example of the recent experience of the recolonisation of the grey wolf in Scandinavia sheds further light on the various ecological and economic mechanisms working. This example demonstrates that the wolf value may be quite modest to justify the wolf recolonisation. This example also demonstrates that the effects of economic forces often are difficult to predict when operating in an interspecies relationship. This indicates that detailed knowledge about the ecology and cost and benefit structure is crucial to carry out a sound recolonised species management policy.

## REFERENCES

Boman, M. and Bostedt, G. 1999. Valuing the wolf in Sweden. Are benefits contingent on the supply?. In R. Brännlund and B. Kristrom (eds.). *Topics in Environmental Economics*. Dordrecht: Kluwer. 157–174.

Boyce, M. 1997. Case study 5: The greater Yellowstone ecosystem. In G. Meffe and R. Carroll. *Principles of Conservation Biology*. Sunderland. MA: Sinauer. 631–642.

Clark, C. 1990. *Mathematical Bioeconomics*. New York: Wiley Interscience.

Dasgupta, P. and Mäler, K. 2003. The economics of non-convex ecosystems: introduction. *Environmental and Resource Economics*. **26**. 499–525.

Freeman, A. M. 2003. *The Measurement of Environmental and Resource Values*. Washington, DC: Resources for the Future.

Graham, K., Beckerman, A. and Thirgood, S. 2005. Human predator-prey conflicts: ecological correlates, prey losses and patterns of management. *Biological Conservation*. **122**. 159–171.

Horan, R. and Bulte, E. 2004. Optimal and open access harvesting of multi-use species in a second-best world. *Environment and Resource Economics*. **28**. 251–272.

Huffaker, R., Bhat, M. and Lenhart, S. 1992. Optimal trapping strategies for diffusing nuisance beaver populations. *Natural Resource Modeling*. **6**. 71–97.

IUCN 1995. Guidelines for re-introductions. The World Conservation Unit: Gland, Switzerland: www.iucn.org/themes/ssc/pubs/policy/reinte.htm.

Krutilla, J. 1967. Conservation reconsidered. *American Economic Review*. **57**. 778–786.

Matson, T., Goldizen, A. and Jarman, P. 2004. Factors affecting the success of translocations of the black-faced impala in Namibia. *Biological Conservation*. **116**. 359–365.

Milner, J., Nilsen, E., Wabakken, P. and Storaas, T. 2005. Hunting moose or keeping sheep? Producing meat in areas with carnivores. *Alces*.

Nilsen, E., Pettersen, T., Gundersen, H., Milner, J., Mysterud, A., Solberg, E., Andreassen, H. and Stenseth, N. C. 2005. Moose harvesting in the presence of wolves. *Journal of Applied Ecology*. **42**. 389–399.

Nunes, P. and van den Bergh, J. 2001. Economic valuation of biodiversity; sense or nonsense?. *Ecological Economics*. **39**. 203–222.

Perrings, C. 2007. Biological invasions and poverty. In A. Kontoleon, U. Pascual and T. Swanson (eds.). *Biodiversity Economics*. Cambridge: Cambridge University Press.

Perzanowski, K., Wanda, O. and Kozak, I. 2004. Constraints for re-establishing a meta-population of the European bison in Ukraine. *Biological Conservation*. **120**. 345–353.

Peterson, R. 1999. Wolf-moose interaction on Isle Royale: the end of natural regulation?. *Ecological Applications*. **9**. 10–16.

Ragozin, D. and Brown, G. 1985. Harvesting policies and non-market valuation in a predator-prey system. *Journal of Environmental Economics and Management*. **12**. 155–168.

Rondeau, D. 2001. Along the way back from the brink. *Journal of Environmental Economics and Management*. **42**. 156–182.

Skonhoft, A. 1995. On the conflicts of wildlife management in Africa. *International Journal of Sustainable Development and World Ecology*. **2**. 267–277.

Skonhoft, A. 2006. The cost and benefit of animal predation; an analysis of Scandinavian wolf re-colonisation. *Ecological Economics*. **58** (4). 830–841.

Skonhoft, A. and Schulz, C. E. 2005. On the economics of ecological nuisance. Working paper. Department of Economics, University of Tromso, Norway.

Steury, T. and Murray, D. 2004. Modeling the reintroduction of lynx to the southern portion of its range. *Biological Conservation*. **117**. 127–141.

Swanson, T. 1994. *The International Regulation of Extinction*. London: Macmillan.

Wabakken, P., Sand, H., Liberg, O. and Bjarvall, B. 2001. The recovery, distribution and population dynamics of wolves on the Scandinavian peninsula. *Canadian Journal of Zoology*. **79**. 189–204.

Wam H. K., Hofstad, O., Nævdal, E. and Sankhayan, P. 2005. A bio-economic model for optimal harvest of timber and moose. *Forest Ecology and Management*. **206**. 207–219.

Zivin, J., Hueth, B. M. and Zilberman, D. 2000. Managing a multiple-use resource; the case of feral pig management in California rangeland. *Journal of Environmental Economics and Management*. **39**. 189–204.

*Part IV*

# Managing agro-biodiversity: causes, values and policies

# 21 On the role of crop biodiversity in the management of environmental risk

*Salvatore Di Falco and Jean-Paul Chavas*

## 1 Introduction

Genetic diversity is the information that is contained in the genes of individual plants, animals and micro-organisms. Species diversity is the diversity of species within which gene flow occurs under natural conditions. Agricultural biodiversity is defined as a component of biodiversity, referring to all diversity within and among species found in crop and domesticated livestock systems, including wild relatives, interacting species of pollinators, pests, parasites and other organisms (Qualset *et al.* 1995; Wood and Lenné 1999; Smale and Drucker 2007).

In managed systems, such as agro-ecosystems, crop genetic resources are the raw materials for modern crop breeding, selection programmes, pest resistance, productivity, stability and future agronomic improvements.[1] A number of studies in the agro-ecology literature suggest that genetic variability within and between crop species confers the potential to resist stress, provide shelter from adverse conditions and increase the resilience and sustainability of agro-ecosystems. Crop biodiversity erosion increases the vulnerability of the crop to biotic and abiotic stresses. Biodiversity reduction promotes build-up of crop pest and pathogen populations. Plot studies show that intercropping would reduce the probability of absolute failure of crop and that crop diversification increases crop income stability (Walker *et al.* 1983).

Therefore, the greater the diversity between and/or within species and functional groups, the greater the tolerance to pests. This is because pests have more ability to spread through crops with the same genetic base (Sumner *et al.* 1981; Altieri and Lieberman 1986; Gliessman 1986; Heisey *et al.* 1997). Further, the performance of different species varies with climatic and other agro-ecological conditions. The agro-ecosystem is subject to stresses caused by inadequate rainfall and soil moisture, randomness of temperature and potential evaporation. All these factors have

[1] In this chapter we will use the terms crop biodiversity, crop genetic diversity or diversity interchangeably.

the potential to shape crop development and variation (Pecetti *et al.* 1992; Loss and Siddique 1994).

Having plants that are functionally similar contributes to resilience and ensures that whatever the environmental conditions there will be plants of a given functional type that thrive under those conditions (Heal 2000) and it allows the agro-ecosystem to maintain productivity over a wider range of conditions (Tilman and Downing 1994; Tilman *et al.* 1996; Naeem *et al.* 1994). Thus, crop biodiversity confers potential resistance to droughts and other environmental stresses and the cost of crop genetic uniformity can potentially be high.

Surprisingly in applied economics, the relationships between crop diversity and both productivity and stability of yields have received much less attention and provided mixed evidence. For instance, Smale *et al.* (1998) estimated the effects on productivity of diversity among modern varieties in the districts of Punjab of Pakistan. This study found that diversity is positively related to the mean of yields and negatively correlated with the variance of yields in rainfed districts. Widawsky and Rozelle (1998), using data from regions of China, tested the impact of rice varietal diversity on the mean and the variance of yields. They found that the number of planted varieties reduces both the mean and the variance of yields, although the variance estimates are not statistically significant. More recently, Di Falco and Perrings (2003, 2005) found diversity to be positively correlated with yields and negatively correlated with revenue variability in two case studies on cereals production in Southern Italy. The common denominator of these studies is that they all use the stochastic production function approach suggested by Just and Pope (1978). They therefore adopt a specification that estimates the role of crop biodiversity on the mean and the variance of yields (or revenues). However, the potential reduction of farmers' welfare is determined not only by higher variance of yields but also by the probability of crop failure (Di Falco and Chavas 2005).

Empirical evidence suggests that most decision makers exhibit decreasing absolute risk aversion (e.g. Binswanger 1981; Chavas and Holt 1996; Chavas 2004). Such farmers are also averse to being exposed to unexpected low returns and are said to exhibit 'downside risk aversion' (Menezes *et al.* 1980; Antle 1987).

Therefore, risk-averse farmers may have an incentive to plant crop species or varieties that reduce the variance of returns. Farmers exhibiting downside risk aversion have an incentive to grow varieties that affect positively the skewness of the distribution of returns. In this context, besides its effect on agricultural productivity, crop genetic diversity can generate benefits by reducing farmers' exposure to both risk and 'downside risk' (Di Falco and Chavas 2005). This can be particularly relevant when production takes place in areas where there is no large technological progress, no irrigation and high weather variability (either temporal or spatial). Under these conditions the production variability may not capture the extent of risk exposure.

The management of environmental risk should involve both a reduction in yield variance and a decrease in exposure to downside risk (e.g. severe drought leading to crop failure). The above arguments suggest that risk assessment needs to go beyond a simple variance assessment to capture also exposures to unfavourable downside risk.

## 2 Analysis

In this section a general methodology to estimate the role of crop genetic diversity on the mean, the variance and the skewness of yield is presented. A farm production technology can be represented mathematically by the following production function:

$$y = g(\boldsymbol{x}, \boldsymbol{v}) \tag{1}$$

where y is output, $\boldsymbol{x}$ is a vector of inputs (e.g. fertiliser, pesticide, labour, crop biodiversity), $\boldsymbol{v}$ is a vector of non-controllable inputs (e.g. weather conditions) and $g(\boldsymbol{x}, \boldsymbol{v})$ denotes the largest feasible output given $\boldsymbol{x}$ and $\boldsymbol{v}$. Given that weather conditions during the growing season are not known ahead of time, $\boldsymbol{v}$ is treated as a random vector, with a given subjective probability distribution. Treating $g(\boldsymbol{x}, \boldsymbol{v})$ as a random variable indicates the need to assess the probability distribution of $g(\boldsymbol{x}, \boldsymbol{v})$. Following Antle (1983), we explore the moment-based approach to this assessment. Consider the following econometric specification for $g(\boldsymbol{x}, \boldsymbol{v})$ (Di Falco and Chavas 2005):

$$\begin{aligned} g(\boldsymbol{x}, \boldsymbol{v}) = {} & f_1(\boldsymbol{x}, \beta_1) + [f_2(\boldsymbol{x}, \beta_2) - f_3(\boldsymbol{x}, \beta_3)^{2/3}]^{1/2} e_2(\boldsymbol{v}) \\ & + [f_3(\boldsymbol{x}, \beta_3)]^{1/3} e_3(\boldsymbol{v}), \end{aligned} \tag{2}$$

where $f_2(\boldsymbol{x}, \beta_2) > 0$ and the random variables $e_2(\boldsymbol{v})$ and $e_3(\boldsymbol{v})$ are independently distributed and satisfy $E[e_2(\boldsymbol{v})] = E[e_3(\boldsymbol{v})] = 0$, $E[e_2(\boldsymbol{v})^2] = E[e_3(\boldsymbol{v})^2] = 1$, $E[e_2(\boldsymbol{v})^3] = 0$, and $E[e_3(\boldsymbol{v})^3] = 1$. It means that the random variables $e_2(\boldsymbol{v})$ and $e_3(\boldsymbol{v})$ are normalised: they are each distributed with mean zero and variance 1. In addition, $e_2(\boldsymbol{v})$ has zero skewness ($E[e_2(\boldsymbol{v})^3] = 0$) while the random variable $e_3(\boldsymbol{v})$ is asymmetrically distributed and has positive skewness ($E[e_3(\boldsymbol{v})^3] = 1$). It follows from (2) that

$$E[g(\boldsymbol{x}, \boldsymbol{v})] = f_1(\boldsymbol{x}, \beta_1), \tag{2a}$$
$$E[(g(\boldsymbol{x}, \boldsymbol{v}) - f_1(\boldsymbol{x}, \beta_1))^2] = f_2(\boldsymbol{x}, \beta_2), \tag{2b}$$
$$E[(g(\boldsymbol{x}, \boldsymbol{v}) - f_1(\boldsymbol{x}, \beta_1))^3] = f_3(\boldsymbol{x}, \beta_3). \tag{2c}$$

The specification (2) provides a convenient representation of the first three central moments of the distribution of $g(\boldsymbol{x}, \boldsymbol{v})$. Indeed, from (2a), the first moment (the mean) is given by $f_1(\mathbf{x}, \beta_1)$. From (2b), the second central moment (the variance) is given by $f_2(\boldsymbol{x}, \beta_2) > 0$. And from (3c), the third central moment (measuring skewness) is given by $f_3(\boldsymbol{x}, \beta_3)$. This provides a flexible representation of the impacts of inputs $\boldsymbol{x}$ on the distribution of output under production uncertainty. In addition, if we treat the distribution of $e_2(\boldsymbol{v})$ and $e_3(\boldsymbol{v})$ as given, then the three moments $f_1(\boldsymbol{x}, \beta_1)$, $f_2(\boldsymbol{x}, \beta_2)$ and $f_3(\boldsymbol{x}, \beta_3)$ are sufficient statistics for the distribution of $g(\boldsymbol{x}, \boldsymbol{v})$ in the specification (2).

Equation (2) can be interpreted as a standard regression model where $f_1(\boldsymbol{x}, \beta_1)$ is the regression line representing mean effects and $\{[f_2(\boldsymbol{x}, \beta_2) - f_3(\boldsymbol{x}, \beta_3)^{2/3}]^{1/2} e_2(\boldsymbol{v}) + [f_3(\boldsymbol{x}, \beta_3)]^{1/3} e_3(\boldsymbol{v})\}$ is an error term with mean zero, variance $f_2(\boldsymbol{x}, \beta_2)$ and skewness $f_3(\boldsymbol{x}, \beta_3)$. By considering explicitly skewness effects, it is a generalisation of the standard stochastic production specification (Just and Pope 1978, 1979; Di Falco and Chavas 2005).

The Just and Pope specification studies the effects of $\boldsymbol{x}$ on the variance of output which are of special interest and help to determine whether inputs are risk increasing (with $\partial f_2(\mathbf{x}, \beta_2)/\partial x > 0$), risk neutral (with $\partial f_2(\mathbf{x}, \beta_2)/\partial x = 0$), or risk decreasing (with $\partial f_2(\mathbf{x}, \beta_2)/\partial x < 0$).

However, in situations where exposure to downside risk is relevant and skewness effects are important, we need to analyse the impact of $\boldsymbol{x}$ on the skewness of the distribution of output. This implies establishing which inputs are downside-risk increasing (with $\partial f_3(\boldsymbol{x}, \beta_3)/\partial x < 0$), downside-risk neutral (with $\partial f_3(\boldsymbol{x}, \beta_3)/\partial x = 0$), or downside-risk decreasing (with $\partial f_3(\boldsymbol{x}, \beta_3)/\partial x > 0$). In this specific case, the question is to identify whether crop biodiversity is positively related to the skewness of the distribution of output and can provide a viable way to manage exposure to downside risk.

## 3 Econometric estimation

Equation (2) can be interpreted as a regression model where the error term exhibits possible heteroscedasticity (given by $f_2(\boldsymbol{x}, \beta_2)$) and skewness (given by $f_3(\boldsymbol{x}, \beta_3)$). Following Antle (1983), the parameters

$(\beta_1, \beta_2, \beta_3)$ in (2) can be consistently estimated using the feasible GLS (Generalised Least Squares). The empirical strategy consists of three steps. First estimate

$$y = f_1(\boldsymbol{x}, \beta_1) + u_1 \tag{3a}$$

yielding $\beta_1^e$, a consistent estimator of $\beta_1$, and the associated error term $u_1^e = y - f(\boldsymbol{x}, \beta_1^e)$. Second, consider the regression models

$$\left(u_1^e\right)^i = f_i(\boldsymbol{x}, \beta_i) + u_i \tag{3b}$$

for i = 2, 3. Applying feasible GLS to (3b) generates consistent estimators of the parameters (Antle, 1983). However, it can be noted that the variance of $u_1$ in (3a) is $f_2(\boldsymbol{x}, \beta_2)$, and the variance of $u_i$ in (3b) is $[f_{2i}(\boldsymbol{x}, \beta_{2ie}) - f_i(\boldsymbol{x}, \beta_{ie})^2]$, i = 2, 3 (see Antle 1983). It follows that both equations (3a) and (3b) exhibit heteroscedasticity. Therefore, in the third step heteroscedasticity needs to be taken into consideration in the estimation of the parameters. Antle (1983) suggests the weighted regression approach to solve it. However, these weights are given by the inverse of the variance of the corresponding error terms. As noted by Antle (1983), the estimated variances for $u_2$ or $u_3$ are not guaranteed to be positive. In situations where the weights are found to be non-positive for some observations, this precludes the use of the weighted regression approach in (3b). In such circumstances, however, unweighted regression remains feasible providing that heteroscedastic-consistent standard errors are used. This provides consistent estimates of (3b).

## 4 Empirical analysis

To illustrate the role of crop biodiversity on farm productivity and in the management of environmental risk, we present in this section some empirical evidence based on a recent study (Di Falco and Chavas 2005). This study is particularly suited for the scope of this chapter for two reasons. First, it uses data from specialised wheat farms. This is particularly important because large parts of published studies use aggregate data (region, district, township level). Second, data are from a drought-prone area where production takes place without irrigation and with limited technological progress (Sicily, Italy). Production takes places in remote and marginal areas (mostly on hills and mountains). The area is characterised by typical Mediterranean weather that is prone to severe droughts. Severe and prolonged droughts deplete soil moisture and adversely affect the productivity of the land. Therefore growing conditions in Sicily can be difficult and farmers are potentially exposed to downside risk. Thus, because of the difficult environmental conditions there may be the possibility that yield will fall below a certain threshold level.

### 4.1 *Wheat and crop biodiversity in Sicily*

Sicily is one of the largest durum wheat producers in Italy, accounting for more than 20 per cent of the national production. Among the farms in the sample, production characteristics are fairly homogenous. Both the cropping technique and technological endowments are similar among farmers. Most Sicilian farmers grow more than one cultivar in their farm. The choice of cultivars is mainly driven by agro-ecological and climatic conditions. Different cultivars provide different quantitative and qualitative responses to different weather conditions. Wheat varieties also differ in protein, gluten concentration and colour. The adoption of newer varieties is increasing rapidly. Newer varieties are typically of shorter stature and often more productive. However, old improved varieties are still widely grown, including *Adamello, Appulo, Simeto* and *Valnova*. Some of these taller varieties have been grown for decades and farmers know their performance well. It is interesting to note that a number of the old improved varieties incorporate genetic material from farmers' varieties or improved varieties used in the 1920s (e.g. *Cappelli*). The use of landraces in the field is currently scant and is mainly driven by local 'niche' markets. However, part of their genetic material is incorporated in some of the varieties that are still widely grown.

Finding a suitable diversity index is not an easy task and there is no single indicator of biodiversity that can adequately capture all interactions between genes and the environment (Smale *et al.* 1998; Brock and Xepapadeas 2003; Goeshl and Swanson 2007). Some previous empirical studies (e.g. Smale *et al.* 1998; Di Falco and Perrings 2005) make use of spatial diversity indices or are based on the count of varieties grown. However, increasing the number of varieties does not necessarily imply expanding the gene pool. Indeed, commercial varieties can be genetically very similar as they are distinguished mostly by their commercial name (Di Falco and Chavas 2005). By the same token, plant breeders seek new traits by crossing progenitors and recombining material. So, breeding activity can expand genetic diversity by incorporating some new genetic material into a commercial variety. This process results in lengthening the pedigree of the variety in order to confer tolerance, resistance and yield potential (Evenson and Gollin 1997; Smale *et al.* 2002). Therefore, in farming systems where varieties are mainly released through breeding programmes, pedigree analysis can be used in order to build more adequate diversity metrics. Pedigree complexity can be used to proxy the extent of genetic variability (Meng *et al.* 1998). Pedigree-based indicators provide a measure of latent genetic diversity, defined as the genetic variation that is manifested when challenged by environmental conditions (Souza *et al.* 1994).

This analysis follows Smale *et al.* (1998) using the number of parental combinations in the pedigree of the variety as genetic diversity indicator. The number of parental combinations is counted only the first time it appears and calculated as average over the varieties grown by the farmer. This provides an indicator that is independent of the number of varieties grown.

### 4.2 Database and econometric specification

This subsection presents the data sources, variables and crop genetic diversity indicator used in the analysis. Data are drawn from a survey conducted by the research centre on wheat *G. P. Ballatore, Enna.*[2] The survey involved sixty durum wheat farms in Sicily during 1999. The farms in the sample were scattered over more than 17,000 km$^2$. Thus, these economic units face very different agro-ecological conditions, temperatures and rainfall.

To control for agro-ecological heterogeneity, farm altitude was inserted as an explanatory variable in the estimating equations. Indeed, at different elevations, temperature and soil quality differ. This can impact soil moisture and crop sensitivity to drought. The average land size is 48 hectares and average durum wheat yield productivity is 2,920 kg/ha. All the relevant data information is summarised in Table 21.1.

In general, pesticide and fertiliser use as well as genetic diversity can impact farm productivity and the higher moments of output distribution. The average productivity effects are captured by $f_1(\boldsymbol{x}, \beta_1)$ in equation (2a) and (3a), where $\boldsymbol{x}$ includes the amount of pesticide and fertiliser use as well as a measure of crop genetic diversity. The variance and skewness effects are captured respectively by $f_2(\boldsymbol{x}, \beta_2)$ and $f_3(\boldsymbol{x}, \beta_3)$ in equations (2b)–(2c) and (3b).

Diversity, pesticides and fertiliser can interact and affect both productivity and risk. However, the direction of these interactions is not clear and the existing literature provides very little evidence about it (Priestley and Bayles 1980; Heisey *et al.* 1997; Widawsky and Rozelle 1998). To explore

[2] Altamore, L. *Valutazione Economica del Processo Produttivo del Grano Dure nelle Principali aree Cerealicole della Sicilia e del Molise, in* G. Fardella, Aspetti tecnici, economici e qualitativi della produzione di grano duro nel Mezzogiorno d'Italia, Stampa Anteprima, 1999 (in English: *Technical, economic and qualitative aspects of durum wheat production in southern Italy*).

Table 21.1 *Sample statistics*

| Variable | Definition | Sample mean | Sample standard deviation | Sample minimum | Sample maximum |
|---|---|---|---|---|---|
| Diversity | 100 * (number of parental combinations in the variety's pedigree) | 1.4 | 0.76 | 0.17 | 4.5 |
| Fertiliser | Quantity of fertiliser used per hectare in kg | 350 | 91 | 200 | 745 |
| Pesticide | Quantity of pesticide used per hectare in litres | 16.1 | 54.9 | 0.25 | 400 |
| Altitude | Farm altitude | 450 | 159 | 200 | 775 |
| Yield | Yield of durum wheat per hectare in 'quintals' | 24 | 6.65 | 14.3 | 44.4 |

Adapted from Di Falco and Chavas (2005).

these issues empirically, the same interaction terms have been included in the regressions. A quadratic specification for $f_i(\boldsymbol{x}, \beta_i)$ is adopted for $\boldsymbol{x}, i = 1, 2, 3$. However, given that some of the coefficients were not statistically different from zero, some of the insignificant variables were dropped. A particular specification involving selected quadratic and interaction effects is presented in Table 21.2.

## 5 Econometric results

The focus of this analysis is on the impact of crop biodiversity on the first three moments of the distributions of durum wheat yield. The flexible-moments approach is, therefore, applied to estimate the role of crop biodiversity, conventional inputs and some relevant interaction terms on the mean, the variance and the skewness of durum wheat yield. The associated econometric estimates are reported in Table 21.2.[3]

Most of the coefficient estimates reported in Table 21.2 are statistically different from zero at the 5 per cent significance level. This applies to mean yield, the variance of yield as well as the skewness of yield. This in turn indicates that crop genetic diversity, pesticide and fertiliser use have significant effects on the first three moments of the distribution of yield.

The estimated mean function in Table 21.2 shows that the coefficients of the linear terms for crop genetic diversity, fertiliser and pesticide are

[3] Estimates were tested for endogeneity by using a residual-based test. Given that we failed to reject the null hypothesis of exogeneity, it appears that the estimates are not adversely affected by endogeneity bias (Di Falco and Chavas 2005).

Table 21.2 *Econometric estimates of mean, variance and skewness of yield*

| Variables | Mean function $f_1(\boldsymbol{x}, \beta_1)$ | Variance function $f_2(\boldsymbol{x}, \beta_2)$ | Skewness function $f_3(\boldsymbol{x}, \beta_3)$ |
|---|---|---|---|
| Constant | −6.5 | 43* | 4813** |
| | (7.5) | (23.9) | (2137) |
| Diversity | 2.5*** | −27.38** | −5828** |
| | (0.5) | (11.7) | (2676) |
| Diversity squared | −0.106*** | – | 484* |
| | (0.027) | | (270) |
| Pesticide | 0.45*** | −1.27** | −5.05** |
| | (0.074) | (0.57) | (2.49) |
| Pesticide squared | −0.001*** | – | – |
| | (0.0002) | | |
| Fertiliser | 0.105*** | −0.077 | −10.26** |
| | (0.0035) | (0.07) | (4.45) |
| Fertiliser squared | −0.0001*** | – | – |
| | (0.00004) | | |
| Diversity* fertiliser | – | – | 10.4** |
| | | | (4.48) |
| Diversity* pesticide | – | 2.13** | – |
| | | (1.02) | |
| Altitude | 0.0015 | 0.096 | 2.11($^a$) |
| | (0.0035) | (0.079) | (1.3) |
| N = 50 | | | |

Significance levels are denoted by one asterisk (*) at the 10% level, two asterisks (**) at the 5 % level, three asterisks (***) at the 1 % level and one a ($^a$) at the 10 % two-sided test

Standard errors are presented in parentheses below the parameter estimates

Adapted from Di Falco and Chavas (2006)

positive and significant. The quadratic terms are negative and statistically significant. Therefore, this indicates that crop biodiversity, fertiliser and pesticide use each have a positive effect on mean yield, but this effect declines with input use. This also indicates that diminishing marginal expected productivity applies to both conventional inputs and crop biodiversity.

Table 21.2 also shows that crop genetic diversity and pesticide use have a significant effect on the variance of yield. Note that, in contrast to

Just and Pope (1979), we find that the effect of fertiliser on production variance is statistically insignificant. The coefficients of the linear terms for diversity and pesticide are both negative and statistically significant at the 5 per cent level. However, the coefficient of the interaction term (diversity * pesticide) is positive and significant. This indicates that the effect of diversity on yield variance varies with pesticide use. That is, crop biodiversity has a negative effect on variance when pesticide use is low, but a positive effect when pesticide use is high. Thus, while both diversity and pesticide have the potential to reduce variance, they behave as substitutes in their risk-reducing effects. Crop biodiversity is most effective in reducing variance when pesticide use is light or moderate. This suggests the presence of strong interaction effects between pest management, ecological management and risk management. Such effects need to be taken into consideration in the design and evaluation of management strategies (Di Falco and Chavas 2005).

All crop biodiversity, fertiliser and pesticide affect significantly the skewness of yield. Both fertiliser and pesticide coefficient estimates are negative. However, fertiliser use has a positive marginal impact on the skewness (evaluated at the sample mean) while pesticide use can increase downside-risk exposure. This is probably due to the low levels of pest infestation in dry environments. Crop biodiversity has an important role in risk management. Indeed, crop biodiversity is positively and significantly correlated with the skewness function. This indicates that increasing diversity hedge against the risk of crop failure and therefore keeping crop genetic diversity in the field is a viable way to manage downside-risk exposure.

## 6 Conclusion

This chapter has focused on the role of crop biodiversity on productivity and environmental risk. We started from the consideration that much of published research on the role of diversity on productivity and risk is conducted by using the stochastic production function suggested by Just and Pope (1978). The adoption of this framework implies that risk effects are captured by the variance of yields. However, when agro-ecological conditions are particularly challenging, weather is highly variable (both spatially and temporally) and technological progress is slow, the mean variance approach might not capture the full extent of risk exposure. Indeed, the exposure to downside risk can be particularly relevant when production takes place in areas with these characteristics. The risk effects, therefore, involve both a reduction in yield variance and a decrease in exposure to downside risk (e.g. severe drought leading to crop failure) and thus risk assessment needs to go beyond a simple variance assessment to capture also exposures to unfavourable downside risk.

In order to illustrate empirically these issues, this chapter has presented a case study that uses data from durum wheat farms from rainfed agriculture in drought-prone areas of Sicily, Italy. The empirical analysis indicates that crop diversity has a potential beneficial role in supporting farm productivity and in managing environmental risk. It has also been found that such diversity can reduce the variability of yields. However, the effect of diversity on yield variance appears to vary with pesticide use. While both diversity and pesticide have the potential to reduce variance, they behave as substitutes in their risk-reducing effects. This finding suggests the presence of strong interaction effects between pest management, ecological management and risk management. Lastly, crop biodiversity is found to be positively correlated with skewness of the distribution of yields. This indicates that diversity can help reduce downside-risk exposure (e.g. the probability of crop failure). Therefore, when unfavourable climatic and agro-ecological conditions expose farmers to particular environmental risks, crop diversity may become an important asset for risk management.

## REFERENCES

Altieri, M. and Liebman, M. 1986. Insect, weed and plant disease management in multiple cropping systems. In C. Francis (ed.). *Multiple Cropping Systems*. New York: Macmillan.

Antle, J. 1983. Testing the stochastic structure of production: a flexible-moment based approach. *Journal of Business and Economic Statistics*. **1**. 192–201.

Antle, J. 1987. Econometric estimation of producers' risk attitudes. *American Journal of Agricultural Economics*. **69**. 509–522.

Binswanger, H. P. 1981. Attitudes toward risk: theoretical implications of an experiment in rural India. *The Economic Journal*. **91**. 867–890.

Brock, W. A. and Xepapadeas, A. 2003. Valuing biodiversity from an economic perspective: a unified economic, ecological and genetic approach. *American Economic Review*. **93**. 1597–1614.

Chavas, J. P. 2004. *Risk Analysis in Theory and Practice*. New York: Elsevier.

Chavas, J. P. and Holt, M. T. 1996. Economic behavior under uncertainty: a joint analysis of risk preferences and technology. *Review of Economics and Statistics*. **78**. 329–335.

Di Falco, S. and Chavas, J. P. 2006. Crop genetic diversity, farm productivity and the management of environmental risk in rainfed agriculture. *European Review of Agricultural Economics*. **33**. 289–314.

Di Falco, S. and Chavas, J. P. 2005. Crop Biodiversity, Farm Productivity and the Management of Environmental Risk in Rainfed Agriculture. Paper presented at the 79th Annual Conference of the Agricultural Economic Society, Nottingham (UK). April.

Di Falco, S. and Perrings, C. 2003. Crop genetic diversity, productivity and stability of agroecosystems: A theoretical and empirical investigation. *Scottish Journal of Political Economy*. **50** (2). 207–216.

Di Falco, S. and Perrings, C. 2005. Crop biodiversity, risk management and the implications of agricultural assistance. *Ecological Economics*. **55**. 459–466.

Evenson, R. E. and Gollin, D. 1997. Genetic resources, international organizations, and improvement in rice varieties. *Economic Development and Cultural Change*. **45** (3). 471–500.

Gliessman, S. R. 1986. Plant interactions in multiple cropping systems. In C. Francis (ed.). *Multiple Cropping Systems*. New York: Macmillan.

Goeshl, T. and Swanson, T. 2007. Designing the legacy library of genetic resources: approaches, methods, and results'. In A. Kontoleon, U. Pascual and T. Swanson (eds.). *Biodiversity Economics*. Cambridge: Cambridge University Press.

Heal, G. 2000. *Nature and the Marketplace: Capturing the Value of Ecosystem Services*. New York: Island Press.

Heisey, P. W., Smale, M., Byerlee, D. and Souza, E. 1997. Wheat rusts and the costs of genetic diversity in the Punjab of Pakistan. *American Journal of Agricultural Economics*. **79**. 726–737.

Just, R. E. and Pope, R. D. (1978). Stochastic representation of production functions and econometric implications. *Journal of Econometrics*. **7**. 67–86.

Just, R. E. and Pope, R. D. 1979. Production function estimation and related risk considerations. *American Journal of Agricultural Economics*. **61**. 276–284.

Loss, S. P. and Siddique, K. H. M. 1994. Morphological and physiological traits associated with wheat yield increases in Mediterranean environment. *Advances in Agronomy*. **52**. 229–276.

Menezes, C., Geiss, G. and Tressler, J. 1980. Increasing downside risk. *American Economic Review*. **70**. 481–487.

Meng, E. C. H., Smale, M., Bellon, M. R. and Grimanelli, D. 1998. Definition and Measurement of Crop Diversity for Economic Analysis. In M. Smale (ed.). *Farmers, Gene Banks, and Crop Breeding*. Boston: Kluwer.

Naeem, S. L., Thompson, J., Lawler, S. P., Lawton, J. H. and Woodfin, R. M. 1994. Declining biodiversity can affect the functioning of ecosystems. *Nature*. **368**. 734–737.

Pecetti, L., Damania, A. B. and Kashour, G. 1992. Geographic variation for spike and grain characteristics in durum wheat germplasm adapted to dry land conditions. *Genetic Resources and Crop Evolution*. **39**. 97–105.

Priestley, R. H. and Bayles, R. A. 1980. Varietal diversification as a means of reducing the spread of cereal diseases in the United Kingdom. *Journal of the National Institute of Agricultural Botany*. **15**. 205–214.

Qualset, C. O., McGuire, P. E. and Warburton, M. L. 1995. Agrobiodiversity: key to agricultural productivity. *California Agriculture*. **49** (6). 45–49.

Smale, M. and Drucker, A. 2007. Agricultural development and the diversity of crop and livestock genetic resources: a review of the economics literature. In A. Kontoleon, U. Pascual and T. Swanson (eds.). *Biodiversity Economics*. Cambridge: Cambridge University Press.

Smale, M., Hartell, J., Heisey, P. W. and Senauer, B. 1998. The contribution of genetic resources and diversity to wheat production in the Punjab of Pakistan. *American Journal of Agricultural Economics*. **80**. 482–493.

Smale, M., Reynolds, M. P., Warburton, M., Skovmand, B., Trethowan, R., Singh, R. P., Ortiz-Monasterio, I. and Crossa, J. 2002. Dimensions of diversity in modern spring bread wheat in developing countries from 1965. *Crop Sciences*. **42**. 1766–1779.

Souza, E., Fox, P. N., Byerlee, D. and Skovmand, B. 1994. Spring wheat diversity in irrigated areas of two developing countries. *Crop Science*. **34**. 774–783.

Sumner, D. R., Doupnik, B. and Boosalis, M. G. 1981. Effects of tillage and multicropping on plant diseases. *Annual Review of Phytopathology*. **19**. 167–187.

Tilman, D. and Downing, J. A. 1994. Biodiversity and stability in grasslands. *Nature*. **367**. 363–365.

Tilman, D., Wedin, D. and Knops, J. 1996. Productivity and sustainability influenced by biodiversity in grassland ecosystems. *Nature*. **379** 718–720.

Walker T. S., Singh, R. P. and Jodha, N. S. 1983. Dimensions of farm level diversification in the semi arid tropics of rural south India. Economic Program Progress report 51. Patancheru, India: ICRISAT.

Widawsky, D. and Rozelle, S. 1998. Varietal Diversity and Yield Variability in Chinese Rice Production. In M. Smale (ed.). *Farmers, Gene Banks, and Crop Breeding*. Boston: Kluwer.

Wood, D. and Lenné, J. M. (eds.). 1999. *Agrobiodiversity: Characterization, Utilization and Management*. Wallingford: CABI.

# 22 Assessing the private value of agro-biodiversity in Hungarian home gardens using the data enrichment method

*Ekin Birol, Andreas Kontoleon and Melinda Smale*

## 1 Introduction

Hungarian agriculture today has a dual structure consisting of large-scale, mechanised farms alongside semi-subsistence, small-scale farms managed with family labour and traditional practices. Dualism has persisted in some form throughout Hungarian history. From 1955 to 1989, during the socialist period of collectivised agriculture, families were permitted to produce for their own needs on small tracts adjacent to their dwellings, commonly known as 'home gardens' (Szelényi 1998; Kovách 1999; Swain 2000; Szép 2000; Meurs 2001; Cros Kárpáti *et al.* 2004). These small-scale farms became refuges for a range of local varieties of trees, crops and livestock breeds, as well as soil micro-organisms. Agricultural scientists describe home gardens as micro-agro-ecosystems that are rich in several components of agro-biodiversity (Már and Juhász 2002; Csizmadia 2004).

Despite the changes engendered by transition to market economy during the past decade, the structure of agriculture remains dualistic,[1] in part because incomplete food markets persist. In addition to lower agricultural incomes, high inflation and unemployment rates, consumers have difficulties obtaining reliable product information and predicting product availability (Feick *et al.* 1993). Search costs and transport costs to the

We gratefully acknowledge the European Union's financial support via the 5th European Framework Biodiversity and Economics for Conservation (BIOECON) project, and funds provided by the International Food Policy Research Institute (IFPRI) and International Plant Genetic Resources Institute (IPGRI). We would like to thank Györgyi Bela, Jeff Bennett, Ágnes Gyovai, László Holly, István Már, György Pataki, the late David Pearce, László Podmaniczky, Timothy Swanson, Eric Van Dusen and Mitsuyasu Yabe for useful comments and fruitful discussions. We would also like to thank Vic Adamowicz, Jeff Bennett, Dietrich Earnhart and Joffre Swait for their very helpful suggestions.

[1] In 1994, less than 0.2 per cent of farms (public, cooperative and private) in Hungary operated 84 per cent of agricultural land, whereas 77 per cent of farms operated less than 4 per cent of land on areas smaller than 0.5 ha (Sarris *et al.* 1999).

nearest food market remain high. The number of hypermarkets in Hungary has grown from only 5 in 1996 to 63 in 2003 (Hungarian Central Statistical Office (HCSO) 2003). A study by the World Health Organisation (WHO 2005) found that these have contributed to the disappearance of the few extant local shops and markets.

Consequently, rural families continue to rely on their own production to meet their food needs and maintain diet quality. In 2001, HCSO reported that one fifth of the population produced agricultural goods for their own consumption and as a source of additional income on 697,336 small family farms with an average size of 591 $m^2$ (HCSO 2001). Szivós and Tóth (2003) estimated that 60 per cent of the households in the lowest income quartiles, most of which are located in rural Hungary, consume food from own production, with a value amounting to 19,277 Ft (€75.3) per month. Szép (2000) found that income in kind generated by part-time agricultural production in home gardens amounted to 14 per cent of total income of the households. Thus, as in other countries with economies in transition, home gardens in Hungary generate substantial private benefits (Wyzan 1996; Seeth *et al.* 1998).

Many expect that as a result of continued economic transition and the nation's accession to the European Union (EU), the dual structure of Hungarian agriculture and the share of home-produced food will eventually disappear (Sarris *et al.* 1999; Vajda 2003; Fertő *et al.* 2004; Weingarten *et al.* 2004). The rural population is expected to continue to decline and age as younger generations migrate to urban areas (Harcsa *et al.* 1994; Sarris *et al.* 1999; Juhász 2001).

If this is the case, private provision of public goods generated by home garden management cannot be sustained in the long run. Although the reformed Common Agricultural Policy (CAP) of the EU aims to promote agro-biodiversity and other public goods generated by agricultural production through multi-functional agriculture (Romstad *et al.* 2000; Lankoski 2000), the contribution of home gardens to multifunctional agriculture in Hungary appears to have been overlooked in other EU and national policies. For example, Hungary's National Rural Development Plan (NRDP) implements several agri-environmental schemes to advance the use of specified farming methods in environmentally sensitive areas (ESAs) (Juhász *et al.* 2000), but so far the role of home gardens within these schemes has not been elucidated.

This chapter identifies the least-cost options for including farming communities in Hungary's agri-environmental schemes, by characterising those communities in which farmers who value agro-biodiversity in their home gardens most are located. We employ the data enrichment method by combining revealed preference and stated preference methods, using survey and choice experiment data collected from 239 farm households across 22 communities in three regions of Hungary. Findings from the combined approach are compared to those obtained separately from the choice experiment and farm household analyses. We conclude that the data enrichment method enables more robust and efficient identification of least-cost farming communities for maintaining agro-biodiversity in Hungarian home gardens. This result is similar to those found in previous studies that have employed the data enrichment method to value non-market, environmental goods (e.g. Adamowicz *et al.* 1994; Earnhart 2001; 2002).

The analysis makes several contributions to the literature on the valuation of biodiversity resources. Perhaps most importantly, it confirms that the data enrichment approach leads to policy recommendations that are distinct from those generated by either revealed preferences or stated preferences alone. Thus, use of one or the other, instead of both, could lead to inappropriate recommendations. Further, in the literature on the valuation of agro-biodiversity, this analysis is the first to combine and compare data for the same households rather than comparing data from different sources (ECOGEN 2005). Third, it contributes to the body of non-market valuation studies applied in the agriculture context by estimating farmers' private valuation of an entire micro-agro-ecosystem. Previous stated preference choice experiment studies have investigated the public's valuation of components of agri-environmental schemes (e.g. Hanley *et al.* 1998; Campbell *et al.* 2006), as well as farmer demand for certain traits of livestock breeds (Scarpa *et al.* 2003a; Scarpa *et al.* 2003b; Ruto 2005) and crop genetic resources (Ndjeunga and Nelson 2005). In addition, previous revealed preference hedonic analysis studies have estimated the value of attributes of crop genetic resources (e.g. Unnevehr 1986; Dalton 2004; Langyintuo *et al.* 2004; Edmeades 2006), and animal genetic resources (e.g. Richards and Jeffery 1995; Jabbar *et al.* 1998; Jabbar and Diedhiou 2003). Fourth, the study presented in this chapter is the first one that combines choice experiment data with farm household data. Other studies have combined choice experiment data with travel cost data, hedonic pricing data or have combined contingent valuation data with travel cost data (Cameron 1992; Adamowicz *et al.* 1994; 1997; Englin and Cameron 1996; Kling 1997; Rosenberger and Loomis 1999; Boxall *et al.* 2002; Earnhart 2001; 2002). Finally, in the relatively scant literature on data enrichment, this analysis is the first related to agro-biodiversity (previous studies have combined data on recreation, environmental amenity, cultural heritage and market goods (e.g. housing market, transportation) (see e.g. Adamowicz *et al.* 1994; 1997; Rosenberger and Loomis 1999; Boxall *et al.* 2002; Earnhart 2001; 2002)). Findings have implications for the design of cost-effective agri-environmental schemes in other EU member Central and Eastern European Countries (CEECs) with similar dual agricultural structures, such as Slovenia and Poland.

The next section describes the rationale for enrichment of data by combining stated and revealed preference valuation methods. The following section summarises data collection methods and data. Section 4 presents the theoretical basis of the approach. Section 5 reports the results of the econometric analyses and the final section states the policy implications and concludes the chapter.

## 2 Rationale for data enrichment by combining revealed and stated preference data

Methods for valuing non-market, public goods are categorised as revealed preference or indirect methods and stated preference or direct methods. Revealed preference methods use actual choices made by consumers in related or surrogate markets, in which the non-market good is implicitly traded, to estimate the value of the non-market good. Stated preference methods have been developed to solve the problem of valuing those non-market goods that have no related or surrogate markets. In these approaches, consumer preferences are elicited directly based on hypothetical, rather than actual, scenarios.

Both stated and revealed preference methods have advantages and disadvantages. Stated preference methods are commonly criticised because the behaviour they depict is not observed in reality (Cummings *et al.* 1986; Mitchell and Carson 1989) and generally fail to take into account certain types of real market constraints (Louviere *et al.* 2000). Nonetheless, these methods provide the only means for estimating the value of public goods that have no related or surrogate markets. In cases where the revealed data are limited and do not encompass the range of proposed quality or quantity changes in the attributes of a public good, stated preference method can be used to cover a much wider range of attribute levels, and hence can be used to consider choices that are fundamentally different from existing ones. Moreover, the choice experiment method, in particular, can be used to measure the value of changing the quantity or quality of multiple attributes of a public good. This stated preference model tends to be more robust than the revealed preference models since a wider and broader array of attributes can be build into choice experiments resulting in rich attribute tradeoff information (Swait *et al.* 1994).

Revealed preference data have high 'face validity' because the data reflect real choices taking into account various constraints on individual decisions, such as market imperfections, budgets and time (Louviere *et al.* 2000). A major drawback of using revealed preference data is that because the attributes and attribute levels of a non-market good, which are explanatory variables, do not vary much over time, their parameters may not be estimated. Hence, the value of changes in the quality or quantity provided of the public good could be difficult to estimate. Moreover, attributes in models estimated from choices in actual settings cannot predict the impact of changing policies (Louviere *et al.* 2000). In other words, revealed preference methods may suffer on the grounds that the new situation (after the change in the quality or the quantity of the non-market good) may be outside the current set of experiences (or outside the data range). Thus, simulation of the new situation would involve extrapolation outside the range used to estimate the model (Adamowicz *et al.* 1994). Moreover, collinearity among multiple attributes is common in revealed preference data, generating coefficients with the wrong signs or implausible magnitudes, and making it difficult to separate attribute effects (Freeman 1993; Greene 1997; Louviere *et al.* 2000; Hensher *et al.* 2005). Separation of these attributes may be necessary, however, in order to accurately represent benefits and costs in policy analysis (Adamowicz *et al.* 1994).

Recent research indicates that combining the stated and revealed preference methods through data fusion, which is also known as the data enrichment method, builds on the strengths and diminishes the drawbacks of each method. The amount of information increases, and findings can be cross-validated (Haab and McConnell 2002). Use of revealed preference data ensures that estimation is anchored in observed behaviour. At the same time, inclusion of stated preference responses to hypothetical changes enables identification of parameters that otherwise would not be identified. Fusing data sources is similar to creating an 'artificial' panel of data, with revealed preference methods generating cross-sectional data about the observed, current choices of consumers facing real market constraints, and stated preference data recording the options the same consumers might choose at another point in time. Revealed preference models are suitable for short-term forecasting of small departures from the current state, whereas a stated preference model is more appropriate to predict structural changes that occur over longer time periods (Louviere *et al.* 2000). The choice experiment method, a stated preference approach, also employs statistical design to eliminate colinearity among the attributes of goods. Overall, combining stated and revealed responses improves the efficiency of estimates of values of the changes in the quality/quantity of the non-market good over time (Haab and McConnell 2002). Thus, the accuracy of welfare measures derived from non-market, public goods are improved through applying the data enrichment method (Adamowicz *et al.* 1994; Earnhart 2001; Haab and McConnell 2002).

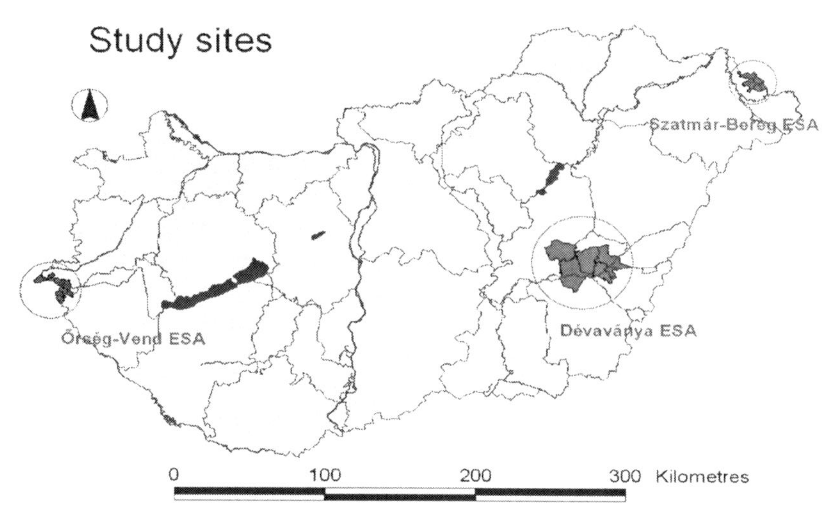


Figure 22.1. Location of the selected ESAs
Source: GIS Laboratory, Institute of Environmental and Landscape Management, Szent István University, Gödöllo, Hungary

The revealed preference approach we employ is a discrete-choice farm household model, similar to the discrete-choice hedonic model applied by Cropper *et al.* (1993) and Earnhart (2001; 2002). We use the choice experiment method as the stated preference approach, for several reasons. First, the purpose of the study is to value the agro-biodiversity components, or multiple attributes, of home gardens. The choice experiment method also reduces some of the biases inherent in the contingent valuation method. Since the survey format is designed to mimic the actual choices of farm households, it is less prone to hypothetical bias. Respondents have a solid understanding of the good being valued in the choice experiment, so that the likelihood of information bias is greatly reduced (Earnhart 2001; 2002; Bateman *et al.* 2003). We also analyse the choice experiment data with a discrete-choice econometric model.

## 3 Data collection

The sample design for the choice experiment and farm household surveys consisted of two stages. In the first stage, secondary data from HCSO (2001) and NRDP were used to select three ESAs (Dévaványa, Őrség-Vend and Szatmár-Bereg) amongst eleven ESAs identified by the NRDP (Figure 22.1). These three ESAs were purposively selected to represent contrasting levels of market development and varying agro-ecologies associated with different farming systems and land-use intensity. In each selected site, pilot agri-environmental schemes were under way and high levels of agro-biodiversity (in terms of crop genetic diversity) have been identified (Már 2002).

Twenty-two communities (five in Dévaványa, eleven in Őrség-Vend and six in Szatmár-Bereg) were included in the sample. Dévaványa, located on the Hungarian Great Plain, is closest to the economic centre of the country of the three ESAs. Soil and climatic conditions are well suited to intensive agricultural production. Populations, areas and population density are relatively high. Labour migration is not a major problem, although the number of inhabitants is stagnating. The unemployment rate in this region (12.4 per cent) is slightly higher than the Hungarian average. Dévaványa is statistically different from the other two ESAs in most indicators of economic development and market integration, including: presence of a train station; distance to the nearest market; number of primary and secondary schools; food markets; and the number of shops and enterprises.

The two isolated ESAs are more similar to each other than either is to Dévaványa. Located in the southwest, Őrség-Vend has a heterogeneous agricultural landscape with poor soil conditions that render intensive agricultural production methods impossible. Communities are very small in area and most are far from towns and markets. Of the three ESAs, Őrség-Vend is the least developed with fewest shops and enterprises. Its small population is declining and ageing, though the unemployment rate of this region is the lowest in the country at 4.8 per cent. Szatmár-Bereg is situated in the northeast, far from the economic centre of the country. The agricultural landscape of this ESA is heterogeneous and climatic conditions are unfavourable. Communities in Szatmár-Bereg are also small, and its declining, ageing population reflects a lack of public investments in infrastructure and employment generation. Roads are of poor quality and the regional unemployment rate is the highest in the country (19 per cent) (National Labour Centre 2000; Juhász *et al.* 2000; Gyovai 2002).

In the second stage of the sample design, all communities within each ESA were sorted based on population sizes, and an initial sample of 1,800 households (600 households per ESA) was sampled randomly from a complete list of all households compiled from telephone books and village maps. A screening survey was sent to all of the 1,800 households to identify all those engaged in home garden management. The response rate to the screening survey was only 13 per cent, but the final sample was augmented through personal visits to listed sample households with the assistance of key informants in each community. A total of 239 farm households (74 in Dévaványa, 81 in Őrség-Vend and 84 in Szatmár-Bereg) were personally interviewed in August 2002 with the choice experiment and household surveys. All households sampled had home gardens, and the findings reported in this chapter are statistically representative of the selected ESAs, as well as other ESAs in Hungary to the extent that they share characteristics.

### *3.1 Choice experiment data*

The most important components of agro-biodiversity managed in Hungarian home gardens were identified with NRDP experts and agricultural scientists, drawing on the results of informal and focus group interviews with farmers in each ESA. This background work resulted in the attributes and levels used in the choice experiment (Table 22.1). Each attribute represents a different component of agro-biodiversity. The total number of crop varieties grown in a home garden of fixed size is an indicator of crop variety richness. In this choice experiment both inter- and infra-species diversity of field crops, trees and vegetables are considered. Crop variety diversity is one of the most crucial components of agro-biodiversity (FAO 1999). Presence of a landrace or a traditional, local variety in the home garden expresses crop genetic diversity.[2] Preliminary molecular analysis and agro-morphological evaluation conducted on bean landrace samples collected from the sampled households' home gardens reveal that the majority of these landraces are distinct and identifiable and contain rare and adaptive traits, and are genetically heterogeneous (Már and Juhász 2002). The traditional method of integrated crop and livestock production represents agro-diversity, or diversity in agricultural management practices (Brookfield and Stocking 1999). Organic production takes place if crops are grown without any industrially produced and marketed chemicals, such as pesticides, herbicides, insecticides, fungicides or soil disinfectants. Previous experiments found that use of organic production methods resulted in soil micro-organism diversity (e.g. Lupwayi *et al.* 1997; Mäder *et al.* 2002). The expected percentage of the annual household food consumption supplied by the home garden, i.e. food self-sufficiency, represents the family's dependence on its own production. The proxy monetary attribute is converted into actual monetary units for each household by using secondary data from HCSO on the regional-level household expenditure on food consumption.

[2] Landraces, or traditional varieties or local varieties, are variants, varieties, or populations of crops, with plants that are often highly variable in appearance, whose genetic structure is shaped by farmers' seed selection practices and management, as well as natural selection processes, over generations of cultivation.

Table 22.1 *Home garden attributes and attribute levels used in the choice experiment*

| Home garden attribute | Definition | Attribute levels |
|---|---|---|
| Crop variety diversity | The total number of different crop species and varieties that are cultivated in the home garden. | 6, 13, 20, 25 |
| Landrace | Whether or not the home garden contains a crop variety that has been passed down from the previous generation and/or has not been purchased from a commercial seed supplier. | Home garden contains a landrace vs. home garden does not contain a landrace |
| Agro-diversity | Integrated crop and livestock production on the home garden, representing diversity in agricultural management system. | Integrated crop and livestock production vs. specialised crop production |
| Organic production | Whether or not industrially produced and marketed chemical inputs are applied in home garden production. | Organic production vs. non-organic production |
| Food self-sufficiency | The percentage of annual household food consumption that it is expected the home garden will supply. | 15 per cent, 45 per cent, 60 per cent, 75 per cent |

Source: Hungarian Home Garden Choice Experiment, Hungarian On-Farm Conservation of Agro-biodiversity Project 2002

Assuming that the following home gardens were the only choices you have, which one would you prefer to manage?

| Home garden characteristics | Garden A | Garden B | |
|---|---|---|---|
| Total number of crop varieties grown in the home garden | 25 | 20 | |
| Home garden has a landrace | No | Yes | |
| Crop production in the home garden is integrated with livestock production | Yes | Yes | Neither home garden A nor home garden B: I will not manage a home garden |
| Home garden crops produced entirely using organic methods | No | No | |
| Expected proportion (in %) of annual household food consumption met through food production in the home garden | 45 | 75 | |
| I prefer to cultivate | Garden A... | Garden B... | Neither garden ...... |

Figure 22.2. Sample choice set

A large number of unique home garden prototypes can be constructed from this number of attributes and levels.[3] Using SPSS Conjoint 8.0 software and experimental design theory, main effects, consisting of 32 pair-wise comparisons of home garden prototypes, were recovered with an orthogonalisation procedure.[4] These were randomly blocked to six different versions, two with six choice sets and the remaining four with five choice sets. In face-to-face interviews, each farmer was presented with five or six choice sets, each containing two home garden management strategies and an option to select neither. The farmers who took part in the choice experiment were those responsible for making decisions in the home garden. Enumerators explained the context in which choices were to be made (a 500 $m^2$ home garden); that attributes of the home garden had been selected as a result of prior research and were combined artificially. Overall, a total of 1,279 choices were elicited from 239 farmers taking part in the choice experiment. An example of a choice set is provided in Figure 22.2.

[3] The total number of home garden prototypes that can be generated is $4^2 * 2^3 = 128$.

[4] Although exclusion of interaction effects in the experimental design may introduce bias into main effects estimations, main effects usually account for more than 80 per cent of the explained variance in a model (Louviere 1988; Louviere *et al.* 2000). Moreover, the aim of this choice experiment was to investigate farmer demand for each home garden attribute independently of the others. As explained in Section 2 above, an advantage of the choice experiment approach relative to revealed preference approaches is that the effects of each attribute on respondents' demand for the good can be separated, avoiding collinearity between the attributes (Adamowicz *et al.* 1994; 1997).

Table 22.2 *Home garden management and agro-biodiversity by ESA*

| Home garden attribute | Dévaványa N = 74 | Őrség-Vend N = 81 | Szatmár-Bereg N = 84 |
|---|---|---|---|
| | | Mean (s.e.) | |
| Home garden area ($m^2$)*** | 460.2 (456) | 1817.6 (3257.6) | 2548.1 (2850.3) |
| Annual food self-sufficiency (€)** | 526.9 (572) | 398.1 (582.6) | 661.2 (676.8) |
| Crop variety diversity*** | 17.2 (9.1) | 28.5 (12.2) | 18.6 (7.3) |
| | | Percentage | |
| Landrace cultivation*** | 23.4 | 53.1 | 55.4 |
| Agro-diversity | 71.6 | 75.3 | 85.9 |
| Organic production* | 16.2 | 17.3 | 4.4 |

Source: Hungarian Home Garden Diversity Household Survey, Hungarian On-Farm Conservation of Agro-biodiversity Project 2002 T-tests and Pearson Chi square tests show significant differences among at least one pair of ESAs (*) at 10 per cent significance level; (**) at 5 per cent significance level and (***) at 1 per cent significance level

### 3.2 *Farm household data*

The farm household data were collected through a structured survey, which gathered information about the households' social and economic characteristics, and their consumption and agricultural production characteristics, including the four components of agro-biodiversity that can be found in their home gardens. Information on the percentage of annual household income spent on food consumption was also collected, and this was converted into actual monetary value of food self-sufficiency provided by the home garden for each household, by using secondary data from HCSO on the regional-level household expenditure on food consumption.

Actual home garden areas, annual food self-sufficiency provided by home gardens and agricultural biodiversity levels found on home gardens surveyed are reported by ESA in Table 22.2. Home gardens in Dévaványa, the most densely populated region and the most favoured ESA in terms of agro-ecological conditions and market and other infrastructure, are the smallest and those in Szatmár-Bereg, the most isolated ESA, are the largest. The annual value of the food self-sufficiency provided by the home garden is the highest in Szatmár-Bereg, the region with the lowest average incomes. Crop variety diversity is significantly higher in Őrség-Vend than in the other two ESAs. In Dévaványa, the percentage of households growing landraces is less than half of that found in the other two. Use of organic production methods in the home garden is similarly represented in Dévaványa and Őrség-Vend. Across the three sites, the majority of households tend livestock along with crops in their homestead plots, with no statistically significant differences. Only 4.4 per cent of farmers in Szatmár-Bereg, the region with the largest home gardens, employ organic practices, which is significantly lower than in the other regions.

## 4 Theoretical framework

The theoretical basis for modelling the choice of farm management strategies in both the farm household and choice experiment analyses is random-utility theory (Luce 1959; McFadden 1974). The farm household chooses a management strategy that determines levels of agro-biodiversity and food self-sufficiency, the attributes of interest. Heterogeneity among farm households, markets and agro-ecological conditions leads to variation in choices. Variation in choices in turn leads to variation in attribute levels.

The overall utility the farm household $n$ derives from management strategy $i$, $U_{in}$, consists of a deterministic component, $V_{in}$, and a random component, $e_{in}$ : $U_{in} = V_{in}(\mathbf{Z_i}, \mathbf{C_n}) + e_{in}$. The deterministic component is modelled as an indirect utility function conditional on $\mathbf{Z_i}$, the vector of management attributes, and $\mathbf{C_n}$, the vector of household, market and agro-ecological characteristics which are specific to individual households and influence utility. Denote $\pi_n(i)$ as the probability that farm household $n$ chooses management strategy $i$ rather than management strategy $j$ among all the feasible management strategies available in the set $H_n$. If the random components are identically and independently distributed (IID) Type I Extreme Value with scale factor $\mu$, then $\pi_n(i)$ is of the logit form:

$$\begin{aligned}\pi_n(i) &= \text{Prob}(V_{in} + e_{in} \geq V_{jn} + e_{jn} : j \in H_n) \\ &= \exp(\mu V_{in}) / \Sigma_{j \in H} \exp(\mu V_{jn}). \qquad (1)\end{aligned}$$

The scale factor $\mu$, which is inversely related to the variance of the error term in the conditional logit model (Ben-Akiva and Lerman 1985), plays a crucial role in the process of combining data. Equation [1] shows that the scale factor and the vector of utility parameters of the estimated model are inseparable and multiplicative. Hence, it is not possible to identify $\mu$ within a single data set and it is generally assumed to equal one. The scale factor associated with any data source, however, affects the values of the estimated parameters. This in turn implies that one cannot directly compare parameters from different data sources, even if the two data sources were generated by the same utility function. Consequently, a statistical test is required to determine if the parameter equality holds between data sets after accounting for scale (i.e. variance) differences (Swait and Louviere 1993; Louviere *et al.* 2000).

The structure of the choice experiment restricts $H_n$ to three. In the farm household analysis, on the other hand, farmers select one specific home garden management strategy from an infinitely large number of management strategies available to them. One means of coping with this mismatch is to select a subset of observations that includes both the strategies chosen by farmers and a fixed number of rejected strategies randomly drawn from the feasible set. Because they take account of both observed and rejected options, regression estimates are consistent and reflect the correct choice model (McFadden 1978; Earnhart 2001; 2002). Parsons and Kealy (1992) show that even a limited number of alternatives, as small as three, is appropriate for randomly drawn opportunity sets in a random utility model, whereas Chattopadhyay (2000) uses two alternatives.

For the revealed preference analysis, we postulate that the feasible set for each farmer consists of all management strategies that have been employed in the community in which the household is located, as elicited by the farm household survey. Given the similarities in farm household-level, agro-ecological and market-related conditions within each community, it seems reasonable to assume that any farm household in any one community could feasibly choose any other management strategy that has already been undertaken in its community. For each farm household in the farm household analysis, the feasible set includes two randomly drawn alternatives and the home garden management strategy undertaken by the household.

## 5 Econometric analysis

As explained above, it is feasible to combine the two discrete-choice models employed in this chapter since they reflect the same process of selecting a home garden management strategy. Both are applications of random utility theory. Each selection model considers the same attributes: four agro-biodiversity components and food self-sufficiency. Since each model is based on random utility theory and can be estimated with the conditional logit model, each can be used for welfare analysis of agro-biodiversity levels. The results of each approach taken alone and the combined approach can be compared (Cropper *et al.* 1993; Adamowicz *et al.* 1994; Earnhart 2001).

### 5.1 *Separate estimation of revealed and stated preference data*

Using the complete data set from all three ESAs, conditional logit models with logarithmic and linear specifications were compared for the two approaches. In both cases the highest value of the log-likelihood function was found for the specification with all of the attributes in linear form. For the population represented by the sample, indirect utility from home garden attributes takes the form

$$V_{in} = \beta_1(Z_{crop\ variety\ diversity}) + \beta_2(Z_{landrace}) + \beta_3(Z_{agro\text{-}diversity}) + \beta_4(Z_{organic}) + \beta_5(Z_{foodself\text{-}sufficiency}) \quad (2)$$

where $\beta_{1-5}$ refer to the vector of coefficients associated with the vector of attributes describing home garden characteristics. The regression equation is estimated with a conditional logit model using full-information maximum likelihood techniques (LIMDEP 8.0 NLOGIT 3.0).

The results of the conditional logit estimation for the stated preference, choice experiment model are reported in Table 22.3. The null hypothesis that the separate effects of ESAs are equal to zero was rejected with a Swait-Louviere log-likelihood ratio test and six degrees of freedom at the 0.5 per cent significance level, based on regressions with the pooled and separate ESA samples. Findings are therefore reported only for the separate ESA samples.

In Dévaványa, where food markets are fully developed, and home gardens are small in size, the stated preferences of farm households for crop variety diversity is statistically insignificant. Nevertheless, the observed richness of crop varieties in Dévaványai home gardens is as high as it is in Szatmár-Bereg, one of the two more remote ESAs. Farmer demand for landraces is also insignificant in this ESA. Few sample farmers in Dévaványa cultivate landraces. The stated demand for agro-diversity is large and significant, owing to complementarity between field crop production and animal husbandry in the home garden. There is also a significant and relatively large stated demand for organic production.

In the isolated ESA of Őrség-Vend, where community-level food markets are lacking, farmers state clearly preferences for home gardens with diverse crop varieties and landraces. No stated demand for organic production is evident in this ESA, perhaps reflecting its poor soil quality,

Table 22.3 *Conditional logit regression of stated preference, choice experiment data*

| | Coefficient (s.e.) | | |
|---|---|---|---|
| Attribute | Dévaványa | Őrség-Vend | Szatmár-Bereg |
| ASC | 0.409* (0.277) | −0.242 (0.247) | 0.392*(0.273) |
| Crop variety diversity | 0.009 (0.011) | 0.018** (0.010) | 0.014*(0.010) |
| Landrace | 0.015 (0.076) | 0.185*** (0.071) | 0.123** (0.069) |
| Agro-diversity | 0.447*** (0.079) | 0.268*** (0.076) | 0.40*** (0.074) |
| Organic production | 0.213*** (0.083) | 0.084 (0.076) | 0.121* (0.075) |
| Food self-sufficiency | $0.183 \times 10^{-5}$*** | $0.347 \times 10^{-5}$*** | $0.356 \times 10^{-5}$*** |
| | $(0.682 \times 10^{-6})$ | $(0.576 \times 10^{-6})$ | $(0.705 \times 10^{-6})$ |
| Sample size | 393 | 436 | 450 |
| $\rho^2$ | 0.092 | 0.103 | 0.167 |
| Log likelihood | −381.99 | −419.52 | −401.01 |

Source: Hungarian Home Garden Choice Experiment, Hungarian On-Farm Conservation of Agro-biodiversity Project 2002. (*)10 per cent significance level; (**)5 per cent significance level; (***) 1 per cent significance level two-tailed tests

even though in reality this ESA supports the highest percentage of home gardens managed with this method.

In Szatmár-Bereg, the other isolated ESA, where market infrastructure is also poor, farm households also state a positive and significant demand for landraces and higher levels of crop variety diversity. Farmers in Szatmár-Bereg also place great importance on agro-diversity. The high unemployment rates in this ESA may render labour-intensive animal husbandry practices less costly in terms of the opportunity cost of time. The coefficient on organic production is positive and significant, even though the sample data indicate that, among the three ESAs, Szatmár-Bereg supports the lowest percentage of home gardens managed with this method.

Across the three ESAs, the level of food self-sufficiency obtained from the home garden contributes positively and significantly to the demand for a hypothetical home garden management strategy with higher levels of this attribute. An alternative specific constant (ASC) was included in the stated preference model to account for the proportion of respondents selecting one or the other of the management strategies offered in the experiment.[5] The sign on the ASC is positive and significant for

[5] While coding the data the ASC were equalled to one when the respondent chose either home garden A or B and to 0 when the respondent chose 'neither home garden' alternative.

Table 22.4 *Conditional logit regression of revealed preference, farm household data*

| | Coefficient (s.e.) | | |
|---|---|---|---|
| Attribute | Dévaványa | Őrség-Vend | Szatmár-Bereg |
| Crop variety diversity | 0.009 (0.023) | −0.381 (0.016) | 0.065*** (0.023) |
| Landrace | −0.167 (0.197) | 0.044 (0.219) | 0.218* (0.169) |
| Agro-diversity | −0.003 (0.175) | 0.175 (0.225) | −0.303 (0.255) |
| Organic production | −0.268* (0.222) | −0.101 (0.199) | 0.029 (0.325) |
| Food self-sufficiency | $0.276 \times 10^{-5}$** | $0.10 \times 10^{-5}$* | $0.380 \times 10^{-5}$** |
| | ($0.133 \times 10^{-5}$) | ($0.619 \times 10^{-6}$) | ($0.167 \times 10^{-5}$) |
| Sample size | 74 | 81 | 84 |
| $\rho^2$ | 0.054 | 0.022 | 0.090 |
| Log likelihood | −66.88 | −77.00 | −73.97 |

Source: Hungarian Home Garden Choice Experiment, Hungarian On-Farm Conservation of Agro-biodiversity Project 2002. (*) 10 per cent significance level; (**) 5 per cent significance level; (***) 1 per cent significance level two-tailed tests

Dévaványa and Szatmár-Bereg, which is puzzling. A negative sign on the ASC coefficient would imply that farmers are highly responsive to changes in choice set quality and make decisions that are closer both to rational choice theory and the behaviour observed in reality (Dhar 1997; Huber and Pinnell 1994). The sign on the ASC coefficient for Őrség-Vend is statistically insignificant, although it carries the expected negative sign.

Table 22.4 reports the conditional logit estimates for the revealed preference, discrete choice, farm household model. The null hypothesis that the separate effects of ESAs are equal to zero was again rejected with a Swait-Louviere log-likelihood ratio test and five degrees of freedom at the 0.5 per cent significance level and only separate regression results are reported. In Dévaványa, the only significant agro-biodiversity attribute is organic production, and its sign is negative, contrary to the result of the stated preference model reported above. In Őrség-Vend none of the agro-biodiversity attributes affects the home garden management choice. Similarly to the results of the stated preference data, Szatmári households are more likely to manage home gardens with higher levels of crop variety diversity and landraces. Across the three ESAs the food self-sufficiency attribute positively and significantly affects the likelihood that the households would choose to manage a home garden, which provides higher levels of food security. Finally, in this model an ASC was not included since in reality all the farmers interviewed manage home gardens.

Overall, the results of the revealed preference models are not highly significant. The parameters of several attribute levels could not be estimated in this model because of collinearity between attributes and lack of variation in the data for some of the attributes (e.g. agro-diversity). Some researchers have argued that since the attributes and variables in the revealed preference data sets are likely to be ill-conditioned (largely invariant and suffer from collinearity) parameter estimates are likely to be biased or statistically insignificant. For this reason, many analysts prefer analyses based on stated preference data. Another alternative is to combine the data sets (Hensher *et al.* 2005), as presented in the next section.

### 5.2 *Joint estimation of revealed and stated preference data*

Swait and Louviere (1993) describe the appropriate steps for joint estimation of data sources and parameter comparison, followed here. First, we estimated the stated and revealed preference models separately to generate the log likelihood values for each data set (reported in Tables 22.3 and 22.4). Second, we concatenated or 'stacked' the two data set matrices and estimated the joint model to obtain a single, shared set of parameters.

The most common econometric approach to use when combining revealed and stated preference data is a two-level nested logit model, known as an 'artificial tree structure' (Hensher and Bradley 1993). Though this structure has no obvious behavioural meaning, it is a convenient statistical model, designed to uncover differences in scale (i.e. variance) between the data sets while estimating model parameters (Louviere *et al.* 2000; Hensher *et al.* 2005).

A nested logit model is a hierarchy of conditional logit models, linked via a tree structure. Conditional logit models underlie the data within each cluster, implying that the assumption of constant variance (scale factors) must hold within each cluster, although they can differ between clusters. By accommodating different scale factors between clusters explicitly, and estimating the scale factor of one data set relative to that of the other, nested logit provides a simple way to accomplish the estimation process required to fuse the revealed preference and stated preference data sources.

In order to illustrate the fusion process undertaken in this chapter, consider a nested logit model with two levels (revealed preference farm household model and stated preference choice experiment model), each with a cluster of three alternative home garden profiles. The choice model in each cluster is conditional logit, so that the scale of each cluster is equal to the inverse of the cluster inclusive value. The cluster inclusive value

Table 22.5 *Conditional logit regression of combined stated and revealed preference data*

| Attributes in the Utility Functions | | | |
|---|---|---|---|
| | Coefficient (s.e.) | | |
| Attribute | Dévaványa | Őrség-Vend | Szatmár-Bereg |
| ASC | $-1.373^{***}$ (0.195) | $-1.356^{***}$ (0.187) | $-1.406^{***}$ (0.189) |
| Crop variety diversity | $0.046^{***}$ (0.009) | $0.053^{***}$ (0.009) | $0.065^{***}$ (0.008) |
| Landrace | 0.062 (0.071) | $0.194^{***}$ (0.068) | $0.164^{***}$ (0.066) |
| Agro-diversity | $0.405^{***}$ (0.075) | $0.287^{***}$ (0.073) | $0.431^{***}$ (0.076) |
| Organic production | $0.223^{***}$ (0.078) | $0.091^{*}$ (0.072) | $0.169^{**}$ (0.077) |
| Food self-sufficiency | $0.442 \times 10^{-5***}$ | $0.415 \times 10^{-5***}$ | $0.276 \times 10^{-5***}$ |
| | $(0.560 \times 10^{-6})$ | $(0.468 \times 10^{-6})$ | $(0.133 \times 10^{-5})$ |
| | Inclusive Value Parameters | | |
| RP1 | 1.00 $(0.3 \times 10^{8})$ | 1.00 $(0.7 \times 10^{15})$ | 1.00 $(1.27 \times 10^{8})$ |
| SP1, SP2, SP3 | Fixed Parameters | | |
| Sample size | 467 | 517 | 534 |
| $\rho^2$ | 0.121 | 0.136 | 0.200 |
| Log likelihood | −450.98 | −491.40 | −469.49 |

Source: Hungarian Home Garden Choice Experiment, Hungarian On-Farm Conservation of Agro-biodiversity Project 2002. (*) 10 per cent significance level; (**) 5 per cent significance level; (***) 1 per cent significance level two-tailed tests

is a parameter estimate used to establish the extent of (in)dependence between linked choices. It is possible to identify only one of the relative scale factors by normalising the inclusive value of the other data set to unity. The nested structure we used assumes that the inclusive value parameters associated with all revealed preference alternatives are equal and fixes the inclusive value parameter of the stated preference at unity. The nested logit model was estimated in LIMDEP 8.0 NLOGIT 3.0 using full-information maximum likelihood techniques.

The results of the combined model are reported in Table 22.5. Similarly to the revealed and stated preference data, the null hypothesis that the separate effects of ESAs are equal to zero was rejected with a Swait-Louviere log-likelihood ratio test and six degrees of freedom at the 0.5 per cent significance level, based on regressions with the pooled and separate ESA samples.

The inclusive value parameter for both of the branches is one for each of the ESAs, implying that the variances are equal. It is not uncommon for the variance structure of the stated and revealed preference data sets to be statistically similar (Adamowicz *et al.* 1997; Hensher *et al.* 2005).

Following Swait and Louviere (1993), we then tested the hypothesis that parameters are equal for the two data sets with the Swait-Louviere log-likelihood ratio test comparing the joint (restricted) and individual (unrestricted) models. We failed to reject the hypothesis that the parameters are equal at the 25 per cent significance level for Szatmár-Bereg and Őrség-Vend and at the 95 per cent significance level for Dévaványa at six degrees of freedom.[6] For all three ESAs, the data are consistent with the hypothesis that the revealed and stated preference data are compatible. By inference, the revealed and stated preference models share the same preference structures.

In Dévaványa, farmers choose to manage home gardens with livestock, organic practices and higher numbers of crop varieties, but without landraces. In Őrség-Vend and Szatmár-Bereg, the economically, geographically and agro-ecologically more marginalised ESAs, farmers choose to manage home gardens with not only these attributes, but also landraces. Across the three ESAs, the level of food self-sufficiency provided by the home garden also contributes positively and significantly to the demand for a home garden management strategy. This finding illustrates the fact that farm households across Hungary still depend on their home gardens for food security and diet quality.

The parameter estimates of the combined model have the same signs as those of the stated preference model, but with the enriched data, two additional factors were identified as statistically significant (crop variety diversity in Dévaványa and organic production methods in Őrség-Vend). The collinearity in the revealed preference model has been reduced by its fusion with the orthogonally designed, stated preference data (Adamowicz *et al.* 1994). As shown by Swait and Louviere (1993), the combined analysis therefore improves the precision and stability of the estimates of model parameters.

The significance levels for almost all of the attributes improved considerably. Estimation of the combined model also enabled estimation of ASCs, which was not feasible with revealed preference data alone (Adamowicz *et al.* 1997). Recall also that the ASC coefficients in the stated preference model were positive or statistically insignificant. The joint model resulted in estimations that are closer both to rational choice theory and to observed behaviour.

In addition, estimation results reveal that actual and hypothetical home garden management decisions are guided by similar decision processes with regard to each of the agro-biodiversity and food self-sufficiency

[6] The degrees of freedom equal the number of parameters in the revealed data model plus the number of parameters in the stated data model minus the number of parameters in the joint model plus one additional degree for the relative scale factor (Louviere *et al.* 2000).

attributes. Finally, the results of the combined data estimations are more efficient than those of the stated preference data, since for each one of the ESA-level regressions the overall fit of the models, as measured with McFadden's $\rho^2$, improves with data fusion. The combined model outperforms the individual models, as found in several other studies (e.g. Earnhart 2001; 2002; Haener *et al.* 2001; Adamowicz *et al.* 1994, 1997).

### 5.3 *Welfare measures of agro-biodiversity management in the home garden*

The choice experiment method and the discrete-choice, farm household-model are both consistent with utility maximisation and demand theory. From the parameter estimates reported in Tables 22.3–22.5 above, welfare measures can be estimated from the conditional logit model using the following formula:

$$CS = \frac{\ln \sum_i \exp(V_{i1}) - \ln \sum_i \exp(V_{i0})}{\alpha} \tag{3}$$

where CS is the compensating surplus welfare measure, $\alpha$ is the marginal utility of income (represented by the coefficient of the food self-sufficiency attribute) and $V_{i0}$ and $V_{i1}$ represent indirect utility functions before and after the change under consideration. The marginal value of change in a single attribute can be represented as a ratio of coefficients, reducing equation (3) to

$$W = -1\left(\frac{\beta_{attribute}}{\beta_{monetary\ attribute}}\right) \tag{4}$$

This part-worth (or implicit price) formula represents the marginal rate of substitution between the monetary attribute (the monetary value of food self-sufficiency) and the agro-biodiversity attribute in question, or the marginal welfare measure (i.e. willingness to accept (WTA) compensation) for a change in any of the agro-biodiversity attributes.[7,8]

[7] Notice, however, that specifying t-ratios or standard errors for these ratios presented in equation [4] is more complex. Each estimate is the ratio of two parameters, each of which is also an estimate surrounded by a range of uncertainty. Even when the t values are 'statistically significant' it does not follow that the ratios are. One approach commonly used to calculate standard errors of the welfare measures involves simulation techniques to establish the empirical distribution of the marginal welfare measure (Bateman *et al.* 2003). One such method is the Wald (Delta) method contained within LIMDEP 8.0 NLOGIT 3.0, which computes values and standard errors for specified linear functions of parameter estimates.

[8] Note that for the effects coded binary attributes (i.e. landrace, agro-diversity and organic production) the implicit price formula becomes $W = -2\left(\frac{\beta_{attribute}}{\beta_{monetary\ attribute}}\right)$ (Hu *et al.* 2004).

Table 22.6 *Willingness to accept compensation welfare measures for each agro-biodiversity attribute per ESA per household per annum*[*]

| Regression | Crop variety diversity | Landrace | Agro-diversity | Organic production |
|---|---|---|---|---|
| *Welfare measures from stated preference, choice experiment regressions* | | | | |
| Dévaványa | – | – | −1841.4 | −878.4 |
| Őrség-Vend | −19.6 | −402.4 | −583 | – |
| Szatmár-Bereg | −13 | −266.4 | −877 | −256.4 |
| *Welfare measures from revealed preference, farm household regressions* | | | | |
| Dévaványa | – | – | – | – |
| Őrség-Vend | – | – | – | – |
| Szatmár-Bereg | −64.8 | −432.8 | – | – |
| *Welfare measures from combined stated preference and revealed preference regressions* | | | | |
| Dévaványa | −40 | – | −715.8 | −394.6 |
| Őrség-Vend | −49.6 | −366 | −540 | −172.2 |
| Szatmár-Bereg | −41.3 | −200.6 | −544.4 | −213.2 |

* Welfare measures are calculated with the Delta method, Wald procedure contained within LIMDEP. – indicates that the Wald procedure resulted in insignificant WTA values for this attribute. Figures are in € ** Wald procedure resulted in insignificant welfare estimates

The implicit prices of each of the agro-biodiversity attributes for each model and ESA are estimated using the Wald procedure (Delta method) in LIMDEP 8.0 NLOGIT 3.0. The WTA compensation values for each agro-biodiversity component are reported in Table 22.6. Signs on the WTA values derived from estimated demand functions can be viewed as a test for theoretical validity, as all of them are negative, implying that the welfare estimates are consistent with theoretical expectations.

For the stated preference data set, neither the crop genetic nor the crop variety diversity component of agro-biodiversity yields significant benefits for farmers in Dévaványa. Agro-diversity and organic production, however, yield the highest benefits to dévaványai farmers compared to those in other ESAs. Farmers in Őrség-Vend value agro-diversity about a third as much as those in Dévaványa, though they do not value organic production methods. Landrace cultivation and crop variety diversity are valued most highly by farmers in Őrség-Vend. In this ESA, an additional landrace benefits farmers twenty times as much as any additional seed type purchased from the shops. In Szatmár-Bereg, the agro-diversity attribute yields the highest benefits to farm households.

The welfare estimates obtained from the revealed preference data analysis are all insignificant with the exception of crop variety diversity and landraces in Szatmár-Bereg. These values are considerably higher than those obtained from the stated preference method. As explained above, the difference between the two models could be explained by the collinearity between the attributes, which appear to confound the calculation of the welfare measures (Earnhart 2001).

Relative to the WTA measures based on each individual data set, combining the revealed and stated data improves benefit valuation and generates estimates that are statistically significant and more robust. Across the ESAs, crop variety diversity is valued most highly by Őrségi farm households, followed by those located in Szatmár-Bereg and Dévaványa, which have similar WTA estimates for this attribute. The highest values for agrodiversity and organic production methods are found among farmers in Dévaványa. Farmers in Őrség-Vend value landraces the most.

The implicit prices for attributes reported in Table 22.6 do not provide CS estimates for the alternative home garden management strategies. Unlike the attributes, the strategies that provide the attributes could be directly supported by programmes such as the NRDP in order to promote on-farm conservation of agro-biodiversity. In order to estimate the CS for farm households, four home garden management scenarios were developed and the total value of private benefits to farm households were calculated for each management scenario.

Scenario 1 represents a home garden with a high level of agrobiodiversity, including 20 crop varieties and at least one landrace, managed with organic production techniques, as well as livestock. The home garden in scenario 2 has a high level of crop biodiversity only, containing 25 crop varieties and at least one landrace. In scenario 3, the home garden is managed with traditional methods of organic production and mixed livestock and crop production, contains fewer (13) crop varieties and at least one landrace. In the final scenario, the home garden has a total of only six crop varieties.[9]

According to the stated preference method, under the high agricultural biodiversity scenarios, the private value that an average farm household appropriates from a home garden is the highest in Dévaványa (Table 22.7). As expected, however, CS scenarios generate very different welfare estimates for the revealed, stated and combined methods. On average, across scenarios and ESAs, the welfare measures derived from the revealed preference method are insignificant for Dévaványa and Őrség-Vend, whereas they overestimate the private values associated with home

[9] Note that in order to estimate overall CS for each home garden management scenario it is necessary to include the welfare measure on ASC, which captures the systematic but unobserved information about farmers' choices.

Table 22.7 *Willingness to accept compensation welfare measures for home garden management scenarios per ESA per household per annum**

| Regression | Scenario 1 – High agro-biodiversity | Scenario 2 – High crop biodiversity | Scenario 3 – Traditional methods | Scenario 4 – Low agro-biodiversity |
|---|---|---|---|---|
| *Welfare measures from stated-preference, choice-experiment regressions* | | | | |
| Dévaványa | −2719.8 | –** | −2719.8 | – |
| Őrség-Vend | −1377.4 | −892.4 | −837.8 | −117.6 |
| Szatmár-Bereg | −1659.8 | −591.4 | −1302.4 | −78 |
| *Welfare measures from revealed-preference, farm household regressions* | | | | |
| Dévaványa | – | – | – | – |
| Őrség-Vend | – | – | – | – |
| Szatmár-Bereg | −1728.8 | −2052.8 | −842.4 | −388.8 |
| *Welfare measures from combined stated-preference and revealed-preference regressions* | | | | |
| Dévaványa | −697.3 | 213.1 | −417.3 | 973.1 |
| Őrség-Vend | −793.4 | −329.2 | −80.2 | 979.2 |
| Szatmár-Bereg | −895.2 | −344.1 | −404.2 | 641.2 |

* Figures are in € *** Welfare measures are calculated with the Delta method, Wald procedure contained within LIMDEP – indicates that the Wald procedure resulted in insignificant WTA values for this attribute

garden management scenarios in Szatmár-Bereg. Those generated by the stated preference method also overestimate the private values relative to the combined approach. The CS calculations for the combined data are more efficient than those estimated by either approach taken singly. The results confirm that farmers in Dévaványa value the high agro-biodiversity and traditional method scenarios positively, but they derive large negative use values from the other two home garden management scenarios. In the isolated ESAs, farmers derive higher levels of utility from home gardens managed with more agro-biodiversity. Farm households in Szatmár-Bereg value home gardens with high agro-biodiversity levels and high crop variety diversity the most, and traditional methods almost as much as those in Dévaványa. These results support the evidence that farmers located in the most economically, geographically and agro-ecologically marginalised communities derive the highest private benefits from the public goods generated by these home gardens. Furthermore, the value estimate for high agro-biodiversity home garden management scenario for Szatmár-Bereg, the region with the lowest average incomes, is similar to the estimations by Szivós and Tóth (2003), who found that 60 per cent of the households in the lowest income quartiles in Hungary consume food from own production with a value amounting to €75.3 per month, i.e. €903 per annum.

## 6 Conclusions and policy implications

This chapter employed a stated preference method (a choice experiment) and a revealed preference method (a discrete-choice, farm household-model) to estimate the private values of agro-biodiversity managed on home gardens to farmers in Hungary. The results of the stated and revealed preference methods are compared, and the data from these two methods are also combined.

The main advance of the combined model presented in this chapter is the use of the stated preference approach to improve the quality of the estimates from the revealed preference approach. In this application, the revealed preference approach is based on a household farm model of on-farm diversity in a discrete choice framework. Combining data sources resulted in more robust and efficient estimates of private values of agro-biodiversity, revealing that fusion of data sources not only enriches the results but also reduces certain problems that are associated with either method. In applied research, combining data sources will have research design and cost implications that need to be assessed.

Still, similar preference functions were retrieved from the revealed preference and stated preference data. This suggests two conclusions. First, farmers can evidently handle hypothetical questions about observable components of agro-biodiversity on their farms. Second, stated preference methods can be valuable, and in some research projects less costly, tools to investigate farmer choices with respect to agro-biodiversity.

In terms of policy implications, the results of this chapter disclosed that there are significant ESA-level differences in private valuation of agro-biodiversity, which should be taken into consideration when designing agri-environmental schemes to support provision of these public goods. Calculation of the compensating surplus values for home gardens with different levels of agro-biodiversity revealed that farmers located in the most marginalised regions, especially those in Szatmár-Bereg, appropriate the highest private values from the agro-biodiversity they manage. Szatmár-Bereg is the least-cost option for agri-environmental schemes that encourage farmers to undertake home garden management practices to support continued conservation of the nation's agro-biodiversity riches.

### REFERENCES

Adamowicz W. L., Swait, J., Boxall, P., Louviere, J. and Williams, M. 1997. Perceptions versus objective measures of environmental quality in combined revealed and stated preference models of environmental valuation. *Journal of Environmental Economics and Management*. **32** (1). 65–84.

Adamowicz, W. L., Louviere, J. and Williams, M. 1994. Combining stated and revealed preference methods for valuing environmental amenities. *Journal of Environmental Economics and Management*. **26**. 271–292.

Bateman, I. J., Carson, R. T., Day, B., Hanemann, W. M., Hanley, N., Hett, T., Jones-Lee, M., Loomes, G., Mourato, S., Ozdemiroglu, E., Pearce, D. W., Sugden, R. and Swanson, S. 2003. *Guidelines for the Use of Stated Preference Techniques for the Valuation of Preferences for Non-market Goods*. Cheltenham: Edward Elgar.

Ben-Akiva, M. and Lerman, S. R. 1985. *Discrete Choice Analysis, Theory and Application to Travel Demand*. Cambridge, MA: MIT Press.

Boxall, P. C., Englin, J. and Adamowicz, W. L. 2002. Valuing undiscovered attributes: a combined revealed-stated preference analysis of North America Aboriginal artifacts. *Journal of Environmental Economics and Management*. **45**. 213–230.

Brookfield, H. and Stocking, M. 1999. Agrodiversity: definition, description and design. *Global Environmental Change*. **9**. 77–80.

Cameron, T. A. 1992. Combining contingent valuation and travel cost data for the valuation of nonmarket goods. *Land Economics*. **68** (3). 302–317.

Campbell, D., Hutchinson, W. G. and Scarpa, R. 2006. Using Discrete Choice Experiments to Derive Individual-Specific WTP Estimates for Landscape Improvements under Agri-Environmental Schemes: Evidence from the Rural Environment Protection Scheme in Ireland. FEEM Working Paper No. 26.2006.

Chattopadhyay, S. 2000. The effectiveness of McFadden's nested logit model in valuing amenity improvements. *Regional Science and Urban Economics*. **30**. 23–43.

Cropper, M., Deck, L., Kishor, N. and McConnell K., 1993. Valuing product attributes using single market data: comparison of hedonic and discrete choice approaches. *Review of Economics and Statistics*. **75** (2). 225–232.

Cros Kárpáti, Z., Gubicza, C., and Ónodi, G. 2004. *Kertségek és kertmûvelõk: Urbanizáció vagy vidékfejlesztés?* Budapest: Mezõgazda.

Csizmadia, G. 2004. Analysis of Small Farm Useful Soil Nutritive Matter from Dévaványa, Őrség-Vend and Szatmár-Bereg Regions of Hungary. Tápiószele: Institute of Agrobotany Working Paper.

Cummings, R. G., Brookshire, D. S. and Schulze, W. D. 1986. *Valuing Environmental Goods: A State of the Arts Assessment of the Contingent Valuation Method*. New Jersey: Rowman and Allanheld.

Dalton, T. A. 2004. Household hedonic model of rice traits: economic values from farmers in West Africa. *Agricultural Economics*. **31**. 149–159.

Dhar, R. 1997. Consumer preference for a no-choice option. *Journal of Consumer Research*. **24**. 215–231.

Earnhart, D. 2001. Combining revealed and stated preference methods to value environmental amenities at residential locations. *Land Economics*. **77** (1). 12–29.

Earnhart, D. 2002. Combining revealed and stated data to examine housing decisions using discrete choice analysis. *Journal of Urban Economics*. **5** (1). 143–169.

ECOGEN. 2005. *Economic Literature on Crop and Livestock Genetic Resources*. Washington, DC: International Food Policy Research Institute and Rome: International Plant Genetic Resources Institute.

Edmeades, S. 2006. *A Hedonic Approach to Estimating the Supply of Variety Attributes of a Subsistence Crop*. Environment and Production Technology Discussion paper No 148. Washington, DC: International Food Policy Research Institute.

Englin, J. and Cameron, T. A. 1996. Augmenting travel cost models with contingent behaviour data: Poisson regression analyses with individual panel data. *Environmental and Resource Economics*. 7. 133–147.

FAO. 1999. *Multifunctional Character of Agriculture and Land*. Maastricht: Conference Background Paper No. 1.

Feick, L. F., Higie, R. A. and Price, L. L. 1993. *Consumer search and decision problems in a transitional economy: Hungary 1989–1992*. Cambridge, MA: Marketing Science Institute, Report No. 93–113.

Fertő, I., Fogács, Cs., Juhász, A., Kürhty, G. 2004. *Regoverning Markets*. Budapest: Country report. http://www.regoverningmarkets.org/docs/country_study_ Hungary_new2005.pdf

Freeman, M. A. 1993. *The Measurement of Environmental and Resource Values: Theory and Methods*. Washington, DC: Resources for the Future.

Greene, W. H. 1997. *Econometric Analysis*. New York: Prentice Hall International.

Gyovai, Á. 2002. Site and sample selection for analysis of crop diversity on Hungarian small farms. In M. Smale, I. Már and D. I. Jarvis (eds.). *The Economics of Conserving Agricultural Biodiversity on-Farm: Research methods developed from IPGRI's Global Project 'Strengthening the Scientific Basis of In Situ Conservation of Agricultural Biodiversity'*. Rome: International Plant Genetic Resources Institute.

Haab, T. C., and McConnell, K. E. 2002. *Valuing Environmental and Natural Resources: the Econometrics of Non-Market Valuation*. Cheltenham: Edward Elgar.

Haener, M., Boxall, P. C. and Adamowicz, W. L. 2001. Modeling recreation site choice: Do hypothetical choices reflect actual behavior? *American Journal of Agricultural Economics*. **83** (3). 629–642.

Hanley, N., MacMillan, D., Wright, R. E., Bullock, C., Simpson, I., Parsisson, D. and Crabtree, B. 1998. Contingent valuation versus choice experiments: estimating the benefits of environmentally sensitive areas in Scotland. *Journal of Agricultural Economics*. **49**. 1–15.

Harcsa, I., Kovách, I. and Szelényi, I. 1994. A posztszocialista átalakulási válság a mezőgazdaságban és a falusi társadalomban. (Postsocialist Transitional Crisis in Agriculture and Rural Society). *Szociológiai Szemle*. **3**. 15–43.

HCSO. 2001 2003. http://www.ksh.hu/pls/ksh/docs/index_eng.html. Budapest.

Hensher, D. A., Rose, J. M. and Greene, W. H. 2005. *Applied Choice Analysis. A Primer*. Cambridge: Cambridge University Press.

Hensher, D. A. and Bradley, M. 1993. Using stated choice data to enrich revealed preference discrete choice models. *Marketing Letters*. **4**. 139–152.

Hu, W., Hünnemeyer, A., Veeman, M., Adamowicz, W. and Srivastava, L. 2004. Trading off health, environmental and genetic modification attributes in food. *European Review of Agricultural Economics*. **31**. 389–408.

Huber, J and Pinnell, J. 1994. *The impact of set quality and choice difficulty on the decision to defer purchase.* Working paper, Durham, NC: Duke University: The Fuqua School of Business.

Jabbar, M., Swallow, B., d'Iteran, G. and Busari, A. 1998. Farmer preferences and market values of cattle breeds of west and central Africa. *Journal of Sustainable Agriculture.* **12**. 21–47.

Jabbar, M. A. and Diedhiou, M. L. 2003. Does breed matter to cattle farmers and buyers? Evidence from West Africa. *Ecological Economics.* **45** (3). 461–472.

Juhász, I., Ángyán, J., Fésûs, I., Podmaniczky, L., Tar, F. and Madarassy, A. 2000. *National Agri-Environment Programme: For the Support of Environmentally Friendly Agricultural Production Methods Ensuring the Protection of the Nature and the Preservation of the Landscape.* Budapest: Ministry of Agriculture and Rural Development, Agri-Environmental Studies.

Juhász, P. 2001. Mezőgazdaságunk és az uniós kihivás (Hungarian Agriculture and the EU Challenge). *Beszélő.* April.

Kling, C. L. 1997. The gains from combining travel cost and contingent valuation data to value nonmarket goods. *Land Economics.* **73** (3). 428–439.

Kovách, I. 1999. Hungary: cooperative farms and household plots. In M. Meurs (ed.). *Many Shades of Red: State Policy and Collective Agriculture.* Boulder, CO: Rowman and Littlefield.

Langyintuo, A., Ntoukam, G, Murdock, L., Lowenberg-DeBoer, J. and Miller, D. 2004. Consumer preferences for cowpea in Cameroon and Ghana. *Agricultural Economics.* **30**. 203–213.

Lankoski, J. (ed.). 2000. *Multifunctional character of agriculture.* Helsinki: Agricultural Economics Research Institute, Research Report No. 241.

Louviere, J. J. 1988. *Analyzing Decision Making: Metric Conjoint Analysis.* Newbury Park, CA: Sage Publications, Inc.

Louviere, J. J., Hensher, D. A., Swait, J. D. and Adamowicz, W. L. 2000. *Stated Choice Methods: Analysis and Applications.* Cambridge: Cambridge University Press.

Luce, D. 1959. *Individual Choice Behaviour.* New York: John Wiley.

Lupwayi, N., Rick, W. and Clayton, G. 1997. Zillions of Lives Underground. *APGC Newsletter.* http://www.pulse.ab.ca/newsletter/97fall/zillion.html.

Mäder, P., Fliessbach, A., Dubois, D., Gunst, L., Fried, P. and Niggli, U. 2002. Soil fertility and biodiversity in organic farming. *Science.* **296**. 1694–1697.

Már, I. 2002. Safeguarding agricultural biodiversity on-farms in Hungary. In M. Smale, I. Már and D. I. Jarvis (eds.). *The Economics of Conserving Agricultural Biodiversity On-farm: Research Methods Developed from IPGRI's Global Project 'Strengthening the Scientific Basis of In Situ Conservation of Agricultural Biodiversity'.* Rome: International Plant Genetic Resources Institute.

Már, I. and Juhász, A. 2002. A tájtermesztésben hasznosítható bab (Phaseolus vulgaris L.) egyensúlyi populációk agrobotanikai vizsgálata (Agrobotanical analysis of bean – Phaseolus vulgaris L. – equilibrium populations suitable for regional land cultivation). Debrecen, Hungary.

McFadden, D. 1974. Conditional logit analysis of qualitative choice behaviour. In P. Zarembka (ed.). *Frontiers in Econometrics.* New York: Academic Press.

McFadden, D. 1978. Modeling the Choice of Residential Location. In A. Karlqvist, L. Lundqvist, F. Snickars and J. W. Weibull (eds.). *Spatial Interaction Theory and Planning Models*. Amsterdam: North-Holland.

Meurs, M. 2001. *The Evolution of Agrarian Institutions: A Comparative Study of Post Socialist Hungary and Bulgaria*. Ann Arbor, MI: University of Michigan Press.

Mitchell, R. C. and Carson, R. T. 1989. *Using Surveys to Value Public Goods: The Contingent Valuation Method*. Baltimore: Johns Hopkins Press.

*National Labour Centre*. 2000. http://www.ikm.iif.hu/english/economy/labour.htm Budapest.

Ndjeunga, J. and Nelson, C. H. 2005. Toward understanding household preference for consumption characteristics of millet varieties: a case study from western Niger. *Agricultural Economics*. **32** (2). 151–165.

Parsons, G. and Kealy, M. J. 1992. Randomly drawn opportunity sets in a random utility model of lake recreation. *Land Economics*. **68**. 93–106.

Richards, T. and Jeffrey, S. 1995. Hedonic pricing of dairy bulls – an alternative index of genetic merit, Department of Rural Economy. Project Report 95-04. Faculty of Agriculture, Forestry, and Home Economics. University of Alberta Edmonton, Canada.

Romstad, E., Vatn, A., Rørstad, P. K. and Søyland, V. 2000. *Multifunctional Agriculture: Implications for Policy Design*. Agricultural University of Norway, Department of Economics and Social Sciences, Report No. 21.

Rosenberger, R. S. and Loomis, J. B. 1999. The value of ranch open space to tourists: Combining observed and contingent behavior data. *Growth and Change*. **30** (3). 366–383.

Ruto, E. 2005. Valuing Animal Genetic Resources: A Choice Modelling Application to Indigenous Cattle in Kenya. Paper presented at the Seventh Annual BIOECON Conference "Economics and the Analysis of Biology and Biodiversity". Kings College, Cambridge. 20–21 September.

Sarris, A. H., Doucha, T. and Mathijs, E. 1999. Agricultural restructuring in central and eastern Europe: implications for competitiveness and rural development. *European Review of Agricultural Economics*. **26**. 305–329.

Scarpa, R., Drucker, A., Anderson, S., Ferraes-Ehuan, N., Gomez, V., Risopatron, C. R. and Rubio-Leonel, O. 2003a. Valuing animal genetic resources in peasant economies: the case of the box keken creole pig in Yucatan. *Ecological Economics*. **45** (3). 427–443.

Scarpa, R., Ruto, E. S. K., Kristjanson, P., Radeny, M., Drucker, A. G. and Rege, J. E. O. 2003b. Valuing indigenous cattle breeds in Kenya: An empirical comparison of stated and revealed preference value estimates. *Ecological Economics*. **45** (3). 409–426.

Seeth, H. T., Chachnov, S., Surinov, A. and Braun, J. v. 1998. Russian poverty: muddling through economic transition with garden plots. *World Development*. **26** (9). 1611–1623.

Swain, N. 2000. Post-Socialist Rural Economy and Society in the CEECs: the Socio-Economic Contest for SAPARD and EU Enlargement. Paper presented at the International Conference: European Rural Policy at the

Crossroads, The Arkleton Centre for Rural Development Research, King's College, University of Aberdeen, Scotland.

Swait, J. and Louviere, J. 1993. The role of the scale parameter in the estimation and comparison of multinomial logit models. *Journal of Marketing Research*. **30**. 305–314.

Swait, J., Louviere, J. and Williams, M. 1994. A sequential approach to exploiting the combined strengths of SP and RP data: application to freight shipper choice. *Transportation*. **21**. 135–152.

Szelényi, I. (ed.). 1998. *Privatising the Land*. London: Routledge.

Szép, K. 2000. The chance of agricultural work in the competition for time: case of household plots in Hungary. *Society and Economy in Central and Eastern Europe*. **22** (4). 95–106.

Szivós, P. and Tóth, I. G. 2003. *Stabilization of the structure of society (Stabilizálódó társadalomszerkezet)*. Budapest: TÁRKI MONITOR Reports.

Unnevehr, L. 1986. Consumer demand for rice grain quality and returns to research for quality improvement in Southeast Asia. *American Journal of Agricultural Economics*. **68**. 634–641.

Vajda, L. 2003. *The view from Central and Eastern Europe, Agricultural outlook forum 2003*, Washington, DC: USDA. http://www.usda.gov/oce/waob/oc2003/speeches/vajda.pdf.

Weingarten, P., Baum, S., Frohberg, K., Hartmann, M. and Matthews, A. 2004. *The future of rural areas in the CEE new member states*. Network of Independent Agricultural Experts in the CEE Candidate Countries. http://europa.eu.int/comm/agriculture/publi/reports/ccrurdev/text_en.pdf

WHO. 2005. The Impact of Food and Nutrition on Public Health: The case for a Food and Nutrition Policy and an Action Plan for the European Region of WHO 2000–2005 and the Draft Urban Food and Nutrition Action Plan. http://www.hospitalitywales.demon.co.uk/nyfaweb/fap4fnp/fap_26.htm

Wyzan, M. 1996. Increased inequality, poverty accompany economic transition. *Transition*. **2** (20). 24–27.

# 23 Agricultural development and the diversity of crop and livestock genetic resources: a review of the economics literature

*Melinda Smale and Adam G. Drucker*

## 1 Introduction

Managing the biological diversity of crop and livestock genetic resources is of fundamental importance: (i) as a means of survival for the world's rural poor (Smale 2006); (ii) as a mechanism for buffering against output losses due to emerging pests and diseases, even in fully commercialised agricultural systems (Heisey *et al.* 1997); (iii) as an input into locally sustainable, indigenous technology systems (Bellon *et al.* 1997; Anderson 2003); (iv) as a means of satisfying the evolving tastes and preferences of consumers as economies change (Evenson *et al.* 1998); and (v), as a biological asset for the future genetic improvement on which the global supply of food and agricultural products depends (Brown 1990; Koo *et al.* 2004).

Geneticists often hypothesise that rare, locally adapted genotypes may be found among the varieties and breeds maintained by farmers in extreme or heterogeneous environments. Some genotypes are thought to contain tolerance or resistance traits that are not only valuable to the farmers who manage them but also to the global genetic resource endowment on which future improvement of crop and livestock depends. The foremost policy challenge is that many of the domesticated landscapes of conservation interest are found in poorer regions of the world, in nations undergoing rapid social and economic change.

Crop and livestock genetic resources are managed sustainably when they satisfy the present needs of farm families while also retaining their genetic integrity for the longer-term needs of society. The 'sustainable management' of genetic resources (GR) is defined as the combined set of actions (and policies) by which a sample, or the whole, of a plant/animal population is subjected to processes of genetic and/or environmental manipulation with the aim of sustaining, utilising, restoring, enhancing and understanding (characterising) the quality and/or quantity of the GR and its products.

The properties of crop and livestock genetic resources mean that managing them in sustainable ways will entail careful husbandry within domesticated landscapes as well as banks and breeding programmes, at local, national and international scales. In this chapter 'agricultural biodiversity' refers to all diversity within and among species found in domesticated crop and livestock systems, including wild relatives, interacting species of pollinators, pests, parasites and other organisms. Domesticated biodiversity (crops, aquaculture fish, livestock), is a consequence of deliberate human intervention, serving both as a production component and as a source for genetic improvement. Located *in-situ*, domesticated biodiversity is linked outside cultivated landscapes with the biodiversity found in protected reserves or maintained in the *ex-situ* collections of breeders and gene banks.

'Crop biodiversity' is the biological diversity of crops, encompassing both phenotypic and genotypic variation, including varieties recognised as agro-morphologically distinct by farmers and those recognised as genetically distinct by plant breeders.[1] Typically, farmers' varieties do not satisfy breeders' or legal definitions of variety because they are heterogeneous, exhibit less uniformity and segregate genetically. Often, farmers' varieties are called 'landraces,' while those bred by professional plant breeders are termed 'modern varieties'. Though definitions and concepts of landraces are numerous in the crop science literature, Harlan (1992) defines them broadly as variants, varieties or populations of crops, with plants that are often highly variable in appearance. The genetic structure of landraces is shaped by farmers' seed selection practices and management, as well as natural selection processes, over generations of cultivation.

Similarly, for livestock, biological diversity encompasses both phenotypic as well as genotypic variation. The Food and Agriculture Organization of the United Nations (FAO 1999, p. 5) defines 'breed' as: 'either a subspecific group of domestic livestock with definable and identifiable external characteristics that enable it to be separated by visual appraisal from other similarly defined groups within the same species; or a group for which geographical and/or cultural separation from phenotypically similar groups has led to acceptance of its separate identity'. A combination of phenotypic studies (including classical morphometric studies, in which morphological characteristics are measured), biochemical (e.g. protein polymorphism, blood group) analyses and, more recently, studies of DNA, are the main sources of data about genetic relationships among

[1] A plant or animal phenotype is the observable manifestation of a genotype. A genotype is determined by its alleles, or types of genes. Genetic segregation occurs when breeding produces multiple, as compared to a single, genotype. An improved variety of wheat or rice, for example, breeds 'true to type' for many generations. Morphology refers to physical characteristics or form.

breeds, varieties and strains (Rege *et al.* 2003). Populations within each species can be classified as wild and feral populations, landraces or primary populations, standardised breeds, selected lines and any conserved genetic material.

As far as the economics of crop and livestock genetic diversity is concerned, all sources of economic value associated with biodiversity, as with other goods and services, emanate from human preferences. In agriculture, most of the value associated with genetic resources is often thought to be related to use, rather than non-use values, although option values can be significant and a study by Cica *et al.* (2003) suggests that existence values can be too. Further, crop and livestock genetic resources are not traded in markets, leading them to be undervalued and invoking challenges for economists. Goods that are not traded in markets tend to be undervalued. Thus, economic theory predicts that farmers who produce goods that are not traded in markets will tend to under-produce them relative to the national, regional or global needs. Policy interventions are therefore necessary to support their production if society's goals are to be met.

Economists have tools that can be of use in designing these policy interventions, and a number of these have been applied in the literature about crop and livestock biodiversity. The economics literature about animal genetic resources (AnGR), conservation and sustainable use has developed rapidly during recent years, although the economics literature about the value of plant genetic resources (PGR) for agriculture has a longer history and is therefore more extensive. This is particularly evident in the context of developing economies, where crop improvement appears to have taken policy precedence. Furthermore, sampling valuable crop landraces for conservation *ex-situ* has been comparatively inexpensive, while the technology for conserving livestock *ex-situ* is not yet operational.

This chapter reviews the literature for crop and livestock components of agro-biodiversity, organised according to research themes or questions.[2] The next two sections review the findings, economic methods and limitations of the research carried out within the economics of crop and livestock genetic diversity, respectively. Then, section 4 concludes by briefly summarising the key economic implications of the findings from the review.

## 2 Crop genetic diversity: economic methods, findings and limitations

This section focuses on crop genetic diversity and provides a summary of the current knowledge about, (i) the marginal value and, (ii) the rate

[2] A companion review can be found in Drucker *et al.* (2005).

of return to improvement of crop GR in commercial agriculture, (iii) the effect of crop biodiversity on productivity, vulnerability and efficiency in agriculture, (iv) the costs and benefits of *ex-situ* conservation, (v) the factors that determine the levels of *in-situ* crop diversity during economic change, (vi) the value of crop genetic diversity to farmers themselves. We now turn to these issues.

### 2.1 *What is the marginal value of a plant species or crop genetic resources?*

Overviews and surveys discussing the sources of economic value in plant genetic resources have been numerous, including Brown (1990), Pearce and Moran (1994) and Swanson (1996). The value of diversity in crop or animal species has been modelled theoretically, supported in some cases by empirical data (e.g. Brown and Goldstein 1984; Weitzman 1993; Polasky and Solow 1995). The values of plant genetic resources and their diversity in crop breeding have been estimated by applying a combination of production economics and forms of hedonic analysis (Evenson *et al.* 1998).

The literature generally concurs that the marginal commercial value expected from an individual plant genetic resource in agricultural use will not be high enough to fund national innovation or conservation efforts at levels desirable for society. The perception that individual plant genetic resources have great commercial value is based largely on anecdotal cases in which substances identified in wild, indigenous plants have generated profits for pharmaceutical companies. Economics research has cast doubt on the likelihood that the willingness to pay for prospecting these resources in the pharmaceutical industry would be sufficient to promote the conservation of their habitats (e.g. Simpson *et al.* 1996). Evidence to suggest that any one landrace or improved variety will generate large commercial returns in agricultural use – and therefore significant benefits through restricting access to it – is even more modest. Though there are instances in which a single plant genetic resource has proved extremely valuable, these cannot be generalised. There are three reasons why economists are sceptical.

The first reason is the process of plant breeding. In plant breeding, numerous genetic resources are continually shuffled and reshuffled in an uncertain search for traits that are well expressed in a crop variety destined for highly differentiated production conditions. Economically important traits are distributed statistically across plant genetic resources, with varying likelihood of encountering useful levels. The traits demanded by societies, such as resistance to plant pests and diseases, and quality attributes preferred by consumers, also change frequently in response to environmental stress and economic changes, keeping plant breeders on a treadmill to surpass past accomplishments. Breeding products (crop varieties) contain many 'ingredients' that are also genetic resources and these products are in turn combined with others to produce the next variety. The marginal contribution of the last resource used may be slight. Attributing value to each ingredient is difficult.

A second reason is the nature of crop production. Changes in productivity that underlie economic benefits from new varieties involve multiple factors in interaction with the seed. A well-known example is the Green Revolution in wheat. The economic benefits associated with the Green Revolution cannot be ascribed solely to the dwarfing genes, the landrace that contained them or the scientist who initially bred them into another cultivar. A number of farm physical, social, economic and policy factors influenced the widespread adoption of those cultivars, generating economic benefits through yield gains. Concurrently, major changes in the growing environments for varieties enhanced those yield gains, such as increased water use, fertiliser application and the expertise of farmers. Production benefits were then transmitted via prices and distributed to society through effects on producers' and consumers' incomes.

A third reason is the existence of substitutes. The same trait may be apparent to one degree or another in many other plant genetic resources. Seed samples of the same genetic resource may also be found in more than one *ex-situ* collection, in more than one political jurisdiction. Even when rare in a given collection, accessions carrying useful traits might be duplicated among seed samples (accessions) in multiple collections. Similarly, though locally rare in farmers' fields, they could be globally abundant.

It is important to remember, however, that the commercial value of plant genetic resources is a relatively small component of their total use value in agriculture. Since other values are not captured well in market prices, public investments in innovation and conservation will continue to be needed to attain socially desirable outcomes. Since the potential usefulness of any single genetic resource is often highly uncertain, and time horizons for developing products from genetic resources are long, private investors typically underinvest in conserving them at the levels needed by society. Tastes and preferences are dynamic; and unforeseen production shocks occur. Thus, the public sector has played and will continue to play a pivotal role in conserving these resources in the foreseeable future.

Notwithstanding that the literature has advanced our theoretical and conceptual understanding of important issues, feasible, cost-effective approaches for valuing multiple components of agricultural biodiversity and services are needed for the design of effective programmes to manage crop genetic resources in sustainable ways.

### 2.2 *What is the rate of return to improvement of crop genetic resources?*

The compendium and state-of-the-art of methods used to assess the economic benefits or productivity gain from the improvement of crop genetic resources are found in Alston *et al.* (1998), Alston *et al.* (2000) and Morris and Heisey (2003). Economists have repeatedly demonstrated that rates of return to investment in plant breeding programmes are high (Alston *et al.* 2000; Evenson and Gollin 2003), documenting the important role of plant genetic resources in the development of world agriculture. The successive, continuous releases of improved varieties by plant breeding programmes, many of them publicly financed, have generated economic returns that far outweigh the costs of investment. Although the marginal benefit that can be attributed to a single gene or genetic resource in plant breeding is likely to represent a relatively small proportion of the total, the productivity benefits accruing to society as a whole and especially to consumers in terms of lower food prices are large relative to the costs of investing in plant breeding. This is particularly true in less advanced agricultural economies where consumers spend a much larger proportion of their budgets on food. Successful innovation has depended on access to a wide range of materials.

Though the methods for attributing the benefits of crop improvement to plant breeding programmes are advanced (Pardey *et al.* 2004), methods for apportioning these benefits among ancestors typically impose unrealistic assumptions – even in highly bred crops. For example, the use of Mendelian rules of inheritance ignores the effects of selection in breeding. In general, estimates are only as reliable as the pedigree that has been recorded. Moreover, assessing the economic benefits from genetic resources in crops that are not highly bred or minor crops would require the applications of other methods since these crops do not have pedigrees.

### 2.3 *What is the effect of crop biodiversity on productivity, vulnerability and efficiency?*

Initial attempts by economists linking genetic diversity to crop productivity were undertaken in a partial productivity framework with a mean-variance production function or simultaneous equation system with cost shares (Smale *et al.* 1998; Widawsky and Rozelle 1998; Meng *et al.* 2003). These studies tested the relationship of crop biodiversity to productivity, yield variability and economic efficiency, particularly in farming systems dominated by modern varieties. So far, hypothesis tests have been inconclusive. Associations are sometimes positive and sometimes negative, and findings are specific to location, time period and cropping system.

Datasets used in these analyses were largely secondary, measured at the township, provincial or regional level, representing farming systems with modern varieties. In addition, most diversity indices were constructed from pedigree data. Indices measuring the rate of variety change and spatial diversity indices were also employed. Meng *et al.* (2003) combined biometric techniques with morphological data to construct spatial diversity indices.

Heisey *et al.* (1997) demonstrated that higher levels of latent genetic diversity in modern wheat varieties would have generated costs in terms of yield losses in some years in the Punjab of Pakistan. In other years, the mix of varieties and their spatial distribution across the region generated both lower overall yields and less diversity than was feasible.

These studies suffer from some important limitations. A more complete theoretical framework of decision making under risk is required to test hypotheses, with multiple outputs and differentiated genetic inputs, estimated structurally where data permit, perhaps including higher moments. To draw generalisations and validate empirical findings, a wider cross-section of case studies conducted in commercially oriented, as well as mixed and or subsistence-oriented systems, is needed. The role of crop genetic diversity in mitigating production and consumption vulnerability in marginal environments has not been adequately explored. The contribution of crop genetic diversity to ecosystem services and resilience has not been investigated empirically.

### 2.4 *What are the costs and benefits of ex-situ conservation?*

In order to estimate the benefits expected from using an additional gene bank accession in crop breeding, studies have employed mathematical programming, Monte Carlo simulations and maximum entropy methods in a search theoretic framework, combined with partial equilibrium estimates of the productivity impact of the bred materials in farmer's fields (Gollin *et al.* 2000; Zohrabian *et al.* 2003). Costs of conserving accessions have been estimated by applying the microeconomic theory of the firm and capital investment decisions (Koo *et al.* 2004). Based on these methods, tools could be developed and directly applied with spreadsheet analysis to gene bank cost data. Other than this literature, sample surveys have been conducted to assess the extent of gene bank utilisation by plant breeders, other scientists and farmers (e.g. Rejesus *et al.* 1996).

The results of these studies suggest that the expected marginal value of exploiting an individual accession in commercial agricultural use justifies the cost of conserving it in a gene bank. The costs of conserving accessions in gene banks are relatively easy to tabulate compared to the expected benefits from the accessions they conserve. If, as is shown in a set of recent studies compiled by Koo *et al.* (2004), the costs of conserving an accession are shown to be lower than any sensible lower-bound estimate of benefits, undertaking the expensive and challenging exercise of benefits estimation is not necessary to justify its conservation. Zohrabian *et al.* (2003) found that the expected marginal benefit from exploring an additional unimproved gene bank accession in breeding resistant varieties of soybean more than justified the costs of acquiring and conserving it. Since the payoff can be large for problems of economic importance when the desired traits are rare, conserving some categories of materials 'untapped' for years can be justifiable; infrequent use of individual accessions by plant breeding programmes does not, in itself, imply that an additional accession will have low value (Gollin *et al.* 2000). Additionally, a recent study of a large national gene bank indicates higher rates of direct utilisation in plant breeding than suggested earlier, secondary use through sharing within and outside respondents' institutions and proportionately higher use rates among respondents in low- and middle-income countries (Smale and Day-Rubenstein 2002).

Several limitations are apparent in this literature. First, cost and benefits estimated from such detailed studies of a few large national and international gene banks cannot be generalised for all gene banks. Second, the range in estimated benefits is extremely sensitive to assumptions concerning the lag until variety release, and the discount rate, or time value of money. Though the statistical theory used in the search models accounts for relative abundance and the genetic differences among accessions with respect to the trait of interest, the range in simulated benefits is too wide for confidence. Finally, the cost analyses distinguish between crops and types of collections, but treat each accession as genetically equivalent.

### 2.5 *Which factors predict variation in crop biodiversity on farms as economies change? Which farmers are most likely to maintain it?*

Seminal approaches that first attempted to answer these interrelated questions (Brush *et al.* 1992; Meng 1997) built on the literature about the adoption of agricultural innovations in developing economies (Feder *et al.* 1985; Feder and Umali 1993). Later, Van Dusen (2000) developed a farmer decision-making model of on-farm crop diversity in the theoretic framework of the agricultural household. Several trait- or attribute-based approaches have been advanced (Edmeades *et al.* 2006; Wale and Mburu 2006). The theoretical model is applied econometrically in a reduced form equation. Dependent variables are diversity indices constructed over optimal choices, as observed on farms. Crop biodiversity is generally treated as an outcome, or indirect choice, of farmer decision-making rather than a deliberate choice.

In these studies, on-farm conservation is defined as the choice by farmers to continue cultivating genetically diverse crops, in the agricultural systems where the crops have evolved historically through processes of human and natural selection (Bellon *et al.* 1997). The premise of recent empirical studies (Meng 1997; Van Dusen 2000; case studies compiled in Smale 2006) is that the highest benefit-cost ratios for on-farm conservation of crop biodiversity will occur where both society and the farmers who maintain it benefit. According to this concept, the highest benefit-cost ratios for on-farm conservation of crop genetic resources will occur where the private benefits or utility farmers earn from managing them as well as the public value associated with their biological diversity are high. In these areas, since farmers are already bearing the costs of maintaining diversity and reveal a preference for doing so, the costs of public interventions to support them will be least. As economic development occurs, a necessary condition for this outcome is that consumers demand products arising from crop biodiversity, so that the costs of maintaining it are paid through the market channel. In locations where both the public value and private value of crop biodiversity are known to be relatively low, there is no need to invest in any form of conservation. Where crop biodiversity is great but farmers derive little private value from it, *ex-situ* conservation is the best strategy. Where there is little crop biodiversity but farmers care a lot about it, there is no need for public investment at all since no value is associated with conservation.

So far, a major aim of case studies about on-farm conservation has been to characterise candidate sites for on-farm conservation, and within these locations, farmers with high probabilities of maintaining it during economic change (Meng 1997; Van Dusen 2000, Birol 2004; Gauchan 2004). Researchers have sought to identify the factors that increase and decrease the likelihood that farmers will continue to manage crop biodiversity, and develop statistical profiles of those most likely to maintain it. These profiles can be used to design targeted programmes in centres of crop biological diversity.

Two of the overriding determinants of crop biodiversity levels on farms are geographical location and environmental heterogeneity, as suggested by theories of population genetics and island biogeography. Further, in most of the studies undertaken in low-income countries, agricultural production is accomplished with limited use of purchased inputs. Farm technology consists largely of family labour, and in some cases, animal traction, in combination with land and soil quality. Where measured, higher numbers of plots, fragments and slopes are positively associated with crop biodiversity on farms. However, the direction of land quality relationships (soil erosion and fertility, moisture content) depends on the context.

Another common determinant is relative isolation from physical market infrastructure, which induces farmers to rely on their own production to meet the food and fodder needs of their families. Nonetheless, the relationship of market development and commercialisation to crop biodiversity on farms appears more complex in these studies when specific market features, other than sheer isolation from physical infrastructure or road density, are disengaged. For example, market participation as a seller enhances the range of endemic banana varieties grown in Uganda, while participating as a buyer has the opposite effect (Edmeades *et al.* 2006). On the hillsides of Ethiopia, different types of markets or road access, such as walking distance from household to road, from the farm to the nearest input shop or dealer, or from the village to the district market, seem to influence the richness (numbers) of varieties grown in different ways (Benin *et al.* 2004; Gebremedhin *et al.* 2006). Markets clearly provide incentives for farmers to grow aromatic quality landraces, but not to grow coarse-grained landraces in Nepal, though farmers preferred coarse-grained landraces for their adaptation to stress and agronomic traits (Gauchan *et al.* 2005a). Seed system characteristics are significant determinants of millet biodiversity at the farm and community levels in southern India (Nagarajan 2005). A fourth is cultural richness and cohesiveness, or cultural autonomy, as these relate to the selection pressures applied on the plant materials. For obvious reasons, this determinant has been more fully analysed in the anthropological and ethnobotanical literature than in the applied economics literature (Brush 2002).

Many of the case study findings suggest that factors associated with economic development may not, in the short-term, detract from intra-crop and in particular inter-crop diversity on farms, whether observed at the farm level or at higher levels of aggregation, such as village, settlement, district or region. Within the poor, marginal environments where most diversity is still found, education of men and women tends to have a positive effect, if at all. Access to animal traction, credit, land and other assets enhance rather than detract from crop biodiversity in most of these studies.

On the other hand, those households currently maintaining crop biodiversity are generally older, regardless of empirical context. As the farm population ages and declines as a proportion of the total, public investments must be made to encourage the retention of local knowledge in crops and varieties – in some form. Labour effects are multiple and counteracting. It is evident that diversification in any form studied is most often associated with relatively labour-intensive production. Non-farm cash transfers and income contribute to sustaining intra- and inter-crop diversity in several of the cases, but the Mexico case (Van Dusen 2006) reveals the negative impact of long-term, international migration.

The Peru case study (Winters *et al.* 2005) illustrates how the rapid uptake of a more remunerative, labour-intensive activity – dairy farming – may lead to the decline of intra-crop diversity. There will often be better ways to relieve poverty than through either the introduction of crop varieties or the diversification of crop varieties.

Last, but not least, statistical profiles of households most likely to sustain crop biodiversity suggest that conservation programmes can be designed to address social equity goals. Though most farmers on the hillsides of Nepal and Ethiopia may be ranked as poor by global standards, targeting the households relatively more likely to maintain valuable landraces in those locations is by no means equivalent to targeting the poor (Gauchan 2004; Benin *et al.* 2004). In Hungary (Birol 2004), targeting the households most likely to maintain agro-biodiversity at least cost is equivalent to targeting the poor, or relatively disadvantaged rural populations.

Though this body of literature is scant, even less well analysed in the published economics literature are species sometimes known as 'orphan crops', which are of minor economic importance globally but have also benefited less from public or private research investments. Too much research has focused on a single crop while treating other crops and economic activities as exogenous or given. Analysis at the household level does not provide sufficient information about diversity in larger biological units, even when explanatory economic variables measured in larger units can be introduced into the equation. Moreover, variation across communities may be more important for programme design than variation within any single community. An economics conceptual framework will need to be developed to relate analyses based on the household model to larger scales of aggregation. Few institutional approaches are apparent in the literature so far (e.g. Bela *et al.* 2005).

## 2.6 *What is the value of crop genetic resources to farmers?*

Predictions from econometric models regarding the last question above represent the preferences farmers reveal for crop varieties and attributes given their production technology, cash expenditures and other constraints. They provide one means of ranking locations, farmers or the sets of varieties according to their private value, in terms of current, direct use. Stated preference methods have been used, with varying degrees of sophistication. Matrix ranking, utility scores and other approaches have been used in focus group and household interviews (Gauchan *et al.* 2005b; Bela *et al.* 2005; Lipper *et al.* 2005). Econometric methods have been applied to data from choice experiments conducted with sample surveys to estimate the value farmers assign to components of agro-biodiversity, including the richness of crop varieties, cultivation of landraces, use of organic methods and integrated crop and livestock production (Birol *et al.* 2004). What have we learned from these studies?

Across a range of crops, national income levels and agro-ecological environments, case studies support the notion that farmers value various dimensions of crop biodiversity (Smale 2005). Yet, the predictions of economic theory are confirmed, even among regions in relatively rich nations, like Hungary (Birol 2004). Farmers in the less productive, most remote regions of this high-income country value agro-biodiversity the most. As the settlements in which farmers reside develop and the physical infrastructure of their markets becomes denser, they will rely less on their home-produced goods for food and the value they ascribe to agro-biodiversity on their farms will diminish. Farmers in southern Italy enjoy an historical endowment of local wheat diversity, producing durum wheat in a challenging environment for controlled, highly articulated and differentiated markets. Durum wheat diversity and crop diversification appear to contribute positively to crop productivity and farmer revenues in southern Italy (Di Falco 2003).

Notwithstanding the use of stated preference valuation studies, additional applications of stated preference methods, with different survey instruments, are needed in order to assess the advantages and disadvantages of this research tool for valuing agro-biodiversity and its components in poorer countries with less literate populations. The well-known limitation of all stated preference approaches is their hypothetical nature compared to revealed preferences, though both stated and revealed preferences have advantages and drawbacks. Combining choice experiment and farm household data analysis could strengthen the reliability of results. In addition, the roles of production and consumption risk are relevant to both revealed and stated preference formulation but have not yet been investigated with a theoretic framework in this literature.

## 3 Livestock genetic diversity: economic methods, findings and limitations

This section turns its attention to livestock genetic diversity and reviews the following issues: (i) the value of livestock GR to farmers, (ii) the costs and benefits of its conservation, (iii) targeting for *in-situ* breed conservation, (iv) the traits which should be addressed in breeding programmes, and (v) the way specific policies influence GR conservation and sustainable use, as well as means for assessing conservation priorities.

### *3.1 What is the value of livestock genetic resources to farmers?*

While choice experiments have been used to determine the economic value of animal genetic resources (AnGR) to farmers, based on an initial identification of potential methodologies, Drucker *et al.* (2001, p. 9) classified a number of additional approaches as appropriate for determining the actual economic importance of a breed, including: aggregate demand and supply; cross-sectional farm and household; market share; and intellectual property rights (IPR) and contracts approaches. Such approaches are not mutually exclusive and they can be used to, respectively: (i) identify the value of a breed to society by measuring consumer and producer surplus (first two approaches); (ii) provide an indication of the current market value of a given breed; and (iii) promote market creation and support for the fair and equitable sharing of AnGR benefits.

However, conventional productivity evaluation criteria are inadequate to evaluate subsistence livestock production and have tended to overestimate the benefits of breed substitution. Adaptive traits and non-income functions form important components of the total value of indigenous breed animals to livestock keepers. Tano *et al.* (2003) and Scarpa *et al.* (2003a, 2003b) valued the phenotypic traits expressed in indigenous breeds of livestock, demonstrating that adaptive traits and non-income functions are shown to form important components of the total value of the animals to livestock keepers. In West Africa, for example, the most important traits for incorporation into breed improvement programme goals were found to be disease resistance, fitness for traction and reproductive performance. Beef and milk production were less important.

In Kenya, Karugia *et al.* (2001) showed that while crossbreeding of dairy cattle has had an overall positive impact on social welfare, under 'traditional' production systems, farm-level performance has been little improved by replacing indigenous zebu with exotic breeds. Using an aggregate demand and supply approach covering both national and farm levels, they argue that conventional economic evaluations of crossbreeding programmes have overestimated their benefits by ignoring subsidies, the increased costs of management such as veterinary support services, and the higher levels of risk and socio-environmental costs associated with the loss of the indigenous genotypes.

In comparing the performance of different genotypes (indigenous goats vs. exotic crosses), Ayalew *et al.* (2003) come to a similar conclusion. The secondary importance of meat and milk production traits in many production systems leads them to argue that conventional productivity evaluation criteria are inadequate to evaluate subsistence livestock production, because they fail to capture non-marketable benefits of the livestock, and the core concept of a single limiting input is inappropriate to subsistence production, as multiple limiting inputs (livestock, labour, land) are involved in the production process. As many of the livestock functions as possible (physical and socio-economic) should thus be aggregated into monetary values and related to the resources used, irrespective of whether these 'products' are marketed, home-consumed or maintained for later use. Evaluation of flock-level productivity indices for subsistence goat production in the eastern Ethiopian highlands shows that indigenous goat flocks generated significantly higher net benefits under improved than under traditional management, challenging the prevailing notion that indigenous livestock do not adequately respond to improvements in the level of management. Furthermore, it is shown that under the subsistence mode of production considered, the premise that crossbred goats are more productive and beneficial than the indigenous goats is wrong.

While the studies reviewed above have been useful to derive policy relevant findings, few examples of the aggregate demand and supply approach or market share approach exist (but see Karugia *et al.* 2001; Drucker and Anderson 2004). This largely reflects the lack of data for breeds other than those that are commercially popular, combined with the difficulties of estimating shadow prices for home labour and forage use. In addition, it should be stressed that the market share approach fails to account for consumer/producer surplus.

### 3.2 *What are the costs and benefits of conservation?*

Drucker *et al.* (2001) categorised a number of methodological tools as being appropriate for determining the appropriateness of *in-situ* conservation programme costs. These included contingent valuation, production loss averted, opportunity cost and least-cost approaches. These approaches are capable of, respectively: identifying society's willingness to pay for AnGR conservation; indicating the magnitude of potential production losses in the absence of maintaining AnGR diversity; identifying the cost of maintaining such diversity; and identifying cost-efficient programmes for the conservation of AnGR.

Recent *in-situ* studies have drawn on the construction of bio-economic models to model conservation costs and the use of contingent valuation techniques to model benefits (Cicia *et al.* 2003; Drucker and Anderson 2004), as well as a range of other techniques borrowed from the economics literature, including that of the plant genetic resources valuation literature. On the benefit side, these have included estimates of market share and production loss averted, while least-cost/opportunity cost calculations have been used on the cost side (Drucker and Anderson 2004; Pattison *et al.* in press). With regard to the latter, such calculations have also been applied within the context of estimating the costs of establishing a safe minimum standard for livestock breed population numbers as part of a conservation programme (Drucker 2006).

The main finding from the above literature is that the costs of implementing an *in-situ* breed conservation programme may be relatively small, both when compared with the size of subsidies currently being provided to the commercial livestock sector and with regard to the benefits of conservation. However, few such conservation initiatives exist and even where the value of indigenous breeds has been recognised and support mechanisms implemented, significant shortcomings can be identified. Similar work regarding the costs and benefits of the *ex-situ* (cryo)conservation[3] of livestock remains limited. Under the assumption that technical feasibility brings cryoconservation of livestock species to within the same level of magnitude as that of plants, extensive conservation efforts would likely be justified on economic grounds.

Specific benefits of livestock diversity conservation that accrue to livestock-keepers are related to the fact that livestock with different agronomic and product characteristics suit a range of local community needs, including the provision of non-output functions. Livestock provide manure to enhance crop yields, and transport for inputs and products, serving also for traction. Where rural financial and insurance markets are not well developed, they enable farm families to smooth variation in income and consumption levels over time. Livestock constitute savings and insurance, buffering against crop failure and cyclical patterns in crop-related income. They enable families to accumulate capital and diversify, serving a range of socio-cultural roles related to status and the obligations of their owners (Anderson 2003). Nevertheless, very limited work on valuing these livestock-keeper level benefits has been carried out. Similarly, the benefits to breeders of the existence of such diversity is also difficult to assess given the focus on improved (exotic) breeds and the failure of a number of crossbreeding programmes based on exotic x indigenous crosses. For society as a whole livestock diversity conservation may generate significant option and existence values but again these have not been valued systematically.

[3] This is the collection and deep-freezing of semen, ova, embryos or tissues which may be used to regenerate animals.

In a developed country case study, Cicia *et al.* (2003) use a bio-economic model and stated preference methods to estimate, respectively, the costs and benefits of establishing a conservation programme for the threatened 'Pentro' horse of Italy. A large positive net present value is associated with the proposed conservation activity (benefit/cost ratio > 2.9).

Even where the value of indigenous breeds has been recognised and support mechanisms implemented, incentives for conservation are inadequate. In an examination of farm animal biodiversity conservation measures and their potential costs in the European Union (EU), Signorello and Pappalardo (2003) report that many breeds at risk of extinction according to the FAO World Watch List are not covered by support payments because they do not appear in national Rural Development Plans. Where they are made, payments do not take into account the different degrees of extinction risk that exist between breeds. Payment levels are in any case inadequate, meaning that it can still remain unprofitable to rear indigenous breeds.

Nevertheless, conservation costs are shown to be relatively small by Drucker (2006) in a number of case studies. The costs of implementing a safe minimum standard are low (depending on the species/breed and location, these range from between approximately €3,000 – €425,000 pa), both when compared with the size of subsidies currently being provided to the livestock sector (<1 per cent of the total subsidy) and with regard to the benefits of conservation (benefit-cost ratio of >2.9). The finding that costs are lowest in the developing country is encouraging, given that an estimated 70 per cent of the livestock breeds existing today are found in developing countries where the risk of loss is highest (Rege and Gibson 2003).

Similar work regarding the costs and benefits of the *ex-situ* (cryo)conservation of livestock remains limited. Gandini and Pizzi (2003) provide a brief review of literature that largely provides information regarding the current situation, which is rapidly changing.

Having reviewed this literature, we now turn to address the potential way ahead. Indeed, although the safe minimum standard approach is shown to have a role to play *in-situ* AnGR conservation, more extensive quantification of the components required to determine costs needs to be undertaken before it can be applied in practice. Such economic valuation needs to cover both the full range of breeds/species being considered, as well as to ensure that as many as possible of the elements making up their total economic value are accounted for. Furthermore, with regard to *ex-situ* AnGR conservation, cryoconservation technologies for livestock are only well-developed for a handful of species. Hence, valuation work in this field has been extremely limited to date, despite the fact that there is need for an integrated conservation approach, which combines a range of available *ex-situ* and *in-situ* options.

### 3.3 *Which farmers should be targeted for participating in in-situ breed conservation programmes? Which farmers are most likely to maintain indigenous breeds?*

A range of stated and revealed preference techniques can be used to relate household characteristics to breed preferences and opportunity costs of production. As in the crops literature cited above, the premise of these studies is that continued conservation of genetic resource diversity on-farm makes most economic sense in those locations where both society and the farmers who maintain it benefit the most. In targeting such households, Mendelsohn (2003) argues that conservationists must first make the case for why society should be willing to pay to protect apparently 'unprofitable' AnGR resources and then must design conservation programmes that will effectively protect what society treasures.

Studies generally conclude that household characteristics play an important role in determining differences in farmer breed preferences. This additional information can be of use in designing cost-effective conservation programmes. The 'least cost' of an *in-situ* conservation programme can be expressed as the cost necessary to raise the comparative advantage of such breeds above that of competing breeds, animals or off-farm activities; and a relatively small investment may suffice to maintain their advantage in a particular farming system. This approach has recently been applied to estimate Creole pig conservation costs in Mexico (Drucker and Anderson 2004; Pattison *et al.* in press) and Boran cattle in Ethiopia (Zander 2006).

Scarpa *et al.* (2003b) show that for Creole pigs in Mexico, the respondent's age, years of schooling, the size of the household and the number of economically active household members were important factors in explaining breed trait preferences. Younger, less educated and lower income households placed relatively higher values on the attributes of indigenous piglets compared to exotics and their crosses (Drucker and Anderson, 2004). Findings by Pattison *et al.* (in press) further corroborate these results. In the context of a 10-year conservation programme designed to increase Creole pig population numbers to a 'not at risk' status, he notes that small, less well-off households would require lower levels or even (in 65 per cent of cases) no compensation at all.

With particular regard to years of schooling, Scarpa *et al.* (2003b) found that it interacted significantly and positively with the need to purchase feed. This suggests that more educated people are less reluctant to buy weaned piglets which require purchased feed (an attribute more closely associated with exotics and their crosses) during rearing.

In the context of the number of economically active household members (a proxy for income, both on and off-farm), Scarpa *et al.* (2003b) show that small households with only one income-earner place relatively more value on piglets that do not require feed purchase, show high disease resistance and need only one bath a week (the latter a proxy for heat tolerance). All these factors are more closely associated with the indigenous breed.

Drucker *et al.* (1999) find no particular pattern regarding breeds, village size and the existence of a commercial pig farm within the village. It is concluded that other factors explain the above average presence of indigenous and crossbreed Creole pigs in particular villages.

In addition to opportunity cost approaches, a number of methodologies exist for identifying priorities at the level of the breeding programme. These include breeding programme evaluation, genetic production function, hedonic and farm simulation model approaches. Described in detail in Drucker *et al.* (2001), these approaches can be used to, respectively: identify the net economic benefits of stock improvement (first two); identify trait values; and model the impact of improved animal characteristics on farm economies. Zander (2006) carried out similar work for the Boran cattle of Ethiopia, characterising households by breed type ownership.

While the above studies reveal that household information can be of use in designing cost-effective conservation programmes, it is also apparent that the methodologies are data-intensive and require integration with participatory rural appraisal approaches.

### 3.4 *Which traits should be addressed in breeding programmes?*

Jabbar *et al.* (1998), using a hedonic approach, showed in a Nigerian case study that although there were some differences in prices that were solely because of breed, most variation in prices was caused by variables such as wither height and girth circumference, which vary from animal to animal within breeds. Variation because of type of animal or month of transaction was also greater than that because of breed.

Assessing livestock keepers' breeding practices and breed preferences in southwest Nigeria, Jabbar and Diedhiou (2003) confirm a strong trend away from trypanotolerant[4] breeds. The best hopes for implementing a

[4] Breeds that are tolerant to trypanosomosis, a disease spread by the tsetse fly and the cause of sleeping sickness in humans.

conservation programme for breeds at risk are likely to be in locations where the breed is still found, disease remains a constraint, the breed is better suited to the farming systems and there are still large markets for the breed.

### 3.5 *How do specific policies influence GR conservation and sustainable use? How can conservation strategies be made cost-efficient? Which breeds should be priorities for conservation?*

Simianer *et al.* (2003) and Reist-Marti (2003) provide one of the few examples of the conceptual development of a decision-support tool in this area. Recognising the large number of indigenous livestock breeds that are currently threatened and the fact that not all can be saved given limited conservation budgets, they elaborate a framework for the allocation of a given budget among a set of breeds such that the expected amount of between-breed diversity conserved is maximised. Drawing on Weitzman (1993) it is argued that the optimum criterion for a conservation scheme is to maximise the expected total utility of the set of breeds, which is a weighted sum of diversity,[5] extinction probabilities and the value of the conserved breeds. In addition, where the models are sufficiently specified and essential data on key parameters are available, the framework can be used for rational decision-making on a global scale. The findings from the application of these methods are interesting.

The current rapid rate of loss of GR diversity is the result of a number of underlying factors. While, in some cases, changes in production systems and consumer preferences reflect the natural evolution of developing economies and markets, in other cases, production systems, breed choice and consumer preferences have been distorted by local, national and international policy. Such distortions may arise from macroeconomic interventions (e.g. exchange and interest rates); regulatory and pricing policy (e.g. taxation, price controls, market and trade regulations); investment policy (e.g. infrastructure development); and institutional policy (e.g. land ownership, GR property rights).

Furthermore, conservation policy needs to promote cost-efficient strategies and this can be achieved through the development of

[5] Note that the measure of diversity used can be based on genetic distances (as in both this and the original Weitzman study) but alternative measures of diversity (e.g. based on the existence of unique attributes of certain breeds – such as trypanotolerance) could also be used. The implications for which breeds should be conserved may well differ depending on how the diversity index is constructed and the overall goal of the conservation programme (conservation of genetic diversity *per se*, maximising the number of unique traits conserved or maximising the livelihood contribution of the livestock diversity conserved).

'Weitzman-type' decision-support tools. Such tools permit the allocation of a given budget among a set of breeds such that the expected amount of between-breed diversity conserved is maximised. For example, applications by Simianer *et al.* (2003) and Reist-Marti (2003) indicate that conservation funds should be spent on only three to nine (depending on different model assumptions) of 23 breeds of African zebu and zenga cattle, and these are not necessarily the most endangered ones.

Since the methodology applied to try to answer the question is dependent on an understanding of individual breed conservation costs, these need to be applied in conjunction with opportunity/least cost and safe minimum standard approaches. As discussed above, the findings from these types of studies have shown that only minimal incentives and interventions may in fact be needed to ensure continued indigenous breed sustainable use, as the costs of implementing an *in-situ* breed conservation programme in certain areas are relatively low. For example, Scarpa *et al.* (2003b) show that the net value that backyard producers place on the Mexican Creole pig is very similar to that of the other breeds. Nevertheless, where opportunity costs for indigenous breed production do exist vis-à-vis the main commercial breeds,[6] compensation payments must be adequate to make the rearing of such breeds profitable.

While the influence of policy factors on livestock diversity is readily discernible in broad terms, little is known about their relative importance. There is a need for such understanding as a first step towards the implementation of policies and market strategies that promote the effective utilisation and conservation of the diverse populations of indigenous livestock breeds. To this end, the development of a number of policy 'decision-support tools' has been proposed, although measures of breed genetic distances and conservation costs are lacking for many species/breeds and no such tools have yet been implemented in practice.

Policy issues related to GR property rights influencing access to and exchange of livestock germplasm are increasingly being discussed in international fora. However, Drucker and Gibson (2003) identify a range of issues that need to be researched before the relative costs and benefits of such an international regulatory instrument for AnGR can be determined. These include an improved understanding of the importance of continued access and trade in livestock germplasm for research and

[6] Note that the existence of such an opportunity cost differential is not always the case – for example, see Ayalew *et al.* 2003. Furthermore, alternatives to compensation approaches also exist. For example, where branding and niche market development have eliminated this opportunity cost, as is the case with Reggiano cattle and parmesan cheese.

development purposes; and the nature of the costs and benefits arising from AnGR research.

## 4 Conclusions

The review offered in this chapter indicates that advances in economic valuation for both crop and livestock GR have eased some methodological/analytical constraints, and that in some respects, data constraints may be more binding. A wide range of tools and analytical approaches has been successfully applied to a number of crops/species and breeds, in a number of production systems and locations. Application of these methods can provide useful estimates of the market and non-market value of variety/breed attributes. Such data are crucial for: (i) identifying trait values in breeding programmes; (ii) demonstrating the benefits, as well as the costs of conservation; (iii) identifying cost-efficient, diversity-maximising, or optimal conservation strategies; and (iv) orienting policies aimed at genetic resources (GR) conservation and sustainable use.

Methodological advances continue to be important in several ways. First, a number of strategic areas have not yet been addressed with adapted tools in either the crops or livestock literature. Second, there are advantages and disadvantages associated with both revealed and stated preference approaches to valuation, so that a combination of approaches will often prove more satisfactory. Still, greater accuracy is likely to come at a price of greater respondent burden and research expenditures. Third, there are obvious limitations to what can be accomplished solely through valuation exercises, since management of genetic resources involves crucial institutional and organisational decisions.

Valuation methodologies need to be incorporated into decision-support tools that can be applied in contexts where they can be used to inform policy decisions and support poor farmers, towards the goal of conserving agro-biodiversity for sustainable use. Finally, economic studies undertaken with respect to one component have treated the goods or services provided by the other component as 'exogenous', or external. Not taking a holistic approach, however, could bias the estimated costs, benefits and policy recommendations in important ways if interactions among biodiversity components are significant. At an operational level, too, there is potential to benefit from a better understanding of the interactions between these components, especially given the fact that interventions often deal with the same people at the community level. Harnessing such interactions may be one of the best ways to conserve regional biodiversity, but will depend on information and insights from a much wider range of literature and disciplines than economics alone.

## REFERENCES

Alston, J. M., Chan-Kang, C., Marra, M. C., Pardey, P. G. and Wyatt, T. J. 2000. A meta-analysis of rates of return to agricultural R and D: ex pede herculem? IFPRI Research Report No. 113. Washington, DC: International Food Policy Research Institute.

Alston, J. M., Norton, G. W. and Pardey, P. G. 1998. *Science under Scarcity: Principles and Practice for Agricultural Research Evaluation and Priority Setting.* Wallingford: CABI International.

Anderson, S. 2003. Animal genetic resources and sustainable livelihoods. *Ecological Economics.* **45** (3). 331–339.

Ayalew, W., King, J. M., Bruns, E. and Rischkowsky, B. 2003. Economic evaluation of smallholder subsistence livestock production: lessons from an Ethiopian goat development program. *Ecological Economics.* **45** (3). 473–485.

Bela, G., Balázs, B. and Pataki, G. 2005. Institutions, stakeholders, and the management of crop genetic sources on Hungarian family farms. In M. Smale (ed.). 251–269.

Bellon, M. R., Pham, J. L. and Jackson, M. T. 1997. Genetic conservation: a role for rice farmers. In N. Maxted, B. Ford-Lloyd and J. G. Hawkes (eds.). *Plant Genetic Conservation: The In Situ Approach.* London, New York: Chapman and Hall. 263–289.

Benin, S., Smale, M., Pender, J., Gebremedhin, B. and Ehui, S. 2004. The economic determinants of cereal crop diversity on farms in the Ethiopian Highlands. *Agricultural Economics.* **31**. 197–208.

Birol, E. 2004. Valuing agricultural biodiversity on home gardens in Hungary: An application of stated and revealed preference methods. PhD Thesis. University College London, University of London.

Brown, G. and Goldstein, J. H. 1984. A model for valuing endangered species. *Journal of Environmental Economics and Management.* **11**. 303–309.

Brown, G. M. 1990. Valuing Genetic Resources. In G. H. Orians, G. M. Brown, W. E. Kunin and J. E Swierzbinski (eds.). *Preservation and Valuation of Biological Resources.* Seattle: University of Washington Press. 203–226.

Brush, S. B., Taylor, J. E. and Bellon, M. R. 1992. Technology adoption and biological diversity in Andean potato agriculture. *Journal of Development Economics.* **39**. 365–387.

Brush, S. B. (ed.). 2002. *Genes in the Fields: On-Farm Conservation of Crop Diversity.* Rome: International Plant Genetic Resources Institute (IPGRI). Ottawa: International Development Research Centre (IDRC). Boca Raton, FL: Lewis Publishers.

Cicia, G., D'Ercole, E. and Marino, D. 2003. Costs and benefits of preserving farm animal genetic resources from extinction: CVM and bio-economic model for valuing a conservation program for the Italian Pentro horse. *Ecological Economics.* **45** (3). 445–459.

Di Falco, S. 2003. Crop Genetic Diversity, Agroecosystem Production and the Stability of Farm Income. Ph.D. dissertation Environment Department, University of York.

Drucker, A. G., Gómez, V., Ferraes-Ehuan, N., Rubio, O. and Anderson, S. 1999. Comparative economic analysis of *criollo*, crossbreed and imported pigs in backyard production of Yucatan, Mexico. FMVZ-UADY, mimeo.

Drucker, A. G., Gómez, V. and Anderson, S. 2001. The economic valuation of farm animal genetic resources: a survey of available methods. *Ecological Economics*. **36** (1). 1–18.

Drucker, A. G. and Gibson, J. P. 2003. A Legal and Regulatory Framework for AnGR? Issues for Consideration. Paper presented at: Workshop on Farm AnGR Legal and Regulatory Framework, Maputo, Mozambique, 19–23 May. FAO/UNDP/GTZ/CTA/SADC.

Drucker, A. G. 2006. An application of the use of safe minimum standards in the conservation of livestock biodiversity. *Environment and Development Economics*. **11**. 77–94.

Drucker, A. G. and Anderson, S. 2004. Economic analysis of animal genetic resources and the use of rural appraisal methods: lessons from South-East Mexico. *International Journal of Sustainable Agriculture*. **2** (2). 77–97.

Drucker, A. G., Smale, M. and Zambrano, P. 2005. *Valuation and Sustainable Management of Crop and Livestock Biodiversity: A Review of Applied Economics Literature*. Washington, DC: International Food Policy Research Institute, Rome: the Systemwide Genetic Resources Program and International Plant Genetic Resources Institute.

Edmeades, S., Karamura, D. and Smale, M. 2005. Demand for cultivar attributes and the biodiversity of bananas in Uganda. In M. Smale (ed.). 97–118.

Evenson, R. E. and Gollin, D. (ed.). 2003. *Crop Variety Improvement and its Effect on Productivity: The Impact of International Agricultural Research*. Wallingford: CABI Publishing.

Evenson, R. E., Gollin, D. and Santaniello, V. (ed.). 1998. *Agricultural Values of Plant Genetic Resources*. Wallingford: CABI, FAO, University of Tor Vergata, and CABI Publishing.

Feder, G., Just, R. E. and Zilberman, D. 1985. Adoption of agricultural innovations in developing countries: a survey. *Economic Development and Cultural Change*. **33**. 255–298.

Feder, G. and Umali, D. L. 1993. The adoption of agricultural innovations: a review. *Technological Forecasting and Social Change*. **43**. 215–239.

FAO. 1999. The global strategy for the management of farm animal genetic resources: Executive brief. Rome: Food and Agriculture Organization. 49 p.

Gandini, G. and Pizzi, F. 2003. In situ and ex situ conservation techniques: financial aspects. In D. Planchenault (ed.). Workshop on Cryopreservation of Animal Genetic Resources in Europe. Proceedings of the Technical Workshop, Paris. 23 February.

Gauchan, D. 2004. Conserving Crop Genetic Resources On-Farm: The Case of Rice in Nepal. University of Birmingham, PhD dissertation.

Gauchan, D., Smale, M. and Chaudhury, P. 2005a. Market-based incentives for conserving diversity on farms: the case of rice landraces in Central Terai, Nepal. *Genetic Resources and Crop Evolution*. **52**. 293–303.

Gauchan, D., Van Dusen, M. E. and Smale, M. 2005b. On farm conservation of rice biodiversity in Nepal: A simultaneous estimation approach. Washington, DC: Environment and Production Technology Division Discussion Paper 144, International Food Policy Research Institute.

Gebremedhin, B., Smale, M. and Pender, J. 2005. Determinants of cereal diversity in villages of Northern Ethiopia. In M. Smale (ed.). 177–191.

Gollin, D., Smale, M. and Skovmand, B. 2000. Searching an *Ex Situ* collection of wheat genetic resources. *American Journal of Agricultural Economics*. **82**. 812–827.

Harlan, J. R. 1992. *Crops and Man (2nd edn)*, Madison, WI: American Society of Agronomy and Crop Science Society of America, Madison.

Heisey, P. W., Smale M., Byerlee, D. and Souza, E. 1997. Wheat rusts and the costs of genetic diversity in the Punjab of Pakistan. *American Journal of Agricultural Economics*. **79**. 726–737.

Jabbar, M., Swallow, B., d'Iteran, G. and Busari, A. 1998. Farmer preferences and market values of cattle breeds of west and central Africa. *Journal of Sustainable Agriculture*. **12**. 21–47.

Jabbar, M. A. and Diedhiou, M. L. 2003. Does breed matter to cattle farmers and buyers? Evidence from West Africa. *Ecological Economics*. **45** (3). 461–472.

Jarvis, D., Sthapit, L. and Sears, L. (ed.). 2000. *Conserving Agricultural Biodiversity In Situ: A Scientific Basis for Sustainable Agriculture*. Rome, Italy: International Plant Genetic Resources Institute.

Karugia, J., Mwai, O., Kaitho, R., Drucker, A., Wollny, C. and Rege, J. E. O. 2001. *Economic Analysis of Crossbreeding Programmes in Sub-Saharan Africa: A Conceptual Framework and Kenyan Case Study. Animal Genetic Resources Research 2*. Nairobi, Kenya: ILRI. 55.

Koo, B., Pardey, P. G. and Wright, B. D. 2004. *Saving seeds: The Economics of Conserving Crop Genetic Resources Ex Situ in the Future Harvest Centres of the CGIAR*. Wallingford: CABI Publishing.

Lipper, L., Cavatassi, R. and Winters, P. 2005. Seeds supply and on farm demand for diversity: a case study of Eastern Ethiopia. In M. Smale (ed.). 233–250.

Mendelsohn, R. 2003. The challenge of conserving indigenous domesticated animals. *Ecological Economics*. **45**. 501–510.

Meng, E. C. H. 1997. Land allocation decisions and in situ conservation of crop genetic resources: The case of wheat landraces in Turkey. Ph.D. dissertation. Davis, CA : University of California.

Meng, E. C. H., Smale, M., Rozelle, S. D., Hu, R. and Huang, J. 2003. Wheat genetic diversity in China: measurement and cost. In S. D. Rozelle and D. A. Sumner (ed.). *Agricultural Trade and Policy in China: Issues, Analysis and Implications*. Ashgate, Burlington, VT, 251–267.

Meng, E. C. H., Smale, M., Bellon, M. R. and Grimanelli, D. 1998. Definition and measurements of crop diversity for economic analysis. In M. Smale (ed.). *Farmers, Gene Banks and Crop Breeding: Economic Analyses of Diversity in Wheat, Maize, and Rice*. Boston, MA: Kluwer Academic Publishers. 19–31.

Morris, M. L. and Heisey, P. W. 2003. Estimating the benefits of plant breeding research: methodological issues and practical challenges. *Agricultural Economics*. **29**. 241–252.

Nagarajan, L. and Smale, M. 2005. Local Seed Systems and Village-Level Determinants of Millet Crop Diversity in Marginal Environments of India. EPTD Discussion Paper 135. Washington, DC: IFPRI.

Pardey, P., Alston, J. M., Chan-Kang, C., Magalhães, E. C. and Vosti, S. 2004. *Assessing and Attributing the Benefits from Varietal Improvement Research in Brazil*. Research Report 136. Washington, DC: International Food Policy Research Institute.

Pattison, J., Drucker, A. and Anderson, S. In press. The cost of conserving livestock diversity? Incentive measures and conservation options for maintaining indigenous Pelón pigs in Yucatan, Mexico. *Tropical Animal Health and Production*.

Pearce, D. and Moran, D. 1994. *The Economic Value of Biodiversity*. London: Earthscan.

Polasky, S. and Solow A. 1995. On the Value of a Collection of Species. *Journal of Environmental Economics and Management*. **29**. 298–303.

Rege, J. E. O. 2003. *In-situ* conservation of farm animal genetic resources. In CIP-UPWARD. *Conservation and Sustainable Use of Agrobiodiversity: A Source Book*. Los Baños, Laguna, The Philippines. 434–438.

Rege, J. E. O. and Gibson, J. P. 2003. Animal genetic resources and economic development: issues in relation to economic valuation. *Ecological Economics*. **45**. 319–330.

Reist-Marti, S., Simianer, H., Gibson, G., Hanotte, O. and Rege, J. E. O. 2003. Weitzman's approach and breed diversity conservation: an application to African cattle breeds. *Conservation Biology*. **17** (5). 1299–1311.

Rejesus, R. M., Smale, M. and Ginkel, M. V. 1996. Wheat breeders' perspectives on genetic diversity and germplasm use: findings from an international survey. *Plant Varieties and Seeds*. **9**. 129–147.

Scarpa, R., Ruto, E. S. K., Kristjanson, P., Radeny, M., Drucker, A. G. and Rege, J. E. O. 2003a. Valuing indigenous cattle breeds in Kenya: an empirical comparison of stated and revealed preference value estimates. *Ecological Economics*. **45** (3). 409–426.

Scarpa, R., Drucker, A. G., Anderson, S., Ferraes-Ehuan, N., Gómez, V., Risopatrón, C. R. and Rubio-Leonel, O. 2003b. Valuing genetic resources in peasant economies: the case of 'hairless, Creole pigs in Yucatan. *Ecological Economics*. **45** (3). 427–443.

Signorello, G. and Pappalardo, G. 2003. Domestic animal biodiversity conservation: a case study of rural development plans in the European Union. *Ecological Economics*. **45** (3). 487–499.

Simianer, H., Reist-Marti, S. B., Gibson, J., Hanotte, O. and Rege, J. E. O. 2003. An approach to the optimal allocation of conservation funds to minimise loss of genetic diversity between livestock breeds. *Ecological Economics*. **45** (3). 377–392.

Simpson, R. D., Sedjo, R. A. and Reid, J. W. 1996. Valuing biodiversity for use in pharmaceutical research. *Journal of Political Economy*. **104**. 163–185.

Smale, M. and Day-Rubenstein, K. 2002. The demand for crop genetic resources: international use of the US National Plant Germplasm System. *World Development*. **30**. 1639–1655.

Smale, M., Hartell, J., Heisey, P. W. and Senauer, B. 1998. The contribution of genetic resources and diversity to wheat production in the Punjab of Pakistan. *American Journal of Agricultural Economics*. **80**. 482–493.

Smale, M. (ed.). 2005. *Valuing Crop Biodiversity: On-Farm Genetic Resources and Economic Change*. Wallingford: CABI Publishing.

Swanson, T. 1996. Global values of biological diversity: the public interest in the conservation of plant genetic resources for agriculture. *Plant Genetic Resources Newsletter*. 1–7.

Tano, K., Kamuanga, M., Faminow, M. D. and Swallow, B. 2003. Using conjoint analysis to estimate farmer's preferences for cattle traits in West Africa. *Ecological Economics*. **45** (3). 393–407.

Van Dusen, M. E. 2000. *In situ* conservation of Crop Genetic Resources in the Mexican *Milpa* System. PhD Thesis University of California at Davis.

Van Dusen, E. 2005. Missing markets, migration and crop biodiversity in the Mexican *milpa* system: A household farm model. In M. Smale (ed.). 192–210.

Wale, E. and Mburu, J. 2006. Demand for attributes and on farm conservation of coffee in Ethiopia. In M. Smale (ed.). 48–62.

Weitzman, M. L. 1993. What to preserve? an application of diversity theory to crane conservation. *The Quarterly Journal of Economics*. **108**. 157–183.

Widawsky, D. and Rozelle, S. D. 1998. Varietal diversity and yield variability in Chinese rice production. In M. Smale (ed.). *Farmers, Gene Banks, and Crop Breeding: Economic Analyses of Diversity in Wheat, Maize and Rice*. Dordrecht: Kluwer Academic Press and CIMMYT. 159–172.

Winters, P., Hintze, L. H. and Ortiz, O. 2005. Rural development and the diversity of potatoes on farms in Cajamarca, Peru. In M. Smale (ed.). 146–161.

Zander, K. K. 2006. Modelling the value of farm animal genetic resources – facilitating priority setting for the conservation of cattle in East Africa. Dissertation. University of Bonn. urn:nbn:de:hbz:5N-09113, URL: http://hss.ulb.uni-bonn.de/diss_online/landw_fak/2006/zander_kerstin

Zohrabian, A., Traxler G., Caudill, S. and Smale, M. 2003. Valuing pre-commercial genetic resources: a maximum entropy approach. *American Journal of Agricultural Economics*. **85**. 429–436.

# Index

Lightning Source UK Ltd.
Milton Keynes UK
12 December 2010

164265UK00001B/110/P